Asheville-Buncombe
Technical Community College
Learning Resources Center
340 Victoria Road
Asheville, NC 28801

Plant Gene Transfer and Expression Protocols

DISCARDED

DEC 1 1 2024

Methods in Molecular Biology™

John M. Walker, Series Editor

55. **Plant Cell Electroporation and Electrofusion Protocols,** edited by *Jac A. Nickoloff, 1995*
54. **YAC Protocols,** edited by *David Markie, 1995*
53. **Yeast Protocols:** *Methods in Cell and Molecular Biology,* edited by *Ivor H. Evans, 1996*
52. **Capillary Electrophoresis:** *Principles, Instrumentation, and Applications,* edited by *Kevin D. Altria, 1996*
51. **Antibody Engineering Protocols,** edited by *Sudhir Paul, 1995*
50. **Species Diagnostics Protocols:** *PCR and Other Nucleic Acid Methods,* edited by *Justin P. Clapp, 1996*
49. **Plant Gene Transfer and Expression Protocols,** edited by *Heddwyn Jones, 1995*
48. **Animal Cell Electroporation and Electrofusion Protocols,** edited by *Jac A. Nickoloff, 1995*
47. **Electroporation Protocols for Microorganisms,** edited by *Jac A. Nickoloff, 1995*
46. **Diagnostic Bacteriology Protocols,** edited by *Jenny Howard and David M. Whitcombe, 1995*
45. **Monoclonal Antibody Protocols,** edited by *William C. Davis, 1995*
44. ***Agrobacterium* Protocols,** edited by *Kevan M. A. Gartland and Michael R. Davey, 1995*
43. **In Vitro Toxicity Testing Protocols,** edited by *Sheila O'Hare and Chris K. Atterwill, 1995*
42. **ELISA:** *Theory and Practice,* by *John R. Crowther, 1995*
41. **Signal Transduction Protocols,** edited by *David A. Kendall and Stephen J. Hill, 1995*
40. **Protein Stability and Folding:** *Theory and Practice,* edited by *Bret A. Shirley, 1995*
39. **Baculovirus Expression Protocols,** edited by *Christopher D. Richardson, 1995*
38. **Cryopreservation and Freeze-Drying Protocols,** edited by *John G. Day and Mark R. McLellan, 1995*
37. **In Vitro Transcription and Translation Protocols,** edited by *Martin J. Tymms, 1995*
36. **Peptide Analysis Protocols,** edited by *Ben M. Dunn and Michael W. Pennington, 1994*
35. **Peptide Synthesis Protocols,** edited by *Michael W. Pennington and Ben M. Dunn, 1994*
34. **Immunocytochemical Methods and Protocols,** edited by *Lorette C. Javois, 1994*
33. ***In Situ* Hybridization Protocols,** edited by *K. H. Andy Choo, 1994*
32. **Basic Protein and Peptide Protocols**, edited by *John M. Walker, 1994*
31. **Protocols for Gene Analysis,** edited by *Adrian J. Harwood, 1994*
30. **DNA–Protein Interactions**, edited by *G. Geoff Kneale, 1994*
29. **Chromosome Analysis Protocols,** edited by *John R. Gosden, 1994*
28. **Protocols for Nucleic Acid Analysis by Nonradioactive Probes,** edited by *Peter G. Isaac, 1994*
27. **Biomembrane Protocols:** *II. Architecture and Function,* edited by *John M. Graham and Joan A. Higgins, 1994*
26. **Protocols for Oligonucleotide Conjugates:** *Synthesis and Analytical Techniques,* edited by *Sudhir Agrawal, 1994*
25. **Computer Analysis of Sequence Data:** *Part II,* edited by *Annette M. Griffin and Hugh G. Griffin, 1994*
24. **Computer Analysis of Sequence Data:** *Part I,* edited by *Annette M. Griffin and Hugh G. Griffin, 1994*
23. **DNA Sequencing Protocols**, edited by *Hugh G. Griffin and Annette M. Griffin, 1993*
22. **Microscopy, Optical Spectroscopy, and Macroscopic Techniques**, edited by *Christopher Jones, Barbara Mulloy, and Adrian H. Thomas, 1993*
21. **Protocols in Molecular Parasitology**, edited by *John E. Hyde, 1993*
20. **Protocols for Oligonucleotides and Analogs:** *Synthesis and Properties,* edited by *Sudhir Agrawal, 1993*
19. **Biomembrane Protocols:** *I. Isolation and Analysis,* edited by *John M. Graham and Joan A. Higgins, 1993*
18. **Transgenesis Techniques:** *Principles and Protocols,* edited by *David Murphy and David A. Carter, 1993*
17. **Spectroscopic Methods and Analyses:** *NMR, Mass Spectrometry, and Metalloprotein Techniques,* edited by *Christopher Jones, Barbara Mulloy, and Adrian H. Thomas, 1993*
16. **Enzymes of Molecular Biology,** edited by *Michael M. Burrell, 1993*
15. **PCR Protocols:** *Current Methods and Applications,* edited by *Bruce A. White, 1993*
14. **Glycoprotein Analysis in Biomedicine,** edited by *Elizabeth F. Hounsell, 1993*
13. **Protocols in Molecular Neurobiology,** edited by *Alan Longstaff and Patricia Revest, 1992*
12. **Pulsed-Field Gel Electrophoresis:** *Protocols, Methods, and Theories,* edited by *Margit Burmeister and Levy Ulanovsky, 1992*
11. **Practical Protein Chromatography,** edited by *Andrew Kenney and Susan Fowell, 1992*
10. **Immunochemical Protocols,** edited by *Margaret M. Manson, 1992*
9. **Protocols in Human Molecular Genetics,** edited by *Christopher G. Mathew, 1991*
8. **Practical Molecular Virology:** *Viral Vectors for Gene Expression,* edited by *Mary K. L. Collins, 1991*
7. **Gene Transfer and Expression Protocols,** edited by *Edward J. Murray, 1991*
6. **Plant Cell and Tissue Culture,** edited by *Jeffrey W. Pollard and John M. Walker, 1990*
5. **Animal Cell Culture,** edited by *Jeffrey W. Pollard and John M. Walker, 1990*

Methods in Molecular Biology™ • 49

Plant Gene Transfer and Expression Protocols

Edited by

Heddwyn Jones

University of Hertfordshire, Hatfield, UK

Asheville-Buncombe
Technical Community College
Learning Resources Center
340 Victoria Road
Asheville, NC 28801

05-0959

Humana Press  **Totowa, New Jersey**

© 1995 Humana Press Inc.
999 Riverview Drive, Suite 208
Totowa, New Jersey 07512

All rights reserved.

No part of this book may be reproduced, stored in a retrieval system, or transmitted in any form or by any means, electronic, mechanical, photocopying, microfilming, recording, or otherwise without written permission from the Publisher. Methods in molecular biology™ is a trademark of the Humana Press Inc.

All authored papers, comments, opinions, conclusions, or recommendations are those of the author(s), and do not necessarily reflect the views of the publisher.

This publication is printed on acid-free paper. ∞
ANSI Z39.48-1984 (American National Standards Institute) Permanence of Paper for Printed Library Materials.

Cover illustration: Fig. 2A from Chapter 25, "*In Situ* Hybridization to Plant Tissue Sections," by Shirley R. Burgess.

Photocopy Authorization Policy:
Authorization to photocopy items for internal or personal use, or the internal or personal use of specific clients, is granted by Humana Press Inc., **provided** that the base fee of US $4.00 per copy, plus US $00.20 per page, is paid directly to the Copyright Clearance Center at 222 Rosewood Drive, Danvers, MA 01923. For those organizations that have been granted a photocopy license from the CCC, a separate system of payment has been arranged and is acceptable to Humana Press Inc. The fee code for users of the Transactional Reporting Service is: [0-89603-321-X/95 $4.00 + $00.20].

Printed in the United States of America. 10 9 8 7 6 5 4 3 2 1

Library of Congress Cataloging in Publication Data

Main entry under title:
Methods in molecular biology™.

Plant gene transfer and expression protocols / edited by Heddwyn Jones.
p. cm. — (Methods in molecular biology™ 49)
Includes index.
ISBN 0-89603-321-X (alk. paper)
1. Plant genetic engineering—Laboratory manuals. 2. Crops—Genetic engineering—Laboratory manuals. 3. Plant gene expression—Laboratory manuals. I. Jones, Heddwyn. II. Series: Methods in molecular biology™ (Totowa, NJ) ; 49.
QK981.5.P567 1995
581.87'322'0724—dc20 95-23429
CIP

Preface

The development of recombinant DNA technology and methods for transferring recombinant genes into plants has brought about significant advances in plant science. First, it has allowed investigation, using reporter genes, into the transcriptional regulation of plant genes—a key to the understanding of the biochemical basis of growth and development in plants. Second, gene transfer technology has facilitated the molecular cloning, by tagging genomic sequences, of important genes (e.g., homeotic genes) whose gene products control the normal pattern of growth and differentiation of plants. Third, overproducing foreign or endogenous proteins in plants can often lead to a better understanding of biochemical and physiological processes. Fourth, gene transfer technology has allowed the improvement of plant agricultural productivity. For example, plants have been engineered with improved viral resistance or the ability to withstand herbicide attack, therefore allowing a more effective use of herbicides to kill weeds. Fifth, there have been recent successes that demonstrate the potential use of plants as biotechnological chemical factories. For example, it is possible to use plants in the production of human antibodies and antigens of medical importance. It has been demonstrated recently that plants can be engineered to produce modified oils and even plastics! This paves the way to redirect agriculture from the production of surplus foods to the production of biotechnological products of industrial importance.

Plant Gene Transfer and Expression Protocols presents a range of protocols for introducing genes into plants, and the analysis of their expression at the mRNA and protein level. The book does not attempt to cover basic nucleic acid manipulation; there are already some excellent books in this area. Moreover, basic DNA technology over the last 10 years or so has been increasingly commercialized and made more amenable to researchers by several biotechnology companies; for example it is now possible to buy a ready-made genomic or cDNA library!

Plant Gene Transfer and Expression Protocols is divided into eight sections, which roughly reflect the flow of biochemical information from DNA to proteins. The first section describes the various expression and reporter gene cassettes that are currently used for studying gene regulation and for the production of recombinant proteins in plants. This is followed by methods for introducing genes into plants using *Agrobacterium* and, in the second section, direct gene transfer methods. The first section includes a chapter that deals with cloning plant genes by T-DNA tagging of *Arabidopsis thaliana.*

The third section of the book deal specifically with the use of reporter genes (e.g., *gus*), and the fourth section deals with determining gene organization and copy number (including the analysis of gene families) by Southern blotting and inverse PCR technology. The fifth and largest section of the book provides protocols for isolating RNA and for the analysis of gene expression at the RNA level using such methods as Northern analysis, RNase protection, RT-PCR, and *in situ* hybridization analysis. Also included in this section are chapters that deal with *Xenopus* oocytes and yeast cells as heterologous hosts for the expression of plant mRNA and cDNA. These systems have recently facilitated the molecular cloning and characterization of various membrane transporters.

The sixth and seventh sections of the book are devoted to methods for studying and manipulating chloroplast and mitochondria, respectively. The final section provides protocols for analyzing recombinant proteins in plants. In this section, protocols are provided for extracting and separating plant proteins by various electrophoretic systems, and for their analysis by Western blotting and ELISA. There is also a chapter in this section on the detection of proteins at the cellular level using immunocytochemical techniques.

Although the emphasis throughout *Plant Gene Transfer and Expression Protocols* is on the analysis of transgenes, the sections on gene organization, RNA, organelles, and protein methodology should also be of value to any researcher who wishes to study an endogenous gene without necessarily resorting to gene transfer technology. Following the previous books of the *Methods in Molecular Biology™* series, emphasis has been placed on generally applicable protocols, with each chapter ending with a Notes section highlighting potential problems and providing tips and alternatives. The protocols are detailed, and with the help of the Notes section, should allow the efficient transfer of these techniques to any laboratory, whether engaged in fundamental or applied aspects of plant science.

I would like to thank the series editor, John Walker, for his help and patience during the preparation of the book and also thank the contributing authors for the excellent material they have provided for this volume.

Heddwyn Jones

Contents

Contributors

PILAR BARCELO • *Biochemistry and Physiology Department, Institute of Arable Crops Research, Harpenden, UK*
SIMON A. BARNES • *Department of Plant Sciences, University of Cambridge, UK; Current address: Laboratory of Plant Molecular Biology, The Rockefeller University, New York, NY*
MARK BELL • *Department of Agricultural Sciences, University of Bristol, UK*
STEFAN BINDER • *Institut für Genbiologische Forschung, Berlin, Germany*
NIGEL W. BLACKHALL • *Department of Life Science, University of Nottingham, UK*
MARGARET I. BOULTON • *Department of Virus Research, Norwich Research Park, Colney, Norwich, UK*
JOHN W. S. BROWN • *Cell and Molecular Genetics Department, Scottish Crop Research Institute, Dundee, Scotland*
SHIRLEY R. BURGESS • *Department of Agricultural Sciences, University of Bristol, UK*
MICHAEL C. CLARKE • *Department of Botany, University of Leicester, UK*
RICHARD COOKE • *Laboratoire de Physiologie et Biologie Moleculaire Vegetale, Universite de Perpignan, France*
IAN S. CURTIS • *Department of Life Science, University of Nottingham, UK*
MICHAEL R. DAVEY • *Department of Life Science, University of Nottingham, UK*
JOHN DAVIES • *Department of Molecular Pathology, Glaxo Research and Development Ltd., Greenford, UK*
MARTINE DEVIC • *Laboratoire de Physiologie et Biologie Moleculaire, Universite de Perpignan, France*
ROGER J. FIDO • *Department of Agricultural Sciences, University of Bristol, UK*
PATRICK GALLOIS • *Laboratoire de Physiologie et Biologie Moleculaire Vegetale, Universite de Perpignan, France*
JOHN GATEHOUSE • *Department of Biological Sciences, University of Durham, UK*
LUTZ GROHMANN • *Institut für Genbiologische Forschung, Berlin, Germany*

François Guerineau • *Department of Botany, University of Leicester, UK*
Gillian A. Hull • *Laboratoire de Physiologie et Biologie Moleculaire, Universite de Perpignan, France*
Heddwyn Jones • *Department of Biosciences, University of Hertfordshire, Hatfield, UK*
Paul A. Lazzeri • *Biochemistry and Physiology Department, AFRC Institute of Arable Crops Research, Harpenden, UK*
Keith Lindsey • *Department of Botany, University of Leicester, UK*
Paulo Marinho • *Laboratoire de Physiologie et Biologie Moleculaire Vegetale, Universite de Perpignan, France*
Anthony J. Miller • *Biochemistry and Physiology Department, AFRC Institute of Arable Crops Research, Harpenden, UK*
E. N. Clare Mills • *Department of Food Molecular Biochemistry, AFRC Institute of Food Research, Norwich, UK*
Michael R. A. Morgan • *Department of Food Molecular Biochemistry, AFRC Institute of Food Research, Norwich, UK*
Paul Muskett • *Department of Botany, University of Leicester, UK*
Johnathan A. Napier • *Department of Agricultural Sciences, University of Bristol, UK*
Remko Offringa • *Institute of Molecular Plant Sciences, Leiden University, Leiden, The Netherlands*
Paul Penon • *Laboratoire de Physiologie et Biologie Moleculaire Vegetale, Universite de Perpignan, France*
Geoffrey W. Plumb • *Department of Food Molecular Biochemistry, AFRC Institute of Food Research, Norwich, UK*
J. Brian Power • *Department of Life Science, University of Nottingham, UK*
Paolo A. Sabelli • *Department of Agricultural Sciences, University of Bristol, UK*
Rod J. Scott • *Department of Botany, University of Leicester, UK*
Charles H. Shaw • *Department of Biological Sciences, University of Durham, UK*
Peter R. Shewry • *Department of Agricultural Sciences, University of Bristol, UK*
Craig G. Simpson • *Cell and Molecular Genetics Department, Scottish Crop Research Institute, Dundee, Scotland*
Jens Stougaard • *Department of Molecular Biology, University of Aarhus, Denmark*
Arthur S. Tatham • *Department of Agricultural Sciences, University of Bristol, UK*

FREDERICA L. THEODOULOU • *Biochemistry and Physiology Department, AFRC Institute of Arable Crops Research, Harpenden, UK*

ANDREW J. THOMPSON • *Department of Biological Sciences, University of Durham, UK*

JENNIFER F. TOPPING • *Department of Botany, University of Leicester, UK*

LAURENCE J. TRUEMAN • *IACR–Rothamsted Experimental Station, Harpenden, UK*

FREDERIQUE VAN DER LEE • *Mogen International nv, Leiden, The Netherlands*

WENBIN WEI • *Department of Botany, University of Leicester, UK; Current address: Centre for Plant Biochemistry and Biotechnology, University of Leeds, UK*

PART I

Agrobacterium-Mediated Transformation

CHAPTER 1

Tools for Expressing Foreign Genes in Plants

François Guerineau

1. Introduction

Since the first reports of tobacco transformation experiments in 1983, a number of fundamental processes, such as gene expression, cell metabolism, or plant development, are being studied using gene transfer experiments. The spectrum of plant species amenable to transformation is continuously widening. This is partly because of the refinement of tissue culture techniques and also because of the development of more and more diverse tools for gene transfer and expression. In this chapter, I will give a list of plasmid constructs containing various components useful for expressing foreign genes in plants: expression cassettes into which genes of interest can easily be inserted, assayable reporter genes that allow accurate quantification of gene expression, selectable marker genes for the selection of transformants, and plant promoters to achieve more specific patterns of gene expression.

2. Expression Cassettes

Efficient expression of foreign genes in transformed plants requires that they are placed under control of a promoter that is active in plant cells. Typical bacterial promoters are not functional in plant cells owing to important differences in the transcription machineries in the two types of organisms. Polyadenylation is also a very important determinant of gene expression. In eukaryotes, mRNAs are polyadenylated in the nuclei before being exported into the cytosol where they are translated. An

From: *Methods in Molecular Biology, Vol. 49: Plant Gene Transfer and Expression Protocols*
Edited by: H. Jones Humana Press Inc., Totowa, NJ

expression cassette will provide a promoter active in plant cells, a polylinker into which a coding sequence can be inserted, and a polyadenylation sequence located downstream of the polylinker. The vectors of all the cassettes described here are small, high-copy number, pBR322 or pUC-derived plasmids encoding ampicillin resistance.

2.1. CaMV 35S Promoter-Based Cassettes

A widely used promoter for expressing foreign genes in plant cells is the promoter directing the synthesis of the cauliflower mosaic virus (CaMV) 35S RNA. This promoter achieves a high level of transcription in nearly all plant tissues. The 35S promoter possesses a transcriptional enhancer located upstream of the TATA box. The duplication of the enhancer results in a higher level of transcription *(1)*. Most of the expression cassettes available contain the 35S promoter linked to the CaMV polyadenylation sequence. All these cassettes differ in their restriction sites upstream and downstream of the promoter and polyadenylation sequences. Also, different strains of CaMV have been used for their construction, the major difference being the presence or absence of an *Eco*RV restriction site between the enhancer sequence and the TATA box. Translation initiation is highly dependent on the sequence surrounding the ATG initiator codon. Some cassettes provide an optimized translation initiator codon context downstream of the promoter sequence and upstream of the polylinker, for the construction of translational fusions.

2.1.1. For Transcriptional Fusions

In the cassettes shown in Fig. 1, no ATG sequence is present between the transcription start and the polylinker sequence. Translation initiation will normally occur at the first ATG codon found in the sequence inserted in the polylinker. As it has been shown that the presence of multiple restriction sites in the untranslated region of mRNAs decreases gene expression *(6)*, the cloning strategy should ensure that as few sites as possible remain upstream of the coding sequence.

2.1.2. For Translational Fusions

The optimal sequence for translation initiation in mammalian cells is CCACC<u>ATG</u>G *(7)*. The consensus sequence around the ATG initiator codons of plant genes was established as AACA<u>ATG</u>G *(8)*. A recent comparison of the effect of these two consensus sequences placed upstream of the β-glucuronidase gene *(gus)* in plant protoplasts has

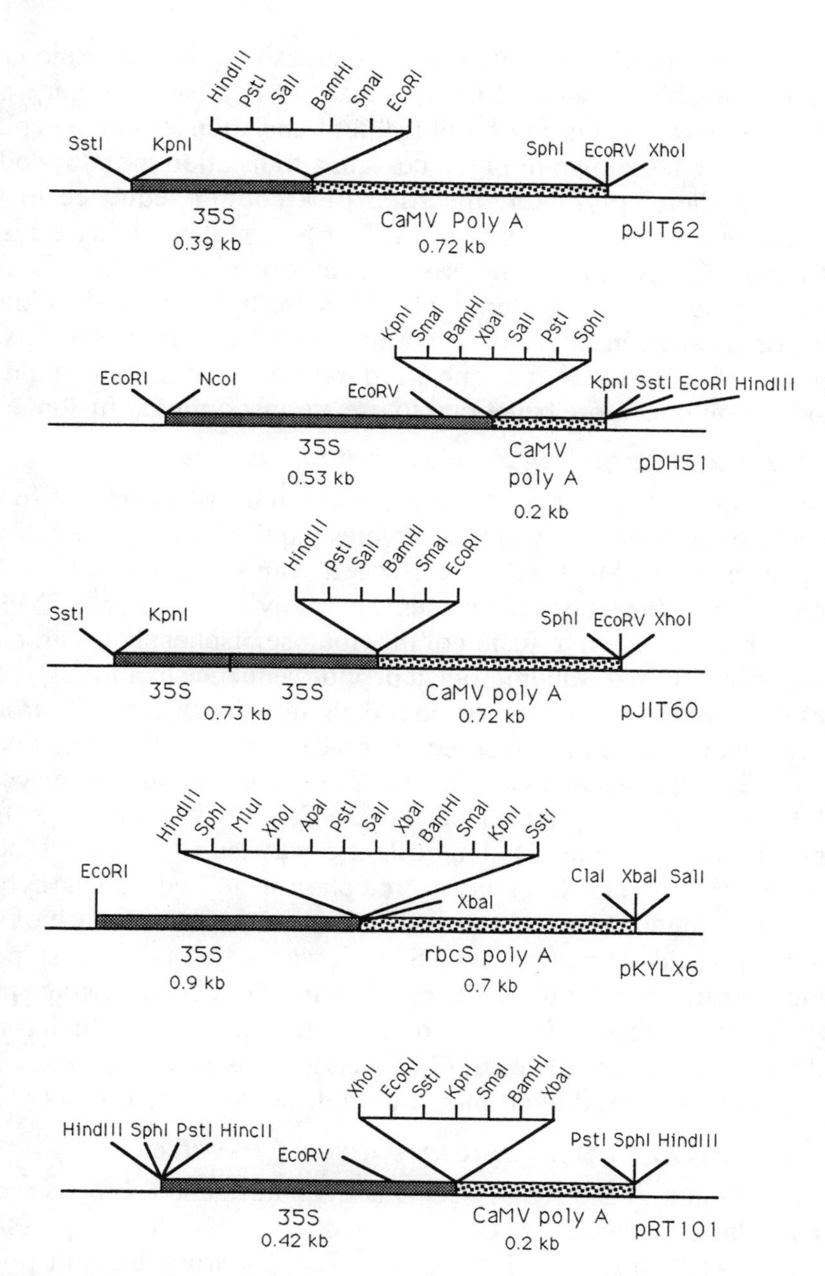

Fig. 1. Maps of CaMV 35S promoter-based expression cassettes for transcriptional fusions. pJIT62 (Guerineau, unpublished), pDH51 *(2)*, pJIT60 *(3)*, pKYLX6 *(4)*, pRT101 *(5)*.

shown that they were equally effective in increasing gene expression *(3)*. This is presumably because of the fact that *gus* is a bacterial gene and does not possess an ATG context optimal for translation initiation in plant cells. The cassettes shown in Fig. 2 contain a translation initiator codon upstream of their polylinker. Insertion of a coding sequence in the polylinker, in frame with the cassette ATG triplet, will result in a translational fusion. Consequently, the protein synthesized in the transformed cells will possess a short N-terminal extension. It is essential to know whether or not such an extension will affect the activity or the stability of the protein. If so, the benefit of enhanced translation initiation would be lost and it would be more beneficial to use a transcriptional fusion.

2.1.3. For Targeting Foreign Proteins to Chloroplasts

Whereas most of the biosynthetic pathways in the plant cell are found in the chloroplasts, very few of the enzymes required are encoded by the chloroplast genome. Most are nuclear-encoded and are imported into the chloroplasts by a transit peptide present at their N-terminus (*see* Chapter 30). It has been shown that fusion of the ribulose bisphosphate carboxylase (RUBISCO) small subunit transit peptide sequence to a foreign protein results in the import of the fusion protein into the chloroplast stroma where the mature protein is released after cleavage from the transit peptide *(10)*. The expression cassette pJIT117 *(11)* contains the sequence of the RUBISCO transit peptide attached to the CaMV 35S promoter with a duplicated enhancer (Fig. 3). This cassette was tested using β-glucuronidase: 17.4% of the GUS activity in protoplasts incubated with the hybrid construct was found in the chloroplast fraction *(11)*. The presence of the 23 first amino acids of mature RUBISCO downstream of the transit peptide would greatly enhance the targeting efficiency *(10)*, but the foreign protein would then be released in the stroma as a fusion protein, which is not suitable for all proteins. The pJIT117 cassette has also been used for importing the bacterial dihydropteroate synthase into chloroplasts *(12)*.

2.2. Plant Promoter-Based Cassettes

The expression of the RUBISCO small subunit gene (*rbcS*) is regulated by light and is tissue-specific (*see* Section 5.4.1.). The expression cassette pKYLX3 *(4)* contains the pea *rbcS-E9* promoter and polyadenylation sequences (Fig. 4). This cassette was able to direct the expression of the chloramphenicol acetyltransferase gene *(cat)* in tobacco calli *(4)*.

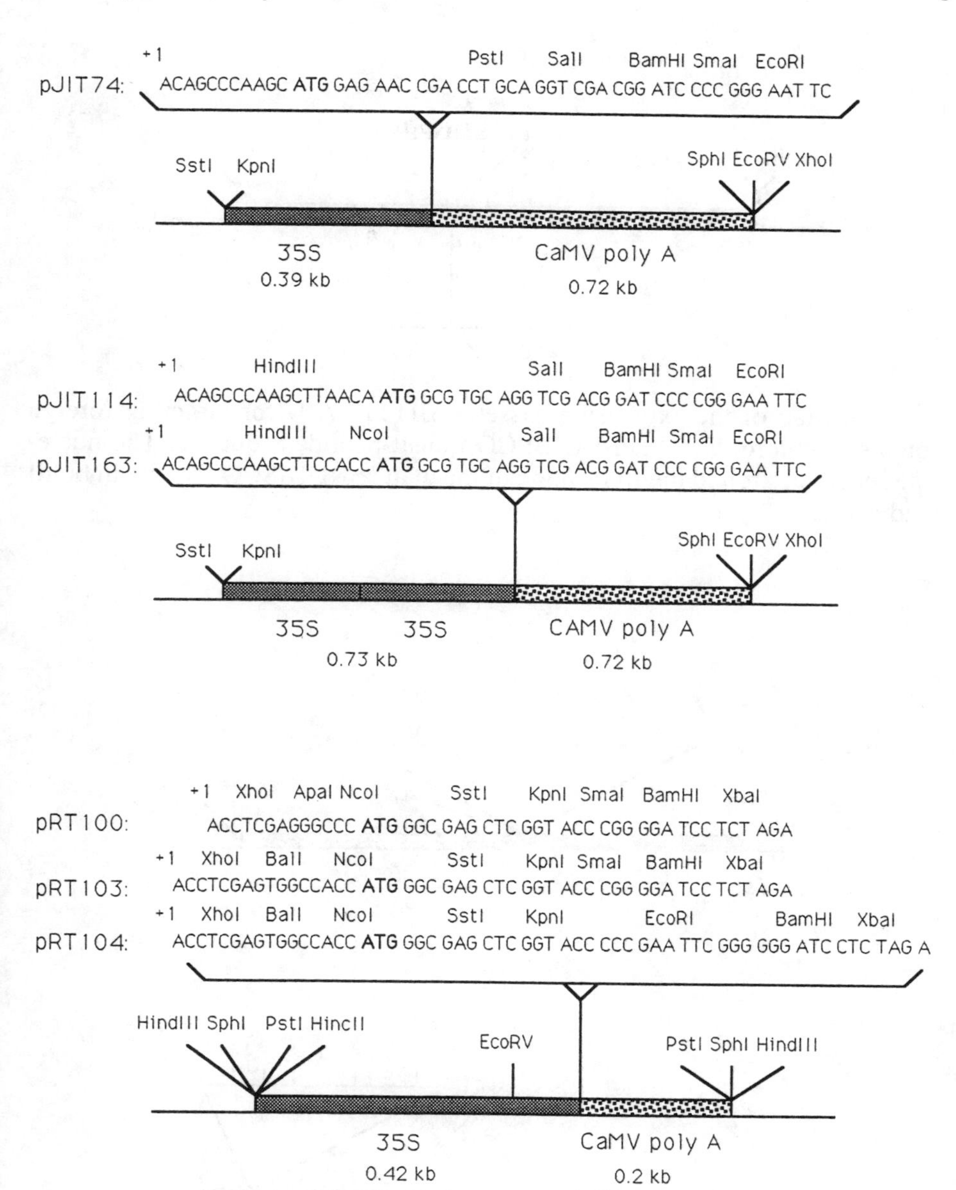

Fig. 2. Maps of CaMV 35S promoter-based expression cassettes for translational fusions. pJIT74 *(9)*, pJIT114 and pJIT163 *(3)*, pRT100, pRT103, pRT104 *(5)*. The translation initiation codons are shown in bold characters. The transcription initiation sites are indicated by +1 above the sequences.

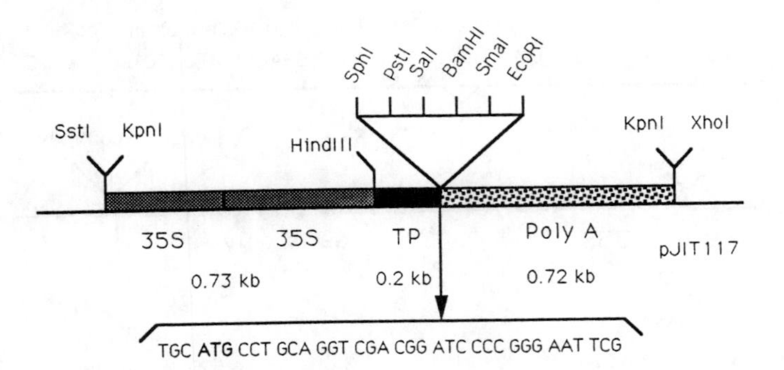

Fig. 3. Map of the expression cassette pJIT117 *(11)* for targeting foreign proteins to chloroplasts. TP, RUBISCO transit peptide sequence. The nucleotide sequence around the first codon of the mature RUBISCO (shown in bold) is indicated.

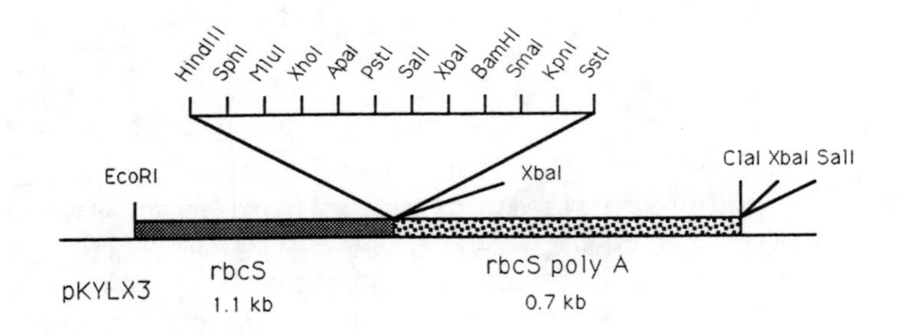

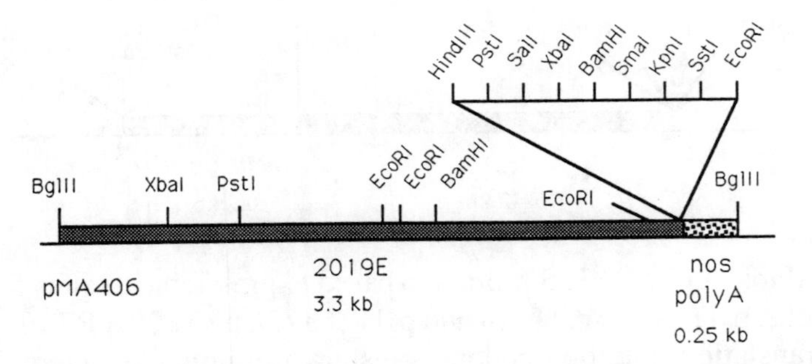

Fig. 4. Maps of two plant promoter-containing expression cassettes. pKYLX3 *(4)*, pMA406 *(13)*. *rbcS*, RUBISCO small subunit; *nos*, nopaline synthase.

The expression of the soybean *Gmhsp17.5-E* gene (also known as *2019E*) is heat-inducible. When a *2019E-gus* gene fusion was electroporated into protoplasts, GUS activity was 10 times higher in protoplasts subjected to a heat shock at 40°C than in protoplasts treated at 29°C *(13)*. The level of expression appeared to be higher than that given by the CaMV 35S promoter. The expression cassette pMA406 contains the *2019E* promoter linked to the polyadenylation sequence of the nopaline synthase *(nos)* gene from *Agrobacterium tumefaciens* (Fig. 4).

3. Reporter Genes

Many studies on plant promoters and on the regulation of gene expression have been made possible by the use of reporter genes. Their main scope is to provide an easy way of assessing gene expression. These genes encode for products which can be quantified using simple biochemical assays. Protocols for the assays are given in Section 3. of this book. Another use for these genes is the detection of transformation events during gene transfer experiments. The expression of a reporter gene can be easily detected in transformants, avoiding the need for more time-consuming characterization.

3.1. The β-Glucuronidase Gene

The β-glucuronidase gene *(vidA or gus)*, which originates from *E. coli* *(14)*, is the most widely used reporter gene in plant molecular biology. Accurate fluorimetric assays or precise histochemical localization of GUS in transgenic tissues are possible *(15)* (*see* Chapter 10). Another interesting property of the enzyme is its ability to tolerate N-terminal extensions *(15)*. Plasmids pBI101-1,-2,-3 provide the three different frames for translational fusions (Fig. 5). Plasmid pJIT166 contains the *gus* gene inserted in the expression cassette pJIT163 *(3)* (Fig. 5). A high GUS activity was recorded in tobacco protoplasts transfected with this plasmid *(3)*. The GenBank and EMBL database accession number for the nucleotide sequence of the *gus* gene is M14641.

3.2. The Firefly Luciferase Gene

The only known substrates for firefly luciferase are ATP and D-luciferin. The extreme specificity of this luminescent reaction is an interesting feature of this reporter gene/assay system. The nucleotide sequence of a luciferase cDNA has been reported *(16)* (accession number M15077). A

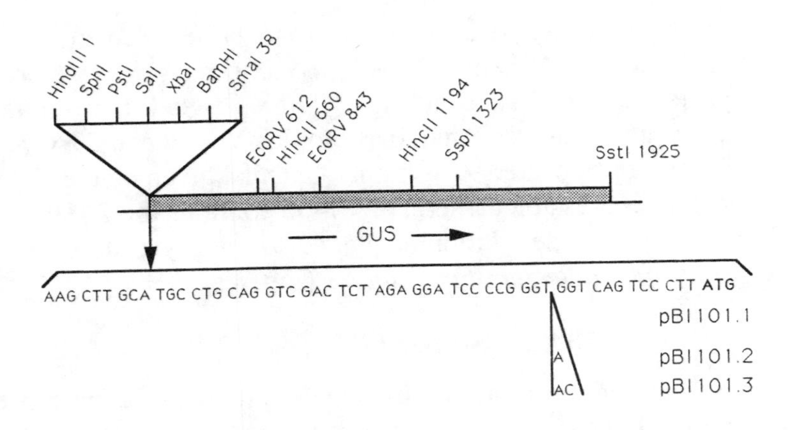

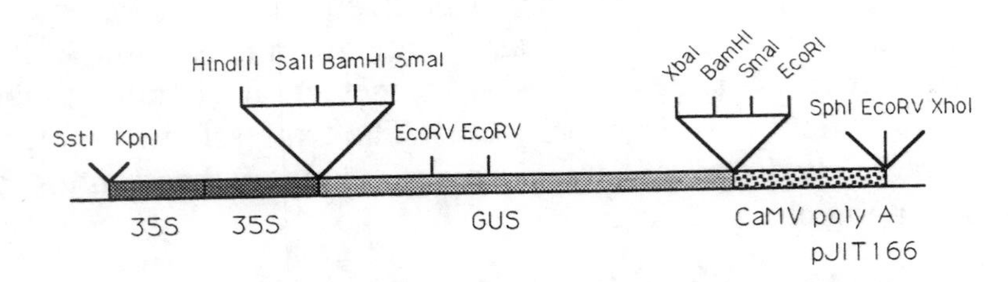

Fig. 5. Maps of the β-glucuronidase *(gus)* coding sequence in pBI101.1, .2, .3 *(15),* and pJIT166 *(3)*. The nucleotide sequence preceding the *gus* translation initiation triplet (shown in bold) is indicated. There are no sites for *Apa*I, *Bgl*II, *Cla*I, *Eco*RI, *Hpa*I, *Kpn*I, *Nco*I, *Sca*I, *Spe*I, *Sst*II, *Stu*I, *Sty*I, *Xho*I in or flanking the *gus* coding sequence in the pBI101 plasmids.

high level of luciferase activity was detected in plants transformed with a 35S-*luc* construct *(17)*. Plasmid pJIT27 *(18)* contains the *luc* coding sequence and pDR100-derived plasmids *(19)* offer other restriction sites for the construction of translational fusions (Fig. 6).

3.3. The Chloramphenicol Acetyltransferase Gene

The most commonly used chloramphenicol acetyltransferase gene *(cat)* originates from transposon *Tn*9 *(20)*. It has been widely used as a reporter gene in mammalian cells and to a lesser extent in plants, owing to the occurrence of the more versatile *gus* gene/assay system. Plasmids pJIT23, pJIT24, pJIT25 (Guerineau, unpublished), and pJIT26 *(9)* carry the *cat* coding sequence in different contexts (Fig. 7). Accession num-

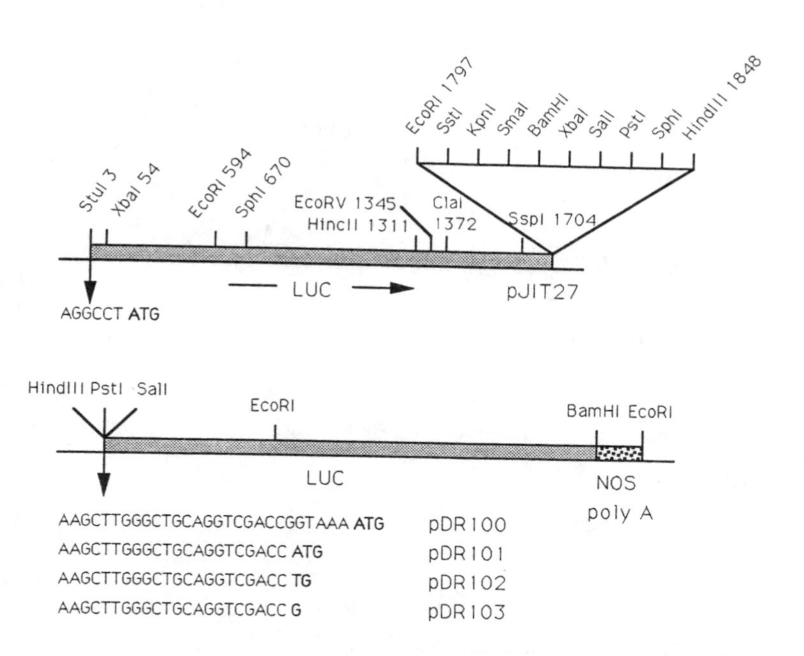

Fig. 6. Maps of the luciferase *(luc)* coding sequence in pJIT27 *(18)* and in pDR plasmids *(19)*. The nucleotide sequences upstream of the luciferase first codon (shown in bold) are indicated. There are no sites for *Apa*I, *Bgl*II, *Hpa*I, *Mlu*I, *Nco*I, *Sca*I, *Spe*I, *Sst*II, *Sty*I, *Xho*I in or flanking the *luc* coding sequence in pJIT27.

bers for the *cat* nucleotide sequence are V00622 and J01841 in the EMBL and GenBank databases, respectively.

3.4. Other Reporter Genes

The *lacZ* gene encoding β-galactosidase (β-GAL) in *E. coli* has been expressed in tobacco crown gall tissues *(21)*. An increase in β-GAL activity up to 20-fold could be detected in some of the transformants. However, the presence of a high endogenous β-GAL activity in plant cells makes this gene inconvenient for sensitive quantification of gene expression. The expression of the neomycin phosphotransferase *(nptII)* (*see* Chapter 12) and phosphinothricin acetyltransferase *(bar)* genes can also be quantified using radiochemical assays *(22,23)*.

4. Selectable Marker Genes

A selectable marker gene is used to recover transformants after a gene transfer experiment. It encodes a protein that confers on transformed cells

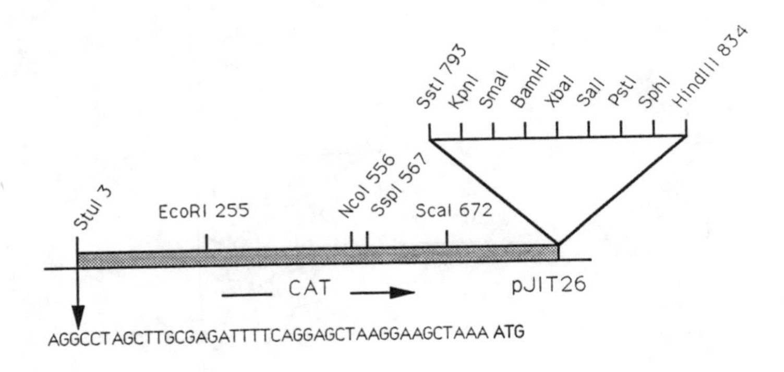

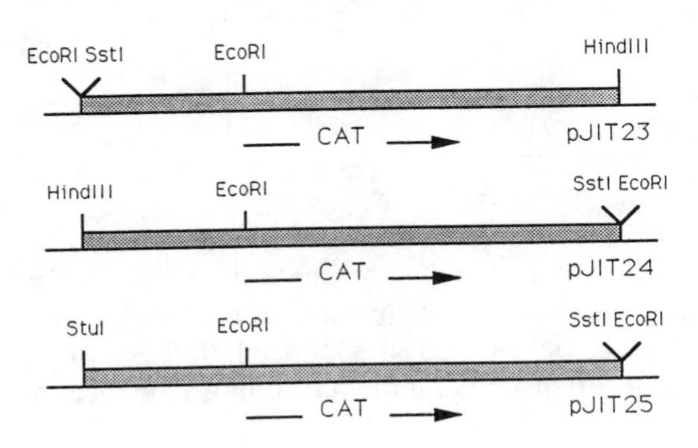

Fig. 7. Maps of the chloramphenicol acetyltransferase gene *(cat)* in pJIT23, 24, 25 (Guerineau, unpublished) and pJIT26 *(9)*. The nucleotide sequence preceding the translation initiation triplet (shown in bold) is indicated. There are no sites for *Apa*I, *Bgl*II, *Cla*I, *Eco*RV, *Hpa*I, *Mlu*I, *Spe*I, *Sst*II, *Xho*I in or flanking the *cat* coding sequence in pJIT26.

the ability to grow on media containing a compound toxic for untransformed cells. Transformants will emerge from the mass of untransformed tissue because of the advantage given by the expression of the resistance gene. The gene product of a selectable marker gene can be a detoxifying enzyme able to degrade the selective agent. Alternatively, it can be a mutated target for the toxic compound. The introduced gene will encode for an enzyme insensitive to inhibition by the selective agent. This enzyme will replace the defective native enzyme in the transformed cells.

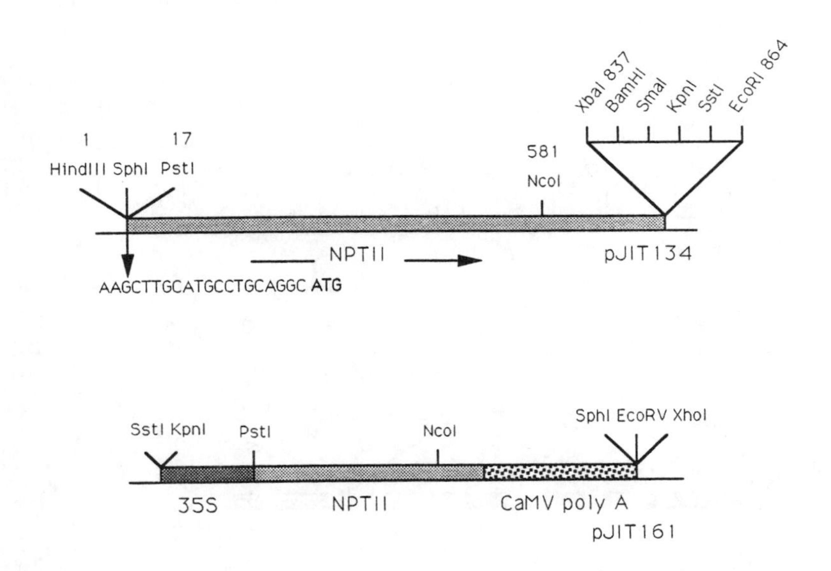

Fig. 8. Maps of the kanamycin resistance gene (*nptII*) contained in pJIT134 and pJIT161 *(9)*. The nucleotide sequence upstream of the translation initiation site (shown in bold) is indicated. There are no sites for *Acc*I, *Apa*I, *Bgl*II, *Cla*I, *Eco*RV, *Hin*cII, *Hpa*I, *Mlu*I, *Sal*I, *Sca*I, *Spe*I, *Ssp*I, *Sst*II, *Stu*I, *Xho*I in or flanking the *nptII* coding sequence in pJIT134.

4.1. Kanamycin Resistance

Some kanamycin resistance genes encode phosphotransferases able to inactivate one or several aminoglycoside antibiotics. The neomycin phosphotransferase gene *(nptII)* from transposon *Tn*5 was the first selectable marker used for plant transformation *(24,25)*. The *nptII* gene fused to the *nos* promoter is a component of many binary vectors and has been used for the recovery of transgenic plants in many species (*see* ref. 26 for review). The *nptII* gene has also been used for plastid transformation *(27)*. A mutated version of this gene, in which the *Pst*I and *Sph*I restriction sites have been removed, is present in pK18 *(28)* (accession number M17626). The coding sequence of this gene has been extracted from pK18, to create pJIT134 *(9)* and placed under control of the CaMV 35S promoter in pJIT161 *(9)* (Fig. 8).

4.2. Hygromycin Resistance

Another gene encoding for a detoxifying enzyme used for plant transformation is the hygromycin phosphotransferase gene (*hpt* or *aphIV*)

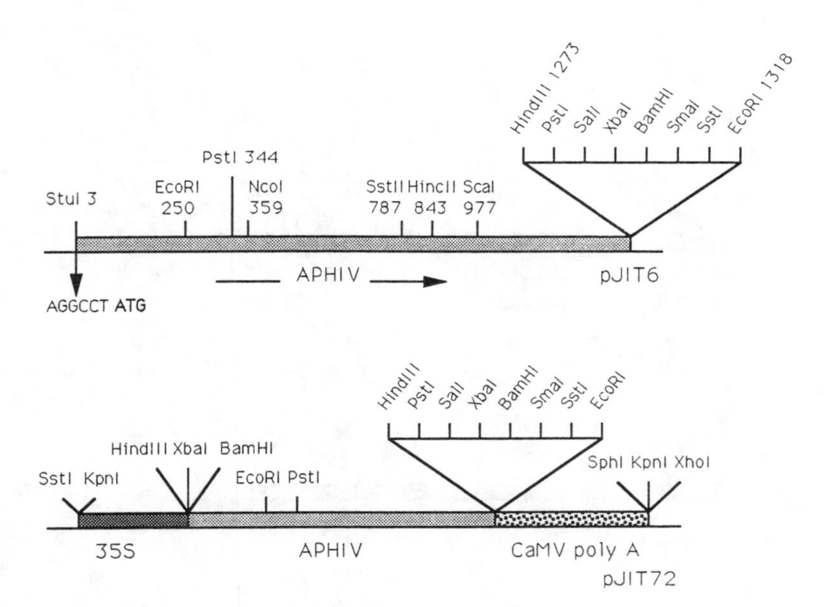

Fig. 9. Maps of the hygromycin resistance gene *(aphIV)* of pJIT6 and pJIT72 *(9)*. The nucleotide sequence preceding the translation initiation triplet (shown in bold) is indicated. There are no sites for *Apa*I, *Bgl*II, *Cla*I, *Eco*RV, *Hpa*I, *Kpn*I, *Mlu*I, *Spe*I, *Sph*I, *Ssp*I, *Xho*I in or flanking the *aphIV* coding sequence in pJIT6.

from *E. coli*. The gene was originally tested for tobacco transformation *(29,30)* and has more recently been used for the transformation of several plant species such as pea *(31)* and maize *(32)*. The coding sequence present in pJIT6 *(9)* (Fig. 9) was recovered from pJR225 *(33)*. It has been cloned downstream of the CaMV 35S promoter to create pJIT72 *(9)* (Fig. 9). The accession number for the hygromycin resistance gene is K01193.

4.3. Streptomycin/Spectinomycin Resistance

The streptomycin resistance gene *(spt)* from transposon *Tn*5 was first developed as a selectable marker for plant transformation *(34)*. More recently, another gene encoding an aminoglycoside-3"-adenyltransferase (*aad*A) has been shown to be a valuable marker gene *(35)*. Transformants expressing either the *spt* or the *aad*A gene form green calli and shoots on selective media containing streptomycin or spectinomycin, whereas untransformed tissues are yellow. This color selection proved to

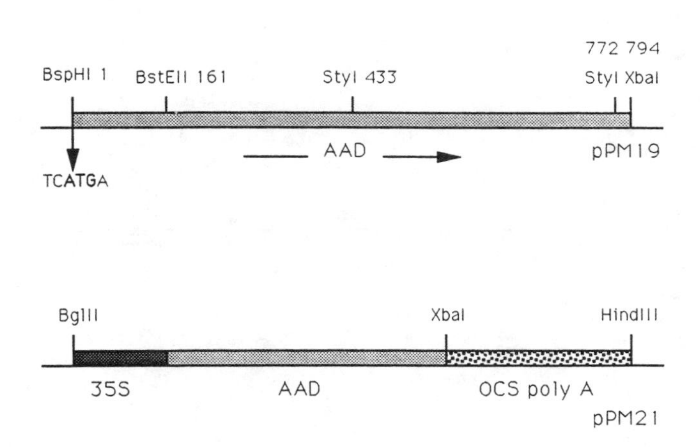

Fig. 10. Maps of the streptomycin/spectinomycin resistance gene *(aad)* present in pPM19 and pPM21 *(35)*. The translation initiator is shown in bold. OCS, octopine synthase. There are no sites for *Acc*I, *Apa*I, *Bam*HI, *Bgl*II, *Cla*I, *Eco*RI, *Eco*RV, *Hinc*II, *Hind*III, *Hpa*I, *Kpn*I, *Mlu*I, *Nco*I, *Pst*I, *Sal*I, *Sca*I, *Sma*I, *Spe*I, *Sph*I, *Ssp*I, *Sst*I, *Sst*II, *Stu*I, *Xho*I in or flanking the *aad* coding sequence in pPM19.

be very useful to monitor transposition events in plants *(36)*. Plasmid pPM21 contains a 35S-*aad*A fusion and the *aad*A coding sequence can be easily extracted from pPM19 *(35)* (Fig. 10). The database accession number for the *aad*A nucleotide sequence *(37)* is X03886.

4.4. Bialaphos Resistance

Bialaphos is an antibiotic consisting of two alanine residues linked to phosphinothricin, a glutamic acid analog that inhibits glutamine synthase. Phosphinothricin is also a chemically synthesized herbicide. The bialaphos resistance gene *(bar)* from *Streptomyces hygroscopicus* encodes a phosphinothricin acetyltransferase that is able to detoxify the herbicide *(38)*. Expression of a 35S-*bar* construct in transgenic tobacco, potato, and tomato plants resulted in a high level of resistance to phosphinothricin and bialaphos in those plants *(23)*. Transformation of oat plants *(39)*, maize *(40)*, and pea *(41)* was recently successful using the *bar* gene as a selectable marker. Plasmid pIJ4104 *(42)* contains a *bar* gene in a convenient context for cloning and pJIT82 *(9)* contains a 35S-*bar* fusion (Fig. 11). The accession number for the sequence of the *bar* gene of pIJ4104 is X17220.

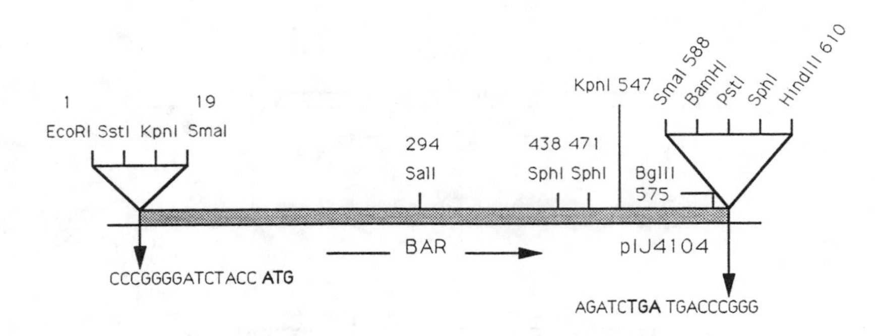

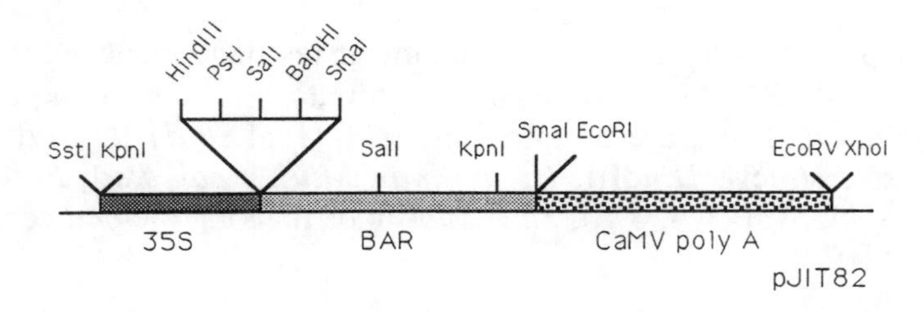

Fig. 11. Maps of the bialaphos resistance gene *(bar)* of pIJ4104 *(42)* and pJIT82 *(9)*. The nucleotide sequences upstream of the translation initiation site (shown in bold) and around the terminator (TGA in bold) are indicated. There are no sites for *Cla*I, *Eco*RV, *Hpa*I, *Mlu*I, *Nco*I, *Sca*I, *Spe*I, *Ssp*I, *Stu*I, *Sty*I, *Xba*I, *Xho*I in or flanking the *bar* coding sequence in pIJ4104.

4.5. Chlorsulfuron Resistance

The target of the herbicide chlorsulfuron is the enzyme acetolactate synthase (ALS). Two mutant *als* alleles, designated as *csr1-1* and *csr1-2*, were isolated from *Arabidopsis thaliana.* An increased level of tolerance to the herbicide was found in tobacco plants transformed with *csr1-1 (43)*. The mutation was shown to have originated from a single base substitution in the *als* coding sequence, making the modified enzyme resistant to inhibition by chlorsulfuron *(44)*. This gene has also been used for flax *(45)* and rice transformation *(46)*. The same mutation introduced into the maize *als* gene has allowed transgenic maize plants to be produced *(47)*. The database accession number for the *csr1-2* sequence is X51514. The *csr1-1* coding sequence can be recovered from pGH1 as a

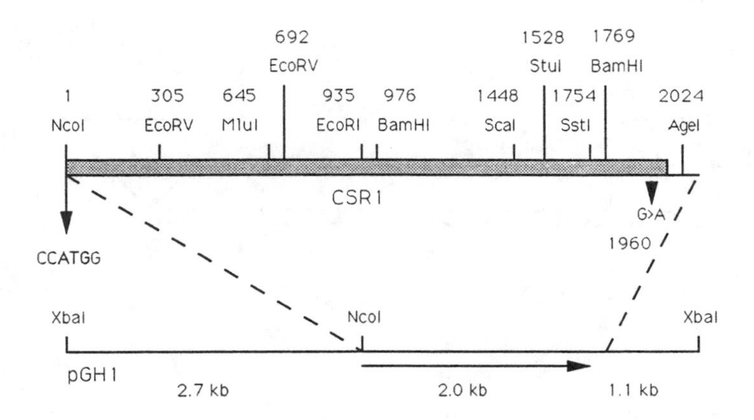

Fig.12. Map of the chlorsulfuron resistance gene *(csr1)* from *Arabidopsis thaliana*, from the translation initiator (shown in bold) to 47 nucleotides downstream of the coding sequence *(44)*. The position of the G to A mutation conferring chlorsulfuron resistance in *csr1-2* is indicated. There are 5 *Hin*dIII sites and no *Apa*I, *Cla*I, *Hpa*I, *Kpn*I, *Pst*I, *Sal*I, *Sma*I, *Spe*I, *Sph*I, *Ssp*I, *Xba*I sites in the *csr1* sequence.

2.02-kb *Nco*I-*Age*I fragment (Fig. 12). pALS used for maize transformation contains the CaMV 35S promoter linked to the *adh1* intron 1, the maize *als* coding sequence, and the *nos* polyadenylation sequence *(47)*, whereas pTRA153 used for rice transformation harbors the 35S promoter linked to the *Arabidopsis csr1* coding sequence and polyadenylation signal *(46)*.

4.6. Sulfonamide Resistance

Asulam is an herbicide that is related to the sulfonamides, a class of chemically synthesized antibacterial compounds. Sulfonamides are inhibitors of dihydropteroate synthase (DHPS), which is an enzyme of the folic acid biosynthetic pathway. Some sulfonamide resistance genes are known to encode a mutated DHPS that is insensitive to sulfonamides. The sulfonamide resistance gene from plasmid R46 has been cloned into pUC19, giving pJIT92 *(48)* (Fig. 13). The coding sequence was inserted into the expression cassette pJIT117 *(11)* (Fig. 3), creating pJIT118 *(12)* (Fig. 13). The chimeric gene was used to transform tobacco leaf explants. Transformants could be selected on asulam or sulfadiazine-containing media *(12)*. The hybrid gene of pJIT119 has also been successfully used

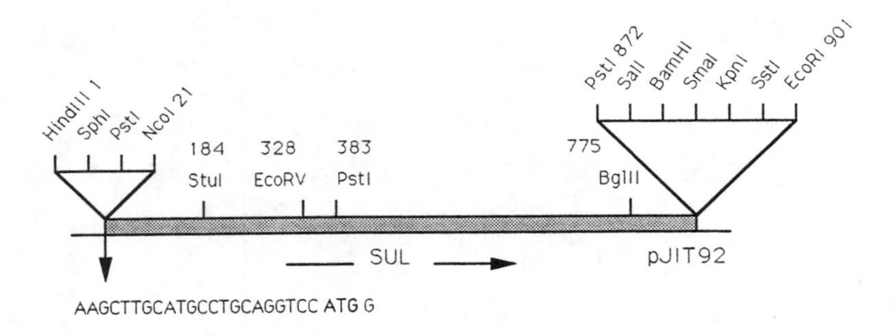

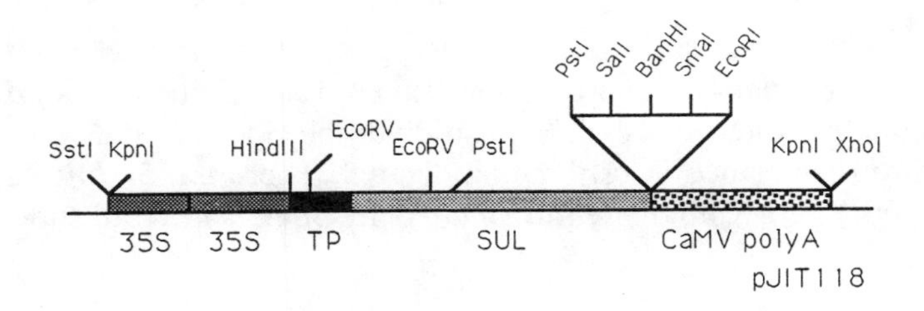

Fig. 13. Maps of the sulfonamide resistance gene *(sul)* of pJIT92 *(48)* and pJIT118 *(12)*. The nucleotide sequence preceding the translation initiation triplet (shown in bold) is indicated. TP, RUBISCO transit peptide. There are no sites for *Apa*I, *Cla*I, *Hpa*I, *Mlu*I, *Sca*I, *Spe*I, *Ssp*I, *Xba*I in or flanking the *sul* coding sequence in pJIT92.

for segregation analysis on transgenic *Arabidopsis thaliana* seedlings (Guerineau, unpublished). The database accession number for the nucleotide sequence of the *sul* gene is X15024.

4.7. Other Selectable Markers

4.7.1. Herbicide Resistance Genes

Similar to what has been observed with the sulfonamide resistance gene conferring asulam resistance, glyphosate resistance was recorded in transgenic plants after targeting a bacterial 5-enolpyruvylshikimate-3-phosphate synthase (EPSPS) to chloroplasts *(49)*. An increased level of tolerance to the herbicide could also be obtained by transformation of plants with a mutant *Petunia epsps* gene *(50)*.

Expression of a *Klebsiella ozaenae* nitrilase gene in transgenic tobacco plants resulted in an increased level of tolerance to the herbicide bromoxynil *(51)*. Similarly, expression in transgenic plants of an *Alcaligenes eutrophus* gene encoding a 2,4-dichlorophenoxyacetate monooxygenase enzyme (DPAM) led to the production of transgenic plants tolerant to 2,4-D *(52)*.

A detoxifying dehalogenase gene from *Pseudomonas putida* could confer on transgenic *Nicotiana plumbaginifolia* an increased level of resistance to 2,2 dichloropropionic acid (2,2 DCPA), the active ingredient of the herbicide Dalapon *(53)*. Direct selection could be applied using 2,2 DCPA.

4.7.2. Other Resistance Genes

Gentamicin is another aminoglycoside that has been used in plant transformation. Gentamicin-3-N-acetyltransferases *(aac[3])* inactivate gentamicin as well as kanamycin and other aminoglycosides. Expression of two of these genes under the control of the CaMV 35S promoter made it possible to select transformants in *Petunia*, tobacco, and other dicotyledonous plant species using gentamicin *(54)*.

Various Gramineae transgenic cell lines could be selected on methotrexate-containing media after transformation with a dihydrofolate reductase *(dhfr)* mouse gene under control of a CaMV 35S promoter *(55)*. The same construct had previously been used for *Petunia* transformation *(56)*.

One of the genes present in transposon *Tn*5 encodes resistance to bleomycin, a DNA damaging compound. Expression of this gene in plant cells resulted in an increase level of resistance to bleomycin *(57)*.

4.7.3. Genes from the Amino-Acid Synthesis Pathways

Two *E. coli* regulatory genes from the aspartate family pathway have recently been tested as selectable marker genes for potato transformation *(58)*. Dihydrodipicolinate synthase *(dhps)* is sensitive to feedback inhibition by lysine but the bacterial enzyme is much less sensitive than its plant counterpart. Plants expressing the bacterial *dhps* were resistant to the toxic lysine analog S-aminoethyl L-cysteine (AEC). Similarly, transfer and expression of the bacterial aspartate kinase (AK) gene conferred tolerance to lysine and threonine, which normally inhibit AK and cause starvation for methionine. In both cases, direct selection of transformants could be achieved and the selection appeared biased in favor of transgenic lines expressing the marker genes at a high level *(58)*.

A *Catharanthus roseus* tryptophan decarboxylase (TD) cDNA placed under control of the CaMV 35S promoter has allowed the selection of tobacco transformants on leaf disks placed on medium containing the toxic tryptophan analog 4-methyl tryptophan *(59)*.

Potential problems associated with the overexpression of enzymes such as DHPS, AK, or TD, might result from the alteration of physiological processes owing to changes in amino acid content. Abnormalities were found in 2 out of 50 tobacco lines expressing *dhps (58)*.

4.7.4. Negative Selectable Marker Genes

Transgenic *Nicotiana plumbaginifolia* constitutively expressing a nitrate reductase gene *(nia)* are killed by chlorate on medium containing ammonium as sole nitrogen source *(60)*. Under these nitrate-free conditions, wild-type plants are not affected by chlorate because the endogenous *nia* gene is not expressed *(60)*.

Cytosine deaminase (CD) converts the nontoxic compound 5-fluorocytosine (5FC) into 5-fluorouracil, which is toxic. CD is not found in eukaryotes. An *E. coli codA* gene encoding CD was fused to the CaMV 35S promoter and transferred to *Arabidopsis*. Untransformed seedlings grew normally when plated on 5FC-containing medium whereas transgenic seedlings died *(61)*.

These negative selection systems might be of interest, for example, in transposon tagging experiments, to eliminate plants not having undergone a transposition event. They could also ease the screening of mutated populations for regulatory mutants.

5. Plant Promoters

The number of plant genes isolated and characterized has dramatically increased in the last few years. The availability of transformation techniques has made it possible to study gene expression in transgenic plants. The use of fusions between promoters and reporter genes has allowed a detailed monitoring of the activity of numerous plant promoters. Some promoters appeared to be active only in certain organs or even cell types in the plant, whereas others were shown to be inducible, that is, activated in the presence of certain nutrients or by special treatments, such as wounding, heat shock, UV light, or pathogen elicitors. Rather than giving a comprehensive account of plant promoters isolated so far, I will focus on a few well characterized promoters having very distinctive patterns of expression. A common feature of these promoters is that all have been used in experiments involving their fusion to reporter genes in transgenic plants, demonstrating their ability to direct the expression of foreign genes in plants in a predictable tissue-specific manner. However, this tissue specificity might not be absolute because of the limit of detection of expression associated with the use of any reporter gene.

5.1. A Root-Specific Promoter

The tobacco *TobRB7* gene was shown to be expressed specifically in roots *(62)*. Homologies with nucleotide sequences of known function suggest that the *TobRB7* gene product might be involved in membrane channeling. The mRNA was not detected in leaves, stems, or shoot meristems. *In situ* hybridizations on root sections showed the presence of the mRNA in root meristems and in the immature central cylinder region. Fusion of the *TobRB7* promoter to the *gus* gene and its expression in tobacco plants resulted in GUS activity being detected exclusively in root tissues. Deletions from the 5'-end of the promoter sequence and subsequent GUS assays on transgenic plants demonstrated that 636 bp upstream of the transcription initiation site was sufficient to direct the expression of the *gus* gene in a root-specific manner. Owing to the deletion of a negative regulatory element located between positions -813 and -636, the GUS activity obtained using the 636 bp promoter was twice as high as that given by the 1.8-kb *TobRB7* upstream sequence. The nucleotide sequence of the whole *TobRB7* genomic clone is given in *(62)* (accession number S45406). The restriction map of the 636-bp promoter sequence is given in Fig.14. This sequence can be recovered as a *Xba*I-*Bam*HI fragment from the *gus* fusion construct *(62)*.

5.2. A Tuber-Specific Promoter

The patatins are a family of proteins found in potato tubers. A number of patatin-encoding genes have been characterized. The upstream sequence of the patatin B33 gene was fused to the *gus* coding sequence. A high specific GUS activity was detected in the tubers of potato plants transformed with the hybrid construct *(63)*. The GUS activity was 100- to 1000-fold higher in tubers than that found in roots, stems, or leaves. Histochemical localization of GUS activity showed that the promoter

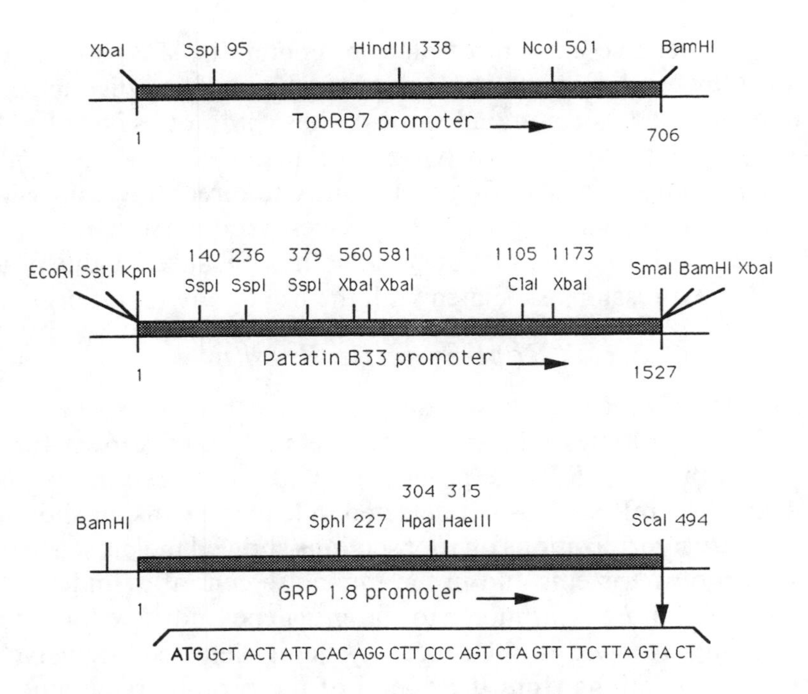

Fig. 14. Maps of a root-specific promoter (TobRB7) *(62)*, of a tuber-specific promoter (patatin B33) *(63)* and of a vascular tissue-specific promoter (GRP 1.8, glycine-rich protein) *(64)*. The sequence extending from the translation initiation site (shown in bold) to the *Sca*I site in the GRP 1.8 construct is indicated. There are no sites for *Acc*I, *Apa*I, *Bgl*II, *Cla*I, *Eco*RI, *Eco*RV, *Hin*cII, *Hpa*I, *Kpn*I, *Mlu*I, *Pst*I, *Sal*I, *Sca*I, *Sma*I, *Spe*I, *Sph*I, *Sst*I, *Sst*II, *Stu*I, *Xho*I in or flanking the TobRB7 promoter shown here. There are no sites for *Acc*I, *Apa*I, *Bgl*II, *Eco*RV, *Hin*cII, *Hin*dIII, *Hpa*I, *Mlu*I, *Nco*I, *Pst*I, *Sal*I, *Sca*I, *Spe*I, *Sph*I, *Sst*II, *Stu*I, *Sty*I, *Xho*I in or flanking the B33 promoter shown here. There are no sites for *Acc*I, *Apa*I, *Bam*HI, *Bgl*II, *Cla*I, *Eco*RI, *Eco*RV, *Hin*dIII, *Kpn*I, *Mlu*I, *Nco*I, *Pst*I, *Sal*I, *Sma*I, *Spe*I, *Ssp*I, *Sst*I, *Sst*II, *Stu*I, *Sty*I, *Xba*I, *Xho*I in the GRP 1.8 promoter shown here.

was active in parenchymatic tissue but not in the peripheral cells of the transgenic tubers. The expression of the patatin B33 gene can be induced in leaves subjected to high concentrations of sucrose *(63)*. The B33 promoter can be recovered on a 1.5-kb *Dra*I fragment. Unique *Eco*RI, *Sst*I, and *Kpn*I sites and *Sma*I and *Bam*HI sites are located respectively 5' and 3' of the promoter sequence in the *gus* fusion construct *(63)* (Fig.14). Its nucleotide sequence (accession number X14483) can be found in *(63)*.

5.3. A Promoter Active in Vascular Tissues

Glycine-rich proteins (GRP) are a class of plant cell wall proteins. Two genes, *GRP 1.0* and *GRP 1.8*, encoding glycine-rich proteins in bean have been isolated on a single genomic clone *(64)*. The *GRP 1.8* promoter was used to drive the expression of the *gus* gene in transgenic tobacco *(65)*. The hybrid gene was shown to be expressed in primary and secondary vascular tissues of roots, stems, leaves, and flowers, during differentiation. The expression was also induced in pith parenchyma cells after excision-wounding of young stems. A promoter fragment of 494 bp containing 427 bp of 5' untranscribed sequence was shown to contain all the information for tissue-specific and wound-inducible expression *(65)*. Deletions in the 5' regulatory sequence of the *GRP 1.8* gene have revealed the existence of two stem elements, one root element and one negative regulatory element *(66)*. The restriction map of the *GRP 1.8* promoter present in the *gus* hybrid construct is shown in Fig. 14. Its nucleotide sequence (accession number X13596) can be found in *(65)* and *(66)*.

5.4. Genes Expressed in Photosynthetic Tissues

5.4.1. A RUBISCO Small Subunit Promoter

The ribulose bisphosphate carboxylase (RUBISCO) small subunit is encoded by a family of nuclear genes *(rbcS)* members of which have been characterized in many species. Their expression, which is light-inducible, is restricted to various photosynthetic tissues. The level and the pattern of expression of members of the RUBISCO gene family have been found to be highly variable *(67)*. When a 1.1-kb fragment containing the tomato *rbcS-3A* promoter was fused to the *cat* coding sequence and transferred to tobacco plants, high level CAT activity was measured in mature leaves *(68)*. In contrast, no or very low expression was detected in roots, stems, flower buds, sepals, petals, stamens, ovaries, or stigmas. The activity in young leaves was approx 10% of that in mature leaves. When the region fused to the *cat* gene was restricted to the 374-bp sequence located upstream of the transcription start, the tissue-specific and light-inducible pattern of expression was maintained, but the level of expression was 5 times lower than that obtained with the full-length promoter. The level of expression given by the full-length *rbcS-3A-cat* fusion was estimated to be 50–70% of that of a CaMV 35S-*cat* construct *(68)*. This is much more than the level of expression given with the pea *rbcS-E9* promoter contained in the expression cassette pKYLX3 (Fig. 4)

(4). The nucleotide sequence of the tomato *rbcS-3A* promoter is found in *(68)*. Its restriction map is shown in Fig. 15.

5.4.2. A Cab Promoter

Genes encoding chlorophyll a/b-binding proteins show patterns of expression similar to those of *rbcS* genes. The promoters of three *Arabidopsis thaliana cab* genes have been cloned upstream of the *cat* coding sequence *(72)*. The *cab-3* promoter appeared to be two to three times stronger than the other two promoters in transgenic tobacco plants. CAT activities were high in green tissues but only weak in roots, stems, and senescing leaves. No activity was found in dark-grown seedlings. The *cab-3* promoter sequence extending 209 bp upstream of the translation start of the *cab-3* gene, driving the expression of the *cat* gene, was sufficient to achieve optimal level of expression, accurate tissue-specificity, and light-induction *(69)*. Note that part of the *cab* coding sequence was also present in the hybrid construct. The restriction map of the *cab3* promoter is shown in Fig. 15. Its nucleotide sequence (accession number X15222) is presented in *(69)*.

5.5. A Flower-Specific Promoter

Chalcone synthase (CHS) is a key enzyme of the flavonoid biosynthetic pathway, which produces compounds that pigment flowers and seed coats and protect plants against pathogens or UV irradiation. Chalcone synthase is encoded by a multigene family, members of which show very different patterns of expression. The *chsA* gene of *Petunia* is expressed primarily in flower tissues where it accounts for 90% of the *chs* mRNA *(70)*. Expression of the *chsA* gene is light-dependent and can be induced by UV light in young seedlings. A DNA fragment containing 805 bp of 5' untranscribed region was fused to the *gus* coding sequence, and the hybrid construct was introduced into *Petunia* plants. It appeared that expression of the hybrid gene occurred in various pigmented and unpigmented cell types of the flower stem, corolla, ovary, anthers, and seed coat *(73)*. Previous gene fusion and deletion experiments had shown that 800-bp of promoter sequence were more efficient at directing the expression of the *cat* gene in transgenic *Petunia* than the whole 2.4-kb upstream sequence *(74)*. The 800-bp *chsA* promoter can be recovered from various plasmids as an *Eco*RI-*Nco*I fragment, the *Nco*I site being created around the ATG initiator codon by site-directed mutagenesis *(70)*.

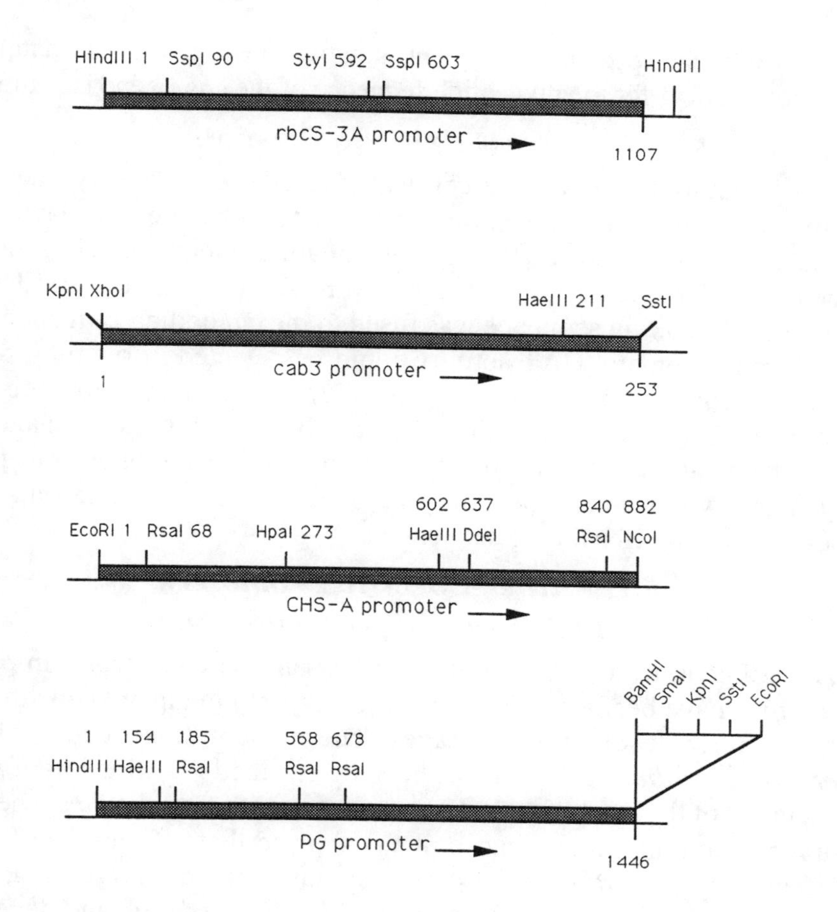

Fig. 15. Maps of promoters active in photosynthetic tissues (rbcS, RUBISCO small subunit) *(68)*, (cab3, chlorophyll a/b-binding protein) *(69)*, in flowers (CHS-A, chalcone synthase) *(70)* or in fruits (PG, polygalacturonase) *(71)*. There are no sites for *Acc*I, *Apa*I, *Bam*HI, *Bgl*II, *Cla*I, *Eco*RI, *Eco*RV, *Hinc*II, *Hpa*I, *Kpn*I, *Mlu*I, *Nco*I, *Pst*I, *Sal*I, *Sca*I, *Sma*I, *Spe*I, *Sph*I, *Sst*I, *Sst*II, *Stu*I, *Xba*I, *Xho*I in the rbcS-3A promoter shown here. There are no sites for *Acc*I, *Apa*I, *Bam*HI, *Bgl*II, *Cla*I, *Eco*RI, *Eco*RV, *Hinc*II, *Hin*dIII, *Hpa*I, *Mlu*I, *Nco*I, *Pst*I, *Sal*I, *Sca*I, *Sma*I, *Spe*I, *Sph*I, *Ssp*I, *Sst*II, *Stu*I, *Sty*I, *Xba*I in or flanking the cab3 promoter shown here. There are no sites for *Acc*I, *Apa*I, *Bam*HI, *Bgl*II, *Cla*I, *Eco*RV, *Hin*dIII, *Kpn*I, *Mlu*I, *Pst*I, *Sal*I, *Sca*I, *Sma*I, *Spe*I, *Sph*I, *Ssp*I, *Sst*I, *Sst*II, *Stu*I, *Xba*I, *Xho*I in the CHS-A promoter shown here. There are no sites for *Acc*I, *Apa*I, *Bgl*II, *Cla*I, *Eco*RV, *Hinc*II, *Hpa*I, *Mlu*I, *Nco*I, *Pst*I, *Sal*I, *Sca*I, *Sph*I, *Sst*II, *Stu*I, *Sty*I, *Xba*I, *Xho*I in or flanking the PG promoter shown here.

The restriction map of this fragment is shown in Fig. 15. The database accession number for the nucleotide sequence of the *chsA* gene is X14591.

5.6. A Fruit-Specific Promoter

Polygalacturonase (PG) is a cell wall degrading enzyme synthesized in ripening tomato fruits. The *pg* gene of tomato has been isolated and the nucleotide sequence of 1.4 kb of upstream sequence has been determined *(71)*. The gene appeared to be expressed only in ripening fruits. When the 5' flanking sequence was fused to the *cat* coding sequence and transferred to tomato, CAT activity could be detected in ripening fruits but not in leaves, roots, or unripe fruits *(71)*. A *Spe*I restriction site was introduced 29 bp upstream of the ATG translation initiator codon and the 1.4-kb sequence containing the *pg* promoter was subcloned into pUC *(71)*, from which it can be easily recovered (Fig. 15). The database accession number for the *pg* sequence is X14074.

5.7. Anther-Specific Promoters

5.7.1. A Tapetum-Specific Promoter

The *A9* gene from *Arabidopsis thaliana* and its counterpart in *Brassica napus* have been shown to be expressed only in tapetal cells during certain anther developmental stages. The nucleotide sequence of the *Arabidopsis thaliana A9* gene has been determined *(75)*. Fusion of various lengths of the *Arabidopsis A9* upstream sequence to the *gus* gene and expression in transgenic tobacco plants showed that a *Hinc*II-*Rsa*I 329-bp fragment was sufficient to direct tapetum-specific expression. The level of expression in the anthers appeared to be very high and developmentally regulated. GUS activity could only be found in the anthers at the stages extending from the beginning of meiosis to the middle of microspore interphase. No activity could be found in pollen, carpels, seeds, or leaves. The active promoter fragment can be easily recovered from pWP70A *(75)* (Fig. 16). The accession number for the *Arabidopsis A9* gene sequence is X61750.

5.7.2. A Pollen-Specific Promoter

An anther-specific gene has been isolated from the tomato genome and its nucleotide sequence has been determined *(76)*. In contrast to the *Arabidopsis A9* gene, the tomato *Lat52* gene was shown to be expressed in pollen grains. A 0.6-kb sequence located upstream of the *Lat52* coding sequence was fused to the *gus* gene and the hybrid construct was

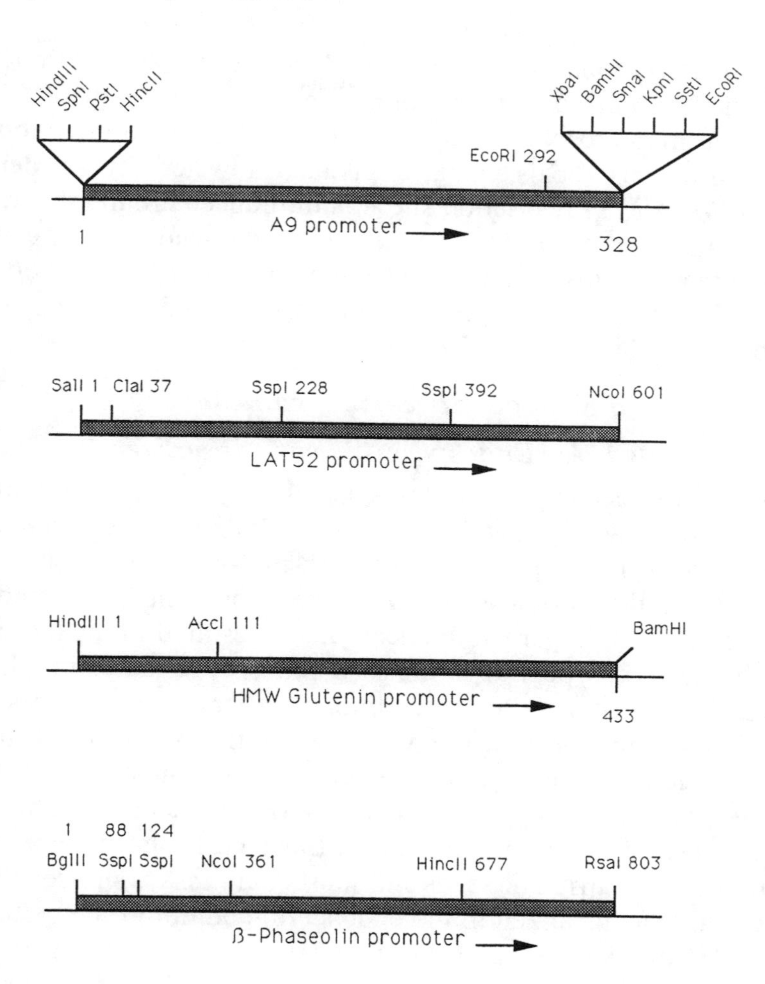

Fig. 16. Maps of promoters specifically expressed in the tapetum (A9) *(75)*, in pollen (LAT52) *(76)*, in endosperm (HMW Glutenin) *(77)*, or in immature embryos (β-phaseolin) *(78)*. There are no sites for AccI, *Apa*I, *Bgl*II, *Cla*I, *Eco*RV, *Hpa*I, *Mlu*I, *Nco*I, *Sal*I, *Sca*I, *Spe*I, *Ssp*I, *Sst*II, *Stu*I, *Sty*I, *Xho*I in or flanking the A9 promoter shown here. There are no sites for *Apa*I, *Bam*HI, *Bgl*II, *Eco*RI, *Eco*RV, *Hin*dIII, *Hpa*I, *Kpn*I, *Mlu*I, *Pst*I, *Sca*I, *Sma*I, *Spe*I, *Sph*I, *Sst*I, *Sst*II, *Stu*I, *Xba*I, *Xho*I in the LAT52 promoter shown here. There are no sites for *Apa*I, *Bgl*II, *Cla*I, *Eco*RI, *Eco*RV, *Hin*cII, *Hpa*I, *Kpn*I, *Mlu*I, *Nco*I, *Pst*I, *Sal*I, *Sca*I, *Sma*I, *Spe*I, *Sph*I, *Ssp*I, *Sst*I, *Sst*II, *Stu*I, *Sty*I, *Xba*I, *Xho*I in or flanking the Glutenin promoter shown here. There are no sites for *Acc*I, *Apa*I, *Bam*HI, *Cla*I, *Eco*RI, *Eco*RV, *Hin*dIII, *Hpa*I, *Kpn*I, *Mlu*I, *Pst*I, *Sal*I, *Sca*I, *Sma*I, *Spe*I, *Sph*I, *Sst*I, *Sst*II, *Stu*I, *Xba*I, *Xho*I in the Phaseolin promoter shown here.

used to transform tobacco, tomato, and *Arabidopsis (79)*. GUS activity was detected in transgenic plants in pollen at the developmental stages extending from microspore mitosis to anthesis. No activity was found in roots, stems, leaves, sepals, petals, or pistils. A low activity was detected in seeds *(80)*. A *Nco*I restriction site was introduced around the translation initiator codon of the *Lat52* gene *(76)* so the whole functional promoter fragment could be recovered from pLAT52-7 *(79)* as a *Sal*I-*Nco*I fragment (Fig. 16). The accession number for the *Lat52* gene nucleotide sequence is X15855.

5.8. Seed-Specific Promoters

5.8.1. An Endosperm-Specific Promoter

Glutenins are seed storage proteins encoded by a multigene family in wheat. High-mol-wt and low-mol-wt glutenin subunits have been identified. The nucleotide sequence of the high-mol-wt glutenin subunit 12 gene located on the chromosome 1D of wheat has been determined *(81)*. Transfer of the whole gene to tobacco plants resulted in the accumulation of the intact polypeptide in the seed endosperm, indicating the correct function of this monocot promoter in transgenic dicots *(77)*. Restriction fragments carrying various lengths of the high-mol-wt glutenin subunit 12 promoter sequence were cloned upstream of the *cat* coding sequence. Expression of the *cat* gene was not detected in the roots, stems, or leaves but only in the seeds of transgenic tobacco plants *(77)*. Dissection of the transgenic seeds and subsequent assays showed that the CAT activity was localized in the endosperm and not in the embryo. It appeared 8 d after anthesis and persisted until seed maturity. A fragment of 433 bp was sufficient to confer endosperm-specific expression. This promoter sequence can be recovered as a *Hin*dIII-*Bam*HI fragment from the pUC-promoter construct *(77)*. Its restriction map is shown in Fig. 16. The accession number for the nucleotide sequence of the gene is X03041.

5.8.2. An Embryo-Specific Promoter

Phaseolins account for a large proportion of the seed storage proteins in beans. Phaseolins accumulate in the cotyledons and their mRNAs are present at high level during the embryo maturation stage preceding seed desiccation and dormancy. The nucleotide sequence of a β-phaseolin gene from *Phaseolus vulgaris* has been determined *(78)*. Fusion of an 0.8-kb fragment from the 5'-flanking region of the gene to the *gus* coding sequence and subsequent transfer of the hybrid gene to tobacco plants resulted in the exclusive expression of the *gus* gene in immature embryos *(82)*. GUS activity was highest in cotyledons. It appeared and increased rapidly 12–17 d after flowering and then remained constant until 25–30 d after flowering. The high GUS activity found in seeds decayed rapidly during the early stages of seed germination: Nearly all activity was lost in seedlings 6 d after seed imbibition. The 0.8-kb promoter region can be recovered as a *Bgl*II-*Rsa*I fragment (Fig. 16). The accession number for its nucleotide sequence is J01263.

Acknowledgments

The author thanks Wyatt Paul, Anna Sorensen, and Rod Scott for their valuable comments on the manuscript, and the BBSRC and the DTI Consortium "Plant Gene Tool Kit" for funding some of the work reported in this chapter.

References

1. Kay, R., Chan, A., Daly, M., and McPherson, J. (1987) Duplication of CaMV 35S promoter sequences creates a strong enhancer for plant genes. *Science* **236,** 1299–1302.
2. Pietrzak, M., Shillito, R. D., Hohn, T., and Potrykus, I. (1986) Expression in plants of two bacterial antibiotic resistance genes after protoplast transformation with a new plant expression vector. *Nucleic Acids Res.* **14,** 5857–5868.
3. Guerineau, F., Lucy, A., and Mullineaux, P. (1992) Effect of two consensus sequences preceding the translation initiator codon on gene expression in plant protoplasts. *Plant Mol. Biol.* **18,** 815–818.
4. Schardl, C. L., Byrd, A. D., Benzion, G., Altschuler, M. A., Hildebrand, D. F., and Hunt, A. G. (1987) Design and construction of a versatile system for the expression of foreign genes in plants. *Gene* **61,** 1–11.
5. Töpfer, R., Matzeit, V., Gronenborn, B., Schell, J., and Steinbiss, H. H. (1987) A set of plant expression vectors for transcriptional and translational fusions. *Nucleic Acids Res.* **15,** 5890.
6. Jones, J. D. G., Dunsmuir, P., and Bedbrook, J. (1985) High level expression of introduced chimaeric genes in regenerated transformed plants. *EMBO J.* **4,** 2411–2418.
7. Kozak, M. (1987) At least six nucleotides preceding the AUG initiator codon enhance translation in mammalian cells. *J. Mol. Biol.* **196,** 947–950.
8. Lütcke, H. A., Chow, K. C., Mickel, F. S., Moss, K. A., Kern, H. F., and Scheele G. A. (1987) Selection of AUG initiation codons differs in plants and animals. *EMBO J.* **6,** 43–48.
9. Guerineau, F. and Mullineaux, P. (1993) Plant transformation and expression vectors, in *Plant Molecular Biology LABFAX* (Croy, R. R. D., ed.), βios Scientific and Blackwell Scientific, Oxford, pp. 121–147.

10. Wasmann, C. C., Reiss, B., Bartlett, S. G., and Bohnert, H. J. (1986) The importance of the transit peptide and the transported protein for protein import into chloroplasts. *Mol. Gen. Genet.* **205,** 446–453.
11. Guerineau, F., Woolston, S., Brooks, L., and Mullineaux, P. (1988) An expression cassette for targeting foreign proteins into chloroplasts. *Nucleic Acids Res.* **16,** 11,380.
12. Guerineau, F., Brooks, L., Meadows, J., Lucy, A., Robinson, C., and Mullineaux, P. (1990) Sulfonamide resistance gene for plant transformation. *Plant Mol. Biol.* **15,** 127–136.
13. Ainley, W. M. and Key, J. L. (1990) Development of a heat shock inducible expression cassette for plants: Characterization of parameters for its use in transient expression assays. *Plant Mol. Biol.* **14,** 949–967.
14. Jefferson, R. A., Burgess, S. M., and Hirsh, D. (1986) β-Glucuronidase from *Escherichia coli* as a gene-fusion marker. *Proc. Natl. Acad. Sci. USA* **83,** 8447–8451.
15. Jefferson, R. A., Kavanagh, T. A., and Bevan, M. W. (1987) *gus* fusions: β-glucuronidase as a sensitive and versatile gene fusion marker in higher plants. *EMBO J.* **6,** 3901–3907.
16. De Wet, J. R., Wood, K. V., Helinski, D. R., and DeLuca, M. (1985) Cloning of firefly luciferase cDNA and the expression of active luciferase in *Escherichia coli. Proc. Natl. Acad. Sci. USA* **82,** 7870–7873.
17. Ow., D. W., Wood, K. V., DeLuca, M., de Wet, J. R., Helinski, D. R., and Howell, S. H. (1986) Transient and stable expression of the firefly luciferase gene in plant cells and transgenic plants. *Science* **234,** 856–859.
18. Mullineaux, P. M., Guerineau, F., and Accotto, G. P. (1990) Processing of complementary sense RNAs of *Digitaria* streak virus in its host and in transgenic tobacco. *Nucleic Acids Res.* **18,** 7259–7265.
19. Riggs, C. D. and Chrispeels, M. J. (1987) Luciferase reporter gene cassettes for plant gene expression studies. *Nucleic Acids Res.* **15,** 8115.
20. Alton, N. K. and Vapnek, D. (1979) Nucleotide sequence analysis of the chloramphenicol resistance transposon *tn*9. *Nature* **282,** 864–869.
21. Matsumoto, S., Takebe, I., and Machida, Y. (1988) *Escherichia coli lacZ* gene as a biochemical and histochemical marker in plant cells. *Gene* **66,** 19–29.
22. McDonnell, R. E., Clark, R. D., Smith, W. A., and Hinchee, M. A. (1987) A simplified method for the detection of neomycin phosphotransferase II activity in transformed plant tissues. *Plant Mol. Biol. Rep.* **5,** 380–386.
23. De Block, M., Botterman, J., Vandewiele, M., Dockx, J., Thoen, C., Gosselé, V., Movva, N. R., Thompson, C., Van Montagu, M., and Leemans, J. (1987) Engineering herbicide resistance in plants by expression of a detoxifying enzyme. *EMBO J.* **6,** 2513–2518.
24. Herrera-Estrella, L., De Block, M., Messens, E., Hernalsteens, J. P., Van Montagu, M., and Schell, J. (1983) Chimeric genes as dominant selectable markers in plant cells. *EMBO J.* **2,** 987–995.
25. Bevan, M. W., Flavell, R. B., and Chilton, M. D. (1983) A chimaeric antibiotic resistance gene as a selectable marker for plant cell transformation. *Nature* **304,** 184–187.

26. Ellis, J. R. (1993) Plant tissue culture and genetic transformation, in *Plant Molecular Biology LABFAX* (Croy, R. R. D., ed.), βios Scientific and Blackwell Scientific, Oxford, pp. 253–285.
27. Carrer, H., Hockenberry, T. N., Svab, Z., and Maliga, P. (1993) Kanamycin resistance as a selectable marker for plastid transformation in tobacco. *Mol. Gen. Genet.* **241,** 49–56.
28. Pridmore, R. D. (1987) New and versatile cloning vectors with kanamycin-resistance marker. *Gene* **56,** 309–312.
29. Waldron, C., Murphy, E. B., Roberts, J. L., Gustafson, G. D., Armour, S. L., and Malcolm, S. K. (1985) Resistance to hygromycin B. A new marker for plant transformation studies. *Plant Mol. Biol.* **5,** 103–108.
30. Van den Elzen, P. J. M., Townsend, J., Lee, K. Y., and Bedbrook, J. R. (1985) A chimaeric hygromycin resistance gene as a selectable marker in plant cells. *Plant Mol. Biol.* **5,** 299–302.
31. Lulsdorf, M. M., Rempel, H., Jackson, J. A., Baliski, D. S., and Hobbs, S. L. A. (1991) Optimizing the production of transformed pea (*Pisum sativum* L.) callus using disarmed *Agrobacterium tumefaciens* strains. *Plant Cell Rep.* **9,** 479–483.
32. Walters, D. A., Vetsch, C. S., Potts, D. E., and Lundquist, R. C. (1992) Transformation and inheritance of a hygromycin phosphotransferase gene in maize plants. *Plant Mol. Biol.* **18,** 189–200.
33. Gritz, L. and Davies, J. (1983) Plasmid-encoded hygromycin B resistance: the sequence of hygromycin B phosphotransferase gene and its expression in *Escherichia coli* and *Saccharomyces cerevisiae*. *Gene* **25,** 179–188.
34. Jones, J. D. G., Svab, Z., Harper, E. C., Hurwitz, C. D., and Maliga, P. (1987) A dominant nuclear streptomycin resistance marker for plant cell transformation. *Mol. Gen. Genet.* **210,** 86–91.
35. Svab, Z., Harper, E. C., Jones, J. D. G., and Maliga, P. (1990) Aminoglycoside-3'-adenyltransferase confers resistance to spectinomycin and streptomycin in *Nicotiana tabacum*. *Plant Mol. Biol.* **14,** 197–205.
36. Jones, J. D. G., Carland, F. M., Maliga, P., and Dooner, H. K. (1989) Visual detection of transposition of the maize element *Activator (Ac)* in tobacco seedlings. *Science* **244,** 204–207.
37. Chinault, A. C., Blakesley, V. A., Roessler, E., Willis, D. G., Smith, C. A., Cook, R. G., and Fenwick, Jr., R. G. (1986) Characterization of transferable plasmids from *Shigella flexneri 2a* that confer resistance to trimethoprim, streptomycin, and sulfonamides. *Plasmid* **15,** 119–131.
38. Thompson, C. J., Movva, N. R., Tizard, R., Crameri, R., Davies, J. E., Lauwereys, M., and Botterman, J. (1987) Characterization of the herbicide-resistance gene *bar* from *Streptomyces hygroscopicus*. *EMBO J.* **6,** 2519–2523.
39. Somers, D. A., Rines, H. W., Gu, W., Kaeppler, H. F., and Bushnell, W. R. (1992) Fertile, transgenic oat plants. *Biotechnology* **10,** 1589–1594.
40. Gordon-Kamm, W. J., Spencer, T. M., Mangano, M. L., Adams, T. R., Daines, R. J., Start, W. G., O'Brien, J. V., Chambers, S. A., Adams, Jr., W. R., Willetts, N. G., Rice, T. B., Mackey, C. J., Krueger, R. W., Kausch, A. P., and Lemaux, P. G. (1990)

Transformation of maize cells and regeneration of fertile transgenic plants. *Plant Cell* **2,** 603–618.
41. Schroeder, H. E., Schotz, A. H., Wardley-Richardson, T., Spencer, D., and Higgins, T. J. V. (1993) Transformation and regeneration of two cultivars of pea (*Pisum sativum* L.) *Plant Physiol.* **101,** 751–757.
42. White, J., Chang, S. Y. P., Bibb, M. J., and Bibb, M. J. (1990) A cassette containing the *bar* gene of *Streptomyces hygroscopicus*: a selectable marker for plant transformation. *Nucleic Acids Res.* **18,** 1062.
43. Haughn, G. W., Smith, J., Mazur, B., and Somerville, C. (1988) Transformation with a mutant *Arabidopsis* acetolactate synthase gene renders tobacco resistant to sulfonylurea herbicides. *Mol. Gen. Genet.* **211,** 266–271.
44. Sathasivan, K., Haughn, G. W., and Murai, N. (1990) Nucleotide sequence of a mutant acetolactate synthase gene from an imidazolinone-resistant *Arabidopsis thaliana* var. Columbia. *Nucleic Acids Res.* **18,** 2188.
45. McHugen, A. (1989) *Agrobacterium* mediated transfer of chlorsulfuron resistance to commercial flax cultivars. *Plant Cell Rep.* **8,** 445–449.
46. Li, Z., Hayashimoto, A., and Murai, N. (1992) A sulfonylurea herbicide resistance gene from *Arabidopsis thaliana* as a new selectable marker for production of fertile transgenic rice plants. *Plant Physiol.* **100,** 662–668.
47. Fromm, M. E., Morrish, F., Armstrong, C., Williams, R., Thomas, J., and Klein, T. M. (1990) Inheritance and expression of chimeric genes in the progeny of transgenic maize plants. *Biotechnology* **8,** 833–839.
48. Guerineau, F. and Mullineaux, P. (1989) Nucleotide sequence of the sulfonamide resistance gene from plasmid R46. *Nucleic Acids Res.* **17,** 4370.
49. Della-Cioppa, G., Bauer, S. C., Taylor, M. L., Rochester, D. E., Klein, B. K., Shah, D. M., Fraley, R. T., and Kishore, G. M. (1987) Targeting a herbicide-resistant enzyme from *Escherichia coli* to chloroplasts of higher plants. *Biotechnology* **5,** 579–584.
50. Shaw, D. M., Horsch, R. B., Klee, H. J., Kishore, G. M., Winter, J. A., Tumer, N. E., Hironaka, C. M., Sanders, P. R., Gasser, C. S., Aykent, S., Siegel, N. R., Rogers, S. G., and Fraley, R. T. (1986) Engineering herbicide tolerance in transgenic plants. *Science* **233,** 478–481.
51. Stalker, D. M., McBride, K. E., and Malyj, L. D. (1988) Herbicide resistance in transgenic plants expressing a bacterial detoxification gene. *Science* **242,** 419–423.
52. Streber, W. R. and Willmitzer, L. (1989) Transgenic tobacco plants expressing a bacterial detoxifying enzyme are resistant to 2,4-D. *Biotechnology* **7,** 811–816.
53. Buchanan-Wollaston, V., Snape, A., and Cannon, F. (1992) A plant selectable marker gene based on the detoxification of the herbicide Dalapon. *Plant Cell Rep.* **11,** 627–631.
54. Hayford, M. B., Medford, J. I., Hoffman, N. L., Rogers, S. G., and Klee, H. J. (1988) Development of a plant transformation selection system based on expression of genes encoding gentamicin acetyltransferases. *Plant Physiol.* **86,** 1216–1222.
55. Hauptmann, R. M., Vasil, V., Ozias-Akins, P., Tabaeizadeh, Z., Rogers, S. G., Fraley, R. T., Horsch, R. B., and Vasil I. K. (1988) Evaluation of selectable markers for obtaining stable transformants in the Gramineae. *Plant Physiol.* **86,** 602–606.

56. Eichholtz, D. A., Rogers, S. G., Horsch, R. B., Klee, H. J., Hayford, M., Hoffmann, N. L., Braford, S. B., Fink, C., Flick, J., O'Connell, K. M., and Fraley, R. T. (1987) Expression of mouse dihydrofolate reductase gene confers methotrexate resistance in transgenic petunia plants. *Somatic Cell Mol. Genet.* **13,** 67–76.
57. Hille, J., Verheggen, F., Roelvink, P., Franssen, H., van Kammen, A., and Zabel, P. (1986) Bleomycin resistance: a new dominant selectable marker for plant cell transformation. *Plant Mol. Biol.* **7,** 171–176.
58. Perl, A., Galili, S., Shaul, O., Ben-Tzvi, I., and Galili, G. (1993) Bacterial dihydrodipicolinate synthase and desensitized aspartate kinase: Two novel selectable markers for plant transformation. *Biotechnology* **11,** 715–718.
59. Goddijn, O. J. M., van der Duyn Schouten, P. M., Schilperoort, R. A., and Hoge, H. C. (1993) A chimaeric tryptophan decarboxylase gene as a novel selectable marker in plant cells. *Plant Mol. Biol.* **22,** 907–912.
60. Nussaume, L., Vincentz, M., and Caboche, M. (1991) Constitutive nitrate reductase: A dominant conditional marker for plant genetics. *Plant J.* **1,** 267–274.
61. Perera, R. J., Linard, C. G., and Signer, E. R. (1993) Cytosine deaminase as a negative selective marker for *Arabidopsis*. *Plant Mol. Biol.* **23,** 793–799.
62. Yamamoto, Y. T., Taylor, C. G., Acedo, G. N., Cheng, C. L., and Conkling, M. A. (1991) Characterization of *cis*-acting sequences regulating root-specific gene expression in tobacco. *Plant Cell* **3,** 371–382.
63. Rocha-Sosa, M., Sonnewald, U., Frommer, W., Stratmann, M., Schell, J., and Willmitzer, L. (1989) Both developmental and metabolic signals activate the promoter of a class I patatin gene. *EMBO J* . **8,** 23–29.
64. Keller, B., Sauer, N., and Lamb, C. J. (1988) Glycine-rich cell wall proteins in bean: gene structure and association of the protein with the vascular system. *EMBO J.* **7,** 3625–3633.
65. Keller, B., Schmid, J., and Lamb, C. J. (1989) Vascular expression of a bean cell wall glycine-rich protein-β-glucuronidase gene fusion in transgenic tobacco. *EMBO J.* **8,** 1309–1314.
66. Keller, B. and Baumgartner, C. (1991) Vascular-specific expression of the bean GRP 1.8 gene is negatively regulated. *Plant Cell* **3,** 1051–1061.
67. Sugita, M. and Gruissem, W. (1987) Developmental, organ-specific, and light-dependent expression of the tomato ribulose-1,5-bisphosphate carboxylase small subunit gene family. *Proc. Natl. Acad. Sci. USA* **84,** 7104–7108.
68. Ueda, T., Pichersky, E., Malik, V. S., and Cashmore, A. R. (1989) Level of expression of the tomato *rbcS-3A* gene is modulated by a far upstream promoter element in a developmentally regulated manner. *Plant Cell* **1,** 217–227.
69. Mitra, A., Choi, H. K., and An, G. (1989) Structural and functional analyses of *Arabidopsis thaliana* chlorophyll a/b-binding protein (*cab*) promoters. *Plant Mol. Biol.* **12,** 169–179.
70. Koes, R. E., Spelt, C. E., and Mol, J. N. M. (1989) The chalcone synthase multigene family of *Petunia hybrida* (V30): differential, light-regulated expression during flower development and UV light induction. *Plant Mol. Biol.* **12,** 213–225.
71. Bird, C. R., Smith, C. J. S., Ray, J. A., Moureau, P., Bevan, M. W., Bird, A. S., Hughes, S., Morris, P. C., Grierson, D., and Schuch W. (1988) The tomato poly-

galacturonase gene and ripening-specific expression in transgenic plants. *Plant Mol. Biol.* **11,** 651–662.
72. An, G. (1987) Integrated regulation of the photosynthetic gene family from *Arabidopsis thaliana* in transformed tobacco cells. *Mol. Gen. Genet.* **207,** 210–216.
73. Koes, R. E., van Blokland R., Quattrocchio, F., van Tunen, A. J., and Mol, J. N. M. (1990) Chalcone synthase promoters in petunia are active in pigmented and unpigmented cell types. *Plant Cell* **2,** 379–392.
74. Van der Meer, I. M., Spelt, C. E., Mol, J. N. M., and Stuitje, A. R. (1990) Promoter analysis of the chalcone synthase (*chs*A) gene of *Petunia hybrida*: a 67 bp promoter region directs flower-specific expression. *Plant Mol. Biol.* **15,** 95–109.
75. Paul, W., Hodge, R., Smartt, S., Draper, J., and Scott, R. (1992) The isolation and characterisation of the tapetum-specific *Arabidopsis thaliana* A9 gene. *Plant Mol. Biol.* **19,** 611–622.
76. Twell, D., Wing, R., Yamaguchi, J., and McCormick, S. (1989) Isolation and expression of an anther-specific gene from tomato. *Mol. Gen. Genet.* **217,** 240–245.
77. Roberts, L. S., Thompson, R. D., and Flavell, R. B. (1989) Tissue-specific expression of a wheat high molecular weight glutenin gene in transgenic tobacco. *Plant Cell* **1,** 569–578.
78. Doyle, J. J., Schuler, M. A., Godette, W. D., Zenger, V., Beachy, R. N., and Slightom, J. L. (1986) The glycosylated seed storage proteins of *Glycine max* and *Phaseolus vulgaris*. Structural homologies of genes and proteins. *J. Biol. Chem.* **261,** 9228–9238.
79. Twell, D., Yamaguchi, J., and McCormick, S. (1990) Pollen-specific gene expression in transgenic plants: coordinate regulation of two different tomato gene promoters during microsporogenesis. *Development* **109,** 705–713.
80. Twell, D., Yamaguchi, J., Wing, R. A., Ushiba, J., and McCormick, S. (1991) Promoter analysis of genes that are coordinately expressed during pollen development reveals pollen-specific enhancer sequences and shared regulatory elements. *Genes Dev.* **5,** 496–507.
81. Thompson, R. D., Bartels, D., and Harberd, N. P. (1985) Nucleotide sequence of a gene from chromosome 1D of wheat encoding a HMW-glutenin subunit. *Nucleic Acids Res.* **13,** 6833–6846.
82. Bustos, M. M., Guiltinan, M. J., Jordano, J., Begum, D., Kalkan, F. A., and Hall, T. C. (1989) Regulation of β-glucuronidase expression in transgenic tobacco plants by an A/T-rich, *cis*-acting sequence found upstream of a french bean β-phaseolin gene. *Plant Cell* **1,** 839–853.

CHAPTER 2

Introduction of Cloning Plasmids into *Agrobacterium tumefaciens*

Charles H. Shaw

1. Introduction

Most experiments utilizing *Agrobacterium tumefaciens* as a vector for the introduction of genes into plant cells commence in *Escherichia coli*. The sheer size and complexity of the Ti-plasmids precludes their direct manipulation. Thus, insertion is usually into a comparatively small binary vector, which is then propagated in *E. coli,* before being introduced into *A. tumefaciens*, where the larger, complementing *vir* plasmid mediates gene transfer to plants. Typically, the binary plasmid will be based on a broad host range replicon, of IncP, IncQ, or IncW derivation, capable of replication in both bacterial hosts. Its construction will have resulted in the binary plasmid possessing the following features:

1. A selectable marker for bacterial cells;
2. A selectable marker for plant cells;
3. A multiple cloning site (MCS) and/or expression site; and
4. The left (LB) and right border (RB) from the Ti-plasmid T-DNA, positioned to define a pseudo T-DNA containing the plant selectable marker and the MCS.

Detailed maps and descriptions of the various expression cassettes, selectable markers, and reporter gene sequences are given in Chapter 1 of this volume *(see also* refs. *1,2)*. Chimeric gene constructs can be readily introduced by standard molecular biological methodology into the cloning sites of various binary (*see* Chapter 6) and intergration vec-

From: *Methods in Molecular Biology, Vol. 49: Plant Gene Transfer and Expression Protocols*
Edited by: H. Jones Humana Press Inc., Totowa, NJ

tors. Descriptions of the use of various *Agrobacterium* plasmid vectors are given in Chapters 3–6 of this volume.

It is clear that a reliable method of introducing plasmids into *A. tumefaciens* is needed. While certain types of plasmid, chiefly IncW derivatives, can be transmitted from *E. coli* to *A.tumefaciens* by IncN plasmids *(3)*, most workers have employed either triparental mating *(2,4)* or electroporation *(5–8)*.

Triparental mating is a conjugation procedure in which pRK2013, a derivative of the IncP plasmid RK2, acts as a helper plasmid, providing the transfer functions to mobilize a wide range of plasmids, including IncP, IncQ, or IncW replicons, from *E. coli* to *A. tumefaciens (2)*. All three strains, donor and helper *E. coli* and recipient *A. tumefaciens*, are mixed and incubated together. This is partly a time-saving measure, avoiding a lengthy two-step conjugation of helper plasmid into donor *E. coli*, followed by the transfer to *A. tumefaciens*. However, it is also a means to avoid any potential incompatibility problems, and instances where the antibiotic resistance marker on pRK2013 (kanamycin) is identical to that of the binary plasmid. In *A. tumefaciens*, pRK2013 is not efficiently maintained, and thus transconjugants can be selected on kanamycin, plus a chromosomal marker to counterselect the donor and helper strains.

Electroporation *(9)* is becoming the method of choice, owing to its high efficiency, convenience, and the fact that some replicons are not so efficiently mobilized by pRK2013. In this method a high voltage pulse, usually generated by capacitor-discharge, is applied to a suspension of cells in a cuvet. It is believed that the pulse induces pores in the cell surface, through which the DNA enters the cell. The method that we employ *(10)*, based on that of Nagel et al. *(6)*, uses the BioRad (Hercules, CA) Gene-Pulser, with Pulse Controller. This gives transformation efficiencies of approx 10^6–10^8/µg DNA, several orders of magnitude greater than the alternative freeze-thaw method *(11)*.

2. Materials

1. Culture Media: *E. coli* and *A. tumefaciens* (*see* Note 1) are grown in LAB M nutrient broth number 2 (Amersham, Braunschweig, Germany) made according to the maker's instructions, and plated on LAB M nutrient agar plus appropriate antibiotics (Sigma, St. Louis, MO) as previously described *(3)*.
2. Dilution of bacteria for plating after triparental mating is done in 10 m*M* Mg_2SO_4.

3. 1 m*M* HEPES/KOH, pH 7.0, diluted in sterile distilled H_2O from a sterile 1*M* stock.
4. 10% filter sterilized glycerol (*see* Note 2).
5. Plasmid DNA for electroporation should be clean but not necessarily ultrapure. Thus it can be prepared by alkaline lysis mini-prep *(12)*.
6. SOC broth *(12)* is a glucose-rich medium used for dilution of electroporated cells: Dissolve 20 g tryptone, 5 g yeast extract, and 0.5 g NaCl, in 950 mL distilled H_2O. Add 10 mL 250 m*M* KCl, adjust pH to 7.0 with 10*N* NaOH and make up to 975 mL. Autoclave, cool, and add 20 mL sterile 1*M* glucose and 5 mL sterile 2*M* $MgCl_2$.

3. Methods

3.1. Triparental Mating

1. Grow 5 mL cultures of each of the *E. coli* donor strain (carrying the binary plasmid), *E. coli* HB101 (pRK2013), and the recipient *A. tumefaciens* strain (*see* Note 3) to exponential phase.
2. Mix 100 µL of the donor, 100 µL of HB101 (pRK2013), and 300 µL of the recipient strain together in a sterile plastic tube.
3. Pipet 100 µL of this mixture onto a nitrocellulose disk placed in the center of a nutrient-agar plate. Incubate overnight at 28°C.
4. Remove the disk and shake in 10 mL of 10 m*M* $MgSO_4$.
5. Spread 100-µL samples of the suspension onto selective media and incubate at 28°C for 48 h to allow for the appearance of transconjugant colonies.
6. Check unselected markers by plating on appropriate media, and check for presence of plasmid by mini-prep *(12)* or Southern blotting on total DNA *(13)*.

3.2. Electroporation

1. Centrifuge 2 × 1.5 mL aliquots of bacteria from an overnight culture of *Agrobacterium tumefaciens* for 30 s at high speed in a microfuge.
2. Resuspend pellets in 0.5 mL of ice-cold 1 m*M* HEPES/KOH, pH 7.0 and recentrifuge.
3. Repeat step 2, twice.
4. Resuspend pellets in 0.5 mL ice-cold 10% glycerol and recentrifuge.
5. Resuspend pellets in 20 µL 10% glycerol and combine the contents of both tubes (40 µL in total).
6. Add plasmid DNA (*see* Note 4) and leave the tube on ice for 2 min.
7. Pulse the DNA–bacteria mixture in an ice-cold 0.2-cm Bio-Rad electroporation cuvet using a Bio-Rad Gene Pulser with Pulse Controller (*see* Note 5).

8. Immediately after the pulse, add 1 mL SOC broth and incubate at 28°C for 4–6 h before plating 100 µL on selective media.

4. Notes

1. All incubations involving *A. tumefaciens* should be at 28°C.
2. The glycerol should be filter sterilized as autoclaving leads to the formation of aldehydes that inhibit electroporation *(14)*.
3. It is best to use a strain of *A. tumefaciens* that carries a chromosomal antibiotic resistance marker, such as rifampicin, to allow counterselection.
4. Best results are achieved at 1µg DNA/40 µL cells.
5. The Gene-Pulser should be set at 25 µF capacitance, 2.5 kV charge, and the Pulse Controller to 400Ω resistance. The time constant was typically about 9 ms in successful transformations.

References

1. Guerineau, F. and Mullineaux, P. (1993) Plant transformation and expression vectors, in *Plant Molecular Biology LabFax* (Croy, R. R. D. C., ed.), Bios, Oxford, pp. 121–147.
2. Herrera-Estrella, L. and Simpson, J. (1988) Foreign gene expression in plants, in *Plant Molecular Biology—A Practical Approach* (Shaw, C. H., ed.), IRL, Oxford, pp. 131–160.
3. Leemans, J., Shaw, C. H., Deblaere, R., De Greve, H., Hernalsttens, J.-P., van Montagu, M., and Schell, J. (1981) Site-specific mutagenesis of *Agrobacterium* Ti-plasmids and transfer of genes to plant cells. *J. Mol. Appl. Genet.* **1,** 149–164.
4. Ditta, G., Stanfield, S., Corbin, D., and Helinski, D. R. (1980) Broad host range DNA cloning system for Gram-negative bacteria: Construction of a gene bank in *Rhizobium meliloti. Proc. Natl. Acad. Sci. USA* **77,** 7347–7351.
5. Wen-jun, S. and Forde, B. G. (1989) Efficient transformation of *Agrobacterium* spp. by high voltage electroporation. *Nucleic Acids Res.* **17,** 8385.
6. Nagel, R., Elliott, A., Masel, A., Birch, R. G., and Manners, J. M. (1990) Electroporation of binary Ti plasmid vector into *Agrobacterium tumefaciens* and *Agrobacterium rhizogenes. FEMS Microbiol. Letts.* **67,** 325–328.
7. Merserau, M., Pazour, G., and Das, A. (1990) Efficient transformation of *Agrobacterium tumefaciens* by electroporation. *Gene* **90,** 149–151.
8. Mozo, T. and Hooykaas, P. J. J. (1991) Electroporation of megaplasmids into *Agrobacterium. Plant Mol. Biol.* **16,** 917–918.
9. Spencer, S. C. (1991) Electroporation technique of DNA transfection, in *Gene Transfer and Expression Protocols* (Murray, E. J., ed.), Humana, Clifton, NJ, pp. 45–52.
10. Palmer, A. C. V. and Shaw, C. H. (1992) The role of *virA and G* phosphorylation in acetosyringone chemotaxis. *J. Gen. Microbiol.* **138,** 2509–2514.
11. Holsters, M., De Waele, D., Depicker, A., Messens, E., Van Montagu, M., and Schell, J. (1978) Transfection and transformation of *Agrobacterium tumefaciens. Mol. Gen. Genet.* **163,** 181–187.

12. Sambrook, J., Fritsch, E. F., and Maniatis, T. (1989) *Molecular Cloning. A Laboratory Manual.* Cold Spring Harbor Laboratory, Cold Spring Harbor, NY.
13. Dhaese, P., De Greve, H., Decraemer, H., Schell, J., and Van Montagu, M. (1979) Rapid mapping of transposon insertion and deletion mutants in the large Ti-plasmids of *Agrobacterium tumefaciens*. *Nucleic Acids Res.* **7,** 1837–1849.
14. Zabarovsky, E. R. and Winberg, G. (1990) High efficiency electroporation of ligated DNA into bacteria. *Nucleic Acids Res.* **18,** 5912.

CHAPTER 3

Leaf Disk Transformation Using *Agrobacterium tumefaciens*-Expression of Heterologous Genes in Tobacco

Patrick Gallois and Paulo Marinho

1. Introduction

Leaf disk transformation of tobacco is a very simple and robust method. It is used with success in many laboratories. The protocol presented here is a simplified version of that of Horsch et al. *(1)*. Basically, it consists of immersing the leaf disks in a liquid culture of *Agrobacterium* carrying the chosen transformation vector. The plant tissue and *Agrobacterium* are then cocultivated on regeneration medium for a period of 2 d at the end of which the leaf disks are transferred to regeneration medium supplemented with an antibiotic to kill the bacteria (cefotaxime), and a selective agent against untransformed plant cells. It takes about 2 mo to obtain rooted plantlets that can be transferred to soil. The protocol presented here works well in our hands with *Nicotiana tabacum* cultivar "petit havana" mutant SR1 *(2)* and *Agrobacterium tumefaciens* strain LBA4404 *(3)* harboring binary vectors conferring kanamycin resistance (100 mg/L). We have also used pBIB-HYG *(4)*, which confers hygromycin resistance (50 mg/L).

The use of transgenic plants has allowed the rapid accumulation of knowledge about the structure and function of plant genes. Tobacco has probably been the species most often transformed, and many experiments involving the introduction of heterologous genes have been carried out.

From: *Methods in Molecular Biology, Vol. 49: Plant Gene Transfer and Expression Protocols*
Edited by: H. Jones Humana Press Inc., Totowa, NJ

These include the introduction of chimeric gene constructs consisting of promoters, which are active in plants, linked to the coding sequences of genes originating from other organisms, for example, bacterial genes which can be used as reporter genes (*see* Chapters 10–12), mammalian genes that code for immunoglobulins *(5)* or viral genes that code for coat proteins *(6)*. Numerous plant genes cloned in dicot or monocot species have been transferred into tobacco under the control of their own promoter or as a gene fusion with a constitutive or inducible promoter. Promoter studies have also been widely carried out, and the general conclusion from such experiments is that, as long as the regulatory cis-acting sequences are included in the integrated foreign gene, its expression pattern is the same as the corresponding transcript in the natural host plant. However, a few important points should be taken into consideration. For most dicot genes, the upstream promoter sequences are sufficient to confer the correct regulation of expression in transgenic tobacco. However, in a few cases transcribed regions of the gene have been shown to be essential for correct regulation (e.g., exonic sequences, 3' end sequences) *(8–11)*.

Expression of monocot genes in tobacco is not as straightforward as expression of dicot genes. Although monocot promoter sequences are usually properly recognized by the tobacco cellular machinery *(12,13)*, there have been some exceptions. For example, insertion of enhancer-like regions from constitutive octopine synthase or CaMV 35S genes upstream of the promoter was necessary for induction of the Maize ADH gene under anaerobic conditions *(14)*. Ueng et al. *(15)* reported that the tissue specificity of a zein genomic clone was lost on introduction into tobacco, as the transcript was detected in seeds (that is normal) but also in leaves, stems, and flowers. Keith and Chua *(16)* have shown that inefficient splicing of the pre-mRNA and inaccurate polyadenylation could lead to reduced stability of monocot mRNAs in tobacco. Second, Matzke et al. *(17)* showed that a prolamin gene was inefficiently translated in sunflower. This explains the observation that expression of monocot storage protein genes in tobacco generally has resulted in a lower expression level than that of the corresponding transcript in the native host plant, and in some cases no foreign protein was detected *(15,17,18)*.

Overexpression or ectopic expression of a given gene may lead to new phenotypes *(19,20)* including lethality. If such an extreme phenotype is suspected, an inducible promoter should be used *(21–23)*.

2. Materials

2.1. Antibiotic Stock Solutions

1. Cefotaxime: 200 mg/mL (Roussel Uclaf sold by Sigma [St. Louis, MO] or preferably local chemist), filter sterilize. Prepare just before use.
2. Kanamycin (Sigma): 100 mg/mL, filter sterilize, keep at –20°C.
3. Streptomycin (Sigma): 50 mg/mL, filter sterilize, keep at –20°C.

2.2. Hormone Stock Solutions

1. 6-Benzylaminopurine (BAP), (Sigma): 10 mg/mL in dimethylsulfoxide (DMSO), no need to sterilize, keep at –20°C. DMSO should be handled under a fume hood.
2. α-Naphthalene acetic acid (NAA), (Sigma): 5 mg/mL in DMSO, no need to sterilize, keep at –20°C. DMSO should be handled under a fume hood.

2.3. Culture Media

1. Tobacco shoot culture medium: 1X Murashige and Skoog salts and vitamins *(24)* (Flow Laboratories, McLean, VA), 30 g/L sucrose, adjust pH to 5.8 with 1*M* KOH, add 10 g/L Bactoagar (Difco, Detroit, MI), autoclave. Pour in Magenta GA7 boxes.
2. *Agrobacterium* culture medium (2YT): 16 g/L bactotryptone (Difco), 10 g/L yeast extract (Difco), 5 g/L NaCl, autoclave, and cool medium to 60°C, add appropriate antibiotics to select for plasmids (e.g., 50 mg/L kanamycin for pBin 19 *(25)* and 200 mg/L streptomycin for pAL4404).
3. Cocultivation medium: 1X Murashige and Skoog salts and vitamins (Flow Laboratories), 30 g/L sucrose, 1 mg/L BAP, adjust pH to 5.8 with KOH 1*M*, add 8 g/L Bactoagar (Difco), and autoclave. Pour into sterile plastic Petri dishes that are 20 mm high (e.g., made by Corning [Corning, NY] or Falcon, Lincoln Park, NJ).
4. Regeneration and selection medium: 1X Murashige and Skoog salts and vitamins (Flow laboratories), 30 g/L sucrose, 1 mg/L BAP, adjust pH to 5.8 with 1*M* KOH , add 8.0 g/L Bactoagar (Difco). Autoclave and cool media to 60°C, add 200 mg/L Cefotaxime to kill off *Agrobacterium* and the appropriate selective agent to select transformed plants depending on the vector used (e.g., for pBin 19, kanamycin at 100 mg/L). Pour into sterile plastic Petri dishes, that are 20 mm high (e.g., made by Corning or Falcon).
5. Rooting medium: 1X Murashige and Skoog salts and vitamins (Flow Laboratories), 30 g/L sucrose, 0.1 mg/L NAA, adjust pH to 5.8 with 1*M* KOH, add 10 g/L Bactoagar (Difco). Autoclave and cool media to 60°C, add 200 mg/L cefotaxime and the appropriate selective agent to select

transformed plants depending on the vector used. Pour in Magenta GA7 boxes (80 mL per box).
6. Soil: We use a 50:50 mix of vermiculite and soil in 100 × 100 × 100 mm plastic pots.
7. Nutritive solution for tobacco grown in soil: Use commercially available nutritive solution for house plants.

3. Methods

3.1. Production of Axenic Tobacco Shoot Culture

Keep an in vitro culture of tobacco plants (*see* Note 1) in GA7 Magenta boxes by subculturing shoots on a regular basis on tobacco shoot culture medium (*see* Section 2.3.1.). This culture is started from seeds sterilized by immersion for 10 min in 11% Domestos (Lever, Warrington, UK) followed by five washes in sterile distilled water. When not in use, the culture can be kept in a cold room for several months. When planning to carry out a transformation experiment, for best results the tobacco plants should be subcultured 5–6 weeks before leaf disk isolation.

3.2. Leaf Disk Infection

1. Start an *Agrobacterium* preculture (*see* Note 2) in 5 mL of 2YT (*see* Section 2.3.2.) with appropriate antibiotics (for pBin 19 use Kanamycin at 50 mg/L *[25]*). Grow overnight at 29°C in an orbital shaker set at 200 rpm.
2. Use 1 mL of the preculture to inoculate a 50 mL *Agrobacterium* culture in 2YT plus antibiotics. Grow for 24 or 36 h depending on the strain (*see* Note 3). For leaf disk inoculation, adjust the culture to $OD_{600} = 1$ with 2YT without antibiotics and pour into a 60 × 100 sterile polypot.
3. Cut dental modeling wax to the size of a Petri dish, immerse into 98% ethanol for 10 min, transfer to a sterile Petri dish, and leave to dry.
4. Using the dental modeling wax as a base, punch out leaf disks from 5–6-wk-old leaves (*see* Note 4) with a sterile 1-cm diameter cork-borer (*see* Note 5). Avoid the primary vein of the leaf or necrotic area. Cut 80 disks per construct (*see* Notes 6 and 7).
5. Transfer all the disks into the *Agrobacterium* culture, shake, and leave for 1 min.
6. Recover the inoculated leaf disks with forceps. Eliminate excess liquid by blotting on sterile filter papers. Circular filter papers the size of a Petri dish are convenient. (Whatman filter papers of 9-cm diameter fit in the lid of a 9-cm petri dish.)
7. Place eight disks upper leaf surface uppermost on a Petri dish containing the cocultivation medium (*see* Section 2.3.3.). Ensure a good contact

between each leaf disk and the culture medium. Seal the Petri dish with Nescofilm. Cultivate for 2 d at 25°C, 16 h daylight (*see* Note 8).

8. After cocultivation, bacterial growth should be visible on the agar at the edge of the leaf disks. Simply transfer the disks to the regeneration and selection medium (*see* Section 2.3.4.). Close the Petri dish with two pieces of cellotape at opposite sides of the dish so as to allow air flow. Low humidity helps the elimination of *Agrobacterium* by antibiotics. Cultivate at 25°C with 16 h daylength for 2 d, then seal the Petri dish with Nescofilm and continue incubation until shoots regenerate.
9. After 1 or 2 wk the leaf disks expand and the center of some disks is lifted away from the selection medium. At that stage, cut the disk in four and press the pieces into the medium in order to ensure a good contact between the plant tissue and the medium.

3.3. Recovery of Transformed Shoots

1. Shoots start to appear about 3 wk after inoculation and generally grow from the edge of the disks or from internal area wounded by the forceps. When the shoots are at least 5-mm long, excise them at the base, avoid taking any callus tissue, and place them individually in Magenta GA7 boxes on rooting medium (*see* Section 2.3.5.). Cefotaxime is kept in the medium to avoid *Agrobacterium* growth (*see* Note 9). The antibiotic used to select the transformants (e.g., kanamycin) is added to the medium to select against escapes (*see* Note 10). To avoid taking clonal shoots representing the same transformation event, it is a good policy to take only one tranformant per leaf disk.
2. Only the plantlets forming roots in the presence of a selective agent should be considered as true transformants; these represent approx 75% of the initial number of green shoots. Once the roots are formed, wash the agar from the root system under a running tap and transfer the plantlets to soil (*see* Section 2.3.6.). To harden the in vitro plantlets, which need to develop a cuticule to control water losses, cover each transformant with a Magenta GA7 box. Leave for 3 d then tilt the box for 1 d before removing it totally. Depending on your particular culture conditions (humidity level) you may need to adjust the period of time required for hardening.
3. Water the plants with commercially available nutritive solution (*see* Section 2.3.7.), following the specifications of the manufacturer.

3.4. Analysis of Transformants

Only DNA analysis using Southern blotting or PCR (*see* Chapters 13 and 14; ref. *26*) will confirm whether regenerants are true transformants. Second, some plants may have integrated the antibiotic resistance gene without the gene or construct of interest that was initially linked to it within the T-DNA. Position effect may also silence the introduced nonselected gene. We found using pBin19 *(25)* that around 40% of the kanamycin resistant plants did not express the reporter gene to be analyzed (*see* Note 7).

The number of independent loci at which integration occurred can be estimated by the segregation ratio of resistant to sensitive seedlings in the progeny. When using kanamycin as a selective agent, sterilize seeds (10 min in 11% Domestos [Lever] and five washes in sterile distilled water) and sow at least 100 of them on tobacco shoot culture medium supplemented with 400 mg/L of kanamycin. Sensitive seedling germinate and bleach within 1 wk. For an interpretation of the different genotype ratios that can be obtained, see Heberle-Bors et al. *(27)*. For a discussion on analysis of first generation transgenic plants and statistics, see Nap et al. *(28)*.

4. Notes

1. We use *Nicotiana tabacum* cv "Petit Havana" mutant SR1 because it stays relatively small when cultured in soil compared to other cultivars. In 10 × 10 × 10 cm pots, this cv. can reach 80–100 cm in height. It also regenerates very well in in vitro culture. Other tobacco cultivars can, however, be transformed with success (e.g., Samsun, Xanthi, W38) using the method described here.
2. We use *Agrobacterium* strain LBA4404 (sold by Clontech, Palo Alto, CA), which, in our hands, works very well. C58 is another strain that we use and, although it has the reputation to transform more efficiently than LBA4404, we have found it more difficult to eliminate from cultures after cocultivation.
3. During growth, the *Agrobacterium* culture becomes turbid and typically forms short chains of white aggregates.
4. Leaves (nonsterile) from young plants grown in soil can also be used as a source material. Sterilize whole leaves with 11% Domestos (bleach) for 10 min and rinse 5X with sterile distilled water. We find in vitro grown plantlets easier to handle since it is sometimes quite difficult to obtain sterile tissue from soil grown plants. In addition, if small wounds are inflicted to leaves during isolation, the bleach attacks the tissue in this area that can no longer be used for the production of leaf disks.
5. Disks are used because they are conveniently generated with a cork borer. Square explants or strips could be used equally well. One cause of death of leaf disks has been found to be the temperature of the cork-borer.

After flaming, the cork-borer should be allowed to cool before excising the leaf disks.

6. In some protocols, leaf disks are precultured on cocultivation medium for 2 d before inoculation. We did not find this step to be necessary for a good transformation frequency. In the same way we did not find it necessary to use a nurse-culture of tobacco cell suspension during cocultivation *(1)*.
7. Allow 80 disks per construct to be on the safe side. Some disks will not regenerate, some shoots will bleach on rooting medium containing the selective agent or will not produce roots. Among the transformants that do form roots (ca. 75% of the initial number in the case of kanamycin selection), some will not express the gene of interest that was cotransferred together with the selectable marker gene. Using pBin19 as a transformation vector, we found that up to 40% of the kanamycin resistant transformants do not express the cotransferred gene; in many of them only part of the gene is present. This is probably owing to the structure of pBin19 where the transferred gene is located between the kanamycin resistance gene and the left border. Finally, position effect owing to the integration site of the T-DNA will greatly influence the level of expression of a given construct, and if levels of expression are to be compared between different constructs or if a high expressor is desirable, at least 10 expressors should be generated for a given construct. Although Stockhaus et al. *(29)* showed in their study that a high copy number of an integrated foreign gene leads to a high level of expression, most studies have shown that there is no such correlation *(13,18,30,31)*.
8. If the transformation goes wrong, it is important to have control plates to refer to in order to solve the problem for the next round. The controls are:
   a. Noninoculated leaf disks on regeneration medium with no selective agent, to check the regeneration medium.
   b. Noninoculated leaf disks on regeneration medium with selective agent, to check the efficiency of antibiotic selection.
   c. Inoculated disks on regeneration medium with Cefotaxime but without the antibiotic selection agent, to check whether the inoculation has damaged the leaf disks thus impairing plant regeneration (optional).
   d. Additionally, it is good practice to test antibiotic selection on rooting media with a few untransformed shoots. With kanamycin at a concentration of 100 mg/L, shoots of up to 2 cm in height can be selected on rooting medium (large untransformed shoots take longer to bleach than small ones).
9. If difficulties are experienced in killing *Agrobacteria* after cocultivation, Augmentin (Beecham Laboratories, Nanterre, France) at 200 mg/L can also be added to media containing Cefotaxime. Prepare just before use.

10. Escapes are plants that regenerate on the selective medium without expressing the selection marker. Antibiotic selection on rooting medium should allow one to discriminate between escapes and true transformants. With the pBin19 vector and kanamycin selection (100 mg/L), we have found that around 10% of the shoots bleach on antibiotic selection in rooting media. It has been shown that kanamycin has poor phloem mobility and can diffuse only over short distances *(32)*, therefore escapes can potentially regenerate from a part of the explant that is not in contact with the gelose medium. Different selective agents give different percentage of escapes; escapes in this context are not, however, to be confused with transformants expressing the selection marker but not expressing the cotransferred gene of interest (*see* Note 7).

References

1. Horsch, R. B., Fry, J. E., Hoffman, N. L., Eichholtz, D., Rogers, S. G., and Fraley, R. T. (1985) A simple and general method for transferring genes into plants. *Science* **227,** 1229–1231.
2. Etzold, T., Fritz, C. C., Shell, J., and Schreier, P. H. (1987) A point mutation in the chloroplast 16S rRNA gene of a streptomycin resistant *Nicotiana tabacum*. *FEBS Lett.* **219,** 343–346.
3. Ooms, G., Hooykaas, P. J. J., van Veen, R. J. M., van Beelen, P., Regenburg-tuink, T. J. G., and Shilperoort, R. A. (1982) Octopine Ti-plasmid deletion mutants of *Agrobacterium tumefaciens* with emphasis on the right side of the T-region. *Plasmid* **7,** 15–29.
4. Becker, D. (1990) Binary vectors which allow the exchange of plant selectable markers and reporter gene. *Nucleic Acids Res.* **18,** 203.
5. During, K., Hoppe, S., Kreuzaler, F., and Schell, J. (1990) Synthesis and self-assembly of functional monoclonal antibody in transgenic *Nicotiana tabacum*. *Plant Mol. Biol.* **15,** 281–293.
6. Powell, P. A., Nelson, R. S., De, B., Hoffmann, N., Rogers, S. G., Fraley, R. J., and Beachy, R. N. (1986) Delay of disease development in transgenic plants that express the tobacco mosaic virus coat protein gene. *Science* **232,** 738.
7. Douglas, C. J., Hauffe, K. D., Ites-Morales, M. E., Ellard, M., Paszkowski, V., Hahlbrock, K., and Dangl, J. L. (1991) Exonic sequences are required for elicitor and light activation of a plant defense gene, but promoter sequences are sufficient for tissue specific expression. *EMBO J.* **7,** 1767–1775.
8. Gallo-Meagher, M., Sowinski, D. A., Elliot, R. C., and Thomson, W. F. (1992) Both internal and external regulatory elements control expression of the pea Fed-1 gene in transgenic tobacco seedlings. *Plant Cell* **4,** 389–395.
9. Dietrich, R. A., Radke, S. E., and Harada J. J. (1992) Downstream DNA sequences are required to activate a gene expressed in the root cortex of embryos and seedlings. *Plant Cell* **4,** 1371–1382.

10. An, G., Mitra, A., Choi, H. K., Costa, M. A., An, K., Thornburg, R. W., and Ryan C. A. (1989) Functional analysis of the 3' control region of the potato wound-inducible proteinase inhibitor II gene. *Plant Cell* **1,** 115–122.
11. Dean, C., Favreau, M., Bedbrook, J., and Dunsmuir, P. (1989) Sequences down-stream of translation start regulate quantitative expression of two petunia rbcs genes. *Plant Cell* **1,** 201–208.
12. Colot, V., Robert, L. S., Kavanagh, T. A., Goldsbrough, A. P., Bevan M. W., and Thomson, R. D. (1987) Localization of sequences in wheat endosperm protein genes which confer tissue-specific expression in tobacco. *EMBO J.* **6,** 3559–3564.
13. Marris, C., Gallois, P., Copley, J., and Kreis, M. (1988) The 5'-flanking region of a barley B hordein gene controls tissue and development specific CAT expression in tobacco plants. *Plant Mol. Biol.* **10,** 359–366.
14. Ellis, J. G., Llewellyn, D. J., Dennis, E. S., and Peacock, W. J. (1987) Maize ADh-1 promoter sequences control anaerobic regulation: addition of upstream promoter elements from constitutive genes is necessary for expression in tobacco. *EMBO J.* **6,** 11–16.
15. Ueng, P., Galili, G., Sapanara, V., Goldsbrough, P. B., Dube, P., Beachy, R. N., and Larkins, B. A. (1988) Expression of a maize storage protein gene in petunia plants is not restricted to seeds. *Plant Physiol.* **86,** 1281–1285.
16. Keith, B. and Chua, N. H. (1986) Monocot and dicot pre-mRNAs are processed with different efficiencies in transgenic tobacco. *EMBO J.* **5,** 2419–2425.
17. Matzke, M. A., Susani, M., Binns, A. N., Lewis, E. D., Rubensteins, I., and Matzke, A. J. M. (1984) Transcription of a zein gene introduced into sunflower using a Ti plasmid vector. *EMBO J.* **3,** 1525–1531.
18. Robert, L. S., Thomson, K. D., and Flawell, R. (1989) Tissue specific expression of wheat high molecular weight glutenin gene in transgenic tobacco. *Plant Cell* **1,** 569–578.
19. Mandel, M. A., Bowman, J. L., Kempin S. A., Ma, H., Meyerowitz, E. M., and Yanofsky, M. F. (1992) Manipulation of flower structure in transgenic tobacco. *Cell* **71,** 133–143.
20. Hirel, B., Marsolier, M. C., Hoarav, A., Hoarav, J., Brangeon, J., Shafer, R., and Verma, D. P. S. (1992) Forcing expression of a soybean root glutamine synthetase gene in tobacco leaves induces a native gene encoding cytosolic enzyme. *Plant Mol. Biol.* **20,** 207–218.
21. Ainley, M. W., Mc Neil, K. J., Hill, J. W., Lingle, W. L., Simpson, R. B., Brenner, M. L., Nagao, R. T., and Key, J. L. (1993) Regulatable endogenous production of cytokinins up to toxic levels in transgenic plants and plant tissues. *Plant Mol. Biol.* **22,** 13–23.
22. Ward, E. R., Ryals, J. A., and Miflin, B. J. (1993) Chemical regulation of transgene expression in plants. *Plant Mol. Biol.* **22,** 361–366.
23. Gatz, C., Frohberg, C., and Wendenburg, R. (1992) Stringent repression and homogeneous de-repression by tetracycline of a modified CaMV 35S promoter in intact transgenic tobacco plants. *The Plant J.* **2,** 397–404.

24. Murashige, T. and Skoog, F. (1962) A revised medium for rapid growth and bioassays with tobacco tissue cultures. *Plant Physiol.* **15,** 473–497.
25. Bevan, M. (1984) Binary *Agrobacterium* vectors for plant transformation. *Nucleic Acids Res.* **12,** 8711–8721.
26. Edward, K., Johnstone, C., and Thompson, C. (1991) A simple and rapid method for the preparation of plant genomic DNA for PCR analysis. *Nucleic Acids. Res.* **19,** 1349.
27. Heberle-bors, E., Charvat, B., Thompson, D., Shernthaner, J. P., Barta, A., Matzke, A. J. M., and Matzke, M. A. (1988) Genetic analysis of the T-DNA insertions into tobacco genome. *Plant Cell Rep.* **7,** 571–574.
28. Nap, J. P., Keizer, P., and Jansen, R. (1993) First generation transgenic plants and statistics. *Plant Mol. Biol. Rep.* **11,** 156–164.
29. Stauckhaus, J., Eckes, P., Blau, A, Schell, J., and Willmitzer, L. (1987) Organ specific and dosage dependent expression of a leaf/stem specific gene from potato after tagging into potato and tobacco plants. *Nucleic Acids Res.* **15,** 3479–3491.
30. Deroles, S. C. and Gardner, R. C. (1988) Expression and inheritance of kanamycin resistance in a large number of transgenic petunias generated by *Agrobacterium* mediated transformation. *Plant Mol. Biol.* **11,** 355–364.
31. Shirsat, A. R., Wilford, N., and Croy, R. R. R. (1989) Gene copy number and levels of expression in transgenic plants of a seed specific gene. *Plant Sci.* **61,** 75–80.
32. Weide, R., Koornneef, M., and Zabel, P. A. (1989) Simple, non-destructive spraying assay for the detection of an active kanamycin resistance gene in transgenic tomato plants. *Theor. Appl. Genet.* **78,** 169–172.

CHAPTER 4

Agrobacterium rhizogenes as a Vector for Transforming Higher Plants

Application in Lotus corniculatus *Transformation*

Jens Stougaard

1. Introduction

Plant transformation using *Agrobacterium rhizogenes* provides features that, for certain purposes, are advantageous compared to the more commonly used *Agrobacterium tumefaciens* techniques. In order to obtain transgenic plants after *A. tumefaciens* mediated transformation with "disarmed" binary vectors, selection of transgenic tissue on antibiotics or herbicides must be possible, and a procedure for regeneration must be worked out *(1)*. In contrast, most plant cells transformed with *A. rhizogenes* develop into transgenic roots that are easily recognized, separated, and cultivated in vitro *(2,3)*. From these root cultures a number of plants regenerate spontaneously or following hormone application *(2–4)*, and, despite the "hairy root phenotype" *(2)* of regenerated plants, *A. rhizogenes*-mediated transformation is useful for recalcitrant species.

The Ri plasmids of the various *A. rhizogenes* isolates carry different T-DNA segments and virulence regions *(5)*. The plant response to infection with a particular strain depends on both the susceptibility and the influence of the T-DNA encoded oncogenes on the plant phytohormone levels and balances. To establish a useful *A. rhizogenes* transformation system, several strains should be tested for the virulence and the severity of the "hairy root phenotype." For *Lotus corniculatus* (bird's-foot tre-

From: *Methods in Molecular Biology, Vol. 49: Plant Gene Transfer and Expression Protocols*
Edited by: H. Jones Humana Press Inc., Totowa, NJ

Table 1
Strategies Using the *Agrobacterium rhizogenes* C58C1 pRi15834 for Transformation of *Lotus corniculatus*

Cotransfer of T-DNA from binary vector	
Cloning vectors:	Suitable broad host range binary vector carrying a kanamycin, spectinomycin/streptomycin, tetracycline, or gentamycin resistance.
Recipient:	*A. rhizogenes* C58C1 pRi15834 carrying a chromosomal rifampicin resistance.
Recombination of integration vectors into the TL-DNA segment	
Cloning vectors:	E. coli plasmids carrying a fragment of TL-DNA for homologous recombination. These plasmids are unable to replicate in *Agrobacterium* and encodes one of the resistance markers mentioned for binary vectors. Examples are pAR5 and pAR1 *(9)*.
Recipient:	*A. rhizogenes* C58C1 pRi15834 carrying a chromosomal rifampicin resistance.
Recombination of pBR322 derived vectors into TL-DNA segments carrying pBR322	
Cloning vectors:	All pBR322 derived plasmids carrying sufficient sequences for recombination and still unable to replicate in *Agrobacterium*. Suitable resistance markers for selection in *Agrobacterium* are mentioned for binary vectors. An example is pIV10 *(11)*.
Recipient:	*A. rhizogenes* C58C1 pRi12 and C58C1 pRi14 *(10)* carrying a chromosomal rifampicin resistance. pRi12 has pBR322 and a 35SGUS reference gene integrated in the TL-segment. pRi14 has pBR322 and a 35SCAT reference gene in the TL-segment.

foil) the C58C1 pRi15834 strain is used extensively and the *Agrobacterium* gene transfer system is exploited in three different strategies.

1. Binary vectors are conjugated *(6,7)* or transformed *(8)* into *A. rhizogenes* and the "binary" T-DNA cotransferred into the plant cells *(9)*.
2. Integration vectors containing a restriction fragment from the TL-DNA are transferred together with the TL-DNA region after homologous recombination into the pRi plasmid *(9)*.
3. A variant of the previous approach uses *A. rhizogenes* acceptor strains where pBR322 integrated into the TL-DNA segment is used as acceptor sequence in recombination *(10,11)* (*see* Table 1).

For many species of the *Leguminosae* family, the combination of in vitro techniques for transformation and regeneration have proven difficult, and transgenic plants have only been obtained with a select group of species and varieties. The *A. rhizogenes* approach generating composite plants with a transgenic root system on an untransformed wild-type shoot is therefore especially useful for the study of root nodule formation on legumes *(10,12,13)*. This "short cut" transformation can also be used to prescreen experiments planned for more time-consuming transformation procedures.

A major interest pursued in transgenic nodules is the mechanisms regulating expression of root nodule expressed genes (nodulins). Typically, the expression of a nodulin gene promoter from one plant species is tested in other transformable plant species. Examination of the data will therefore, first of all, allow an evaluation of the conservation of the regulatory mechanisms. Provided the temporal and spatial expression is conserved, compared with the known expression pattern of the endogenous gene from the parent plant, further experiments defining components of this conserved mechanism can be attempted, for example, by cis-analysis of the promoter. Possible deviations from known expression patterns, arising, for example, from the absence of negative control circuits in the heterologous plant host leading to ectopic expression of the promoter, may also be informative and contribute information on the regulatory mechanisms. Where possible, studies in the homologous system can reveal components of the regulatory mechanism that are not conserved or not operational in the heterologous system. Analysis of gene expression in both a homologous and a heterologous system can therefore, when carefully deciphered, provide a detailed and general knowledge of the regulatory mechanisms. Some caution should, however, be observed when "hairy roots" or derived plants are used as transgenic tissue. Expression of the rol, aux, and possibly other oncogenes, influences the phytohormone concentrations *(14)*, and changes the hormonal sensitivity of the plant cells *(14,15)*. The expression patterns of plant genes responding to plant phytohormone signals may therefore be complex in "hairy roots." In this situation, binary plasmids containing biochemically characterized rol gene(s) could be used to study the hormone regulation *per se (14,16)*, but generally it is advisable to consider alternative transformation procedures, and to compare expression patterns with the expression of the endogenous wild-type plant gene localized by *in situ* hybridization or other histochemical methods. Ultimately, the "hairy root" phenotype could be eliminated in the progeny of primary transformants by genetic segregation of a binary T-DNA from the Ri T-DNA(s).

In the following sections, procedures for *A. rhizogenes*-mediated transformation and regeneration of *Lotus corniculatus* are outlined, together with protocols for nodulating composite plants.

2. Materials

2.1. Tissue Culture Media

1. L-S medium: Half-concentration of B5 medium without sucrose. Dissolve 1.94 g of ready made B5 (Flow Laboratories [McLean, VA], cat. no. 26-130-22), adjust pH to 5.5, add 12 g Noble agar (Difco, Detroit, MI) and water to 1 L, then autoclave. Medium can be remelted once in a microwave oven.
2. L medium: Half-concentration B5 medium. Dissolve 1.94 g of B5, 10 g sucrose, adjust pH to 5.5, add 9 g Noble agar, and water to 1 L, then autoclave. Medium can be remelted once in a microwave oven.
3. LW medium: Half-concentration B5 with 0.01% Tween-20. Dissolve 1.94 g B5, 10 g sucrose, add 100 µL Tween-20, 8 mL of 1*M* NaH_2PO_4 pH 5.5, adjust volume to 1 L, then autoclave.

2.2. Buffers

1. SP plus Tween 0.01%: For 1 L dissolve 8.5 g NaCl, 7.0 g K_2HPO_4, 3.0 g KH_2PO_4, add 100 µL Tween 20, adjust pH to 7.2, autoclave.
2. 2% hypochlorite, 0.02% Tween 20: For 20 mL mix 2.7 mL of 15% sodium hypochlorite solution with autoclaved water in a sterile universal container, add 4 µL Tween 20, shake gently, and use the same day.

2.3. Media for Growth of Bacteria

1. YB medium for *Rhizobium loti* inoculum: Dissolve 0.4 g yeast extract, 0.5 g K_2HPO_4, 0.2 g $MgSO_4 \cdot 7H_2O$, 0.1 g NaCl, adjust pH to 6.8 and volume to 1 L, autoclave.
2. YMA medium for *R. loti* plates: Add 2 g mannitol and 15 g agar to 1 L of YB medium, autoclave.
3. LB medium for *Agrobacterium* and *Escherichia coli*: Mix 5 g yeast extract, 10 g bacto tryptone, 5 g NaCl, adjust pH to 7.0, and volume to 1 L, autoclave.
4. LA medium for *Agrobacterium* and *E. coli* plates: Add 14 g of agar to 1 L of LB before autoclaving.

5. B&D plant growth medium *(17)*: Five milliliters of four stock solutions, A, B, C, and D, are mixed with 10 L of water. Stock A contains 294.1 g $CaCl_2 \cdot 2\ H_2O$ per L. Stock B contains 136.1 g KH_2PO_4 per L. Stock C contains 6.7 g Fe-citrate per L, heat to dissolve. Stock D contains; 123.3 g $MgSO_4 \cdot 7H_2O$, 87.0 g K_2SO_4, 0.338 g $MnSO_4 \cdot H_2O$, 0.247 g H_3BO_3, 0.288 g $ZnSO_4 \cdot 7H_2O$, 0.1 g $CuSO_4 \cdot 5H_2O$, 0.056 g $CoSO_4 \cdot 7H_2O$, 0.048 g $Na_2MoO_4 \cdot 2H_2O$ per L. Stocks are stored in brown bottles at 4°C. To avoid precipitation of B&D medium, autoclave B separately and add to medium afterward.

2.4. Antibiotics and Fungicides

1. Cefotaxime at 50 mg/mL: For 1 L of medium, dissolve 500 mg in 10 mL H_2O, filter sterilize, and use the same day.
2. Kanamycin at 100 mg/mL: Dissolve 3.2 g in 25 mL H_2O, filter sterilize (*see* Note 4). Store in aliquots at –20°C.
3. Rifampicin at 50 mg/mL: Dissolve 1 g of rimactan (Ciba Geigy, Basel, Switzerland) in 20 mL of water, filter sterilize. Store in aliquots at –20°C.
4. Spectinomycin at 100 mg/mL: Dissolve 2 g of spectinomycin in 20 mL of water, filter sterilize. Store in aliquots at –20°C.
5. Carbendazime at 30 mg/mL: Dissolve 300 mg carbendazime in 10 mL of DMSO, store in aliquots at –20°C.

2.5. Miscellaneous

1. Seeds of *L. corniculatus* "Rodéo" are supplied by Cooperative Agricole Mathieu (Gers, France).
2. Pouches: Northrup King Co. (Minneapolis, NM).
3. Nitrocellulose filters: Millipore (Bedford, MA) cat. no. GSWP047S0.
4. Universal containers: Gibco BRL (Gaithersburg, MD) cat. no. 3-64238.
5. Plant containers: Magenta Plast (Roskilde, Denmark) cat. no. 10197.
6. Hypodermic needles: Becton Dickinson (Dublin, Ireland) (27G3/4) 0.4 × 19, cat. no. 002200110.
7. Carbendazime: Reidel de Haën AG (Hannover, Germany).

3. Methods

3.1. Triparental Mating of Binary Vectors into Agrobacterium (Fig. 1)

1. Streak the *E. coli* strain containing the binary vector and the *E. coli* strain carrying the helper plasmid(s) onto fresh selective LA plates (Fig. 1, kanamycin). Incubate overnight at 37°C.
2. Inoculate 5 mL of liquid medium (LB) with *Agrobacterium* at high density, grow overnight at 28°C without selection.

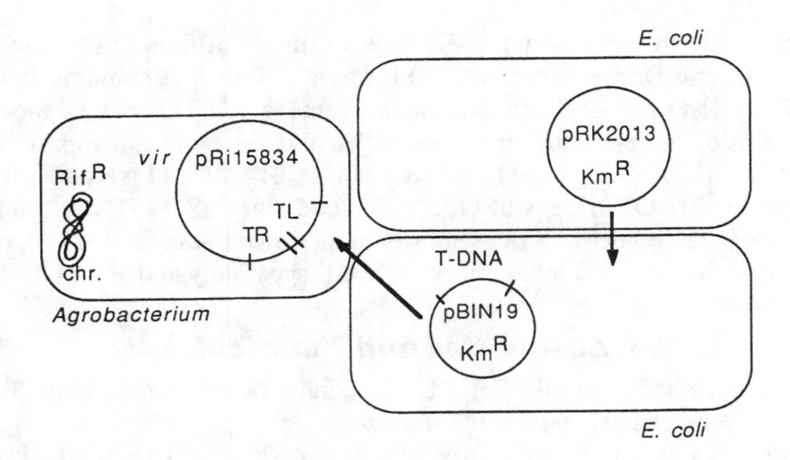

Fig. 1. Triparental mating of the binary vector pBIN19 into *Agrobacterium rhizogenes* C58C1 pRi15834. The pRK2013 helper plasmid is not stably established in *Agrobacterium (7)*. *Agrobacterium rhizogenes* transconjugants containing the binary vector are selected on rifampicin, a chromosomal (chr) marker, and kanamycin. The *vir* region of the pRi plasmid contains the virulence genes.

3. Start liquid culture (LB) of each of the *E. coli* strains containing binary vector and the helper plasmid(s). Scrape from selective plates using a toothpick, inoculate at high density, and grow at 37°C without selection for 1 h.
4. Arrange nitrocellulose filters on well dried solid LA plates and spot 50 µL of the *Agrobacterium* overnight culture onto the mating filters. Spot 200 µL of an *E. coli* culture containing a binary vector and 200 µL of the *E. coli* carrying helper plasmid(s) on top of the *Agrobacterium*. As controls, spot *Agrobacterium* and the *E. coli* strains on separate filters.
5. Leave the plates in a vertical flow bench, without lids, until the liquid has dried from the filters, incubate at 28°C overnight.
6. Use forceps to move the filters into universal containers with 4 mL of SP Buffer. Tighten the lids and vortex until the bacteria are washed off the filter and form a suspension in the buffer.
7. Plate 10–20 µL of the bacterial suspension onto LA plates selecting for the *A. rhizogenes* C58C1 chromosomal rifampicin resistance marker and the marker on the binary vector (Fig. 1, kanamycin). Incubate at 28°C for 2 d. (*See* Fig. 1 and Table 2 for conditions and concentrations of antibiotics.)

Table 2
Concentrations of Antibiotics for the Selection of *Agrobacterium* C58C1 pRi15834 and Derived Strains (*see* Note 4)

Rifampicin	100 µg/mL
Kanamycin	300 µg/mL
Spectinomycin	100 µg/mL
Streptomycin	100 µg/mL
Gentamycin	100 µg/mL
Tetracycline	1–2 µg/mL

8. Restreak transconjugants on new selective LA plates before use (Fig. 1, rifampicin and kanamycin) (*see* Note 1).

3.2. Conjugational Transfer of Integration Vectors into Agrobacterium (Fig. 2)

1. In this procedure the *E. coli* containing the integration vector and the *E. coli* carrying the helper plasmid(s) are crossed before mating to the *Agrobacterium* strain. Scrape the strains from selective plates and start 3-mL liquid cultures with high inoculum, incubate at 37°C for 3 h. Start the cross by transferring 1.5 mL of a *recA E. coli* containing the integration vector and 0.3 mL of *E. coli* carrying helper plasmid(s) to the same tube. Incubate for 1 h at 37°C without shaking. Plate 100 µL from the cross and each of the controls on LA plates selecting for integration vector and helper plasmid(s) (Fig. 2; spectinomycin, kanamycin, tetracycline). Incubate overnight at 37°C.
2. The colonies appearing after selection for both integration vector and helper plasmid(s) are used to start the *E. coli* donor culture for the *Agrobacterium* mating. The procedure is the filter procedure used for mating of binary vectors (*see* Section 3.1.), except that 200 µL of *E. coli* culture and 50 µL *Agrobacterium* is used for the mating, and that 50 to 100 µL are plated onto selective LA plates from the SP buffer suspensions (Fig. 2; rifampicin, spectinomycin). (*See* Fig. 2 for a schematic representation and Table 2 for concentrations of antibiotics.)

3.3. Infection of Seedling Hypocotyls (Fig. 3)

1. Take lids off individual 9-cm Petri dishes and arrange the dishes at an angle using the lids as a support. Cast the agar slopes pouring approx 20 mL of L-S medium into the lowest part of the plates.

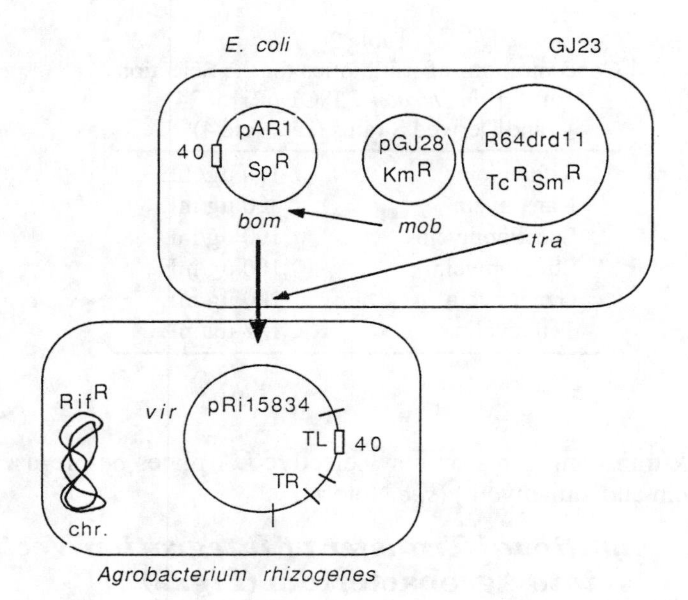

Fig. 2. Conjugational transfer of pAR1 *(9)* using the *E. coli* helper strain GJ23 containing the helper plasmids pGJ28 and pR64*drd*11 *(6)*. *Agrobacterium rhizogenes* C58C1 pRi15834 transconjugants with pAR1 recombined into the TL-region through the *Eco*RI fragment 40 are selected on rifampicin and spectinomycin. The *E. coli* donor is counterselected with rifampicin. In this plasmid helper system the *mob* functions interact with the *bom* site of the plasmid to be conjugated, the *tra* functions mediate the transfer. The pR64*drd*11 and pGJ28 plasmids are normally not maintained in *Agrobacterium (6)*. The *vir* region of the pRi plasmid contains the virulence genes. The rifampicin resistance marker is on the chromosome (chr) of C58C1.

2. Surface sterilize seeds in 20 mL of 2% hypochlorite, 0.02% Tween. Shake for 20 min and rinse the seeds in 6 × 20 mL of autoclaved water. Transfer the seeds onto a sterile filter paper in a Petri dish, add 3–5 mL of sterile water and incubate at 26° overnight.
3. Arrange 10–12 seeds at the top of the agar slope in the L-S plates. Pack 10 plates together using transparent tape. Incubate plates at 26°C continuous light (Philips TLD 30W/33 tube) for 5 d.
4. Streak the *Agrobacterium rhizogenes* strains for the transformation onto fresh selective medium on d 3 after the start of seed germination. Plates for histidine auxotrophic strains *(10)* are supplemented with 100 µg/mL of histidine. Incubate at 28°C.

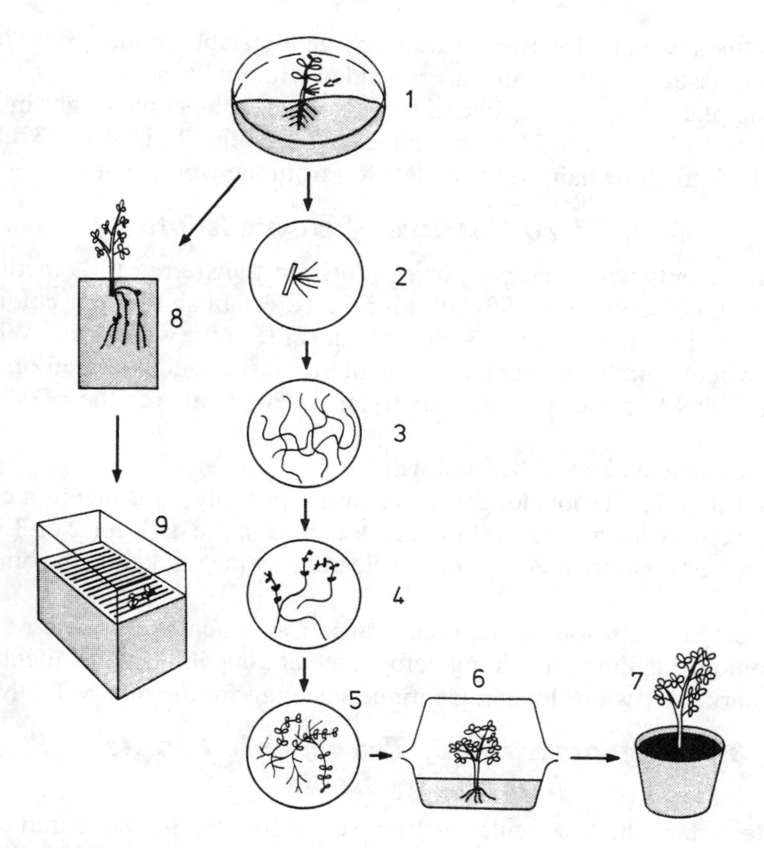

Fig. 3. Illustration of the *Lotus corniculatus* transformation and regeneration procedure. **(1)** Hairy roots emerging (arrow) from wounded hypocotyl of an *L. corniculatus* seedling infected with *A. rhizogenes*. **(2)** Initiation of root culture after transfer of stem fragment with hairy roots. Incubation in the dark. **(3)** Root culture established from an individual hairy root. Incubation in the dark. **(4)** Small green calli and small shoots appearing on the root culture after transfer into light. **(5)** Elongated shoots still attached to the root culture. **(6)** A shoot from a root culture is transferred for rooting in a plant container. **(7)** Rooted plant is transferred to a pot. **(8)** "Short-cut transformation"; the normal root system is removed and the hairy root is inoculated with *Rhizobium*. **(9)** Rack system used for pouches.

5. Seedlings are "wound-site" infected on d 5 after germination using a hypodermic needle. A needle is fitted onto a 2-mL syringe and dipped into a freshly grown *Agrobacterium* colony on a selective plate. The hypoco-

tyls are then wounded in two positions by gently stabbing the needle into the plant tissue. Control plants are wounded without bacteria.

6. Pack the plates together again and keep wounded plants overnight in the dark at 21°C. Incubate at 21°C continuous light (Philips TLD 30W/33 tube) for 2–3 wk to allow hairy roots to develop from the wound sites.

3.4. In Vitro Culture of Hairy Roots

1. Stem fragments with emerging hairy roots are transferred to L medium containing cefotaxime at 300–500 µg/mL. To establish the root cultures effectively, transfer 1–2 cm of stem on either side of the wound site. When binary vectors are used, select for the plant marker gene located on the vector T-DNA, for example, kanamycin at 50µg/mL for the pGV941 T-DNA *(9)*.
2. Incubate in the dark at 26°C for 2 wk.
3. Transfer individual roots longer than 3 cm, separately, to L medium containing cefotaxime at 300–500 µg/mL. Incubate in the dark for 2 or 3 wk. If binary vectors were used, continue selecting for the marker on the binary T-DNA.
4. Remove agar from root system and transfer as much as possible of the roots onto L medium containing cefotaxime at 300–500 µg/mL. Incubate in the dark for 1 wk or longer. Continue selecting for the binary T-DNA.

3.5. Regeneration of Transgenic Plants from Hairy Roots

1. Transfer established root cultures from step 4 (or step 3) above into continuous light (Philips TLD [Einhoven, Holland] 30W/33 tube) at 26°C. A small part of the root culture is subcultured on L medium (with selection) and kept in the dark as root culture.
2. Leave the root cultures in the light without changing the medium. Spontaneous shoot formation is first visible as green spots on the roots. Shoots will then develop from the spots and continue to grow into proper elongated shoots.
3. Shoots of 3–4 cm in lengths are cut off, if possible together with a bit of root, and rooted for 2 wk on 100 mL of L medium in plant containers. Kanamycin at 50 µg/mL can be used as extra selection for binary T-DNA.
4. Transfer plants to pots containing a nitrogen free support, water with B&D medium supplemented with 5 m*M* KNO_3 (*see* Note 3), and inoculate with 2 mL of a *Rhizobium loti* culture grown for 3–4 d in YB medium. Cover with a plastic bag.
5. Remove plastic bags to harden the plants and continue to water with B&D.

3.6. Nodulation of Composite Plants (Fig. 3)

1. Start transformation as described in Section 3.3. (*see* Note 2).
2. Remove the main root from plants with hairy roots emerging at the wound site and transfer 5–8 shoots to a plant container containing 100 mL LW medium plus 500 µg/mL of cefotaxime. Take care to submerge the plants. Wash the plants by shaking gently for 2–3 d at 26°C continuous light. For histidine auxotraphic strains *(10)*, include 10 µg/mL of histidine in this wash only.
3. Change the media and continue to wash for another 2–3 d, then wash the plants in six changes of autoclaved water. Collect and autoclave media and water before disposal.
4. Place two shoots with hairy roots into opposite sites of a pouch containing 20 mL of B&D with 2 m*M* KNO_3 (*see* Note 3). Inoculate with 1 mL of a *Rhizobium loti* culture previously grown for 3–4 d in YB medium. Pouches are then arranged close together in a rack system covered with a lid (Fig. 3, step 9). Autoclaved water is poured into the rack to limit evaporation from pouches. Pouches must be close together for support and to keep the opening closed around the plants.
5. Move the racks to growth cabinets with 16 h/8 h day/night cycle, 246 µmol/m^2/s, 20°C/16°C temperature regime, 70% humidity. Nodules will appear after 4–5 wk. Pick nodules when they have reached the appropriate developmental stage. The general lifetime of nodules grown in pouches is lower then for nodules grown in pots.

4. Notes

1. Transformation procedures for *Agrobacterium* are available *(8,18)*, and binary vectors can, using these procedures, be transferred into *Agrobacterium* faster than when using the triparental mating system. The combined frequency of transformation and recombination may, however, limit the use of direct transformation for transfer of integration vectors *(6,8)*.
2. Nodulation of composite plants taken directly from the L-S plates is faster than nodulation of washed and decontaminated plants. Provided the *Agrobacteria* can be contained in the growth conditions used and the plant reporter gene constructs are designed to avoid expression in the *Agrobacteria*, this approach can be used to speed up the nodulation. With the decontaminated plants, the efficiency of the washing procedure can be tested by plating homogenized plant material on selective LA plates. Composite plants can, of course, also be grown in jars, pots, or Petri dishes. Composite plants grown in pouches under lid requires light of high intensity, 246 µmol/m^2/s. If necessary, the fungicide carbendazime can be included at 30 µg/mL in all tissue culture and washing media.

3. Some nitrate is necessary for the recovery of plants after tissue culture, but too much nitrate will inhibit nodulation. The level of nitrate should be balanced so that the plants turn slightly yellow just prior to onset of nitrogen fixation. Under the conditions described, 5 L of B&D plus 5 m*M* KNO_3 is used (once) for 22 pots and 2 m*M* KNO_3 is used in the pouches.
4. The C58c1 pRi15834 strains are very sensitive toward tetracycline. It might be necessary to test the toxicity of the tetracycline batch used and to make several selective plates, increasing the tetracycline concentration stepwise by 0.5 µg/mL. Note also that most commercial kanamycin preparations have a purity of approx 80%.

References

1. Handberg, K. and Stougaard, J. (1992) *Lotus japonicus*, and autogamous, diploid legume species for classical and molecular genetics. *Plant J.* **2,** 487–496.
2. Tepfer, D. (1984) Transformation of several species of higher plants by *Agrobacterium rhizogenes:* sexual transmission of the transformed genotype and phenotype. *Cell* **37,** 959–967.
3. Mugnier, J. (1988) Establishment of new axenic hairy root lines by inoculation with *Agrobacterium rhizogenes*. *Plant Cell Rep.* **7,** 9–12.
4. Petit, A., Stougaard, J., Kühle, A., Marcker, K. A., and Tempé, J. (1987) Transformation and regeneration of the legume *Lotus corniculatus*: a system for molecular studies of symbiotic nitrogen fixation. *Mol. Gen. Genet.* **207,** 245–250.
5. Birot, A.-M., Bouchez, D., Casse-Delbart, F., Durand-Tardif, M., Jouanin, L., Pautot, V., Robaglia, C., Tepfer, D., Tepfer, M., Tourneur, J., and Vilaine, F. (1987) Studies and uses of the Ri plasmids of *Agrobacterium rhizogenes*. *Plant Physiol. Biochem.* **25,** 323–335.
6. Van Haute, E., Joos, H., Maes, M., Warren, G., Van Montagu, M., and Schell, J. (1983) Intergenic transfer and exchange recombination of restriction fragments cloned in pBR322: a novel strategy for the reversed genetics of the Ti plasmid of *Agrobacterium tumefaciens*. *EMBO J.* **2,** 411–417.
7. Ditta, G., Stanfield, S., Corbin, D., and Helinski, D. R. (1980) Broad host range DNA cloning system for Gram-negative bacteria: construction of a gene bank of *Rhizobium meliloti*. *Proc. Natl. Acad. Sci. USA* **77,** 7347–7351.
8. Holsters, M., de Waele, D., Depicker, A., Messens, E., van Montagu, M., and Schell, J. (1978) Transfection and transformation of *Agrobacterium tumefaciens*. *Mol. Gen. Genet.* **163,** 181–187.
9. Stougaard, J. Abildsten, D., and Marcker, K. A. (1987) The *Agrobacterium rhizogenes* pRi TL-DNA segment as a gene vector system for transformation of plants. *Mol. Gen. Genet.* **207,** 251–255.

10. Hansen, J. Jørgensen, J.-E., Stougaard, J., and Marcker, K. A. (1989) Hairy roots—a short cut to transgenic root nodules. *Plant Cell Rep.* **8,** 12–15.
11. Ramlov, K., Laursen, N. B., Stougaard, J., and Marcker, K. A. (1993) Site-directed mutagenesis of the organ-specific element in the soybean leghemoglobin *lbc3* gene promoter. *Plant J.* **4,** 577–580.
12. Quandt, H.-J., Pühler, A., and Broer, I. (1993) Transgenic root nodules of *Vicia hirsuta*: a fast and efficient system for the study of gene expression in indeterminate-type nodules. *MPMI* **6,** 699–706.
13. Cheon, C-III, Lee, N.-G., Siddique, A. B. M., Bal, A. K., and Verma D. P. S. (1993) Roles of plant homologs Rab1p and Rab7p in the biogenesis of the peribacteroid membrane, a subcellular compartment formed *de novo* during root nodule symbiosis. *EMBO J.* **12,** 4125–4135.
14. Schmülling, T., Fladung, M., Grossmann, K., and Schell, J. (1993) Hormonal content and sensitivity of transgenic tobacco and potato plants expressing single *rol* genes of *Agrobacterium rhizogenes* T-DNA. *Plant J.* **3,** 371–382.
15. Shen, W. H., Petit, A., Guern, J., and Tempé J. (1988) Hairy roots are more sensitive to auxin than normal roots. *Proc. Natl. Acad. Sci. USA* **85,** 3417–3421.
16. Vilaine, F. and Casse-Delbart, F. (1987) A new vector derived from *Agrobacterium rhizogenes* plasmids: a micro-Ri plasmid and its use to construct a mini-Ri plasmid. *Gene* **55,** 105–114.
17. Broughton, W. J. and Dilworth, M. (1971) Control of leghemoglobin synthesis in snake beans. *Biochem. J.* **125,** 1075–1080.
18. Mattanovich, D., Ruker, F., Machado, A. D. C., Laimer, M., Regner, F., Steinkellner, H., Himmler, G., and Katinger, H. (1989) Efficient transformation of *Agrobacterium* spp. by electroporation. *Nucleic Acids Res.* **17,** 6747.

CHAPTER 5

Agrobacterium-Mediated Transformation of *Arabidopsis thaliana*

Application in T-DNA Tagging

Jennifer F. Topping, Wenbin Wei, Michael C. Clarke, Paul Muskett, and Keith Lindsey

1. Introduction

A number of methods have been developed for the isolation of plant genes, including the differential screening of cDNA libraries, the use of heterologous probes, and the use of protein sequence data to generate oligonucleotides for library screening. These have been well-reviewed. The advent of plant transformation techniques has further increased the tools available for the functional analysis of isolated genes, such as for the introduction of promoter deletion–reporter gene fusions, for the overexpression of the gene product or for generating null expressors by antisense RNA technology. More recently, transformation techniques have been exploited to develop new approaches to isolate plant genes, and these methods can be considered as being of two types: insertional mutagenesis and gene trapping.

Insertional mutagenesis involves the disruption of genes by DNA sequences, and can be achieved either by the random integration of transposable elements or of T-DNA sequences following transformation by *Agrobacterium tumefaciens*. If the inserting DNA should disrupt a gene essential for a given developmental, biochemical, or physiological pathway, then a mutant phenotype may be detectable in plants homozygous

From: *Methods in Molecular Biology, Vol. 49: Plant Gene Transfer and Expression Protocols*
Edited by: H. Jones Humana Press Inc., Totowa, NJ

for the mutation. In species that do not have active endogenous transposable elements, functional heterologous transposon systems, such as those originating from maize *(1)*, can be introduced by stable transformation. Insertional mutagenesis, like other mutagenesis systems, is most usefully exploited in diploid species in which a recessive mutation is less likely to be masked by functional redundancy conferred by the presence of multiple wild-type alleles, such as would be found in polyploids.

Gene trapping techniques allow the identification of native genes that are expressed in spatially or temporally regulated patterns, without the requirement for generating a mutant phenotype. The rationale is to transform populations of plants with a reporter gene that has either a weak promoter *(2,3)* or no promoter *(2,4–7)*. In the former case, a so-called "enhancer trap," the reporter would be expected to be activated by nearby enhancer elements associated with native genes close to, and either upstream or downstream of, the site of T-DNA insertion. Enhancers that regulate the expression pattern of genes may therefore modulate the expression of the reporter, generating a functional tag of the enhancer. In the second case, the activation of a promoterless reporter gene (a "promoter trap") would only be expected to occur if the T-DNA had integrated downstream of a native gene promoter, to generate either a transcriptional or translational gene fusion. Again, the expression pattern of the reporter gene should reflect the expression of the tagged gene, since it is under the transcriptional control of the same promoter. In contrast to the enhancer trap system, a functional promoter trap is expected necessarily to create a gene disruption, which may result in a mutant phenotype in progeny homozygous for the T-DNA insert. However, the approach may also be used to tag genes in polyploid species *(4)*, since detection of the tag can be achieved by the assay of reporter gene activity alone.

Both insertional mutagenesis and gene trapping can facilitate the isolation of genes, by virtue of the physical linkage of known DNA sequences to unknown genomic sequences. This paves the way for gene isolation by either inverse PCR techniques, using transposon or T-DNA sequences as primers *(4)*, by plasmid rescue *(8)*, or by library screening, using the known DNA sequences as probes *(9)*.

A useful model plant in which to carry out mutagenesis screens is *Arabidopsis thaliana (10,11)*. It is naturally self-fertilizing, produces abundant seed within 6 or 8 wk after germination, and grows vigorously in artificial light and in vitro. It has recently received a large amount of attention for molecular genetic studies: It has a small genome with a low content of repetitive DNA, which facilitates gene cloning by map-based techniques, such as chromosome walking, library screening, and PCR *(12)*. It is only very recently that functional transposable elements have been identified in the *Arabidopsis* genome *(13)*, and with this one exception the application of insertional mutagenesis or gene trapping as strategies to isolate genes has required the development of stable transformation systems.

There are two broad approaches to transform plants: vector-free or direct gene transfer systems, and those that involve the use of vectors, notably involving *Agrobacterium*-mediated gene transfer *(14)*. Both systems have been developed for *Arabidopsis*, but we will only consider in any detail those methods that utilize infection by *Agrobacterium tumefaciens*. (For recent coverage of protoplast-based direct gene transfer techniques, *see* ref. *15)* There are three explant tissues of *Arabidopsis* most commonly adopted for infection by *A. tumefaciens*: intact seeds *(16)*, cotyledons *(17)*, and roots *(18)*. The seed transformation system involves no tissue culture steps, and has been used to generate approx 13,000 independent transformants to date *(19)*, but is not easily reproduced in other laboratories. The cotyledon and root transformation protocols have been more widely used. We *(20)* have modified the root transformation system of Valvekens et al. *(19)* such that, in our hands, the frequency of recovery of transformants is increased up to 100-fold; we have generated approx 2500 independent transformants using this method, and it has been used successfully in other laboratories.

In this chapter we describe:

1. Our modified root transformation protocol;
2. The vector we use for promoter trapping and insertional mutagenesis;
3. The screening strategy we use for identifying tagged genes; and
4. The use of inverse PCR to isolate T-DNA flanking sequences.

2. Materials

2.1. Plant Transformation

1. Bleach solution: Five percent (v/v) commercial bleach containing 0.05% (v/v) Tween 20.
2. 70% (v/v) ethanol.
3. Silver thiosulfate stock solution (1.25 mg/mL): Add dropwise a solution of silver nitrate (2.5 mg/mL) to an equal volume of a solution of sodium thiosulfate pentahydrate (14.6 mg/mL). Filter sterilize.

4. Germination medium (GM): Half strength Murashige and Skoog medium (ref. *21*; Sigma [St. Louis, MO] M5519), 10 g/L sucrose. Adjust pH to 5.8 with 1*M* KOH, add 8 g/L Difco (Detroit, MI) Bacto agar, autoclave (121°C, 15 min) to sterilize, add 5 mg/L silver thiosulfate solution (filter sterilized) before plating out.
5. Stock solution of 4 mg/mL 2,4-dichlorophenoxyacetic acid (2,4-D) in dimethylsulfoxide (DMSO).
6. Stock solution of 0.4 mg/mL kinetin in DMSO.
7. Callus-inducing medium (CIM): Gamborg's B5 medium (*22*; Sigma), 0.5 g/L MES, 20 g/L glucose. Adjust pH to 5.8 with 1*M* KOH, add 8 g/L Difco Bacto agar, autoclave (121°C, 15 min) to sterilize. After autoclaving, add 5 mg/L silver thiosulfate solution (filter sterilized), 0.5 mg/L 2,4-D, 0.05 mg/L kinetin.
8. Stock solution of 200 mg/mL vancomycin (filter sterilized).
9. Stock solution of 50 mg/mL kanamycin sulfate (filter sterilized).
10. Stock solution of 20 mg/mL 2-isopentenyladenine (2-iP) in DMSO.
11. Stock solution of 1.2 mg/mL indole-3-acetic acid (IAA) in DMSO.
12. Shoot-inducing medium (SIM): Gamborg's B5 medium (Sigma), 0.5 g/L MES, 20 g/L glucose. Adjust pH to 5.8 with 1*M* KOH, add 8 g/L Difco Bacto agar, autoclave (121°C, 15 min). After autoclaving, add 850 mg/L vancomycin, 35 mg/L kanamycin sulfate, 5 mg/L 2-iP, 0.15 mg/L IAA.
13. Shoot overlay medium (SOM): As SIM, but replacing the agar with 8 g/L low melting point agarose (SeaPlaque, FMC, Rockland, ME); autoclave (121°C, 15 min).
14. Shoot elongation medium (SEM): Half strength Murashige and Skoog Medium (Sigma), 10 g/L sucrose. Adjust pH to 5.8 with 1*M* KOH, add 8 g/L Difco Bacto agar, autoclave (121°C, 15 min).
15. Compost: From Russell et al. (*23*): Irish moss peat (Joseph Bently & Sons, Barrow-on-Humber, UK), John Innes potting compost No. 3, horticultural potting grit (Joseph Bently & Sons), coarse vermiculite (Vermiperl Medium Grade, Silverperl Ltd., Gainsborough, UK). Mix in a ratio of 2:2:2:1, autoclave (121°C, 30 min).
16. LB (Luria-Bertani) medium: 10 g/L bacto-tryptone, 5 g/L bacto-yeast extract, 10 g/L NaCl. Adjust pH to 7.5 with 1*M* NaOH, autoclave (121°C, 15 min).
17. Filter sterilized stock of rifampicin: 20 mg/mL in methanol.
18. *Agrobacterium* culture medium: Sterile LB (Luria-Bertani) medium, 50 mg/L kanamycin sulfate (or other appropriate selective agent), 100 mg/L rifampicin (or other appropriate selective agent).
19. *Agrobacterium* culture dilution medium: Gamborg's B5 salts solution (Sigma), 20 g/L glucose, 0.5 g/L MES. Adjust pH to 5.7 with 1*M* KOH, autoclave (121°C, 15 min).

20. Sterile double-distilled water (autoclave 121°C, 15min).
21. Micropore® gas-permeable tape (3M, Loughborough, UK).
22. 9-cm Petri dishes (Falcon 3003).
23. Nylon or stainless steel mesh (100 μm pore diameter).
24. Sterile filter paper.
25. Polypot containers (SC020, Northern Media Supply Ltd., Loughborough, UK).
26. Gas-permeable transparent SunCap® film (C6920, Sigma).
27. Glass culture tubes (C5916, Sigma).
28. Sterile perlite (autoclave 121°C, 15 min).
29. Sterile Gamborg's B5 salt solution (autoclave 121°C, 15 min).
30. Aracon® tubes (Beta Developments, Gent).

2.2. Promoter Trap Vector pΔgusBin19

This vector *(2)* is characterized by a selectable *nptII* gene near the T-DNA right border, under the transcriptional control of the *nos* promoter and polyadenylation sequences; while at the left border is a promoterless *uid*A (*gus*A) coding region (encoding β-glucuronidase; *24*) plus *nos* polyadenylation sequence (Fig. 1). The 5' end of the *gus*A gene is adjacent to the left border and has its own translation initiation codon. Since there are translational stop codons in the left border region in all reading frames, gene fusions generated on integration are expected to be primarily transcriptional rather than translational, and this is observed (unpublished data).

2.3. Screening for Tagged Genes

1. Materials for GUS enzyme assay and histochemistry are presented in the chapter by Hull et al., this volume.
2. Kanamycin selection plates: Half strength Murashige and Skoog medium, 10 g/L sucrose, adjust pH to 5.8 with 1*M* KOH, add 8 g/L Difco Bacto agar. Autoclave (121°C, 15 min) to sterilize. Add 25 mg/L kanamycin sulfate (filter sterilized) before plating out.

2.4. Inverse PCR

1. 5X ligation buffer: 250 m*M* Tris-HCl pH 7.4, 50 m*M* $MgCl_2$, 50 m*M* dithiothreitol, 5 m*M* ATP.
2. T4 ligase: 1 U/μL (BRL/Gibco, Middlesex, UK).
3. 10X PCR buffer (contains 1.5 m*M* Mg^{2+}; Promega, Southampton, UK).
4. 2 m*M* dNTP.
5. Primers (100 ng/mL): *Seq 1*: 5' CTGAATGGCGAATGGCGC 3'; *nested 2*: 5' CAGGACGTAACATAAGGG 3'; *nested 3*: 5' GACTGGCATGAACTTCG 3'; and *nptII*: 5' GTCATAGCCGAATAGCCTC 3' (*see* Fig. 1).

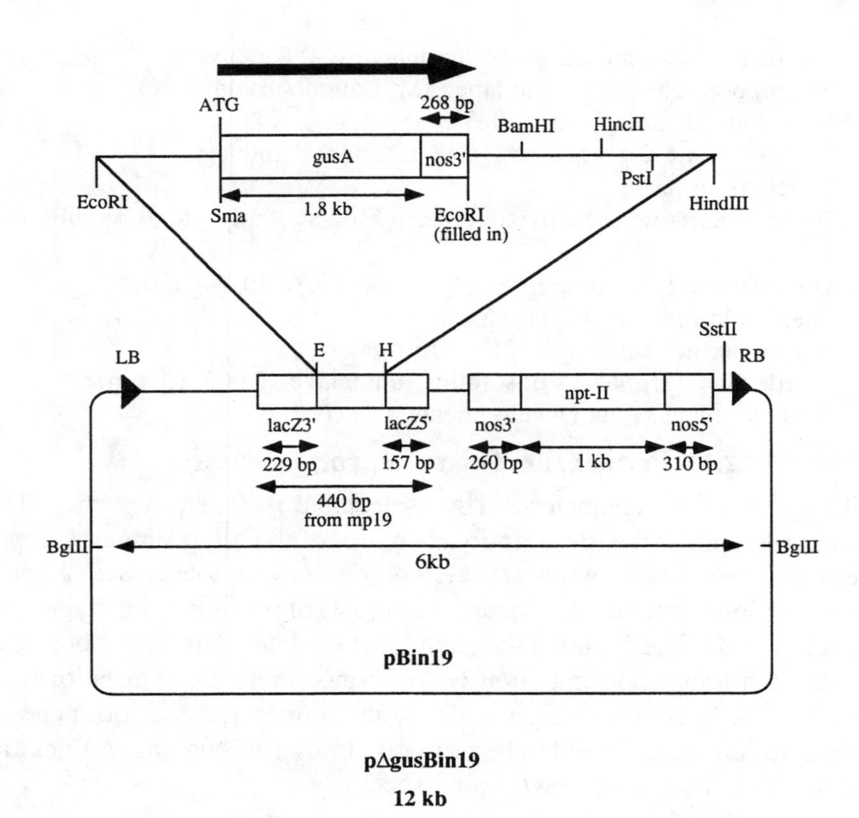

Fig. 1. The promoter trap vector pΔgusBin19 *(2)*.

6. 25 m*M* $MgCl_2$.
7. *Taq* polymerase: 0.5 U/0.1 μL (Promega, Southampton, UK).

3. Methods

3.1. Seed Sterilization and Seedling Growth

1. Vernalize *Arabidopsis* seeds at 4°C for 1 wk before sterilization, to maximize germination frequency.
2. Surface sterilize the seeds by
   a. Immersion in 70% (v/v) ethanol for 30 s; then remove the ethanol with a Pasteur pipet; then
   b. Immerse the seeds in 5% (v/v) commercial bleach solution containing 0.05% Tween 20 as a surfactant for 15–20 min.
3. Remove bleach solution with a Pasteur pipet and wash the seeds six times in sterile double distilled water.

4. Suspend the seeds in a drop of sterile water and, using a pipet, transfer them to the surface of GM agar plates. Spread out the seeds evenly (approx 40 per 9-cm Petri dish) and seal the plates with Micropore tape.
5. To germinate the seeds, incubate for 2 d in dim light (approx 22°C) and then for a further 3–4 wk in a standard light regime for cultures (at least 25–50 μmol/m^2/s) to allow root growth.

3.2. Root Explantation and Inoculation with Agrobacterium

1. When the seedlings are approx 3 wk old and have produced abundant (but not yet greening) roots (*see* Note 1), gently remove them from the agar with forceps and excise the root systems.
2. Place the intact root systems on the surface of agar plates containing CIM, and incubate at approx 20–22°C for 3 d.
3. Transfer the roots to a sterile Petri dish and cut into segments of approx 0.5 cm in length.
4. Grow up a liquid suspension of *Agrobacterium* cells at 29°C on a rotary shaker (200 rpm) for 48 h, then dilute with sterile *Agrobacterium* dilution medium, to a final optical density of 0.1 (600 nm).
5. Suspend the root explants in 20-mL *Agrobacterium* culture medium and incubate at room temperature for 2 min. Drain off the bacterial suspension by placing the roots on a nylon or stainless steel mesh (100 μm pore diameter); remove excess liquid by blotting the roots on sterile filter paper.
6. Transfer the roots to CIM agar plates, and culture for 48–72 h.
7. Wash the roots (over a nylon/steel mesh) with sterile *Agrobacterium* culture dilution medium to remove excess bacterial growth (*see* Note 2), and blot the roots dry on sterile filter paper.

3.3. Selection and Regeneration of Transformants

1. Suspend the root explants in molten SOM, at a temperature of about 30°C. Use the roots of approx 40 seedlings (approx 0.1 g tissue, corresponding to approx 300 individual root segments) per 10 mL medium.
2. Pour 10 mL of the root/SOM on to the surface of agar plates (9 cm) containing 40 mL SIM. Ensure that the root explants are evenly dispersed. Seal the plates with gas-permeable Micropore tape (*see* Note 3), and incubate at approx 25°C in fluorescent light of at least 25–50 μmol/m^2/s.
3. From approx 3 wk, green (putatively transformed) calluses develop on the root explants, and kanamycin-resistant shoots will regenerate over a period of several weeks.
4. When the shoots have expanded leaves, transfer to SEM (20 mL) in either 60 mL Polypots sealed with gas-permeable transparent SunCap film, or glass culture tubes sealed with cotton wool bungs or with SunCap film (*see*

Note 4). Under these conditions, at least 80% of the transformed shoots are expected to flower and set seed (*see* Notes 5 and 6).

3.4. Bulking Up of T2 Plants

1. Fill Petri dishes with a layer of sterile perlite, and add sterile Gamborg's B5 salt solution until the perlite has absorbed the liquid to its full capacity, but no more.
2. Plate out and vernalize the seeds (approx 20 per plate) from T1 plants and culture for 2 d in dim light, and subsequently under full illumination, at approx 22°C (*see* Note 7).
3. When the seedlings are approx 1–2 cm tall, transfer to sterile, moist compost. Grow for 1 wk in high humidity in a micropropagator (approx 20°C), and gradually reduce the local humidity by increasing ventilation over the following week.
4. When the seedlings are 3–5 cm high, cover the seedlings with Aracon tubes (Beta Developments, Gent) and collect seed over the following few months (*see* Note 8).

3.5. Screening for GUS Fusion Expression and Mutants

Screening for tagged genes in transformants containing pΔgusBin19 comprises the identification of lines exhibiting GUS fusion activity and mutant phenotypes that cosegregate with the T-DNA insert. Depending on the developmental process being studied, we may screen for GUS activity either by histochemistry, or in some instances (such as in embryos that may require dissection) by fluorimetry, which is sensitive and allows the rapid screening of relatively large organs (e.g., siliques) followed by the more detailed histochemical analysis of "positives" identified fluorimetrically *(25)*. Methods for GUS enzyme assay and histochemistry are presented in Chapter 10, and will not be repeated here.

The method of screening for mutant phenotypes depends on the type of mutant (developmental, biochemical, physiological) of interest. Confirmation that the mutation is owing to the insertion of a T-DNA involves the demonstration of cosegregation of the T-DNA and the mutant phenotype in progeny from selfing and outcrossing with wild-type lines. This is achieved by determining the linkage of, for example, a kanamycin-resistance phenotype, conferred by the T-DNA, and of T-DNA sequences, determined by Southern hybridization analysis, and the mutant phenotype. Final proof is demonstrated by the complementation of the mutant phenotype by retransformation with a wild-type allele of the mutant gene, cloned from a wild-type library using T-DNA flanking sequences as a probe. Kanamycin resistance/mutant cosegregation analysis is carried out as follows.

1. Plate out approx 200 sterilized T2 seeds on both kanamycin selection plates and either SEM (Section 2.1.14.) or sterile perlite/compost (*see* Section 3.4. steps 1–3). A T-DNA present as a single active locus will segregate 3 kan^r: 1 kan^s in the progeny. The mutant phenotype, if determined by a single recessive mutation, will appear in 25% of the seedlings on the SEM media/perlite/compost that lack kanamycin. If the mutant is embryonic lethal, all the viable progeny will be phenotypically wild-type (heterozygous for the mutation or wild-type).
2. Collect T3 seed from the mutants grown up on nonselective media (assuming they are not embryonic- or seedling-lethal or sterile) and plate out approx 200 on kanamycin selection plates (*see* Note 9). If the mutant phenotype and T-DNA are linked, 100% of the seedlings will be kanamycin resistant (homozygous for the *nptII* gene) *and* 100% mutant phenotype (homozygous for the mutant gene). Kanamycin-resistant but phenotypically wild-type T3 seedlings in turn should segregate, on selfing, 25% kanamycin resistant mutants (homozygous for T-DNA and mutant gene); 50% kanamycin resistant wild-type phenotypes (heterozygous for T-DNA and mutant gene); and 25% kanamycin sensitive wild-type phenotypes (lacking both T-DNA and mutant gene). Southern blots should be carried out to confirm this.
3. To determine whether an embryonic-lethal mutation is caused by a T-DNA, collect T3 seed from a kanamycin-resistant plant (which is expected to be heterozygous for both the mutant gene and T-DNA) and plate it out on kanamycin selection plates. Twenty-five percent of the seeds should fail to germinate (homozygous for the embryonic lethal mutation); of the germinating seeds, 66% should be kanamycin resistant wild-type phenotypes (heterozygous for T-DNA and mutant gene) and 33% should be kanamycin sensitive wild-type phenotypes (lacking both T-DNA and mutant gene).
4. By outcrossing lines putatively homozygous for both T-DNA and mutant gene to wild-types, further evidence of cosegregation can be obtained. To outcross:
   a. Remove sepals, petals, and anthers from unopened buds of the female parent (*see* Note 10).
   b. Remove all other buds, to prevent self-fertilization.
   c. Leave the emasculated buds for about 2 d to allow the stigmata to become receptive to pollen, and then pollinate with dehiscing anthers removed from the male parent.

5. Collect F1 seed (which is expected to be heterozygous for both the mutant gene and T-DNA) and plate it out on kanamycin selection plates. Twenty-five percent of the seedlings are expected to be kanamycin resistant mutants (homozygous for T-DNA and mutant gene); 50% kanamycin resistant wild-type phenotypes (heterozygous for T-DNA and mutant gene); and 25% kanamycin sensitive wild-type phenotypes (lacking both T-DNA and mutant gene; *see* Note 11). Southern blots should be carried out to confirm this.

3.6. Inverse PCR Amplification of T-DNA Flanking Sequences

To successfully amplify T-DNA flanking sequences by IPCR, it is essential to identify restriction fragments containing T-DNA sequences and plant flanking DNA, such that the expected PCR product is no greater than approx 2.5–3 kb. Larger products are difficult to amplify. Purified genomic DNA from transgenic plants is analyzed by Southern hybridization following digestion with a range of restriction enzymes designed to generate appropriate T-DNA/plant DNA border fragments. These fragments are then religated under conditions designed to generate monomeric circles *(4,26)*.

1. Ligate digested DNA fragments (2 µg in 40 µL) in 200 µL 5X ligation buffer plus 14 µL T4 ligase, made up to a final volume of 1 mL with sterile distilled water.
2. Extract 500 µL ligated DNA with phenol/chloroform, precipitate the DNA in ethanol and resuspend to a concentration of 20 ng/µL.
3. Fifty Nanogram DNA is used as substrate for IPCR amplification, in the following reaction mixture: 2.5 µL 10X PCR buffer, 1 µL each primer (100 ng/mL), 2.5 µL 25 m*M* $MgCl_2$, 0.4 µL *Taq* polymerase (0.5 U/0.1 µL); made up to 17 µL with sterile distilled water (*see* Note 12).
4. Pipet the reaction mixture into an Eppendorf tube, cover with 1 drop of paraffin oil, and place the tube in a thermal cycler, holding the temperature at 80°C. To start the reaction, add 2.5 µL 2 m*M* dNTP.
5. Carry out PCR amplification over 30 cycles using the following conditions: Denaturation at 95°C for 4 min during the first cycle and for 1 min in subsequent cycles; primer annealing at 55°C, 2 min; extension by *Taq* polymerase at 72°C for 10 min during the first cycle and for 3 min in subsequent cycles; denaturation at 95°C, 1 min; and final extension at 72°C, 10 min (*see* Notes 13–15).
6. Separate the reaction products in a 1% (w/v) agarose thin gel and stain with ethidium bromide to visualize the DNA (Fig. 2).

IPCR of AtVT-1

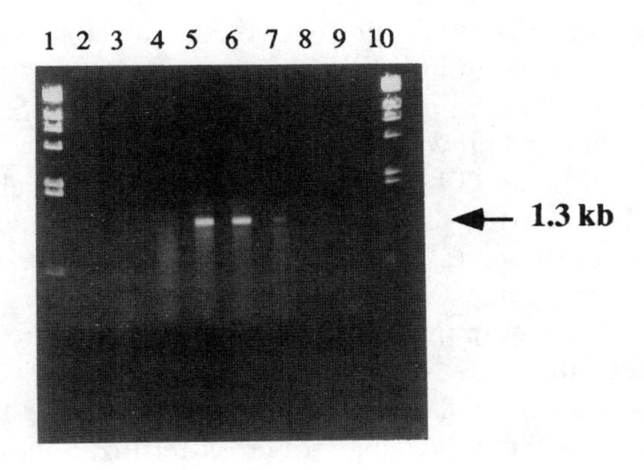

1. Lambda markers
2. empty
3. 17 pg pGUS
4. 500 ng AtVT-1 DNA, 2.0 mM Mg
5. 250 ng AtVT-1 DNA, 2.0 mM Mg
6. 125 ng AtVT-1 DNA, 2.0 mM Mg
7. 125 ng AtVT-1 DNA, 1.5mM Mg
8. empty
9. buffer
10. Lambda markers

Fig. 2. Amplification of T-DNA flanking sequences from a transgenic line, AtVT-1, by inverse PCR (W. Wei and K. Lindsey, unpublished).

4. Notes

4.1. Arabidopsis *Transformation*

1. Roots that are beginning to green are too old to respond to the regeneration conditions.
2. Following both inoculation with *Agrobacterium* and washing with *Agrobacterium* culture dilution medium, remove as much excess liquid as possible from the roots. This is to prevent an overgrowth of the roots by the bacteria.
3. Micropore tape allows gas exchange, preventing the accumulation of moisture inside the Petri dishes.

4. Both SunCap film and cotton wool bungs allow gas exchange, which is assumed to prevent both ethylene accumulation (which may inhibit shoot expansion) and to reduce humidity in the culture vessels (which prevents anther dehiscence and consequently seed set).
5. Seed set is also sensitive to the presence of kanamycin in the shoot elongation medium, and so antibiotics should be omitted at this stage.
6. Do not put more than two or three shoots in one polypot, as this acts to increase humidity and prevent seed set.
7. Primary transformants (T1 plants) may set only relatively few seeds in vitro, typically about 100 seeds. Germinability of T1 seeds (T2 seedlings) is also commonly poor (a typical viability is 70%, but this is variable between transgenic lines). We observe a higher frequency of T2 seedling survival if seeds are germinated on perlite rather than directly in compost or on agar medium.
8. Aracon tubes allow individal transgenic lines to be grown up separately in close proximity, without the danger of cross-fertilization or of seed loss.

4.2. Screening for GUS Fusion Expression and Mutants

9. In view of the low numbers and poor germinability of T1 seeds, genetic analysis on the transformants (e.g., antibiotic resistance segregation) is carried out on T3 and subsequent generations.
10. Emasculation and cross-pollination of *Arabidopsis* flowers is carried out under a dissecting microscope.
11. Kanamycin sensitive seedlings bleach, and fail to produce either true leaves or roots, on kanamycin selection plates.

4.3. Inverse PCR Amplification of T-DNA Flanking Sequences

12. To amplify genomic sequences flanking the T-DNA left border, we have used the *nested 2* and *nested 3* primers, followed by a confirmatory reamplification with *seq 1* and *nested 3*; and for right border flanking sequences, we have used *nested 2* and *nptII*.
13. The concentration of DNA and Mg^{2+} used in the amplification reaction is critical (Fig. 2), and it is advisable to test empirically a range of concentrations of each in preliminary experiments.
14. For other primers than those described here, the cycling conditions must be optimized. The extension time is estimated to be 1 min per 1 kb of expected amplification product.
15. If feasible, use a test plasmid at single copy levels mixed with untransformed genomic DNA as a positive control to optimize the reaction conditions.

Acknowledgments

The authors gratefully acknowledge financial support from AFRC, SERC, and CEC for their work on gene tagging in *Arabidopsis*.

References

1. Dean, C., Sjodin, C., Page, T., Jones, J., and Lister, C. (1992) Behaviour of the maize transposable element *Ac* in *Arabidopsis thaliana. Plant J.* **2,** 69–81.
2. Topping, J. F., Wei, W., and Lindsey, K. (1991) Functional tagging of regulatory elements in the plant genome. *Development* **112,** 1009–1019.
3. Goldsbrough, A. and Bevan, M. (1991) New patterns of gene activity in plants detected using an *Agrobacterium* vector. *Plant Mol. Biol.* **16,** 263–269.
4. Lindsey, K., Wei, W., Clarke, M. C., McArdle, H. F., Rooke, L. M., and Topping, J. F. (1993) Tagging genomic sequences that direct transgene expression by activation of a promoter trap in plants. *Transgenic Research* **2,** 33–47.
5. André, D., Colau, D., Schell, J., van Montagu, M., and Hernalsteens, J.-P. (1986) Gene tagging in plants by a T-DNA insertion mutagen that generates APH (3')II-plant gene fusions. *Mol. Gen. Genet.* **204,** 512–518.
6. Koncz, C., Martini, N., Mayerhofer, R., Koncz-Kalman, Z., Körber, H., Rédei, G. P., and Schell, J. (1989) High frequency T-DNA-mediated gene tagging in plants. *Proc. Natl. Acad. Sci. USA* **86,** 8467–8471.
7. Kertbundit, S., De Greve, H., DeBoeck, F., van Montagu, M., and Hernalsteens, J.-P. (1991) *In vivo* random β-glucuronidase gene fusions in *Arabidopsis thaliana. Proc. Natl. Acad. Sci. USA* **88,** 5212–5216.
8. Yanofsky, M. F., Ma, H., Bowman, J. L., Drews, G. N., Feldmann, K. A., and Meyerowitz, E. M. (1990) The protein encoded by the *Arabidopsis* homeotic gene *agamous* resembles transcription factors. *Nature* **346,** 35–39.
9. Koncz, C., Mayerhofer, R., Koncz-Kalman, Z., Nawrath, C., Reiss, B., Rédei, G. P., and Schell, J. (1990) Isolation of a gene encoding a novel chloroplast protein by T-DNA tagging in *Arabidopsis thaliana. EMBO J.* **9,** 1337–1346.
10. Rédei, G. P. (1975) *Arabidopsis* as a genetic tool. *Ann. Rev. Genet.* **9,** 111–127.
11. Rédei, G. P. and Koncz, C. (1992) Classical mutagenesis, in *Methods in Arabidopsis Research* (Koncz, C., Chua, N. H., and Schell, J., eds.), World Scientific, Singapore, pp. 16–82.
12. Meyerowitz, E. M. (1989) Arabidopsis, a useful weed. *Cell* **56,** 263–269.
13. Tsay, Y.-F., Frank, M. J., Page, T., Dean, C., and Crawford, N. M. (1993) Identification of a mobile endogenous transposon in *Arabidopsis thaliana. Science* **260,** 342–344.
14. Lindsey, K. and Jones, M. G. K. (1990) Selection of transformed plant cells, in *Plant Cell Line Selection: Procedures and Applications* (Dix, P. J., ed.), VCH Verlagsgesellschaft, Weinheim, pp. 317–339.
15. Damm, B. and Willmitzer, L. (1991) *Arabidopsis* protoplast transformation and regeneration, in *Plant Tissue Culture Manual: Fundamentals and Applications* (Lindsey, K., ed.), Kluwer, Dordrecht, pp. A7: 1–20.

16. Feldmann, K. A. and Marks, M. D. (1987) *Agrobacterium*-mediated transformation of germinating seeds of *Arabidopsis thaliana*: a non-tissue culture approach. *Mol. Gen. Genet.* **208,** 1–9.
17. Schmidt, R. and Willmitzer, L. (1991) *Arabidopsis* regeneration and transformation (leaf and cotyledon explant system), in *Plant Tissue Culture Manual: Fundamentals and Applications* (Lindsey, K., ed.), Kluwer, Dordrecht, pp. A6: 1–17.
18. Valvekens, D., van Lijsebettens, M., and van Montagu, M. (1992) *Arabidopsis* regeneration and transformation (root explant system), in *Plant Tissue Culture Manual: Fundamentals and Applications* (Lindsey, K., ed.), Kluwer, Dordrecht, pp. A8: 1–17.
19. Forsthoefel, N. R., Wu, Y., Schulz, B., Bennett, M. J., and Feldmann, K. A. (1992) T-DNA insertion mutagenesis in *Arabidopsis*: prospects and perspectives. *Aust. J. Plant Physiol.* **19,** 353–366.
20. Clarke, M. C., Wei, W., and Lindsey, K. (1992) High frequency transformation of *Arabidopsis thaliana* by *Agrobacterium tumefaciens*. *Plant Mol. Biol. Reporter* **10,** 178–189.
21. Murashige, T. and Skoog, F. (1962) A revised medium for rapid growth and bioassays with tobacco tissue cultures. *Physiol. Plant* **15,** 473–497.
22. Gamborg, O. L., Miller, R. A., and Ojima, K. (1968) Nutrient requirements of suspension cultures of soybean root cells. *Exp. Cell Res.* **50,** 151–158.
23. Russell, J., Fuller, J., Wilson, Z., and Mulligan, B. (1991) Protocol for growing *Arabidopsis*, in *Arabidopsis: The Compleat Guide* (Flanders, D. and Dean, C., eds.), AFRC PMB *Arabidopsis* Programme, Norwich, UK.
24. Jefferson, R. A., Kavanagh, T. A., and Bevan, M. W. (1987) GUS fusions: β-glucuronidase as a sensitive and versatile gene fusion marker in higher plants. *EMBO J.* **6,** 3901–3907.
25. Topping, J. F., Agyeman, F., Henricot, B., and Lindsey, K. (1994) Identification of molecular markers of embryogenesis in *Arabidopsis thaliana* by promoter trapping. *Plant J.* **5,** 895–903.
26. Collins, F. S., and Weissman, S. M. (1984) Directional cloning of DNA fragments at a large distance from an initial probe: a circularized method. *Proc. Natl. Acad. Sci. USA* **81,** 6812–6816.

CHAPTER 6

Agrobacterium-Mediated Transfer of Geminiviruses to Plant Tissues

Margaret I. Boulton

1. Introduction

1.1. Agroinfection

Agrobacterium-mediated transfer of cloned DNA copies of infectious viral genomes to plant tissues has been termed "agroinoculation" or "agroinfection" (for review, *see* ref. *1*). The technique has been used successfully for a wide variety of virus groups, including the caulimoviruses and badnaviruses (genomes of double-stranded [ds] DNA) *(2–4)*, the geminiviruses (ssDNA genomes) *(5–7)*, the phloem-restricted luteoviruses (ssRNA genomes) *(8)*, as well as for viroids (ssRNA) *(9)*. For viruses with RNA genomes and viroids, cloned cDNA copies of the viral genome are required. Agroinfection of viruses (or viroids) with circular genomes requires the insertion of tandem multimeric copies of the genome between the T-DNA borders of an *Agrobacterium* binary vector, but for RNA viruses with linear genomes a cDNA copy of the genome must be inserted between a promoter (usually, the cauliflower mosaic virus 35S promoter [CaMV 35S]) and a transcription termination signal (e.g., the nopaline synthase terminator) *(8)*.

The technique relies on the natural T-DNA transfer mechanism of *Agrobacterium tumefaciens* or *Agrobacterium rhizogenes*; disarmed (nononcogenic) strains are equally as efficient as wild-type bacteria and any T-DNA binary vector may be used. Agroinfection is not a technique for transformation of the plant, only for the transfer of viral genetic information that leads to a virus infection. Infection is produced by a number

From: *Methods in Molecular Biology, Vol. 49: Plant Gene Transfer and Expression Protocols*
Edited by: H. Jones Humana Press Inc., Totowa, NJ

of mechanisms. Copies of linear RNA virus genomes are transcribed from the CaMV 35S promoter to initiate the RNA replication *(8)*. In other cases, viral genomes are released either by homologous recombination between repeated sequences or by replicative release using the viral origin *(10)*.

The technique is applicable to most plant species and is limited only by the host range of the virus or the *Agrobacterium* strain; small plants, tissue explants (e.g., leaf disks), and germinating seedlings may be infected *(11–14)*. The agroinoculation technique has proven especially useful when the virus or its genome cannot be manually inoculated to the plants, thereby allowing the in vivo assay of cloned material. Furthermore, agroinoculation almost always increases infection efficiency even when manual inoculation is possible *(6,7)*.

For further information on *Agrobacterium*, T-DNA binary vectors and leaf disk transformation, refer to Chapters 1–5, this volume.

1.2. Geminiviruses

Geminiviruses infect a wide variety of dicotyledonous (dicot) and monocotyledonous (monocot) plants, they have a small genome size (2.6–3.0 kb), and they replicate to high copy number in the nucleus of infected cells (for reviews, *see 15–17*). The availability of dsDNA intermediates of replication facilitates cloning of wild-type or mutated genomes. Geminiviruses are characterized by one (monopartite) or two (bipartite, DNAs A and B) genomes. The bipartite genome viruses (e.g., African cassava mosaic virus [ACMV] and tomato golden mosaic virus [TGMV]) infect dicots, whereas the majority of the monopartite genome viruses (e.g., maize streak virus [MSV] and wheat dwarf virus [WDV] infect monocots. Exceptions to this include beet curly top virus (BCTV) and tomato yellow leaf curl virus (TYLCV) both of which infect dicots. For the bipartite geminiviruses, DNA A replicates autonomously but DNA B is required for systemic infection of plants *(18)*.

1.3. Uses of Agroinfection

1.3.1. The Study of Agrobacterium-Plant Interactions

Until recently, the host range of *Agrobacterium* (assessed by the presence of tumors on inoculated plants) was thought to be limited to dicots and a very few monocots *(19,20)*. The demonstration *(5)* that MSV DNA, which is not manually transmissible to maize, could produce a systemic infection following agroinoculation suggested that *Agrobacterium* could also interact with a member of the Poaceae, although tumors were not formed. The disease symptoms, produced as a result of amplification of the transferred MSV genomes, provided a sensitive marker for T-DNA transfer. Using the geminiviruses MSV and WDV, and the badnavirus rice tungro bacilliform virus, T-DNA transfer has been shown to occur in a wide range of cereals and grasses *(3,11,14,21)*. DNA transfer was found to be bacterial strain-specific *(11,21)* and the *vir* gene determinants of T-DNA transfer to maize were mapped *(12, 22)*. Agroinfection may also be used to show which *Agrobacterium* strains interact with dicot plant species or cultivars; this may prove particularly useful as a preliminary to transformation of species for which transformation has proven difficult.

1.3.2. The Study of Viral Replication, Gene Function, and Pathogenicity

The use of the sensitive agroinfection technique and mutated viral genomes allows the identification of genes essential for replication and/or symptom expression of the virus *(17,23)*. Replacement of the ACMV coat protein coding sequence with other viral genes such as the BCTV coat protein gene or other ACMV or TGMV DNA B genes has resulted in the identification of factors determining virus insect vector specificity and virus movement *(24–26)*. Geminivirus gene expression has also been investigated; for example, the promoter of the coat protein gene has been shown to be regulated by other viral gene products *(27)*.

1.3.3. Agroinfection of Geminivirus Vectors: Limitations and Uses

The expression of foreign genes from a high copy number viral (episomal) replicon using agroinfection of plants offers the potential of elevated levels of expression without the need for production of transgenic plants or protoplasts. However, it is important to be aware that the technique does not produce a stably transformed plant and replicons are not transferred to a second generation of plants; to ensure a high copy number the vector must maintain its ability to spread throughout the infected plant.

The coat protein gene on DNA A of the bipartite geminiviruses (Fig. 1) is dispensable for virus infection *(28,29)*, and may be replaced with other genes *(30,31)*. However, to obtain systemic infection of the plant, both

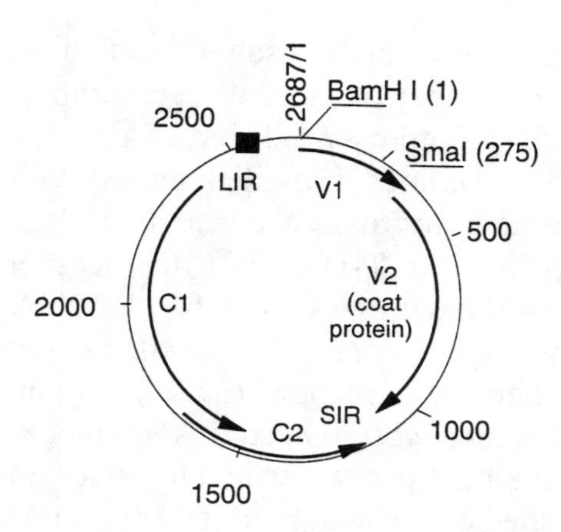

Fig. 1. The genome map of maize streak virus. The four open reading frames (V1, V2, C1, C2; arrowed arcs) have homology with those found in other cereal-infecting geminiviruses. The position of the conserved stem-loop structure is indicated by the infilled rectangle. SIR = small intergenic region, LIR = large intergenic region. The positions of the restriction enzyme sites used for cloning of the partial multimeric copy of the genome are shown, nucleotide numbering is according to ref. *(41)*.

DNA A and DNA B must be inoculated. During systemic infection, a proportion of the molecules may revert to the size of a wild-type molecule by deletion of the foreign gene sequences *(32,33)*. The size constraints can be overcome if leaf disks are agroinoculated as (in this case) viral spread is not required *(34)*. Leaf disk transformation is described in Chapter 3.

Disruption of the coat protein gene of the monopartite viruses (Fig. 1) prevents systemic infection of plants *(7,23–25)*, but vectors based on MSV have been agroinoculated to plants, resulting in expression of GUS *(37)* in restricted areas around the inoculation site in the leaves of young seedlings. Agroinfected plants produced 5–10 times more GUS-positive spots than plants inoculated with an inoculum comprising only the GUS cassette inserted into the binary vector *(37)*. For MSV, foreign DNA may be inserted into the small intergenic region (SIR, Fig. 1); again, the virus is unable to infect the plant systemically, but if the foreign gene cassette includes a strong promoter, the expression levels are equivalent to those obtained with the coat protein replacement vector *(37)*.

Geminivirus vectors may be used for the analysis of basic plant molecular biological mechanisms. The *Activator (Ac)*-transposase-mediated excision of the *Dissociation (Ds)* element was analyzed in maize plants following the agroinfection of an MSV genome containing *Ds (38,39)*. The *Ds* was inserted into the SIR and the excision event was illustrated by plants showing systemic symptoms.

Plant pre-mRNA splicing has been studied following leaf disk agroinfection of TGMV coat protein replacement vectors containing plant intron sequences. Leaf disk nuclei accurately processed the introns of both dicots and monocots *(40)*.

1.4. Requirements for Agroinfection of Plants with Geminiviruses

Agroinfection of plants with geminiviruses can be divided into three steps:

1. Cloning of a tandem multimeric copy of the viral genome and its transfer to an *Agrobacterium* binary vector;
2. Transformation of *Agrobacterium* with the vector; and
3. Agroinoculation of plants.

The first step will be described using MSV strain N *(41)* as an illustration. The cloning of a partially repeated copy of the genome into pBIN19 *(42)* will be described and is shown in Fig. 2. Step (2) is described in Chapter 2. The third step varies according to the host plant used. The following methodology will describe the inoculation of young seedlings of dicots (Section 3.2.) and monocots (Section 3.3., Fig. 3A) and a technique suitable for agroinoculation of germinated cereal seeds (Section 3.4., Fig. 3B). Agroinfection of leaf disks of dicot plants may be done as described in Chapter 3, this volume.

2. Materials

1. *Agrobacterium* strain containing the geminivirus multimeric construct(s). If this is not available, materials 2, 3, and 4 will be required.
2. DNA samples of a cloned full-length copy of the viral genome.
3. An *Agrobacterium* binary vector (e.g., pBIN19).
4. *Agrobacterium* strain pGV3850 *(43)*.
5. Sterilized AB broth and agar *(44)* (*see* Note 8).
6. Appropriate antibiotic solutions, e.g., kanamycin sulfate (*see* Note 7). Some antibiotics are light sensitive, most are stable at –20°C. All should be handled with care (use gloves and avoid inhaling the powder) as they may be toxic and/or allergenic.

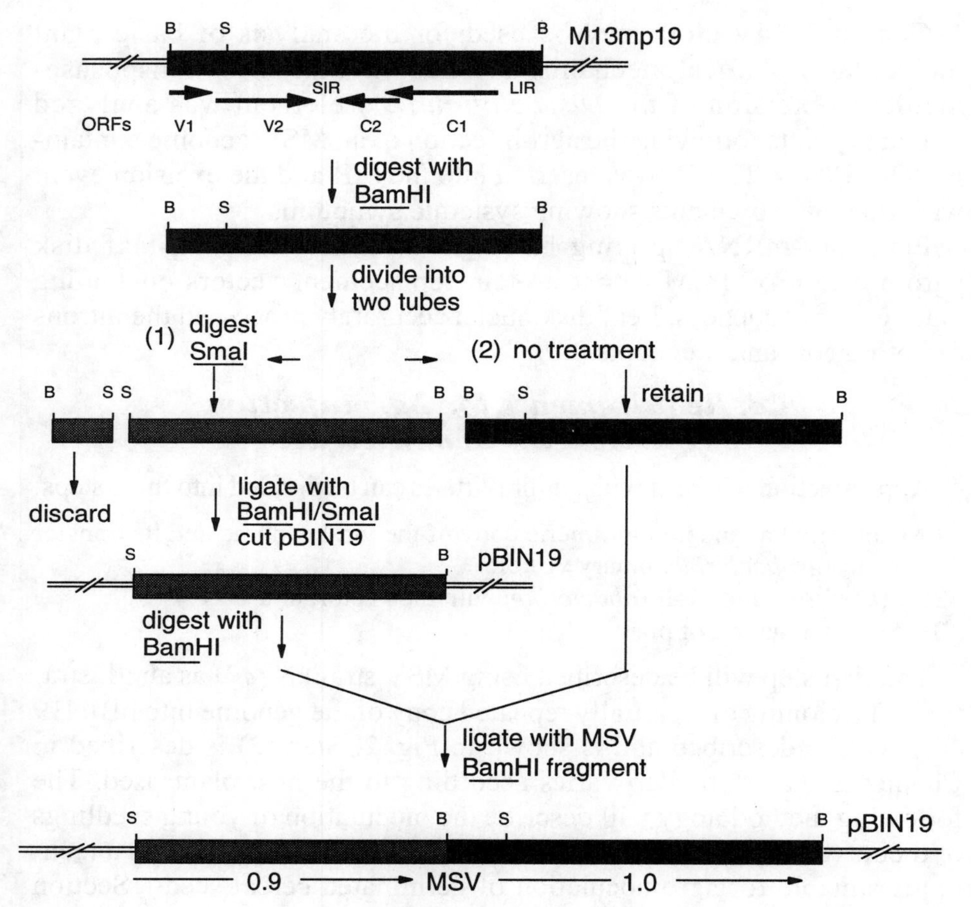

Fig 2. Construction of an MSV partial multimer (1.9 copies of the genome) in pBIN19. A schematic representation of the genome organization of MSV is shown at the top. The genome has been linearized at the *Bam*HI site (B), the *Sma*I site (S) used for the cloning is also shown above the genomic fragment, open reading frames are below. LIR = large intergenic region, SIR = small intergenic region.

7. Sterilized glycerol (10 mL).
8. Seeds or seedlings of the species to be inoculated.
9. Plant pots and peat-based compost.
10. Toothpicks and/or fine sewing needles (Section 3.2.2.).
11. Hamilton syringe (50 µL volume, Sections 3.3. and 3.4. only).

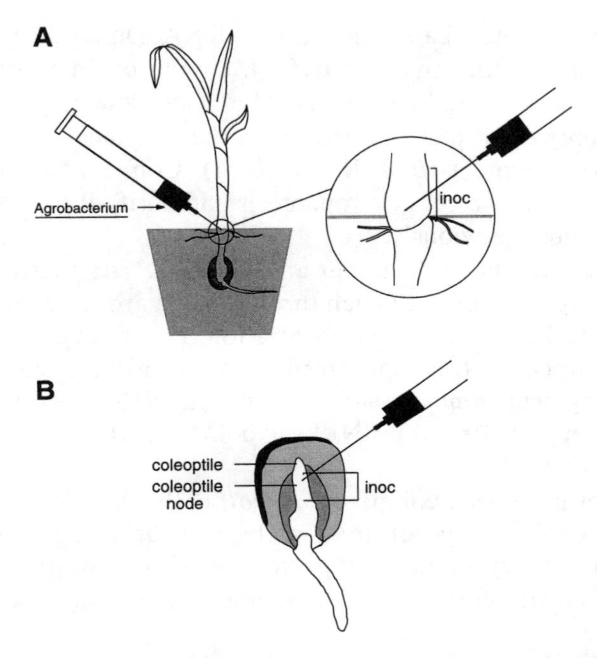

Fig. 3. The area for successful agroinoculation of maize plants **(A)** or 3-d-old germinated seeds **(B)**. Multiple (approx four) injections should be done into the center of the tissue, within the designated area ("inoc").

12. Domestos or similar hypochlorite solution containing wetting agents (Section 3.4.).
13. Appropriate containment facilities for the growth of agroinoculated plant material (*see* Note 14).

3. Methods

3.1. Cloning of a Multimeric Copy of MSV into the Binary Vector

A flow diagram of the cloning procedure is shown in Fig. 2.

1. Digest the cloned MSV DNA with the appropriate restriction enzyme to release a full-length fragment of the genome from the cloning vector. For MSV-Ns this could be *Bam*HI (*see* Note 1).
2. Separate the fragment from the vector by agarose gel electrophoresis. Visualize the fragment using ethidium bromide staining and U-V irradiation and excise it from the agarose. Purify the fragment from the agarose using your preferred technique (*see* Note 2).

3. Retain approximately half of the purified DNA. Digest the remainder with an appropriate restriction enzyme (for MSV this could be *Sma*I). Separate and purify the selected fragment as before (*see* Note 3).
4. Digest approx 5 μg pBINl9 DNA with the enzymes used to isolate the small MSV fragment (*Bam*HI and *Sma*I). Confirm the success of the digestion by agarose gel electrophoresis and purify the vector from agarose as described previously.
5. Assess the amount of fragment and vector DNAs recovered by electrophoresing an aliquot of each through an agarose gel and comparing it with a marker of known concentration. It is not necessary to calculate the concentration spectrophotometrically. Ligate the partial genome fragment (*Bam*HI/*Sma*I) with the pBIN19 vector using T4 DNA ligase and approx 100 ng pBIN19 and a 4*M* excess of the MSV fragment (approx 10 ng).
6. Prepare, or purchase, competent *E. coli* cells (*see* Note 4). Transform the cells with the ligation mix, plate on L broth agar containing 50 μg/mL kanamycin sulfate, 40 μg/mL IPTG and 40 μg/mL X-gal and select the (white colored) recombinant clones. Confirm the presence of the insert.
7. Prepare DNA from a selected recombinant clone. Digest approx 4 μg with the enzyme used to prepare the MSV genomic fragment (*Bam*HI) and dephosphorylate with calf intestinal phosphatase to prevent religation of the vector (*see* Note 5). Extract the digest once with an equal volume of phenol-chloroform to inactivate the phosphatase and collect the DNA by ethanol precipitation.
8. Ligate approx 100 ng of the vector DNA and 10 ng of the full-length (monomeric) fragment.
9. Transform competent *E. coli* cells and select the recombinant colonies on kanamycin-containing medium; IPTG and X-gal need not be added to the agar as all the colonies will be white. Confirm the presence of the multimeric insert, most simply by digesting DNA purified from the colony with enzymes present in the multiple cloning site of pBIN19 that flank the MSV insert and are not found within the MSV genome, e.g., *Xba*I and *Eco*RI. This digestion will release the insert that may be sized by comparison with molecular weight markers following agarose gel electrophoresis (*see* Note 6).
10. Transform *Agrobacterium* with the selected recombinant using one of the techniques outlined in Chapter 2, this volume. A disarmed *A. tumefaciens* strain based on pTiC58, e.g., pGV3850 *(43)* has proved suitable for a very wide range of plants tested by the author.

3.2. Agroinoculation of Dicots with Bipartite Geminiviruses

3.2.1. Preparation of the Agrobacterium Inoculum

1. Two days prior to the inoculation date, prepare the bacterial suspensions. Prepare two sterilized 250-mL Erlenmeyer flasks each containing 25 mL AB broth (and the appropriate antibiotic for maintenance of the binary vector, e.g., kanamycin sulfate). Inoculate one flask with a colony of *Agrobacterium* containing the geminivirus vector based on DNA A, and the other with a colony containing the multimeric copy of DNA B. Incubate, shaking at 28°C, for approx 48 h until a cloudy culture is obtained (*see* Notes 7 and 8).
2. Transfer both cultures to a single sterile tube, mix, and centrifuge at approx 3000*g* for 10 min at room temperature (*see* Note 9).
3. Dispose of the supernatant. Immediately prior to inoculation, resuspend the pellet in an equal volume of sterile distilled water (SDW) (*see* Note 10).

3.2.2. Plant Growth and Inoculation Procedure

1. Sow seeds of the chosen species and prick out the young seedlings into a peat-based compost in approx 7-cm pots. Grow under appropriate conditions until the first three leaves are visible, but not fully expanded. The stems of the plants should have a diameter of at least 2 mm (*see* Notes 11 and 12).
2. Inoculate the seedlings by stabbing the stem using a sharp, narrow toothpick that has been dipped into the *Agrobacterium* solution. The point of the pick should reach the center of the stem but not pass through to the far side (*see* Note 13).
3. Continue to grow the plants under appropriate conditions (*see* Note 14), and examine daily. After approx 10–20 d symptoms should be apparent on the newly developed leaves (*see* Note 15). Sample the leaf tissue as appropriate for the required analysis. If DNA copy number analyses are required, a proportion of the tissue may be stored at –20°C prior to DNA extraction.

3.3. Agroinoculation of Monocot Seedlings with MSV (or Other Cereal-Infecting Geminiviruses)

1. Sow seeds of the chosen species (e.g., maize or wheat) putting three seeds into peat-based compost in approx 10-cm pots; the seeds should be lightly covered with compost. Grow the plants under appropriate conditions until the third leaf (maize) or fourth leaf (wheat) emerges. The stems of the plants should have a diameter of at least 2 mm (*see* Note 16).

2. Prepare the *Agrobacterium* inoculum as described in Section 3.2.1. except that only one flask of culture is required for geminiviruses with a monopartite genome.
3. Use a Hamilton syringe to inoculate the plants by injecting the base of the plant just above the root cap area. A total volume of 25 μL *Agrobacterium* culture should be injected (5 × 5 μL injections) into the stem within 1 cm of the base. The needle should reach the center of the stem but not pass through to the far side (*see* Note 17; Fig. 3A).
4. Continue to grow the plants under appropriate conditions. After approx 7–20 d symptoms should be apparent on the newly developed leaves (*see* Notes 14 and 18). Sample the leaf tissue as appropriate for the required analysis (*see* Note 19). If DNA analyses are required, a proportion of the tissue may be stored at –20°C prior to extraction.

3.4. Agroinoculation of Germinated Monocot Seeds

1. Day 1: Surface sterilize maize seeds by gentle agitation in a closed flask containing 20% Domestos (Lever Brothers) for 30 min. Rinse seeds six times with sterile distilled water (SDW). Place seeds on filter paper (moistened with SDW) in 9-cm Petri dishes (max 10 seeds/dish). Incubate at 25°C in the dark for 3 d.
2. Day 2: Prepare the *Agrobacterium* inoculum as described in Section 3.2.1., except that only one flask of culture is required for geminiviruses with a monopartite genome.
3. Day 4: The shoots of the seedlings should be between 0.5–5 cm. Inoculate the seedlings around the coleoptilar node (*see* Note 20; Fig. 3B) using a Hamilton syringe. Inject each seedling with 20 μL of culture divided between four separate injections. After inoculation, plant the seedlings in pots and grow as described in Section 3.3.
4. Sample the leaf tissue as appropriate for the required analysis (*see* Note 21). If DNA analyses are required, a proportion of the tissue may be stored at –20°C prior to extraction.

4. Notes

4.1. Cloning of the Multimeric Virus Genome

1. All molecular biology protocols described in this section are described in ref. *(45)*. The procedure outlined here is for the monopartite genome viruses. For the bipartite viruses each genome component, DNAs A and B, must be cloned into separate binary vectors. In all cases, to facilitate the cloning, an enzyme that cuts the virus genome only once and is present as a unique site in the multiple cloning region of the binary vector should be chosen. This will depend on the nucleotide sequence of the virus and the

binary vector selected. Binary vectors with multiple cloning sites and color selection to aid identification of recombinant colonies are recommended; pBIN19 is widely and freely available to all researchers.

2. A 1% agarose gel is suitable for most fragments; the choice of buffer depends on the technique for isolation of the DNA from the agarose. In addition to the techniques outlined in ref. *(45)* there are many commercial kits available, all of which should give a successful result. Minimize the length of time that the gel is illuminated with U-V irradiation as this may cause breakage of the DNA.
3. Again, to simplify the procedure, whenever possible choose an enzyme with a unique site in the virus genome and also in the multiple cloning region of the binary vector. If this is impossible, choose an enzyme that produces a blunt end, as otherwise the procedure for obtaining the fragments will have to be modified to produce a blunt end. Blunt-ended molecules may be cloned into *Sma*I sites. The fragment size may be from 10–99% of the entire genome. The presence of the viral origin of replication *(ori)* within this fragment may enhance the infectivity of the construct. The *ori* is situated within the loop of the conserved stem loop of the geminiviruses.
4. To prevent loss of the repeated viral insert from the vector, recombination within the *E. coli* cells should be minimized by the use of *recA*⁻ strains (derivatives of DH5α or JM83 are widely available).
5. It is important that dephosphorylation of the vector is efficient otherwise during the subsequent ligation it will religate without the monomer inserts. As all the colonies will contain inserts (either partial genomic copies or partially repeated copies), color selection cannot be used to identify the required colonies. A high background of religated vector will complicate the selection of clones with multimeric inserts. If the ligation/transformation control (vector plus ligase, without the insert) has a high background, close to that of the test sample, then the dephosphorylation procedure should be repeated.
6. If suitable restriction enzymes cannot be found, the presence of a multimeric insert can usually be confirmed using enzymes with sites only within the virus genome that will produce diagnostic fragments when the monomeric genome fragment is present. Rapid screening of colonies may also be performed using PCR techniques and universal primers or oligonucleotides that have the virus sequence. Theoretically it is possible that the monomeric fragment may insert in either orientation with respect to the partial genome fragment. In the author's experience (based on approx 200 clones), all multimeric inserts will be in the tandem orientation, probably owing to instability of inverted repeat sequences in *E. coli*. However,

it is simple to confirm the orientation of the monomeric insert by restriction enzyme digestion.

4.2. Agroinoculation of Dicots with Bipartite Geminiviruses

7. The host range of the *Agrobacterium* strain should include the plant species used. This can be determined by a literature search of transformed plant species. If the information is not available for the chosen plant, the disarmed strains pGV3850 *(43)* and A281 *(46)* are successful for a wide range of plants and should be tried first. If infectivity is not obtained, the appropriate *Agrobacterium* strain may be found by screening the plants for tumor formation using wild-type strains *(19)*. For members of the Poaceae (cereals and grasses) that do not form tumors, some information is available *(11,21)*. Transfer of the binary vector to alternative *Agrobacterium* strains is described in Chapter 2. All binary vectors tested to date are suitable for agroinfection studies. To maintain the binary vector in the *Agrobacterium,* always include antibiotics appropriate to the binary plasmid in the culture medium.
8. Incubation temperatures from 25–30°C can be used. At higher temperatures the Ti plasmids of some strains of *Agrobacterium* may be lost, resulting in an inability of the bacterium to transfer DNA. If a shaking incubator at this temperature is not available, the bacteria may be streaked thickly on AB agar containing antibiotics and incubated at 28°C for approx 48 h. Collect the bacteria by scraping a flat-ended spatula across the surface of the agar and transfer them to a sterile bijou for resuspension. To ensure a high rate of agroinfection over a long time period, store the *Agrobacterium* culture in glycerol. The 48-h shake culture should be mixed 1:1 (v/v) with sterile glycerol and immediately pipeted into storage containers (e.g., Eppendorf tubes) and frozen at –70°C. The high concentration of glycerol facilitates the sampling of the frozen culture. Bacteria stored in this way, without thawing, have remained viable for seven years in the author's laboratory. Bacteria repeatedly subcultured on plates may lose their infectivity. The use of AB medium is not obligatory, most *Agrobacterium* strains will grow in richer media.
9. Both genomic components (DNAs A and B) of the bipartite geminiviruses are required for systemic infection of plants. Pooling the two cultures facilitates their coinoculation. For leaf disk inoculation (*see* Chapter 3) with a bipartite geminivirus only the DNA A-based vector component needs to be used.
10. All supernatants of cultures containing bacteria and virus vectors should be disposed of either by autoclaving or treatment with a proprietary bacte-

ricide. The *Agrobacterium* inoculum may be left in SDW for up to 4 h without affecting its infectivity. Some antibiotics, e.g., spectinomycin and streptomycin (required for the maintenance of some binary vectors), cause "bleaching" or severe chlorosis of leaves if small amounts remain in the culture medium used for inoculation. When using these antibiotics, the bacterial pellet should be washed with AB broth alone.

11. The plant should be a host for the geminivirus on which the vector is based. This may be determined by searching the literature for the appropriate virus. For many viruses host range information is available from CMI-AAB reference sources *(47)*. As virus–host interactions are complex, wherever possible use the same cultivars of the species as those described in the literature.
12. Older plants may be used but in many species the susceptibility of plants to virus infection decreases with increasing plant age.
13. The toothpick (or equivalent) used should be as fine as possible to prevent killing small plants. For small seedlings with thin stems a fine (sewing) needle is ideal for inoculation. In this case, approx 10 μL of inoculum is pipetted into the area where a petiole joins the stem and the inoculum forced into the stem by repeated small stabs. It is not necessary to introduce the entire inoculum volume. For plants that have little stem (e.g., *Arabidopsis* the inoculum may be pipeted onto the leaves and introduced by repeated small stabs with a fine sewing needle. This inoculation procedure works with many dicot plants whose stems are too thin to inoculate but it results in damage to the inoculated leaf that may make symptom analysis difficult.
14. The plants should be grown under the containment conditions stipulated for plants infected with pathogen sequences inserted into *Agrobacterium*. The requirements are likely to differ in each country.
15. The timing of symptom appearance will vary depending on the virus and host chosen. Symptoms will be visible on leaves that emerge after the inoculation procedure and, in many cases, on the inoculated leaves.

4.3. Agroinoculation of Monocot Seedlings with MSV (or other Cereal-Infecting Geminiviruses)

16. Most cereals may be grown at 25°C with a photoperiod of 16 h; plants are normally ready for infection 10–14 d after sowing. Older plants may be used, although infection rates may be reduced. When inoculating older wheat plants, ensure that all the tillers are injected.
17. No air bubbles should be present in the syringe; if the inoculum is too concentrated to draw up into the syringe it may be diluted until the process is facilitated. Only a limited inoculation region is successful for agroinfection

of monocot seedlings. Fig. 3A shows this region for maize. The equivalent region should be used for other cereals and grasses. Support the stem during inoculation with the fingers of the noninoculating hand. To increase the chances of a successful inoculation with plants with thick stems (e.g., maize) the last inoculation should be down through the center of the stem until a resistance is felt; at this point inject the remainder of the inoculum. Frequently a proportion of the inoculum will not remain in the plant; this does not matter. If the seeds have been sown too deeply, reveal the inoculation site by gently wiping away the soil on one side of the plant using a gloved finger. Disturb the roots as little as possible.

18. Timing of symptom appearance depends on the virus, the host plant, and the growth conditions. For MSV and maize this is typically 10 d. Symptoms will not be apparent when reporter genes have been inserted either into the coat protein or SIR regions of the genome (*see* Section 1.). If the assay is dependent on inserted sequences being removed from the genome, for example transposon excision from the small intergenic region *(38,39)*, symptoms can be the marker for the excision event.
19. Tissue to be sampled: Inoculation of pot-grown seedlings is particularly useful where disease symptoms act as the reporter. Inoculation of germinated seedlings (Section 3.3.), although less straightforward, is more appropriate for reporter gene assays.

4.4. Agroinoculation of Germinated Monocot Seeds

20. The inoculation point for germinated seedlings of monocots is shown in Fig. 3B. Only inoculations done within this region will result in infection.
21. When virus spread is prevented by the insertion of foreign genes, replicating virus or viral DNA may be obtained from the first three leaves emerging after inoculation, even though no symptoms will be seen. The inoculation sites appear as small holes in these leaves.

References

1. Grimsley, N. (1990) Agroinfection. *Physiologia Plantarum* **79,** 147–153.
2. Grimsley, N. H., Hohn, B., Hohn, T., and Walden, R. (1986) Agroinfection, an alternative route for viral infection of plants by using the Ti plasmid. *Proc. Natl. Acad. Sci. USA* **83,** 3282–3286.
3. Dasgupta, I., Hull, R., Eastop, S., Poggi-Pollini, C., Blakebrough, M., Boulton, M. I., and Davies, J. W. (1991) Rice tungro bacilliform virus DNA independently infects rice after *Agrobacterium*-mediated transfer. *J. Gen. Virol.* **72,** 1215–1221.
4. Medbury, S. L., Lockhart, B. E. L., and Olszewski, N. E. (1991) Properties of *Commelina* mottle virus's complete DNA sequence, genomic discontinuities and transcript suggest that it is a pararetrovirus. *Nucleic Acids Res.* **18,** 5505–5513.
5. Grimsley, N. H., Hohn, T., Davies, J. W., and Hohn, B. (1987) *Agrobacterium*-mediated delivery of maize streak virus into maize plants. *Nature* **325,** 177–179.

6. Hayes, R. J., Coutts, R. H. A., and Buck, K. W. (1988) Agroinfection of *Nicotiana* spp. with cloned DNA of tomato golden mosaic virus. *J. Gen. Virol.* **69,** 1487–1496.
7. Briddon, R. W., Watts, J., Markham, P. G., and Stanley, J. (1989) The coat protein of beet curly top virus is essential for infectivity. *Virology* **172,** 628–633.
8. Leiser, R.-M., Ziegler-Graff, V., Reutenauer, A., Herrbach, E., Lemaire, O., Guilley, H., Richards, K., and Jonard, G. (1992) Agroinfection as an alternative to insects for infecting plants with beet western yellows luteovirus. *Proc. Natl. Acad. Sci. USA*. **89,** 9136–9140.
9. Gardner, R., Chonoes, K., and Owens, R. (1986) Potato spindle tuber viroid infections mediated by the Ti plasmid of *Agrobacterium tumefaciens. Plant Mol. Biol.* **6,** 221–228.
10. Stenger, D. C., Revington, G. N., Stevenson, M. C., and Bisaro, D. M. (1991) Replicational release of geminivirus genomes from tandemly-repeated copies: evidence for rolling circle replication of a plant viral DNA. *Proc. Natl. Acad. Sci. USA* **88,** 8029–8033.
11. Boulton, M. I., Buchholz, W. G., Marks, M. S., Markham, P. G., and Davies, J. W. (1989) Specificity of *Agrobacterium*-mediated delivery of maize streak virus DNA to members of the Gramineae. *Plant Mol. Biol.* **12,** 31–40.
12. Boulton, M. I., Raineri, D. M., Davies, J. W., and Nester, E. W. (1993) Identification of genetic factors controlling the ability of *Agrobacterium* to transfer DNA to maize, in *Advances in Molecular Genetics of Plant-Microbe Interactions* (Nester, E. W. and Verma, D. P. S., eds.), Kluwer, The Netherlands, pp. 73–78.
13. Grimsley, N., Jarchow, E., Oetiker, J., Schlaeppi, M., and Hohn, B. (1991) Agroinfection as a tool for the investigation of plant-pathogen interactions, in *Plant Molecular Biology 2* (Herrmann, R. G. and Larkins, B., eds.), Plenum, New York, pp. 225–238.
14. Marks, M. S., Kemp, J. M., Woolston, C. J., and Dale, P. J. (1989) Agroinfection of wheat: a comparison of *Agrobacterium* strains. *Plant Sci.* **63,** 247–256.
15. Davies, J. W. and Stanley, J. (1989) Geminivirus genes and vectors. *Trends. Genet.* **5,** 77–81.
16. Lazarowitz, S. G. (1987) The molecular characterization of geminiviruses. *Plant Mol. Biol. Rept.* **4,** 177–192.
17. Lazarowitz, S. G. (1992) Geminiviruses: genome structure and gene function. *Crit. Rev. Plant Sci.* **11,** 327–349.
18. von Arnim, A., Frischmuth, T., and Stanley, J. (1993) Detection and possible functions of African lease cassava mosaic virus DNA B gene products. *Virology* **192,** 264–272.
19. De Cleene, M. and DeLey, G. (1976) The host range of crown gall. *Bot. Rev.* **42,** 389–466.
20. De Cleene, M. (1984) The susceptibility of monocotyledons to *Agrobacterium tumefaciens. Phytopath. Z.* **113,** 81–89.
21. Boulton, M. I. and Davies, J. W. (1990) Monopartite geminiviruses: markers for gene transfer to cereals. *Aspects of Appl. Biol.* **24,** 79–86.
22. Raineri, D. M., Boulton, M. I., Davies, J. W., and Nester, E. W. (1993) VirA, the plant-signal receptor, is responsible for the Ti plasmid-specific transfer of DNA to maize by *Agrobacterium. Proc. Natl. Acad. Sci. USA* **90,** 3549–3553.

23. Stanley, J. (1991) The molecular determinants of geminivirus pathogenesis. *Semin. Virol.* **2,** 139–149.
24. Briddon, R. W., Pinner, M. S., Stanley, J., and Markham, P. G. (1990) Geminivirus coat protein gene replacement alters insect specificity. *Virology* **177,** 85–94.
25. Etessami, P., Callis, R., Ellwood, S., and Stanley, J. (1988) Delimitation of essential genes of cassava latent virus DNA 2. *Nucleic Acids Res.* **16,** 4811–4829.
26. von Arnim, A. and Stanley, J. (1992) Inhibition of African cassava mosaic virus systemic infection by a movement protein from the related geminivirus tomato golden mosaic virus. *Virology* **187,** 555–564.
27. Sunter, G. and Bisaro, D. (1992) Transactivation of geminivirus AR1 and BR1 gene expression by the viral AL2 gene product occurs at the level of transcription. *Plant Cell* **4,** 1321–1331.
28. Stanley, J. and Townsend, R. (1986) Infectious mutants of cassava latent virus generated *in vivo* from intact recombinant DNA clones containing single copies of the genome. *Nucleic Acids Res.* **14,** 5981–5998.
29. Brough, C. L., Hayes, R. J., Morgan, A. J., Coutts, R. H. A., and Buck, K. W. (1988) Effects of mutagenesis *in vitro* on the ability of cloned tomato golden mosaic virus DNA to infect *Nicotiana benthamiana* plants. *J. Gen. Virol.* **69,** 503–514.
30. Ward, A., Etessami, P., and Stanley, J. (1988) Expression of a bacterial gene in plants mediated by infectious geminivirus DNA. *EMBO J.* **7,** 1583–1587.
31. Hayes, R. J., Petty, I. T. D., Coutts, R. H. A., and Buck, K. W. (1988) Gene amplification and expression in plants by a replicating geminivirus vector. *Nature* **334,** 179–182.
32. Hayes, R. J., Coutts R. H. A., and Buck, K. W. (1989) Stability and expression of bacterial genes in replicating geminivirus vectors in plants. *Nucleic Acids Res.* **17,** 2391–2403.
33. Elmer, S. and Rogers, S. G. (1990) Selection for wild type derivatives of tomato golden mosaic virus during systemic infection. *Nucleic Acids Res.* **18,** 2001–2006.
34. Hanley-Bowdoin, L., Elmer, J. S., and Rogers, S. G. (1988) Transient expression of heterologous RNAs using tomato golden mosaic virus. *Nucleic Acids Res.* **16,** 10,511–10,528.
35. Boulton, M. I., Steinkellner, H., Donson, J., Markham, P. G., King, D. I., and Davies, J. W. (1989) Mutational analysis of the virion-sense genes of maize streak virus. *J. Gen. Virol.* **70,** 2309–2323.
36. Woolston, C. J., Reynolds, H. V., Stacey, N. J., and Mullineaux, P. M. (1989) Replication of wheat dwarf virus DNA in protoplasts and analysis of coat protein mutants in protoplasts and plants. *Nucleic Acids Res.* **17,** 6029–6041.
37. Shen, W. H. and Hohn, B. (1994) Amplification and expression of the β-glucuronidase gene in maize plants by vectors based on maize streak virus. *Plant J.* **5,** 227–236.
38. Shen, W.-H. and Hohn, B. (1992) Excision of a transposable element from a viral vector introduced into maize plants by agroinfection. *Plant J.* **2,** 35–42.
39. Shen, W.-H, Das, S., and Hohn, B. (1992) Mechanism of *DsI* excision from the genome of maize streak virus. *Mol. Gen. Genet.* **233,** 388–394.

40. McCullough, A. J., Lou, H., and Schuler, M. A. (1991) *In vivo* analysis of plant pre-mRNA splicing using an autonomously replicating vector. *Nucleic Acids Res.* **19,** 3001–3009.
41. Mullineaux, P. M., Donson, J., Morris-Krsinich, B. A. M., Boulton, M. I., and Davies, J. W. (1984) The nucleotide sequence of maize streak virus DNA. *EMBO J.* **3,** 3063–3068.
42. Bevan, M. (1984) Binary *Agrobacterium* vectors for plant transformation. *Nucleic Acids Res.* **12,** 8711–8721.
43. Zambryski, P. J., Joos, H., Leemans, J., Van Montagu, M., and Schell, J. (1983) Ti plasmid vector for the introduction of DNA into plant cells without alteration of their normal regeneration capacity. *EMBO J.* **2,** 2143–2150.
44. Cangelosi, G. A., Best, E. A., Martinetti, G., and Nester, E. W. (1991) Genetic analysis of *Agrobacterium. Methods Enzymol.* **204,** 384–397.
45. Sambrook, J., Fritsch, E. F., and Maniatis, T. (1989) *Molecular Cloning: A Laboratory Manual.* Cold Spring Harbor Laboratory, Cold Spring Harbor, NY.
46. Montoya, A. L., Chilton, M.-D., Gordon, M. P., Sciaky, D., and Nester, E. W. (1977) Octopine and nopaline metabolism in *Agrobacterium tumefaciens* and crown gall tumor cells: role of plasmid genes. *J. Bacteriol.* **129,** 101–107.
47. Murant, A. F. and Harrison, B. D. (eds.) *CMI/AAB Descriptions of Plant Viruses.* Sets 1–19 (1970–1989) Association of Applied Biologists, Warwick, UK.

PART II

Direct Gene Transfer

CHAPTER 7

Stable Transformation of Barley via Direct DNA Uptake

Electroporation- and PEG-Mediated Protoplast Transformation

Paul A. Lazzeri

1. Introduction

Transformation technology has great potential for application in many areas of fundamental and applied plant science; the introduction and expression of transgenes will facilitate new approaches in genetics, biochemistry, and physiology and stimulate progress in these fields. In addition, gene transfer techniques will allow the genetic manipulation of crop plants toward improved agronomic characteristics. Indeed, in those plant species that are amenable to *Agrobacterium*–mediated transformation the use of transgenic plants has already had a significant impact in a number of areas of research, and genetically engineered plants with various different modified traits are undergoing field trials or are actually coming to market.

A barrier to more widespread use of genetic manipulation techniques, however, is the difficulty of transforming some of the major crop plant species. Foremost among the groups of crops recalcitrant to genetic manipulation are the cereal and grass species (*Poaceae*). Nevertheless, because of the agronomic importance of poaceous plants there is great interest in transforming these species.

In recent years, however, cereal transformation has been achieved by several different techniques of direct gene transfer. The first successful

From: *Methods in Molecular Biology, Vol. 49: Plant Gene Transfer and Expression Protocols*
Edited by: H. Jones Humana Press Inc., Totowa, NJ

method was protoplast transformation *(2)*, and subsequently particle bombardment *(3)*, cell electroporation *(4)*, and vortexing tissues with silicon carbide "whiskers" *(5)* have all yielded transgenic cereal plants. Each of these methods has advantages and disadvantages in terms of efficiency, the types of tissue that can be targeted, and the specialization/cost of equipment needed (*see* ref. *6* for review).

Protoplast transformation has the advantages that very many cells can be handled with ease, allowing many independent transformants to be produced, that the selection of transformants is relatively simple and, with PEG-mediated transformation, that no specialized equipment is needed. The major disadvantage is the difficulty of regenerating plants from poaceous protoplasts. This is a long-standing problem, but in recent years there has been real progress and plants have been recovered from protoplasts of most of the major poaceous crops *(6)*, and transgenic plants have been obtained in maize *(7)* and rice *(8)*. In species in which embryogenic cell suspensions can routinely be obtained, e.g., japonica or indica rice, protoplast transformation can be an efficient and economic method for producing numbers of transformants, and, even in species where regeneration from protoplasts is difficult, the method gives the possibility of producing many independent transformed lines suitable for construct testing, gene expression analysis, or biochemical or physiological analyses at the cellular level.

In barley, two groups have reported the regeneration of green plants from protoplast cultures *(9–11)*, and in the latter experiments fertile plants that set seed were recovered. There have also been a number of experiments using protoplast transformation as a research tool. Protoplasts from a number of different tissues have been employed to analyze expression of hormone-regulated promoter elements, to examine chloroplast targeting, and to compare different heterologous promoter sequences *(12–16)*. Stable transformation of barley was first achieved via PEG-mediated DNA uptake into nonregenerable suspension protoplasts *(17)*, and, recently, transgenic plants have also been produced by this method (H. Funatsuki et al., personal communication).

The protocols given here are for the isolation, culture and transformation of barley suspension protoplasts. With protopasts of the cultivar Dissa absolute transformation efficiencies (based on total cells treated) of $2.5 - 10 \times 10^{-5}$, and relative transformation efficiencies (based on dividing cells) of $5 - 20 \times 10^{-4}$ routinely have been obtained. The protocols have been developed with Dissa cell suspensions, but with very little modification have proved suitable for other *H. vulgare* cultivars, for wild *Hordeum* species (*H. bulbosum*, *H. marinum*, and *H. murinum*), for tritordeum (*H. chilense* x × *T. durum* amphiploid), and for wheat suspension protoplasts.

2. Materials

2.1. Suspension Culture Medium

1. Composition of L1 basal medium: Macrosalts: 750 mg/L NH_4NO_3; KNO_3, 1750; KH_2PO_4, 200; $MgSO_4 \cdot 7H_2O$, 350; $CaCl_2 \cdot 2H_2O$, 450. Make 10X stock, autoclave, store at ~5°C. Microsalts: 15 mg/L $MnSO_4 \cdot H_2O$, H_3B0_4, 5; $ZnSO_4$, 7.5; KI, 0.75; $Na_2MoO_4 \cdot H_2O$, 0.25; $CuSO_4 \cdot 5H_2O$, 0.025; $CoCl_2.6H_2O$, 0.025; Make 1000X stock, autoclave, store at ~5°C. Iron: 37 mg/L Na_2EDTA; $FeSO_4 \cdot 7H_2O$, 28. Make 100X stock, autoclave, store at ~5°C. Vitamins: 1 mg/L ascorbic acid; Ca-pantothenate, 0.5; folic acid, 0.2; myo-inositol, 100; nicotinic acid, 1; *p*-aminobenzoic acid, 1; pyridoxine-HCl, 1; riboflavin, 0.1; thiamine-HCl, 10. Make 200X stock, aliquot, store at –20°C. Amino acids: 750 mg/L glutamine; proline, 150; asparagine, 100. Make 40X stock, aliquot, store at –20°C. Sugars: 50 g/L maltose; pH: 5.7; osmolarity: ~300 mOsm.

   After combination of all components and bringing up to final volume, sterilize medium by passage through 22-μ*M* membrane filter. Leave at room temperature for 5–7 d (to be sure of sterility), and then store at 2–5°C until use (safe storage life 1–2 mo).
2. L1D2 suspension medium: Composition as basal medium above, with the addition of 2 mg/L 2,4–D (*see* Section 2.1.3.) before filter sterilization.
3. 2,4-D (2,4-dichlorophenoxyacetic acid) stock solution: 1 mg/mL in double-distilled water.

2.2. Protoplast Isolation Solutions

1. Washing (LW) solution: Macro- and microsalts and amino acids as L1 basal medium and in addition 100 g/L sorbitol, 5 g/L glucose, pH 5.7 (osmolarity ~700 mOsm). Autoclave to sterilize, store at 2–5°C.
2. Enzyme solution: 1% w/v Onozuka RS cellulase (Yakult), 0.05% w/v pectolyase Y23 (Seishin), dissolved in LW solution, pH 5.7, ~725 mOsm. Stir solution at least 2 h to ensure all protein is dissolved, store at –20°C in 40-mL aliquots. Filter sterilize directly before use, discard unused solution.

2.3. Protoplast Transformation Solutions

1. C100S solution: 15 g/L $CaCl_2 \cdot 2H_2O$ (~100 m*M*), 1g/L MES, 70 g/L sorbitol, pH 5.7, ~650 mOsm.

2. C100S/PEG 1500 solution: 40% w/v Merck (Darmstadt, Germany) PEG 1500, dissolved in C100S solution, pH 7.0.
3. Electroporation buffer: 14 g/L potassium aspartate (~80 m*M*), 1 g/L calcium gluconate (~5 m*M*), 2.4 g/L HEPES (10 m*M*), 75 g/L sorbitol, pH 7.2, ~700 mOsm. Sterilize these solutions by filtration, store at –20°C, in 2–5-mL aliquots.
4. Plasmid DNA solution: Sterile (alcohol precipitated) plasmid DNA, dissolved in autoclaved 10:1 TE buffer (10 m*M* Tris–HCl, pH 8.0, 1 m*M* EDTA, pH 8.0) at concentration 1 µg/µL. Store at –20°C, in 500-µL aliquots.

2.4. Protoplast Culture Media

1. Liquid culture medium: L1D0.5G medium; L1 basal medium, with 0.5 mg/L 2,4-D and with 90 g/L glucose instead of maltose. Make at 1X concentration, filter sterilize, and store at 2–5°C (1–2 mo).
2. Solid (agarose) culture medium: L1D0.5 medium; L1 basal medium, with 0.5 mg/L 2,4-D and 180 g/L maltose (~700 mOsm). Make at 2X concentration, filter sterilize, and store at 2–5°C. To prepare medium, mix this 2X solution 1:1 with molten (~50°C) 2X SeaPlaque agarose solution (*see* Section 2.4.3.).
3. SeaPlaque Agarose solution: 2X concentration is 2.5% (w/v) in double-distilled water. Make volumes between 50 and 100 mL, autoclave to sterilize.

2.5. Transient GUS Expression Assay Solution

X-Gluc solution: X-Gluc (5-bromo-4-chloro-3-indolyl β-D-glucuronic acid) dissolved in methylcellusolve (ethylene glycol monomethyl ether) at 50 mg/mL; add this to phosphate buffer solution (50 m*M* sodium phosphate buffer, pH 7.0, 0.5 m*M* potassium ferricyanide, 0.5 m*M* potassium ferrocyanide, 0.1% Triton X-100) to give final X-Gluc concentration of 1 mg/mL. Aliquot in 1–10-mL vol, and store at –20°C.

2.6. Selection of Transformed Microcalli

1. G418 sulfate (geneticin) (Gibco, Gaithersburg, MD) stock solution: 1 mg/mL in double-distilled water. Sterilize by filtration, store at –20°C, in 10 mL aliquots.
2. L1D2.5 callus growth medium: L1 basal medium, with 30 g/L maltose, 2 mg/L 2,4-D, and 0.4% w/v (final concentration) Sigma (St. Louis, MO) 1A agarose. Make 2X concentration medium solution, filter sterilize, and prepare medium by mixing 1:1 with molten, sterile, 2X agarose solution (*see* Section 2.6.3.).

3. Sigma 1A agarose solution: 2X concentration is 0.8% w/v in double-distilled water. Make in 250–500-mL vol, autoclave to sterilize.
4. L1D2.5/G25 selection medium: As L1D2.5 callus growth medium, but with 25 mg/L G418 added before filter sterilization.
5. L1D0.5B1 regeneration medium: L1 basal medium, with 30 g/L maltose, 0.5 mg/L 2,4-D, and 1 mg/L BAP (*see* Section 2.6.6.), solidified with 0.4% w/v Sigma 1A agarose.
6. BAP (6-benzylaminopurine) stock solution: 1 mg/mL in double-distilled water.
7. L1D0.5B1G selection medium: as L1D0.5B1 medium (*see* Section 2.6.5.) with G418 sulfate added as appropriate (5–15 mg/L).
8. L1-plantlet growth medium: L1 basal medium, with 30 g/L maltose, solidified with 0.4% w/v Sigma 1A agarose.
9. L1G5 selection medium: As L1 plantlet growth medium (*see* Section 2.6.8.), with 5 mg/L G418 sulfate.

3. Methods

3.1. Protoplast Isolation

The following protocol is for the transformation and culture of protoplasts isolated from barley cell suspensions. Such suspensions are generally established from immature embryo or anther cultures *(10,11,17)*. A strict regime for suspension maintenance (e.g., subculture timing, growth conditions) is essential, as the "health/quality" of the protoplasts (and hence of the donor suspension) is the single most important factor deciding the success and reproducibility of subsequent manipulations.

1. Maintain suspensions in L1D2 medium by biweekly subculture (*see* Note 1). At subculture, transfer 1–2 g (fwt) cells to a 9-cm plastic Petri dish and add 7–10 mL enzyme solution per g cells (*see* Note 2). Seal the dish.
2. Place dishes on a rotary shaker set at 30 rpm, at 25°C. Check the digestion hourly, by microscopic observation, digestion time is usually 2–3 h.
3. Sieve digestion mixture through 100-, 50-, and 25-μm sieves and wash through with LW (washing) solution.
4. Using a pipet (*see* Note 3), distribute the protoplast suspension in centrifuge tubes, and spin for 5 min at 50*g*.
5. Pour supernatants off pellets, then immediately disperse them in residual liquid by gentle shaking. Wash protoplasts by adding 10 mL LW solution to the tubes and then centrifuge as in Section 3.1.4. and again disperse pellets.
6. Resuspend protoplasts in LW and pool the whole isolation in a known volume (20–50 mL) of the solution in a flask. Swirl gently to distribute the

protoplasts evenly and take a sample to count the density in a hemocytometer (Fuchs-Rosenthal type). Calculate total protoplast yield.
7. Either continue immediately with the transformation procedure, or store pooled protoplasts at 4–7°C until use (*see* Note 5).

3.2. Protoplast Transformation

Procedures are given both for stable transformation and for transient gene expression in barley protoplasts, and methods for gene transfer via PEG treatment or via electroporation are given. Where methods diverge alternatives are given.

3.2.1. PEG-Mediated Transformation

1. Distribute aliquots of the protoplast suspension among centrifuge tubes such that individual tubes contain 1–4 × 10^6 protoplasts. Centrifuge at 50*g* for 5 min, pour off supernatants, and disperse pellets.
2. Add 1 mL LW solution to the untransformed control tube and add 0.2 mL C100S solution to the mock transformation tube and to the transformation tubes (*see* Note 6).
3. Add 50 μL plasmid DNA solution (conc. 1 μg/μL) to all the transformation tubes (*see* Note 7) and 50 μL "control" DNA (e.g., salmon sperm DNA) to the mock transformation tube(s). Shake tubes gently.
4. Add 0.6 mL C100S/PEG 1500 solution, dropwise, with gentle shaking, to the transformation and mock transformation tubes (*see* Note 8). Allow transformation mixtures to stand for 10 min, with gentle shaking at intervals.
5. Fill all tubes to 10 mL with LW solution (*see* Note 9), taking care that the washing solution dilutes the viscous PEG mixtures rather than floating above them.
6. Centrifuge the tubes at 50*g* for 7 min, discard supernatants, and disperse pellets. Continue as in Section 3.3. (transient expression) or Section 3.5. (stable transformation) below (*see* Note 10).

3.2.2. Electroporation

1. Distribute aliquots of the protoplast suspension among centrifuge tubes such that individual tubes contain 1–4 × 10^6 protoplasts. Centrifuge at 50*g* for 5 min, pour off the supernatants, and disperse the pellets.
2. Suspend protoplast pellets in 300 μL electroporation buffer and transfer protoplast suspensions to 4 mm electrode-distance electroporation cuvets (*see* Note 11).
3. Add 25 μL of plasmid DNA solution to transformation cuvets, and 25 μL of "control" DNA solution to the mock transformation cuvet(s), and shake cuvets gently to mix the DNA and cells and place cuvets on ice for 10 min.

4. Pulse parameters: For transient expression use a capacitance of 960 μF and voltage of 150V (this gives a field strength of 375 V/cm with 0.4 cm electrode spacing). Under these conditions, pulse duration should be between 80 and 90 ms. For stable transformation, use a capacitance of 960 μF and a voltage between 90 and 110V (this gives field strengths between 225 and 275 V/cm). Pulse durations should be between 150 and 200 ms.
5. Immediately after pulsing, return cuvets to ice and allow to stand for 10 min.
6. Remove the protoplast suspension from the cuvets with a Pasteur pipet and for transient expression assays, transfer the suspension to a 1.5–2.5 mL volume of L1D0.5G liquid culture medium, in tissue culture quality Petri dishes, rinsing the last cells from the cuvet with culture medium. Continue as in Section 3.3. below. For stable transformation experiments, transfer the suspension to test tubes, rinse all the cells from cuvets with LW solution, and make the test tube volumes up to 10 mL with LW solution. Centrifuge tubes at 50*g* for 7 min, discard supernatants, and disperse the pellets. Continue as in Section 3.5. below.

3.3. Enzyme Assays for Transient Gene Expression

1. Suspend protoplasts in liquid L1D0.5G medium at density of 1–2 × 10^6/mL and transfer to tissue culture quality plastic Petri dishes. Incubate the cultures at 25°C in dim light.
2. Harvest protoplasts (cells) after 36–48 h incubation. Scrape adhering cells from the floors of the Petri dishes with a spatula and pipet the suspended cells into a 2.2-mL Eppendorf tube. Rinse the dishes with LW medium to collect any cells remaining, pipeting these into the tube.
3. Centrifuge samples at 15,000 rpm at 4°C for 10 min. Discard supernatants and store Eppendorf tubes containing pellets at –70°C until transient gene expression analysis by standard protocols (e.g., NPT *[18]*, GUS *[19]*) (*see* Chapters 10 and 12, this volume).

3.4. Histochemical Assay for Transient GUS Expression

1. Suspend protoplasts in a 2-mL vol of a 1:1 mixture of 2X L1D0.5 protoplast medium and 2X (2.5% w/v) SeaPlaque agarose at 42°C, using a widemouth pipet. Gently pipet the protoplast suspension into a 6-cm tissue culture quality plastic Petri dish.
2. Seal the dishes, and incubate at 25°C in dim light.
3. After 36–72 h incubation, open the dishes and cut the agarose into quarters with a scalpel. Drip 0.5 mL of X-Gluc phosphate buffer solution onto the center of the dish where the two cuts through the agarose bisect. Tilt the dish so that the X-Gluc solution covers the upper and lower surfaces of the agarose.

4. Reseal dishes using two layers of Nescofilm and incubate at 37°C overnight.
5. Inspect dishes for the presence of blue cells under a stereoscopic microscope, using transmitted light viewing capacity.

3.5. Stable Transformation

1. Suspend treated protoplasts in volumes of a 1:1 mixture of 2X L1D0.5 protoplast medium and 2X (2.5% w/v) SeaPlaque agarose to give protoplast densities between 0.25 and 1×10^6/mL. Pipet 2.5 mL aliquots of the suspended protoplasts into 6-cm plastic Petri dishes (tissue culture quality).
2. Allow agarose to solidify and seal the dishes with Nescofilm. Incubate at 25°C, under dim light or in darkness.
3. Monitor culture development. After 10 d, calculate plating efficiency, i.e., number of microcalli formed/number of protoplasts originally plated.
4. Apply selection pressure after ca. 12 d culture. Cut the agarose in the Petri dishes into four segments and transfer these to a 9-cm Petri dish. Pipet 5 mL of L1D2 suspension medium into each dish and add 180 μL of sterile G418 sulfate stock solution (the final antibiotic amount is therefore 0.18 mg in a total agarose-and-medium vol of 7.5 mL; this gives a final concentration of 25 mg/L). Culture the microcalli from the untreated control with antibiotic selection, and culture the mock-transformed microcalli without selection (*see* Notes 12 and 13).
5. Place dishes on a rotary shaker at 30 rpm, at a temperature of 25°C, under dim light. After 10 d, remove old medium (using a narrow-mouth pipet, so as not to take any free-swimming microcalli) and replace with 5 mL of fresh medium, with or without 25 mg/L G418 as appropriate. Return dishes to shaker.
6. After 20 d liquid selection culture, pick any calli >1 mm in diameter and transfer them onto the surface of solid L1D2.5/G25 medium. Calli showing sustained growth after two 25-d passages on this medium are almost certainly transformants; this can be verified by assays for the activity of marker genes and subsequently by Southern analysis to demonstrate integration of the transferred gene(s) (*see also* Chapter 14 on the use of inverse PCR in transgene screening and analysis).

3.6. Plant Regeneration

1. For regeneration from transformed calli, transfer them to regeneration medium containing 15 mg/L G418 sulfate (L1D0.5B1G15; *see* Section 2.6.). Incubate the dishes at 20–25°C under continuous light (ca. 1000 lux). Culture control calli on regeneration medium without G418 sulfate.
2. Subculture calli at ca. 20-d intervals on L1D0.5B1G15 medium, selecting for compact yellow embryogenic tissue (*see* Notes 14 and 15). Transfer

calli with germinating somatic embryos (or shoots) to hormone-free (L1-G15) medium contained in larger vessels, and culture under higher light intensities (16 h, 6000 lux).

3. Transfer rooted plantlets to a peat/soil mix and grow under greenhouse conditions.
4. Subject plants to further analysis to ascertain transgene copy numbers (Section 4.) and gene expression levels (*see* Chapters 5–8).

4. Notes

4.1. Suspension Culture, Protoplast Isolation

1. Suspensions are cultured in 190-mL plastic vessels (Greiner 967161) or Erlenmeyer flasks, shaken at 100 rpm in dim light at 25°C. At subculture ($3^1/_2$-d intervals), the medium is pipeted-off and cells are harvested, leaving 2.5–3 g fresh weight per vessel. Ten milliliters of fresh L1D2 medium are then added to the vessel, and the harvested cells used for protoplast isolation.
2. The standard enzyme solution contains sorbitol as osmoticum; however, enzyme activity is increased when KCl is used as osmoticum. This reduces digestion time and increases yield from hard-to-digest suspensions, but more care must be taken to avoid "overdigestion," resulting in damage to protoplasts.
3. Use wide-mouth pipets (reduces shearing forces) for all transfers of protoplast suspensions.
4. If protoplast pellets are not dispersed quickly after centrifugation, protoplasts will form aggregates that are later difficult to separate.
5. Isolated protoplasts may be stored up to 24 h at 4–7°C and will still give good plating efficiencies, although there is some loss of viability.

4.2. PEG-Mediated Transformation

6. Two control treatments are usually made—untreated protoplasts and a mock transformation—in which plasmid DNA containing a selectable or scorable marker gene is replaced by other plasmid DNA or by salmon sperm or calf thymus DNA (sheared to approximately the size of the transforming plasmid by sonication or repeated passage through an 18-gage needle). A further control, the addition of plasmid DNA, but without a transformation treatment, may also be made.
7. DNA concentration is an important factor; both transient and stable transformation efficiencies are improved by a high DNA:protoplast ratio. To avoid having to use large amounts of plasmid, the volume of the transformation mixture is kept as low as possible.
8. PEG concentration is also a major factor affecting transformation efficiency. In the standard protocol, the final PEG concentration is ca 25%.

Higher PEG concentrations (30–35%) may be useful in transient expression experiments, but these levels significantly reduce protoplast plating efficiencies, which is a disadvantage in stable transformation experiments.

9. In liquid protoplast cultures, glucose is used instead of maltose in the medium as protoplasts float in the latter sugar and are therefore difficult to collect by centrifugation.
10. In stable transformation experiments, plating efficiency is calculated by finding the mean number of microcalli in a known area of the culture dish (e.g., the field of view for a 10X objective of an inverted microscope) and using this figure to calculate the total number of microcalli present. Plating efficiency is expressed as the percentage of protoplasts plated that give rise to viable microcalli (*see also* Chapter 8).

4.3. Electroporation

11. The procedures given have been developed using a BioRad (Richmond, CA) GenePulser electroporation apparatus, equipped with a Capacitance Extender and using BioRad electroporation cuvets with 4-mm electrode spacing. Other electroporation equipment will differ in output and type of pulse produced, and therefore some optimization of electrical parameters may be necessary to achieve comparable results (*see also* Chapter 8).

4.4. Selection

12. The timing of selection is critical. When selection pressure is applied too early the stress will be too great for small, sensitive microcalli and no resistant colonies will be recovered. When selection is applied too late many large microcalli will survive and will subsequently be very difficult to kill, even using very high antibiotic concentrations. Plating efficiency also affects selection efficiency; very high microcallus densities require higher antibiotic levels for control. As a guide, selection pressure can generally be applied when the majority of microcalli contain about 8–20 cells.
13. Antibiotic concentration is also a factor that may be varied to suit particular conditions. If selection is applied at the right time good discrimination can be achieved with 25 mg/L G418, but for late selection, double this concentration may be required for control. If the untreated control microcalli are selected and the mock-transformed calli left unselected, then both the efficiency of the selection pressure and the viability of treated protoplasts may be assessed.

4.5. Regeneration

14. Frequently, somatic embryos do not appear during the first passage on regeneration medium, but selective subculturing enriches for embryogenic

tissue and allows small groups of embryogenic cells to proliferate and subsequently to regenerate.

15. If it is seen that the maintenance of selection pressure inhibits embryogenesis in transformed cultures (by comparison with nontransformed, nonselected cultures) it may be necessary to reduce the G418 concentration, or remove selection pressure completely to allow regeneration to proceed.

Note Added in Proof

Since the completion of this chapter, clear evidence for the transformation of rice by *Agrobacterium* has been published *(20)*, this result suggests that further effort should be devoted to the development of *Agrobacterium*—mediated transformation procedures for other major cereals.

References

1. Grimsley, N., Hohn, B., Ramos, C., Kado, C., and Rogowsky, P. (1989) DNA transfer from *A* to *Zea Mays* or *Brassica* by agroinfection is dependent on bacterial virulence functions. *Mol. Gen. Gene.* **217,** 309–316.
2. Rhodes, C. A., Pierce, D. A., Mettler, I. J., Mascarenhas, D., and Detmar, J. J. (1988) Genetically transformed maize plants from protoplasts. *Science* **240,** 204–207.
3. Gordon-Kamm, W. J., Spencer, T. M., Mangano, M. L., Adams, T. R., Daines, R. J., Start, W. G., O'Brien, J. V., Chambers, S. A., Adams, W. R., Willets, N. G., Rice, T. B., Mackey, C. J., Krueger, R. W., Kausch, A. P., and Lemaux, P. G. (1990) Transformation of maize cells and regeneration of fertile transgenic plants. *Plant Cell* **2,** 603–618.
4. D'Halluin, K., Bonne, E., Bossut, M., De Beuckeleer, M., and Leemans, J. (1992) Transgenic maize plants by tissue electroporation. *Plant Cell* **4,** 1495–1505.
5. Wang, K., Frame, B., Drayton, P. R., Bagnall, S., Jiao, S., Lewnau, C., Wilson, M., Dunwell, JM., and Thompson, J. A. (1994) Abstract, Keystone Symposium: Improved Crop and Plant Products Through Biotechnology, Park City, U, 9–16 January 1994.
6. Lazzeri, P. A. and Shewry, P. R. (1994) Biotechnology of Cereals, in *Biotechnology and Genetic Engineering Reviews* (Tombs, M. P., ed.), Intercept Ltd., Andover, MD, pp. 79–145.
7. Golovkin, M. V., Abraham, M., Morocz, S., Bottka, S., Feher, A., and Dudits, D. (1993) Production of transgenic maize plants by direct DNA uptake into embryogenic protoplasts. *Plant Sci* **90,** 41–52.
8. Shimamoto, K., Terada, R., Izawa, T., and Fujimoto, H. (1989) Fertile transgenic rice plants regenerated from transformed protoplasts. *Nature* **338,** 274–276.
9. Yan, Q., Zhang, X., Shi, J., and Li, J. (1990) Green plant regeneration from protoplasts of barley (*Hordeum vulgare* L.). *Kexue Tongbao* **35,** 1581–1583.
10. Jähne, A., Lazzeri, P. A., and Loerz, H. (1991) Regeneration of fertile plants from protoplasts derived from embryogenic cell suspensions of barley (*Hordeum vulgare* L.) *Plant Cell Rep.* **10,** 1–6.

11. Funatsuki, H., Loerz, H., and Lazzeri, P. A. (1992) Use of feeder cells to improve barley protoplast culture and regeneration. *Plant Sci.* **85,** 179–187.
12. Chibbar, R. N., Kartha, K. K., Datla, R. S. S., Leung, N., Caswell, K., Mallard, C. S., and Steinhauer, L. (1993) *Plant Cell Rep.* **12,** 506–509.
13. Jacobsen, J. V. and Close, T. J. (1991) Control of transient expression of chimaeric genes by gibberellic acid and abscisic acid in protoplasts prepared from mature barley aleurone layers. *Plant Mol. Biol.* **16,** 713–724.
14. Lee, B. T., Murdoch, K., Topping, J., Kreis, M., and Jones, M. G. K. (1989) Transient expression in aleurone protoplasts isolated from developing caryopses of barley and wheat. *Plant Mol Biol.* **13,** 21–39.
15. Lee, B. T., Murdoch, K., Topping, J., Jones, M. G. K., and Kreis, M. (1991) Transient expression of foreign genes introduced into barley endosperm protoplasts by PEG-mediated transfer or into intact tissue by microprojectile bombardment. *Plant Sci.* **78,** 237–246.
16. Teeri, T. H., Patel, G. H., Aspegren, K., and Kauppinen, V. (1989) Chloroplast targeting of neomycin phosphotransferase II with a pea transit peptide in electroporated barley mesophyll protoplasts. *Plant Cell Rep.* **8,** 187–190.
17. Lazzeri, P. A., Brettschneider, R., Luehrs, R., and Loerz, H. (1991) Stable transformation of barley via PEG-induced direct DNA uptake into protoplasts. *Theor. Appl. Genet.* **81,** 437–444.
18. McDonnell, R. E., Clark, R. D., Smith, W. A., and Hinchee, M. A. (1987) A simplified method for the detection of *neomycin phosphotransferase II* activity in transformed plant tissues. *Plant Mol. Biol. Reptr.* **5,** 380–386.
19. Jefferson, R. A. (1987) Assaying chimeric genes in plants: the GUS fusion system. *Plant Mol. Biol. Reptr.* **5,** 387–405.
20. Hiei, Y., Ohta, S., Komari, T., and Kumashiro, T. (1994) Efficient transformation of rice (*Oryza sativa* L) mediated by *Agrobacterium* and sequence analysis of the boundaries of the T-DNA. *Plant J.* **6,** 271–282.

Chapter 8

Gene Transfer into Plant Protoplasts by Electroporation

Heddwyn Jones

1. Introduction

Introduction of foreign genes into plant cells can be achieved by a variety of methods, including direct transfer into protoplasts using chemical (*1–3; see also* Chapter 7, this volume) and electrical methods *(4–6)*. These methods can be used to study genes both transiently *(1,5,7)* and when stably integrated in regenerated transgenic plants *(2–4,6)*. The electrical method (electroporation) is a simple and rapid procedure and involves applying electrical pulses to a suspension of protoplasts and DNA placed between electrodes in a suitable cuvet. The electrical pulses induce the formation of transient pores in the plasmalemma allowing DNA to enter the cell and nucleus. The method has been used to introduce genes into protoplasts isolated from a range of different species and seems to be a universal method of gene transfer into prokaryotic and eukaryotic cells (an application of the technique to transform *Agrobacteria* is covered in Chapter 2, this volume).

Several electroporation machines are available commercially and offer different pulse types; square-wave-type pulsing machines (e.g., Hoefer Pro-Genetor pulse controller) vs and capacitor discharge (exponentially decaying) pulse machines (e.g., BioRad [Richmond, CA] Gene Pulser) are available. It is not yet clear which one of these machine types is best for gene transfer, in any case, optimization of the electrical settings in these machines should allow efficient gene transfer into potentially any

From: *Methods in Molecular Biology, Vol. 49: Plant Gene Transfer and Expression Protocols*
Edited by: H. Jones Humana Press Inc., Totowa, NJ

protoplast cell type. When optimizing gene transfer into a particular protoplast type for the first time, published conditions that work for a similar protoplast type are a useful starting point; the voltage settings can then be altered for efficient transfer. The best and quickest way of monitoring uptake and expression of genes is to optimize using a reporter gene such as *cat* or *gus* under the control of a strong promoter (e.g., CaMV 35S promoter). Monitoring transiently in this way can be done 48 h after pulsing, which is much quicker than having to score the number of antibiotic resistant calli and/or regenerated plants that can take weeks to months.

A range of electroporation buffers/media have been successfully used for gene transfer, and an ideal medium should lead to minimal damage to the cell during the critical period when the pores are induced. Pulsing can be carried out either at room temperature or at 4°C (cells previously placed on ice prior to pulsing). The form of plasmid DNA used for transformation can be linear or supercoiled, and the working concentration range should be 1–10 µg/mL. In promoter deletion studies it is important that same number of moles of plasmid are electroporated in order to compare promoter activity from one construct to the next. The addition of carrier DNA is optional. For at least some protoplast types (if not all) it is possible to cotransfect plasmids carrying different genes. It is thus possible to regenerate transgenic plants after pulsing with one plasmid carrying an antibiotic selectable marker (e.g., kanamycin resistance; *see* Chapter 12 this volume), and the other plasmid carrying a foreign transgene of interest. Cell number and density is another parameter that needs to be considered for successful gene transfer and should be in the range 0.5–10^6 cells/mL, which is sufficient for both transient gene expression studies and for the recovery of a sufficient number of regenerated transgenic plants. The latter will obviously depend on the plating efficiency of the species under study.

This chapter describes a method for transferring genes into potato protoplasts for transient expression studies and has been used to study the expression of homologous potato genes *(9)*. The method can also be used for the recovery of transgenic potato plants and represents an alternative to the use of *Agrobacterium* as a vector for transforming dicot plant species. For a more general account of the use of electroporation for gene transfer into monocot and dicot species I recommend that the reader consult the article by Shillito and Potrykus *(8)*.

2. Materials

1. In vitro shoot cultures of potato plants (e.g., Maris piper, Desiree) were grown on the salts, vitamins, and sugars of Murashige and Skoog medium (supplied as a powder, Sigma) with 20 g/L sucrose and 0.05 mg/L 6-benzylaminopurine (BAP).
2. Macronutrients (stock × 100): To approx 750 mL of distilled water dissolve 0.190 g KNO_3, 0.044 g $CaCl_2 \cdot H_2O$, 0.037 g $MgSO_4 \cdot 7H_2O$ and 0.017 g KH_2PO_4. Make up to 1 L and store at 4°C.
3. Micronutrients (stock A × 100): 3.7 mg Na_2 EDTA, 2.8 mg $FeSO_4 \cdot 7H_2O$. Make up to 1 L and store at 4°C.
4. Micronutrients (stock B × 1000): 0.6 mg H_3BO_3, 2.0 mg $MnCl_2 \cdot 4H_2O$, 0.9 mg $ZnSO_4 \cdot 7H_2O$, 0.08 mg KI. Make up to 1 L and store at 4°C.
5. Micronutrients (stock C × 10,000): 0.030 mg Na_2 $MoO_4 \cdot 2H_2O$, 0.003 mg $CuSO_4 \cdot 5H_2O$, 0.003 mg $CoSO_4 \cdot 7H_2O$. Make up to 1 L and store at 4°C.
6. Vitamins (stock × 1000): 0.05 mg thiamine HCl, 0.20 mg glycine, 0.50 mg nicotinic acid, 0.05 mg pyridoxine HCl, 0.05 mg folic acid, 0.005 mg biotin. Make up to 1 L and store at 4°C.
7. 1-Napthaleneacetic acid (NAA) stock: Dissolve 20 mg in 5 mL of ethanol (warm to dissolve) and make up to 100 mL with sterile distilled water.
8. 6-Benzylaminopurine (BAP) stock: Dissolve 20 mg in 5 mL of 0.5*N* HCl and make up to 100 mL.
9. A. Medium: To approx 900 mL of distilled water add 10 mL of the macronutrients stock, 10 mL of micronutrient stock, 1 mL of micronutrient stock B, 0.1 mL of micronutrient stock C, 1 mL of vitamin stock, 10 mg myoinositol, 10 mL NAA stock and 2.5 mL of BAP stock. Adjust pH to 5.6 and volume to 1 L. Sterilize by auoclaving.
10. Preplasmolysis solution: To approx 750 mL of distilled water add 10 mL of the macronutrient stock, 81.97 g of mannitol and 0.86 g $CaCl_2 \cdot 2H_2O$. Make up to 1 L and adjust the pH to 5.6. Sterilize by autoclaving.
11. Protoplast wash solution: To approx 750 mL of distilled water add 10 mL of the macronutrient stock, 93.5 g of mannitol and 0.86 g $CaCl_2 \cdot 2H_2O$. Make up to 1 L and adjust pH to 5.6. Sterilize by autoclaving and adjust the osmolality to 570 mOsm.
12. Percoll dilution solution: To approx 750 mL of distilled water add 10 mL of the macronutrient stock, 135 g of mannitol and 0.86 g $CaCl_2 \cdot 2H_2O$. Make up to 1 L, autoclave, and adjust the osmolality to 815 mOsm.
13. Enzyme solution: To 750 mL of distilled water add 10 mL of the macronutrient stock, 15 g Meicelase (Meiji Seika Kaisha, Tokyo, Japan), 1 g Pectolyase Y23 (Seishim Pharmaceuticals, Tokyo, Japan), 0.86 g $CaCl_2 \cdot 2H_2O$ and 89.6 g mannitol. Make up to 1 L and adjust the pH to 5.6. Sterilize by filtration and store frozen in 50-mL aliquots.

14. CLG medium: To approx 750 mL of distilled water add 4.7 g KNO_3, 1.1 g $CaCl_2 \cdot 2H_2O$, 0.95 g $MgSO_4 \cdot 7H_2O$, 0.43 g KH_2PO_4, 25 mL of micronutrient stock A, 2 mL of micronutrient stock B, 0.2 mL of micronutrient stock C, 2.5 mL of the vitamin stock, 1.14 g myo-inositol, 0.1 g glutamine, 51 g sucrose, 46 g mannitol, 0.585 g 2-(*N*-morpholino) ethanesulfonic acid, 5 mL NAA stock, and 2 mL BAP stock. Adjust pH to 5.6, make up to 1 L, and filter sterilize.
15. Electroporator: Bio-Rad gene pulser or equivalent.
16. Pulse medium: To 750 mL of distilled water add 0.7456 g KCl and 93.893 g mannitol, adjust the pH to 7.2, and make up to 1 L. Sterilize by autoclaving.
17. Purified plasmid DNA carrying the gene of interest.

3. Methods

1. Remove approx 1 g of leaf material from 3–4-wk-old cultures and precondition overnight in approx 100 mL of A. medium at 4°C in the dark (*see* Note 1).
2. Slice the preconditioned leaves into small pieces and transfer to approx 30 mL of preplasmolysis solution and incubate at room temperature for 30 min.
3. Transfer the preplasmolyzed shoots to approx 25 mL enzyme solution and shake at 40 rpm on an orbital shaker at room temperature for 4 h in the dark.
4. Pass the resulting protoplast suspension (*see* Note 2) through a 50-μ sieve, then pass through a 38-μ sieve.
5. Transfer 10-mL aliquots of the protoplast suspension to test tubes and centrifuge at 80*g* for 5 min.
6. Remove the supernatant and resuspend each pellet in 1.5 mL of protoplast wash solution.
7. Make up a 30% Percoll solution by mixing 3 mL of sterile Percoll and 7 mL of Percoll dilution solution.
8. Layer 1.5 mL of protoplast suspension over 2.5 mL of 30% Percoll (*see* Note 3).
9. Centrifuge at 100*g* for 5 min.
10. Remove the band of protoplasts at the Percoll–wash solution interface and dilute each tube of protoplasts with 9 mL protoplast wash solution.
11. Centrifuge at 80*g* for 5 min.
12. Remove the supernatant and resuspend the pellet in 10 mL of protoplast wash solution.
13. Centrifuge at 80*g* for 5 min.
14. Resuspend the protoplast in pulse medium at a density of 1×10^6 cells/mL.

15. Add purified plasmid(s) carrying the gene of interest at a concentration of 1 µg/mL (*see* Note 4).
16. Pulse the cells at 225 V/cm using an RC pulse duration of 85 ms. Optimization may be required around these values (*see* Note 5).
17. Transfer the transformed protoplasts to 10 mL CLG medium and culture in the dark at 26°C for 48 h (*see* Note 6).
18. Harvest the protoplast by centrifugation at 80*g* for 5 min and resuspend in the appropriate buffer for CAT or GUS assays depending on the reporter gene used.
19. Gene expression is quantified by GUS or CAT assays according to the protocols outlined in Chapters 10 and 11, respectively.

4. Notes

1. All manipulations involved in protoplast isolation and transformation must be performed under sterile conditions under a laminar flow hood. All test tubes, scalpels, and so on must be autoclaved before use.
2. Protoplasts are very fragile and must be manipulated very gently, particularly when pipeting (use wide mouth pipets).
3. The protoplast suspension is best layered over the 30% Percoll by holding the test tube almost horizontally, and adding the protoplast suspension slowly using a Pasteur pipet.
4. Plasmid DNA can be isolated, for example, using the alkali lysis method of Birnboim and Doly *(10)*. Transient gene expression can be achieved using supercoiled plasmid DNA and there is no need to add carrier DNA. For stable plant transformation, however there is evidence that restricted linearized plasmid DNA increases the efficiency of transformation, moreover, stable transformation efficiency can be further improved by the addition of carrier DNA (e.g., calf thymus DNA) mixed at a ratio of five times the amount of plasmid DNA *(8)*.
5. Conditions causing cell death are generally needed for successful transformation of the remaining cells. As mentioned in the Introduction, when optimizing gene delivery, start by altering voltage and pulse duration; other parameters can then be altered (if required) like pulsing medium, temperature, and DNA concentration.
6. If bacterial contamination is a persistent problem add the following antibiotics to the CLG medium: 250 mg/L ampicillin, 5 mg/L gentamycin, and 5 mg/L tetracycline.

References

1. Lorz, H., Baker, B., and Schell, J. (1985) Gene transfer to cereal cells mediated by protoplast transformation. *Mol. Gen. Genet.* **199,** 178–182.
2. Paszkowski, J., Shillito, R. D., Saul, M. W., Mandak, V., Hohn, T., Hohn, B., and Potrykus, I. (1984) Direct gene transfer to plants. *EMBO J.* **3,** 2717–2722.

3. Potrykus, I., Saul, M. W., Petruska, J., Paszkowski, J., and Shillito, R. D. (1985) Direct gene transfer of cells of a graminaceous monocot. *Mol. Gen Genet.* **199,** 183–188.
4. Fromm, M., Taylor, L. P., and Walbot, V. (1986) Stable transformation of maize after gene transfer by electroporation. *Nature* **319,** 791–793.
5. Fromm, M., Taylor, L. P., and Walbot, V. (1985) Expression of genes transferred into monocot and dicot plant cells by electroporation. *Proc. Natl. Acad. Sci. USA* **82,** 5824–5828.
6. Shillito, R. D., Saul, M. W., Paszkowski, J., Muller, M., and Potrykus, I. (1985) High efficiency direct gene transfer to plants. *Bio/technology* **3,** 1099–1103.
7. Howard, E. A., Walker, J. C., Dennis, E. S., and Peacock, W. J. (1987) Regulated expression of an alcohol dehydrogenase 1 chimeric gene introduced into maize protoplasts. *Planta* **170,** 535–540.
8. Shillito, R. D. and Potrykus, I. (1987) Direct gene transfer to protoplasts of dicotyledenous and monocotyledenous plants by a number of methods, including electroporation. *Methods Enzymol.* **153,** 313–336.
9. Jones, H., Ooms, G., and Jones, M. G. K. (1989) Transient gene expression in electroporated *Solanum* protoplasts. *Plant Mol. Biol.* **13,** 503–511.
10. Birnboim, H. C. and Doly, J. (1979) A rapid alkaline extraction procedure for screening recombinant plasmid DNA. *Nucleic Acids Res.* **7,** 1513–1523.

CHAPTER 9

Transformation of Cereals by Microprojectile Bombardment of Immature Inflorescence and Scutellum Tissues

Pilar Barcelo and Paul A. Lazzeri

1. Introduction

The cereals are one of the groups of crops more recalcitrant to transformation. Since, with the exception of rice *(1)*, cereals have not been transformed by *Agrobacterium*, and highly regenerable protoplast systems are difficult to obtain, these species have remained untransformed for a much longer time than the major dicot species. It was not until 1988 and 1989 that the first transgenic cereal plants (maize *[2]*; rice *[3]*) were produced by the use of direct DNA-transfer into protoplasts. This method, however, has not yet given rise to transgenic wheat or barley plants, species in which the protoplast systems established are not as regenerative and stable as those developed for rice or maize.

As an alternative to protoplast transformation, particle bombardment was first described as a method for the production of transgenic plants in 1987 *(4)*. The technique was relatively quickly shown to be applicable in maize and rice transformation *(5,6)* and also in wheat *(7)*. In all these cases, regenerable callus or suspension cultures were used as targets, which limits broad application because the production of maintainable embryogenic cultures still presents difficulties in many genotypes or species. This problem has been approached by the use of regenerable primary explants, generally immature embryos, as targets for delivery *(8,9)*.

From: *Methods in Molecular Biology, Vol. 49: Plant Gene Transfer and Expression Protocols*
Edited by: H. Jones Humana Press Inc., Totowa, NJ

Recently, several groups have targeted immature embryo (scutellum) tissue in wheat and barley and have recovered transgenic plants *(10–14)* with efficiencies of about 0.2–0.3 plants per bombardment in wheat and one plant per bombardment in barley (typically 20–30 embryos).

In the *Triticeae*, a second primary explant source with high regeneration potential is immature inflorescence tissue *(15)* and in tritordeum bombardment of inflorescence cultures has yielded transgenic plants at an efficiency of 0.5 plants per bombardment *(16)*. This development, giving a choice of target tissues, should broaden the spectrum of cultivars that can be transformed; in some cases, genotypes that show poor regenerative capacity at the immature embryo level are responsive at the inflorescence level and *vice versa*.

We present here procedures for the transformation of tritordeum and wheat by particle bombardment of either immature inflorescence or immature scutellum cultures.

2. Materials

2.1. Plant Material

This protocol concerns the bombardment of immature embryos from wheat and immature inflorescences of the fertile cereal amphiploid, tritordeum ($H^{ch}H^{ch}AABB$). This amphiploid was synthesized by A. Martin *(17)* by doubling of the chromosomes of hybrid plants obtained by embryo rescue from crosses between a wild *Hordeum (Hordeum chilense)* and *durum* wheat cultivars. To produce explants for bombardment, grow tritordeum or wheat plants under 16 h photoperiod with temperatures of 6–18°C during the day and 14°C at night. Winter wheat cultivars require a vernalization period of 8 wk before being transferred to these conditions.

2.2. Culture Media

1. Macrosalts for L3 (10X stock): 2.0 g NH_4NO_3, 17.5 g KNO_3, 2.0 g KH_2PO_4, 3.5 g $MgSO_4·7H_2O$, 4.5 g $CaCl_2·2H_2O$. Make up to 1 L with distilled water. Store at 4°C after autoclaving (*see* Note 1).
2. Macrosalts for MS medium (10X stock): 16.5 g NH_4NO_3, 19.0 g KNO_3, 1.7 g KH_2PO_4, 3.7 g $MgSO_4·7H_2O$, 4.4 g $CaCl_2·2H_2O$. Make up to 1 L with distilled water. Store at 4°C after autoclaving (*see* Note 1).
3. Use 200 mL of the above stock solutions for making 1 L of 2X concentration L3 or MS culture medium.
4. Microsalts for L3 medium (1000X stock): 13.4 g $MnSO_4$, 5.0 g H_3BO_3, 7.5 g $ZnSO_4·7H_2O$, 0.75 g KI, 0.25 g $Na_2MoO_4·2H_2O$, 0.025 g $CuSO_4·5H_2O$,

0.025 g $CoCl_2 \cdot 6H_2O$. Make up to 1 L with distilled water. Store at 4°C after filter sterilizing through a 22-µm membrane filter.

5. Microsalts for MS medium (1000X stock): 10.0 g $MnSO_4$, 6.2 g H_3BO_3, 5.8 g $ZnSO_4 \cdot 7H_2O$, 0.8 g Kl, 0.25 g $Na_2MoO_4 \cdot 2H_2O$, 0.025 g $CuSO_4 \cdot 5H_2O$, 0.025 g $CoCl_2 \cdot 6H_2O$. Make up to 1 L with distilled water. Store at 4°C after filter sterilizing through a 22-µm membrane filter.
6. Use 2 mL of the above stock solutions for making 1 L 2X concentration L3 or MS culture medium.
7. Iron solution for L3 and MS medium (500X stock): 18.65 g Na_2EDTA, 13.90 g $FeSO_4 \cdot 7H_2O$. Make up to 1 L with distilled water (*see* Note 2). Store solution protected from the light at 4°C.
8. Use 4 mL of the above stock solution for making up 1 L 2X concentration L3 or MS culture medium.
9. Vitamins/inositol solution for L3 medium (250X stock): 50.0 g inositol, 2.5 g thiamine HC1, 0.25 g pyridoxine HCl, 0.25 g nicotinic acid, 0.25 g Ca-pantothenate, 0.25 g ascorbic acid, 0.125 g choline chloride, 0.0625 g folic acid. Make up to 1 L with distilled water (*see* Note 3). Store solution in 8-mL aliquots at –20°C.
10. Vitamins/inositol solution for MS medium (250X stock): 25.0 g inositol, 0.025 g thiamine HCl, 0.125 g pyridoxine HCl, 0.125 g nicotinic acid, 0.5 g glycine. Make up to 1 L with distilled water. Store solution in 8-mL aliquots at –20°C.
11. Use 8 mL of the above stock solutions for making 1 L 2X concentration L3 or MS culture medium.
12. Amino acid solution for L3 and MS medium (40X stock): 30.0 g L-glutamine, 6.0 g L-proline, 4.0 g L-asparagine. Make up to 1 L with distilled water (*see* Note 4). Store solution in 50- and 25-mL aliquots at –20°C.
13. Use 50- or 25-mL of the above stock solution for making 1 L 2X concentration L3 or MS culture medium, respectively.
14. Use 60 g maltose for making 1 L 2X concentration L3 culture medium.
15. Use 60 g sucrose for making 1 L 2X concentration MS culture medium.
16. 2,4-Dichloroacetic acid (2,4-D): Prepare a stock solution by dissolving 2,4-D in a few drops of 1*M* NaOH and bring to a concentration of 1 mg/mL with distilled water. Store in 4 mL aliquots at 20°C. Add appropriate volumes of 2,4-D stock to culture medium (*see* Section 2.3.).
17. Geneticin disulfate (G418): Prepare a stock solution at a concentration of 10 mg/mL. Sterilize by filtration (22 µm pore size) and store in 10-mL aliquots at –20°C. Use 10 mL stock solution per 1 L 2X concentration L3 or MS selection medium (*see* Note 5 and Section 2.3.).
18. Sigma S1A agarose: Prepare 2X concentrated solution (8 g/L). Sterilize by autoclaving, store at room temperature (*see* Section 2.3.).

2.3. Medium Preparation

Combine appropriate volumes of macrosalts, microsalts, iron, vitamins/inositol, and amino acid stock solutions, add 60 g of sugar and the appropriate hormone or selection agent as specified below, adjust pH to 5.6–5.8 sterilize by filtering through a 22-μm membrane filter and store at 4°C. The medium is ready for mixing 1:1 with molten 2X agarose solution (final agarose concentration 0.4%) and pouring.

1. Induction medium: Make L3 induction medium for inflorescence by combining L3 macrosalts, L3 microsalts, iron solution, L3 vitamins/inositol, L3 amino acids, and maltose, as detailed in Section 2.2., with the addition of 2,4-D at a final concentration of 2 mg/L (L3D2) culture medium. Make embryo induction medium MS by combining MS macrosalts, MS microsalts, iron solution, MS vitamins/inositol, MS amino acids, and sucrose as detailed in Section 2.2., with the addition of 2,4-D at a final concentration of 1 mg/L (MSD1) culture medium.
2. Regeneration medium: Make L3 and MS regeneration media (L3D0 and MSD0) by combining the components listed above for induction medium without addition of hormones (*see* Note 6).
3. Selection medium: Make L3 and MS selection media (L3D0G50 and MSD0G50) by combining the components listed for L3 and MS regeneration media with the addition of G-418 (Geneticin disulfate) at a final concentration of 50 mg/L of culture media.

2.4. Bombardment Solutions

1. Gold particle solution: Gold particles (Bio-Rad, Richmond, CA or Heraeus, Karlsruhe, Germany) ranging in diameter from 0.4 to 1.2 μm are prepared following the protocol supplied with the Bio-Rad PDS 1000 He particle gun. Suspend 40 mg of the dry gold particles in 1 mL of 100% ethanol, and sonicate until particles are clearly dispersed. Centrifuge briefly, remove supernatant, and resuspend in ethanol. Repeat twice. Centrifuge briefly, remove supernatant, and resuspend gold in 1 mL of sterile distilled water. Repeat the process once more and resuspend finally in 1 mL of sterile distilled water. Aliquot in 50-μL volumes, vortexing between aliquots to ensure an equal distribution of particles. The gold particle suspension may be stored at room temperature.
2. Spermidine: Prepare a 0.1*M* spermidine free-base (Sigma, St. Louis, MO) solution in sterile distilled water. Spermidine is supplied as a 0.92 g/mL density solution. Take 15.8 μL of the solution and add sterile distilled water to give a total volume of 1 mL. Store at –80°C in 25-μL aliquots (*see* Note 7).

3. $CaCl_2$ solution: Prepare a 2.5*M* $CaCl_2$ solution. Sterilize by filtration (22 μm) and store at ~4°C or –20°C in 55-μL aliquots.
4. Plasmid DNA: Prepare plasmid DNA using a (Qiagen Inc., Chatsworth, CA) kit (tip 500). Redissolve in sterile TE buffer (10:1, Tris:EDTA), pH 8.0, at a final concentration of 1 μg/μL (*see* Note 8).

2.5. Transgene Expression Assay Solutions

1. GUS-Buffer: The GUS buffer contains 1 m*M* X-Gluc (5-bromo- 4-chloro-3-indolyl-β-D-glucuronide, cyclohexylammonium salt [Sigma or Biosynth AG, Staad, Switzerland]), 100 m*M* sodium phosphate buffer pH 7.0, 0.5 m*M* potassium ferricyanide, 0.5 m*M* potassium ferricyanide, and 0.1% Triton X-100.
2. Stock solution A: 0.5*M* sodium-phosphate buffer pH 7. This solution contains 58 mL of 1*M* Na_2HPO_4, 42 mL of 1*M* NaH_2PO_4, and 100 mL distilled water (check that the pH of the solution is 7). Store at –20°C in 20-mL aliquots.
3. Stock solution B: 50 m*M* potassium ferricyanide. Store at –20°C in 1-mL aliquots.
4. Stock solution C: 50 m*M* potassium ferrocyanide. Store at –20°C in 1-mL aliquots.
5. For preparing 100 mL of GUS-buffer, dissolve 50 mg X-Gluc in 1 mL dimethylformamide or methylcellulose. Add to this solution the following: 20 mL of solution A, 1 mL of both solutions B and C, and 100 μL Triton X-100. Bring the volume to 100 mL with distilled water, vortex, filter sterilize (22 μm), and store at –20°C in 2-mL aliquots.
6. Fluorometric GUS assay solutions: All solutions required for the fluorometric gus assay are prepared according to *(21)*. Details are also given in Chapter 10 of this volume.
7. *nptII* Assay solutions: All solutions required for the *in situ* gel assay are prepared according to *(22)*. Details are also given in Chapter 12 of this volume.

3. Methods

3.1. Sterilization and Dissection of Immature Inflorescences and Embryos

1. Harvest tillers containing inflorescences ranging in length from 0.5 to 1.5 cm by cutting under the last node of the tiller. Make another cut ~6 cm above the node and seal with Nescofilm (*see* Notes 9 and 10).
2. For immature embryos, remove caryopses from spikelets containing embryos of approx 0.7–1.5 mm in length (*see* Note 10).

3. Sterilize both explants for 15–30 min in 10% hypochlorite solution and rinse twice with sterile distilled water.
4. Dissect inflorescences out under sterile conditions and cut them into ~1–2 mm pieces *(15)*.
5. Dissect embryos aseptically under a stereo dissecting microscope and remove the embryonic axis prior to culture.

3.2. Preparation of Explants for Bombardment

1. Place cut inflorescence explants or dissected embryos (with scutellum tissue facing up) in the center of a 9-cm Petri dish containing induction medium (L3D2 or MSD1), covering a circle of ~2–2.5 cm in diameter. Culture them at 26°C in the dark for 1–3 d prior to bombardment (*see* Note 11).
2. Approximately 3–6 inflorescences or 20–25 embryos are used per bombardment. Owing to the known variation between explants, one control plate per bombarded plate should be used. Control plates are bombarded with gold particles without plasmid DNA (*see* Note 12).

3.3. Coating Microcarriers

1. Gold particles prepared as described in Section 2.4. are coated with plasmid DNA following a protocol modified from the original Bio-Rad procedure: Take 50 μL aliquot of gold particles (2 mg particles) and sonicate until deagglomeration occurs and the particles are clearly in suspension.
2. Add 5 μL of plasmid DNA (1 μg/μL) and vortex immediately to assure good contact of the DNA with the particles.
3. Very quickly add 50 μL of 2.5*M* $CaCl_2$ and 20 μL of 0.1*M* spermidine to the particles and vortex for 1 min. Precipitation of the DNA onto the particles will occur during the vortexing.
4. Discard the supernatant after a short centrifugation and wash the particles with 250 μL of 100% ethanol. Redissolve in 85 μL 100% ethanol and use 5 μL of the gold suspension per bombardment (117 μg particles per shot) (*see* Note 13).

3.4. Loading Macrocarriers

1. Place a macrocarrier in a macrocarrier holder. Pipet 5 μL of coated microcarriers onto the center of the macrocarrier and place it inside a sterile Petri dish (out of the clean bench; *see* Note 12) to allow ethanol evaporation. Following the evaporation of the ethanol coated particles will be attached to the macrocarrier.
2. Monitor the degree of agglutination of loaded macrocarriers under an inverted microscope. Those macrocarriers containing more than five clumps of particles should be discarded. Once the ethanol has completely evaporated, place the macrocarrier, with coated microcarriers facing down, on top of the macrocarrier launch assembly.

3.5. DNA Delivery

1. Place target cultures, prepared as explained in Section 3.2., into the chamber.
2. Place a rupture disk (650–1550 psi) into the rupture disk retaining cap (*see* Note 15). Operate the particle gun according to the instructions supplied with gun PDS 1000/He.

3.6. Culture, Selection, and Plant Regeneration

1. After bombardment, distribute treated explants over the surface of the medium (L3D2 or MSD1) in the original dishes and culture for a further 2–3 weeks at 26°C in darkness. This period of culturing should be as long as it is necessary to induce somatic embryogenesis in the explants.
2. After induction, transfer explants to selection medium (L3D0G50 or MSD0G50) and culture under 26°C and 16 h light (*see* Note 16).
3. After 3–4 wk of selection, subculture surviving explants, which form embryogenic structures or shoots, onto fresh selection medium (L3D0G50 or MSD0G50) and culture them for a period of 3–4 wk to allow shoot regeneration.
4. Transfer regenerated plantlets to Magenta boxes containing selection medium (L3D0G50 or MSD0G50) for successive 4-wk passages until it is clear that the regenerated plants continue to grow under selection conditions.
5. Transfer the antibiotic selected plantlets to regeneration medium (L3D0 or MSD0) for further development.
6. Transfer the plantlets to soil, in pots, and grow them to maturity under normal greenhouse conditions. Before transfer to soil, the plantlets should have a good root system. Those plantlets which have not been able to form roots may be immersed in an IBA (1 mg/mL) solution for 1 min and then cultured further in Magenta boxes until roots appear (*see* Note 17).

3.7. Transgene Expression Assays

3.7.1. Histochemical GUS Assay

1. Two days after bombardment, take a sample of approx 5 explants from both bombarded and control plates and immerse them in X-Gluc buffer (50–60 µL per well of a microtiter plate).
2. Assess blue staining after incubation overnight at 37°C plus 1 d at room temperature.
3. The same procedure may be used to assay GUS expression in leaf pieces and other tissues of putatively transformed plants. For GUS assays on green tissues, chlorophyll should be removed after staining by incubating the leaf segments in absolute ethanol. The length of the incubation time

depends on the age of the leaf and it should last until all the chlorophyll has been totally extracted. Explants may be stored in 70% ethanol or in 1*M* Tris-HCl buffer with a higher pH than 7.

3.7.2. Fluorometric GUS Assay

Fluorometric GUS assays are performed in leaf tissue of putative transgenics as described by *(21)* and detailed in Chapter 10 of this volume.

3.7.3. Assay for NPTII Enzyme Activity

NPTII activity is determined after electrophoresis of protein extracts in nondenaturating polyacrylamide gels by the *in situ* gel assay described by *(22)* and Chapter 12 of this volume.

4. Notes

1. Dissolve $CaCl_2$ before mixing with the other components.
2. This solution can be bought ready-made from Sigma. To make up the solution, dissolve each component separately by stirring and heating. Once the components are dissolved, mix together and stir until the color of the solution changes from light green to yellow-brown. Allow to cool before storage at 4°C. This solution does not need sterilization for storage.
3. Dissolve folic acid in KOH before mixing with the other components.
4. Dissolve L-glutamine at pH 9 before mixing with the other components.
5. Pour selection plates when needed and do not use plates more than 1-mo-old.
6. Embryogenic tritordeum and wheat cultures regenerate plantlets in hormone-free medium but the addition of zeatin (cis-trans isomers mixture) at concentrations of 5–10 mg/L increases the number of plants regenerated.
7. Spermidine solution is very hygroscopic and oxidizable and is supplied stored under argon, therefore it is necessary to store it at –80°C once opened. We suggest you aliquot the original spermidine solution in 15.8-μL aliquots. Each aliquot will be used to prepare 1 mL of 0.1*M* solution. Discard aliquots after use.
8. A number of combinations of scorable/selectable marker and promoters have to date been used for cereal transformation *(18)*. In our experiments on tritordeum transformation *(16)* we have mainly used two plasmids, containing the *uidA* or *neo* genes as scorable or selectable markers, respectively. The 7.4 kb plasmid pAct1-DGus contains the *uidA* gene as scorable marker under the control of the actin1-D promoter fragment from rice *(19)*. The plasmid pCal-neo of 4.7 kb contains the *neo* gene as selectable marker gene driven by the CaMV 35S promoter and the Adh1 intron from maize *(20)*. In all experiments, cotransformation with a 1:1 mixture of the two plasmids was carried out, and both gave good levels of expression.

9. Inflorescences are always located above the last node of the tiller, therefore by cutting under the node the part of the tiller harvested will contain the inflorescence. Both the node and the Nescofilm will prevent the penetration of hypochlorite into the inner region of the tiller where the explant is located. Place harvested tillers in water during the preparation procedure to avoid desiccation.
10. In order to use inflorescence or embryo explants for the production of transgenic plants, it is first necessary to determine the regenerative potential of these explants as well as the developmental stages with the highest regeneration capability. These experiments have to be repeated for each new variety or species to be transformed. For cereals, we recommend that you study inflorescences ranging from 0.4 to ~2 cm and embryos from 0.5 to 1.5 mm in length. Younger or older explants, however, may still be able to regenerate plants, but not as efficiently.
11. Preculture of the explants for 1–6 d prior to bombardment increases the number of transgenic plants produced from bombarded tritordeum inflorescences *(16)*.
12. We observe that nonbombarded explants show lower regeneration potential than bombarded nonselected explants. Therefore, it is necessary to use bombarded explants without DNA as controls for bombardment experiments.
13. This protocol provides enough DNA-coated particles for 16 bombardments, but owing to alcohol evaporation a maximum of 10 bombardments usually can be achieved. We recommend that you perform the bombardments as quickly as possible to avoid increasing the amounts of particles in the later shots.
14. Vibrations from the clean bench and exposure to high humidity during the ethanol evaporation will increase the agglutination of particles and will result in a reduction of transformation efficiency.
15. Five different rupture disks have been used for the production of transgenic plants from immature inflorescence tissues. Disks of 650, 900, 1100, 1350, and 1550 psi may be used. It is advisable to test a range of rupture disks when a new species or even new genotypes are being targeted. The distances between the rupture disk, macrocarrier, stopping screen, and target cells can be controlled as desired. The following distances were found to be optimal for the recovery of transgenic plants from bombarded immature inflorescences and embryos: 2.5 cm between rupture disk and macrocarrier, 0.8 cm between macrocarrier are stopping screen, and 5.5 cm between stopping screen and target tissue.
16. The selectable agent is used at a concentration that is known to fully inhibit growth of nontransformed explants.

17. When plantlets are transferred from Magenta boxes to pots, they need to be maintained at high humidity for 1–2 wk to allow them to harden.

References

1. Hiei, Y., Ohta, S., Komari, T., and Kumashiro, T. (1994) Efficient transformation of rice (*Oryza sativa* L) mediated by *Agrobacterium* and sequence analysis of the boundaries of the T-DNA. *Plant J.* **6,** 271–282.
2. Rhodes, C. A., Pierce, D. A., Mettler I. J., Mascarenhas, D., and Detmer, J. J. (1988) Genetically transformed maize plants from protoplasts. *Science* **240,** 204–207.
3. Shimamoto, K., Terada, R., Izawa, T., and Fujimoto, H. (1989) Fertile transgenic rice plants regenerated from transformed protoplasts. *Nature* **338,** 274–276.
4. Sanford, J. C., Klein, T. M., Wolf, E. D., and Allen, N. (1987) Delivery of substances into cells and tissues using a particle bombardment process. *J. Part. Sci. Technol.* **5,** 27–37.
5. Gordon-Kamm, W. J., Spencer, T. M., Mangano, M. L, Adams, T. R., Daines, R. J., Start, W. G., OBrien, J. V., Chambers, S. A., Adams, W. R., Jr., Willets, N. G., Rice, T. B., Mackey, C. J., Krueger, R. W., Kausch, A. P., and Lemaux, P. G. (1990) Transformation of maize cells and regeneration of fertile transgenic plants. *Plant Cell* **2,** 603–618.
6. Cao, J., Duan, X., McElroy, D., and Wu, R. (1992) Regeneration of herbicide resistant transgenic rice plants following microprojectile-mediated transformation of suspension culture cells. *Plant Cell Reps.* **11,** 586–591.
7. Vasil, V., Castillo, A. M., Fromm, M. E., and Vasil, I. K. (1992) Herbicide resistant fertile transgenic wheat plants obtained by microprojectile bombardment of regenerable embryogenic callus. *Bio/Technology* **10,** 667–674.
8. Christou, P., Ford, T. L., and Kofron, M. (1991) Production of transgenic rice (*Oryza sativa* L.) plants from agronomically important indica and japonica varieties via electric discharge particle acceleration of exogenous DNA into immature zygotic embryos. *Bio/Technology* **9,** 957–962.
9. Koziel, M. G., Beland, G. L., Bowman, C., Carozzi, N. B., Crenshaw, R., Crossland, L., Dawson, J., Desai, N., Hill, M., Kadwell, S., Launis, K., Lewis, K., Maddox, D., McPherson, K., Meghji, M. R., Merlin, E., Rhodes, R., Warren, G. W., Wright, M., and Evola, S. V. (1993) Field performance of elite transgenic maize plants expressing an insecticidal protein derived from *Bacillus thuringiensis*. *Bio/Technology* **11,** 194–200.
10. Weeks, T., Anderson, O. D., and Blechl, A. E. (1993) Rapid production of multiple independent lines of fertile transgenic wheat *(Triticum aestivum)*. *Plant Physiol.* **102,** 1077–1084.
11. Vasil, V., Srivastava, V., Castillo, A. M., Fromm, M. E., and Vasil, I. K. (1993) Rapid production of transgenic wheat plants by direct bombardment of cultured immature embryos. *Bio/Technology* **11,** 1553–1558.
12. Becker, D., Brettschneider, R., and Lörz, H. (1994) Fertile transgenic wheat from microprojectile bombardment of scutellar tissue. *Plant J.* **5,** 299–307.
13. Nehra, N. S., Chibbar, R. N., Leung, N., Caswell, K., Mallard, C., Steinhauer, L., Baga, M., and Kartha, K. K. (1994) Self-fertile transgenic wheat plants regener-

ated from isolated scutellar tissues following microprojectile bombardment with two distinct gene constructs. *Plant J.* **5,** 285–297.

14. Wan, Y. and Lemaux, P. G. (1994) Generation of large numbers of independently transformed fertile barley plants. *Plant Physiol.* **104,** 37–48.
15. Barcelo, P., Vazquez, A., and Martin, A. (1989) Somatic embryogenesis and plant regeneration from tritordeum. *Plant Breeding* **103,** 235–240.
16. Barcelo, P., Hagel, C., Becker, D., Martin, A., and Lorz, H. (1994) Transgenic cereal (tritordeum) plants obtained at high efficiency by microprojectile bombardment of inflorescence tissue. *Plant J.* **5,** 583–592.
17. Martin, A. and Sanchez-Monge, E. (1982) Cytology and morphology of the amphiploid *Hordeum chilense* x *Triticum turgidum* conv. *durum. Euphytica* **31,** 261–267.
18. Lazzeri, P. A. and Shewry, P. R. (1993) Biotechnology of cereals, in *Biotechnology and Genetic Engineering Reviews*, vol. 11 (Tombs M. P., ed.), Intercept Ltd., Andover, MD, pp. 79–145.
19. McElroy, D., Zhang, W., Cao, J., and Wu, R. (1990) Isolation of an efficient Actin promoter for use in rice transformation. *Plant Cell* **2,** 163–171.
20. Callis, J., Fromm, M., and Walbot, V. (1987) Introns increase gene expression in cultured maize cells. *Genes Dev.* **1,** 1183–1200.
21. Jefferson, R. A. (1987) Assaying chimeric genes in plants: The GUS gene fusion system. *Plant Mol. Biol. Rep.* **5,** 387–405.
22. Schreier, P. H., Seftot, E. A., Schell, J., and Bohnert, H. J. (1985) The use of nuclear encoded sequences to detect the light regulated synthesis and transport of a foreign protein into plant chloroplast. *EMBO J.* **4,** 25–32.

PART III

Use of Reporter Genes

Chapter 10

The β-Glucuronidase *(gus)* Reporter Gene System

Gene Fusions; Spectrophotometric, Fluorometric, and Histochemical Detection

Gillian A. Hull and Martine Devic

1. Introduction

1.1. The gus *Reporter Gene*

In 1987, Richard Jefferson et al. *(1)* demonstrated the application of a new reporter gene system in transgenic plants. The reporter gene was the *uidA* gene of *Escherichia coli* that encodes the enzyme β-glucuronidase (GUS). Since then, the *uidA* gene (commonly referred to as the *gus* gene) has become one of the most widely used reporter genes in plant molecular biology. (For a detailed description of the *gus* gene, *see* Chapter 1.)

The most frequent use of the *gus* gene is as a reporter gene for promoter analysis, in both transient assays and in stably transformed plants. It has been used to identify promoter elements involved in many aspects of regulation of gene expression, such as tissue specific and developmental regulation *(2,3)*, hormonal regulation *(4–6)*, response to wounding *(7)*, and photoregulation *(8)*.

The *gus* gene has also been used as a reporter gene in promoter trapping studies (*9–11* and Chapter 5 of this volume), as a marker gene for the development of plant transformation procedures *(12,13)*, and to study the mechanism of *Agrobacterium tumefaciens*-mediated transformation *(14)*. For the latter, a modified *gus* gene containing a plant intron inserted in the coding region was used *(15)*. The presence of the intron does not

From: *Methods in Molecular Biology, Vol. 49: Plant Gene Transfer and Expression Protocols*
Edited by: H. Jones Humana Press Inc., Totowa, NJ

affect the level of *gus* gene expression in transformed plant cells, but prevents expression of the *gus* gene in *Agrobacterium*, because of the absence of eukaryotic splicing machinery in prokaryotes. This allows the detection of GUS activity in plant cells during the early stages of transformation, without problems of background owing to expression of the reporter gene in *Agrobacterium*.

The advantages of the *gus* reporter gene system are discussed in detail in the review by Jefferson and Wilson *(16)*, and are briefly summarized herein.

1. β-Glucuronidase catalyzes the hydrolysis of a wide variety of β-glucuronides. Substrates are commercially available that allow GUS activity to be measured easily and quantitatively by spectrophotometric and fluorometric methods.
2. Endogenous β-glucuronidase activity is absent from most higher plants, allowing the detection of low levels of GUS activity in transformed tissues. (Possible sources of background activity are discussed in Note 1.)
3. The tissue-specific localization of GUS activity can be visualized in a histochemical assay using the indigogenic substrate X-gluc (5-bromo-4-chloro-3-indolyl β-D-glucuronic acid).
4. β-Glucuronidase is a stable enzyme that can be assayed in simple buffers over a range of pH values (5.0–9.0).
5. β-Glucuronidase can tolerate large N-terminal and C-terminal fusions. Kavanagh et al. *(17)* reported that an addition of 53 amino acids is tolerated with no significant drop in activity, and that β-glucuronidase fusions are still active (albeit at a reduced level) when a 126 amino acid extension is added. This property has been utilized to study the function of signal peptides by generating GUS fusion proteins targeted to the chloroplast *(17)*, the mitochondria *(18)*, and the endoplasmic reticulum of plant cells *(19)*. Restrepo et al. *(20)* showed that large C-terminal fusions are also tolerated, and targeted *gus* gene expression to the nucleus of plant cells by fusing the coding sequence of a nuclear located protein of the Tobacco Etch Potyvirus to the 3' end of the *gus* gene coding sequence *(20)*. This approach allows precise visualization of cell-specific *gus* gene expression, and was used by Twell *(21)* to show vegetative cell-specific expression in pollen.

1.2. *gus* Gene Fusions

A number of vectors designed for the construction of *gus* gene fusions are described in Jefferson *(22)*, and can be purchased from Clontech (Palo Alto, CA). Among these is the pBI.101 series in which the *gus* gene coding sequence is cloned in the binary plant transformation vector pBin19 *(23)*, flanked by the lac Z polylinker sequence at the 5' end, and the 3' termination sequence of the nopaline synthase *(nos)* gene of *A. tumefaciens* at the 3' end. This allows DNA sequences to be inserted upstream of the *gus* gene, generating gene fusions that can be transferred directly into plants by *Agrobacterium*-mediated transformation.

Two types of *gus* gene fusions can be constructed: transcriptional or translational gene fusions. In transcriptional gene fusions the site of the fusion is located within untranslated DNA sequences. Transcriptional fusions are often generated in promoter analysis experiments, for example, insertion of a promoter fragment, the 3' end of which is located within the transcribed but untranslated region of the gene, upstream of the ATG initiation codon of the *gus* gene. In this case, transcription of the *gus* gene is under the control of the introduced promoter, and translation is initiated at the ATG codon of the *gus* gene. Similarly, the *nos* ter sequence could be excised and fragments of 3' untranslated DNA inserted downstream of the *gus* termination condon.

In translational fusions, the site of the fusion is located within the coding regions of the genes. When constructing translational gene fusions, it is essential that the correct reading frame is conserved across the junction of the two coding regions. There are three plasmids in the pBI.101 series (pBI.101, pBI.101.2, and pBI.101.3) that provide three different reading frames, relative to the cloning sites, for the construction of N-terminal translational fusions.

DNA fragments to be cloned upstream (or downstream) of the *gus* gene can be generated by several methods; by utilization of existing restriction sites, by exonuclease III or Bal 31 digestion, or by polymerase chain reaction (PCR). Once cloned into the *gus* cassette it is important to check the junction of the gene fusion by sequencing. A primer that anneals to the 5' end of the *gus* gene coding region is available for this purpose from Clontech. If PCR has been used, the sequence of the entire amplified fragment should be checked.

In our laboratory, constructs are generated and checked in a series of plasmids derived from pBI.101, pBI.102, pBI.101.3 in which the promoterless *gus* cassette has been transferred to pUC19. The pUC-based constructs can be used directly in transient expression experiments or the promoter::*gus*::*nos* ter fusion can then be excised and transferred to pBin19 for stable transformation experiments. The pUC-based constructs are more convenient to manipulate and sequence than those in pBin19 owing to the large size and low copy number of pBin19.

This chapter includes protocols for quantitative measurement of GUS activity by fluorometry, and for histochemical localization of GUS activity in plant tissues (*see* Note 2). If you encounter specific problems that are not mentioned in the Notes section, or if you require more detailed information on the methods, recommended reading includes the review articles of Jefferson *(24)*, Jefferson and Wilson *(16)*, and the book, *GUS protocols: Using the GUS Gene as a Reporter of Gene Expression (25)*, which collects the valuable experience of several confirmed research groups.

2. Materials

2.1. Fluorometric Assay

1. Plant tissue: Freshly harvested tissue should be assayed immediately, or frozen in liquid nitrogen and stored at –80°C.
2. Grinding of plant tissue: Micropestles (Treff, Polylabo, Paris, France), or prechilled mortar and pestle, or prechilled glass homogenizer (*see* Note 6).
3. Chemicals: 4-methylumbelliferyl β-D glucuronide (4-MUG) (Sigma, St. Louis, MO); 4-methylumbelliferone (4-MU) (Sigma); BioRad protein assay kit (BioRad, Richmond, CA).
4. Equipment: Fluoroskan II and microtiter plates (Labsystems, Helsinki, Finland).
5. Computer software: Deltasoft™ for fluorescence. ELISA analysis for the Macintosh™ with interface for the fluoroskan readers (BioMetallics, Inc., Princeton, NJ).
6. Spectrophotometer.
7. Disposable 1.6-mL cuvets (Polylabo, France).
8. GUS extraction buffer: 50 m*M* Tris-HCl, pH 7.0, 0.1% Triton X100, 10 m*M* EDTA, 3 m*M* dithiothreitol. Aliquot and store at –20°C.
9. Substrate: 2 m*M* 4-MUG made up in GUS extraction buffer. Aliquot and store in the dark at -20°C.
10. Standards: 10 pmol/μL, 100-pmol/μL, and 1 nmol/μL 4-MU made up in GUS extraction buffer. Store in the dark at –20°C.
11. Bovine serum albumin: (Bio-Rad) 1 mg/mL made up in distilled water. Aliquot and store at –20°C.

2.2. Histochemical Detection

1. X-Gluc: 5-bromo-4-chloro-3-indolyl-β-D-glucuronic acid (X-Gluc) is purchased as cyclohexylammonium salt from Biosynth AG (Skokie, IL). The powder is stored dry at –20°C. To prepare the solution, dissolve 5–10 mg X-Gluc (1–2 m*M*) in 100 μL dimethylformamide under a fume hood. Add

to 10 mL of 100 m*M* sodium-phosphate buffer, pH 7.0. Wrap the tube in aluminum foil to protect from light and store at 4°C for 1 wk. However, it is recommended to prepare fresh X-Gluc solution for each assay.
2. Glutaraldehyde (Fluka).
3. 3-Aminopropyltriethoxy-silane (Tespa) (Sigma).
4. Oxidation catalyst solutions: Stock I: 165 mg potassium ferricyanide, $K_3Fe(CN)_6$ (Sigma) in 10 mL H_2O (50 m*M*). Stock II: 212 mg potassium ferrocyanide, $K_4Fe(CN)_6.3\ H_2O$ (Sigma) in 10 mL H_2O (50 m*M*). Add 100 μL of each stock solution to 10 mL X-Gluc solution for a final concentration of 0.5 m*M*. The stock solutions can be stored for several weeks at 4°C.
5. 70% ethanol/H_2O.
6. Fixative solutions: The solutions are freshly prepared. Prefixation: 1% Formaldehyde in 200 m*M* sodium-phosphate buffer pH 7.0. Solution A: 50 mL ethanol, 5 mL acetic acid, 10 mL 37% formaldehyde, and 35 mL water. Solution B: 4% Glutaraldehyde (stock solution: 25% glutaraldehyde, stored at 4°C) in 200 m*M* sodium-phosphate buffer pH 6.8. Wear gloves and work under a fume hood. Sodium-phosphate buffer 200 m*M*, pH 6.8 for washing.
7. Coating of microscopy slides: 2% TESPA (3-aminopropyltriethoxy-silane, stock solution stored at 4°C) in acetone. Wear gloves and work under a fume hood.

3. Methods

3.1. Fluorometric Assay

The protocol given herein has been used to measure β-glucuronidase (GUS) activity using the substrate 4-methylumbelliferyl β-D glucuronide (4-MUG) in transgenic tobacco and *Arabidopsis thaliana*. β-Glucuronidase catalyzes hydrolysis of the substrate 4-MUG to give the products D-glucuronic acid and 4-methylumbelliferone (4-MU). The latter can be assayed fluorometrically using excitation and emission wavelengths of 365 and 455 nm, respectively. The rate of accumulation of 4-MU is linearly related to the concentration of GUS enzyme and can thus be used as a measure of GUS activity *(16)*.

The protocol given herein is for assaying GUS activity in seeds of transgenic *Arabidopsis*; however, it has been adapted for use with a variety of tissues, including seeds, siliques, leaf, stem, root, and floral tissues from transgenic tobacco and *Arabidopsis* generated for promoter analysis and promoter trapping studies.

The protocol is designed for use with a Fluoroskan II fluorometer, connected to a Macintosh™ computer, which we use in conjunction with the software system Deltasoft (*see* Note 3). (For detailed studies of the sensitivity of β-glucuronidase assays using the Fluoroskan II system, *see* refs. 26 and 27.) Assays are performed in microtiter plates incubated at 37°C in the Fluoroskan II, which is programmed to take readings at regular intervals. This permits continuous kinetic analysis of large numbers of samples (up to 96 per microtiter plate). An alternative protocol for discontinuous (endpoint) measurement of fluorescence, in which the fluorescence value of aliquots of a reaction mixture is measured at regular intervals of time, is also provided.

The kinetic method has several advantages over endpoint assays. An obvious advantage is that it is much less time consuming; once the components of the reaction have been mixed in the microtiter plate and the Fluoroskan II has been programmed, the reaction can be left unattended. Furthermore, errors owing to imprecise timing of the time points (e.g., when large numbers of samples are being assayed simultaneously), and varying background levels between wells in the microtiter plate are avoided.

The Deltasoft program calculates the rate of accumulation of fluorescent product in fluorescence units per minute (fl.u/min). The rates are then converted into pmol 4-MU/min using a calibration curve of fl.u against concentration 4-MU (*see* Note 4). The protein content of each sample is then determined and the rates are standardized and expressed as pmol 4-MU/μg protein/min (*see* Note 5).

3.1.1. Kinetic Fluorometric Assay

1. Aliquot approx 2–5 mg (50–200 seeds) of *Arabidropsis* seeds into a 1.5-mL Eppendorf tube and grind thoroughly in a 150 μL GUS extraction buffer using a motorized micropestle. Remember to assay nontransformed seeds as a negative control. Seeds transformed with the *gus* gene under the control of a constitutive promoter can be used as a positive control if required. After grinding, add 150 μL of GUS extraction buffer and mix by vortexing. If adapting this method to other tissues, refer to Note 6.
2. Centrifuge the homogenate in an Eppendorf centrifuge, at maximum speed, for 5 min, then place on ice.
3. Switch on the Fluoroskan II, as it requires at least 30 min to warm up.
4. Aliquot 100 μL of substrate solution into each well. Prepare sufficient wells to assay each sample in duplicate (*see* Note 7).
5. Add 95 μL of GUS extraction buffer to each well, followed by 5 μL of extract supernatant (final concentration of 4-MUG 1 m*M*). Mix by pipeting. In the blank control wells, add 100 μL of GUS extraction buffer only

(*see* Note 8). In the calibration wells, add 100 μL GUS extraction buffer containing 100 pmol, 500 pmol, 1 nmol, 5 nmol, and 10 nmol of 4-MU. The final volume in all the wells is 200 μL. The amount of extract used in the assay will depend on the level of GUS activity in the sample being tested, and is determined empirically. The Deltasoft program displays kinetic plots of fl.u/min for each well. The amount of extract used should be such that a linear response is obtained between at least four successive readings. The fl.u scale can be varied to accommodate high- and low-level GUS activities. The relationship between fl.u and concentration of 4-MU is linear up to approx 2500 fl.u, which corresponds to a 200 μ*M* solution of 4-MU (40 nmol in 200 μL). The lower limit of detection of the Fluoroskan II is 200 n*M* 4-MU (40 pmol in 200 μL).

6. Set the temperature of the plate reader to 37°C and place the microtiter plate in the Fluoroskan II. Measure the absolute fluorescence values of the calibration wells using the *Read Endpoint* option. These values will be used to plot a calibration curve of fl.u against concentration 4-MU. Then program the Fluoroskan II to measure the rate of accumulation of 4-MU using the *Read Kinetic* option. When expression is relatively high, the plate can be incubated in the Fluoroskan II for periods of up to 3 h, with readings taken every 2–15 min, depending on the length of the incubation. When incubations of more than 3 h are required, it is advisable to incubate the plate (sealed to limit evaporation) in a 37°C incubator and to measure the absolute fluorescence values at regular intervals. Using the 4-MU calibration curve, convert the rate of GUS activity from fluorescence units per minute into pmol 4-MU per minute.

3.1.2. Endpoint Fluorometric Assay

1. In an Eppendorf tube, mix 100 μL of sample extract with 400 μL of 1 m*M* 4-MUG made up in GUS extraction buffer.
2. Place the Eppendorf tube at 37°C. After approx 2 min, remove a 100-μL aliquot of the reaction mix and add this to an equal volume of 0.2*M* Na_2CO_3 (stopping reagent). This timepoint is $T = 0$. Repeat the above step four more times at regular intervals. The frequency of the timepoints depends on the level of activity of the sample. As a first assay try 10, 20, 40, and 60 min. If the readings in the fluorometer are off scale take readings at closer time intervals and/or use less extract in the reaction. If the level of expression is low, incubate longer and/or increase the amount of extract in the reaction.
3. Measure the fluorescence of the stopped reactions, plot against time, and calculate the rate. *Note:* The highly alkaline stopping reagent results in significant amplification of the fluorescence *(16)*.

3.1.3. Protein Assay

1. First prepare a calibration curve using bovine serum albumin (BSA) standards. In 1.6-mL disposable cuvets aliquot 1, 2, 4, 6, 8, 10, and 12 µL of BSA from an aqueous solution of 1 mg/mL. Add 5 µL of GUS extraction buffer and then make up the volume to 800 µL with distilled water. (Some constituents of the GUS extraction buffer may absorb light at 595 nm, and/or increase background by reacting with the reagent; therefore a volume of GUS extraction buffer equal to the volume of extract to be assayed is added to the BSA standards.) Add 200 µL of Bradford reagent, mix by inversion, and leave at room temperature for 5–60 min. Measure the absorbance at 595 nm and plot absorbance against amount of protein (µg).
2. At the same time, place 5 µL of extract in a 1.6-mL disposable cuvet. Make the volume up to 800 µL with distilled water and add 200 µL of Bradford reagent. Mix by inversion, and leave at room temperature for 5–60 min. Measure the absorbance at 595 nm and determine the protein concentration by interpolation from the calibration curve. If the absorbance falls outside the calibration curve, adjust the amount or concentration of extract assayed, making sure that the volume of GUS extraction buffer in the BSA standards is equal to the volume of extract assayed.
3. Standardize samples by expressing GUS activities as pmol 4-MU/µg protein/min (*see* Note 9).

3.2. Histochemical Detection

Our experience in histochemical detection of GUS activity comes from T-DNA tagging studies in which we are looking for genes that are active during embryogenesis. The T-DNA construction used in these studies contains the open reading frame of the GUS reporter gene with its initiation codon orientated toward the left or right border, so that its expression might be driven by endogenous plant promoters *(10)*. In general, the levels of GUS activity resulting from the transcription and translation of this chimeric mRNA are low. As a result, the following protocols are adapted for the detection of low levels of GUS activity in seeds. Alternative methods of visualization of GUS activity are discussed in Notes 10 and 11.

The chemical reaction that allows the histochemical detection of GUS activity occurs in two steps, only the first of which requires the β-glucuronidase enzymatic activity. β-Glucuronidase catalyzes the hydrolysis of the substrate X-Gluc and liberates a molecule of 5-bromo-4-chloro-indoxyl (XH). This form of indoxyl is in equilibrium with its tautomer, 5-bromo-4-chloro-y-indoxyl (X). Neither form is colored. In the presence of oxygen, the soluble indoxyls X and XH are dimerized to form insoluble diX-indigo and diXH-leucoindigo, respectively, resulting in the blue coloration (*see* Note 12).

3.2.1. GUS Localization on Whole Samples

3.2.1.1. Staining Procedure

Fresh plant tissues are immersed in X-Gluc solution and vacuum infiltrated for 5 min to improve the penetration of the substrate. The vacuum should be applied and released slowly in a desiccator. The samples are incubated in the dark at 37°C from 1 h to overnight or until the staining is satisfactory. We tried replacing the phosphate buffer with a corresponding Tris buffer at the same pH, but the results were not as good. Pigments, mainly chlorophylls and carotenoids, are removed by several washes of 70% ethanol until the nonstained tissues are white.

The formation of the diX-indigo precipitates is slow, and diffusion of the soluble indoxyls into neighboring cells before dimerization occurs is possible. Therefore, cells not expressing β-glucuronidase might also exhibit a blue coloration. The addition of potassium ferricyanide and ferrocyanide, at a final concentration of 0.1–0.5 m*M*, to the X-Gluc solution prevents the diffusion of the indoxyls by accelerating the dimerization. This results in a more accurate localization of GUS activity.

In preliminary experiments, it is preferable to omit potassium ferricyanide and ferrocyanide from the reaction in order to get an idea of the intensity of staining. When the GUS reporter gene is under the control of a strong promoter, the staining can "leak" into neighboring cells and the addition of potassium ferricyanide and ferrocyanide is necessary to achieve a true localization of the sites of β-glucuronidase activity. However, potassium ferricyanide and ferrocyanide also have an inhibitory effect on the first reaction, i.e., the enzymatic catalysis of the hydrolysis of X-Gluc. In the case of weak staining, the addition of the oxidation catalysts may completely abolish the reaction.

Whole seedlings, leaves, roots, stems, or flowers can be stained following the above protocol without pretreatment (*see* Notes 13 and 14). For the staining of seeds, it is necessary to dissect the fruit. Open the silique along the raplum to expose the seeds and immerse the opened silique completely in the X-Gluc solution. Our experience with *Arabidopsis* shows that, before the maturation stage, the substrate can easily penetrate into the seed and the embryo. After incubation, remove the seeds from the silique for observation of the localization of GUS activity by light microscopy. The embryos can be excised by gently squeezing the seeds between a slide and a coverslip. During the desiccation stage, the seed coat becomes impermeable. It is therefore necessary to puncture the seed with a needle or better, to excise the embryo as described earlier before staining to ensure the penetration of the substrate.

3.2.1.2. Photography

The samples can be photographed using a camera mounted on a binocular or a light microscope. For large samples, we use a WILD binocular microscope with a zoom lens ranging from 6.3X to 32X. Alternatively, a 35-mm camera on a stable stand with macro lenses or extension tubes can be used, preferably in conjunction with optic fibers. The type of films used, daylight or tungsten, depends on the light source. However, in our hands, tungsten films, even with adequate lighting, always have a tendency to a give a uniform blue background that is incompatible with good localization of the indigo dye. We recommend halogen lamps with a daylight film of 64 ASA.

Often, weakly staining samples are better visualized when photographed on a black background. One simple solution is to place the object on a piece of black paper; however, when illuminated, the black background will have a brownish tint. It is preferable to place a glass plate (e.g., a microscopy slide) on a cylinder of approx 4 cm diameter and 5 cm high with a side illumination. A perfect black background will be produced by the shadow inside the cylinder.

Side illumination of a wet object can result in a shiny appearance, sometimes with undesirable reflection of the light. To avoid this effect, the sample can be mounted between a slide and coverslip that are held slightly apart by blue-tac. The space between the slide and the coverslip is then filled with water.

For the photography of small samples, e.g., immature seeds and young embryos of *Arabidopsis*, or for a more detailed observation of the pattern of GUS expression in a tissue, we use a Zeiss light microscope equipped with bright-field and dark-field optics. The maximum magnification that can reasonably be used on a nonsectioned sample is 20X. This is owing to the fact that the depth of field of a microscope is extremely reduced so that the picture can be focused in only one plane. In our hands, the best pictures are obtained with a 64 ASA tungsten film, without a blue filter.

Staining of whole sample gives valuable and rapid information of organ specificity, and some indication of tissue specificity. To achieve precise tissue- and cell-specific localization of β-glucuronidase activity, thin sections of the sample are required.

3.2.2. GUS Localization on Sections

3.2.2.1. Sections Sliced by Hand

Fresh tissues (e.g., stem) can be sliced directly by hand using a razor blade. If necessary, the tissue can be held between two pieces of polystyrene during sectioning. When the material cannot be sliced manually, it requires embedding into a support: paraffin or resin. To maintain the tissue structure during all the embedding steps, the sample must be fixed.

3.2.2.2. Fixation

To give a protocol for fixation, embedding and sectioning that would be suitable for all plant tissues is impossible. We propose two different fixation solutions—Solution A, for most tissues (to be tried first); and Solution B, as a starting protocol for more recalcitrant tissues such as mature seeds. If you already have experience with your particular tissue, you should try to adapt your protocol to the principal requirement of retaining GUS activity as will be mentioned later.

You are facing a dilemma: On the one hand, good fixation is required to maintain a good structure of the sample, and, on the other hand, some components of the fixative solutions, especially the glutaraldehyde, have an inhibitory effect on β-glucuronidase activity. Preliminary results of staining the whole sample as described earlier, will indicate which aspect you should optimize, i.e., fixation vs enzyme activity. In general, it is better to stain before the fixation. De Block and Debrouwer *(28)* have developed an artifact-free *in situ* β-glucuronidase localization that avoids fixation.

If the staining is strong, it can be performed on material prefixed in 1% formaldehyde in 200 m*M* Na-phosphate buffer, pH 7.0, for up to 30 min (vacuum infiltration is recommended). The sample is then incubated at 37°C in the X-Gluc solution for no longer than the time required for suitable coloration. Finally, the sample is fixed in either Solution A at room temperature or Solution B at 4°C for 4 h and rinsed twice with Na-phosphate buffer. To allow a better penetration of the fixatives, the object can be cut into small pieces of 5 mm in length. If the staining is weak, the prefixation step should be omitted. The structure will not be as well preserved, but it should give enough information to determine the pattern of GUS expression. The sample is now ready for embedding.

3.2.2.3. Embedding and Sectioning

The choice of the medium for embedding depends entirely on the equipment for sectioning at your disposal and your personal experience. Good results have been obtained with paraffin and with synthetic resins such as LRWhite, Historesin, and methacrylate.

3.2.2.4. Cryostat Sections

In this case, plant tissues are sectioned first and stained afterward, as the presence of crystals of the indigo dye in the frozen sample will tear the tissue during sectioning. No fixation is required before sectioning, so there is full retention of GUS activity in the sample. Also, the problem of substrate penetration is alleviated. It is an ideal method for enzyme histochemistry. However, the drawback of this method is that it is difficult to obtain good sections of less than 5 μm, and to achieve the same quality sections as with a sample embedded in resin. Also, it is not easy to orientate the samples in the embedding solution prior to sectioning.

1. Prechill the cryostat chamber to between –20°C and –35°C.
2. Deposit a drop of embedding solution (Reichert-Jung) onto the sample holder and place small pieces of fresh or frozen plant tissue in the droplet.
3. Transfer the sample holder into the cryostat chamber and allow it to cool down.
4. Sections of 5–10 μm are recuperated directly onto a precoated microscopy slide (*see* Section 3.2.2.4.).
5. When dry, stain the sections with X-Gluc in a humidity chamber for 30 min to 12 h at 37°C.
6. The sections are now ready for observation and can be mounted permanently with EUKITT (O. Kindler Gmbh and Co., Polylabo, Paris, France) after dehydration in ethanol.

3.2.2.5. Coating of Microscopy Slides

This treatment makes the slides sticky so that the sections will adhere strongly.

1. Arrange the slides in a slide holder and immerse in 2% TESPA in acetone for 3 min.
2. Wash 2X in acetone and then 2X in water.
3. Air dry under a cover of filter paper to prevent dust deposition.

4. Notes

1. β-Glucuronidase generally is absent in most plant species. However, several reports have mentioned endogenous GUS activity in nontransformed plants, notably in the male gametophyte *(16,29)*. In most cases, this intrinsic GUS activity can be eliminated without reducing the introduced GUS activity when the pH is elevated to 9 *(30)*. Alternatively, the addition of methanol in the X-Gluc solution is reported to suppress endogenous GUS activity *(31)*, however, this has not been tested in our laboratory. Alwen et al. *(32)* have described an endogenous β-glucuronidase activity that is apparently widespread in the plant kingdom and that has an optimum activity at pH 5.0. The fluorometric and histochemical assays described here are all carried out at neutral pH, which is the optimum pH for bacterial β-glucuronidase, and we have not been troubled by endogenous activity. However, a sample of tissue of nontransgenic plants should always be assayed as a control of background activity.

   GUS activity found in nontransgenic plants can also result from endophytic bacteria or fungi. For histochemical assays, Jefferson and Wilson *(16)* recommend the addition of sodium azide at 0.02% or chloramphenicol at 100 μg/mL to the X-Gluc solution to prevent proliferation of these organisms during long incubations.
2. GUS activity can also be assayed using the substrate *p*-nitrophenyl β-D-glucuronide (PNPG). The product of this reaction, *p*-nitrophenyl, is measured spectrophotometrically at A_{605}. Although less sensitive than the fluorometric assay, the spectrophotometric assay is simple and cheap *(16)*. For detailed protocols for this assay, *see* refs. *16* and *26*.
3. The Deltasoft software program provides two options for reading the microtiter plates: *Read Endpoint* and *Read Kinetic. Read Endpoint* performs a single reading of the microtiter plate, and measures the absolute fluorescence in each well. Using *Read Kinetic* the microtiter plate is incubated at 37°C in the Fluoroskan II, and readings are taken at preselected intervals over a chosen period of time. A kinetic plot is displayed for each well and after the final reading has been taken a curve is fitted by linear regression, and the rate is calculated.

   The kinetic plots can be edited, i.e., aberrant points can be masked and the scale of the axes can be changed. In this way, specific regions of the curve can be selected for analysis. Alternatively, the *Maximum Rate* option can be used that calculates the maximum rate between a specified number of points.

   A template file can be created for each plate file (raw data) by entering the position of blanks, standards, and unknown samples. The template file and the plate file are superimposed and the resulting template + plate file is used for data analysis.

A choice of five different curve fitting options are available to fit a curve to the standard concentration vs fluorescence values. Deltasoft can then directly interpolate the concentration of unknown samples from the standard curve. Finally, a statistics option is also available for analysis of results.

Note: There are faults inherent in the Deltasoft software that have apparently been rectified in a new updated version that is now available; however, we have not yet tested this software in our laboratory.

4. In our experiments, fluorescence units are calibrated against pmol 4-MU. For each batch of GUS assays a calibration plate was prepared containing 4-MU standards ranging from 100 pmol/200 µL to 10 nmol/200 µL (eight wells for each concentration). A calibration curve of the mean values was automatically plotted using the Deltasoft *Curve Fit* option. 4-MU Standards were also included on each sample plate and were found to fall within the values of the calibration curve. It is also possible to calibrate the Fluoroskan II with standard dilutions of purified GUS enzyme *(26)*. In this case, Deltasoft can directly convert the rates of activity from fl.u/min into U GUS/min.
5. GUS Activity can alternatively be standardized to DNA content, fresh weight of tissue, and, in some cases, cell number. *See* the review by Jefferson and Wilson *(16)* for a discussion of the merits and drawbacks of the various methods.
6. Most plant tissues can be homogenized in Eppendorf tubes using a micropestle attached to a motor; this allows rapid grinding of large numbers of samples. Large tissue samples (such as whole tobacco leaves or tobacco flowers), or tissues that are difficult to homogenize (such as roots), are homogenized using prechilled mortar and pestles. Tobacco microspores/pollen grains are homogenized in a prechilled glass homogenizer. The volume of buffer depends on the amount of tissue and the method of grinding. Generally, use approx 400 µL of buffer if grinding in a mortar and pestle or glass homogenizer, and 100–200 µL of buffer if using an Eppendorf tube (volumes greater than this will overflow from the Eppendorf tube during grinding). The extract supernatant can be diluted afterwards if necessary.
7. Each GUS assay is performed in duplicate and the mean and variation of replica samples are automatically calculated using the Deltasoft statistics program. Protein assays are also performed in duplicate, and the mean value is used to standardize the GUS activities.
8. Blank controls consisting of substrate and GUS extraction buffer only are also included to provide a measure of spontaneous breakdown of the substrate during the incubation.

9. Breyne et al. *(26)* report that a high concentration of protein (greater than 50 μg in 250 μL) in the assay may inhibit GUS activity. In the protocol given here for *Arabidopsis* seeds, 5 μL of extract generally contains between 1–30 μg of protein.
10. Antibodies directed against β-glucuronidase are commercially available. We tested one of them (Clontech, Eugene, OR) without success. GUS Antibodies from two sources (Clontech and Molecular Probes) have been tested independently by Jean Claude Caissard *(33)* using ELISA assay, Western blot, and immunolocalization. These preparations were found to be impure, i.e., contaminated with nonspecific antibodies that recognize proteins other than GUS.
11. Caissard et al. *(34)* have developed a method for ultrastructural detection of GUS activity. The diX-indigo precipitates are preserved during preparation of samples for electron microscopy, however, electron dense precipitations not owing to indigo crystals can also occur. Therefore, the composition of the microcrystals should be verified by means of X-ray microanalysis.
12. After years of "blues," let us see "la vie en rose"! Biosynth AG, in collaboration with Richard Jefferson, has developed two new substrates: magenta-beta-D-glcA and salmon-beta-D-glcA. They can improve the detection of GUS in cases of difficult backgrounds. However, X-Gluc: (glcA) is still recommended for routine staining procedures.
13. Owing to the poor penetration of X-Gluc in some samples, the staining is not always reproducible. In case of problems, it is possible to add nonionic detergents to the X-Gluc solution (e.g., 0.05% Triton X100 or 0.1% Tween 20) to decrease surface tension. By analyzing many samples, even in difficult conditions of staining, a good idea of the general pattern of GUS expression in the tissue should be obtained.
14. A nondestructive assay for *Arabidopsis* seedlings has been described by Martin et al. *(35)*. The seedlings are transferred to a liquid medium containing 1 m*M* X-Gluc and incubated at 22°C. The plants are rescued by transfer to solid medium.

References

1. Jefferson, R. A., Kavanagh, T. A., and Bevan M. W. (1987) GUS fusions: β-glucuronidase as a sensitive and versatile gene fusion marker in higher plants. *EMBO J.* **6(13),** 3901–3907.
2. Twell, D., Yamaguchi, J., and McCormick, S. (1990) Pollen-specific gene expression in transgenic plants: coordinate regulation of two different tomato gene promoters during microsporogenesis. *Development* **109,** 705–713.
3. Dietrich, R. A., Radke, S. E., and Harada, J. J. (1992) Downstream DNA sequences are required to activate a gene expressed in the root cortex of embryos and seedlings. *Plant Cell* **4(11),** 1371–1382.

4. Marcotte, W. R., Russell, S. M., and Quatrano, R. S. (1989) Abscisic acid responsive sequences from the Em gene of wheat. *Plant Cell* **1,** 969–976.
5. Rogers, J. C. and Rogers, S. W. (1992) Definition and functional implications of gibberellin and abscisic acid *cis*-acting hormone response complexes. *Plant Cell* **4(11),** 1443–1451.
6. Li, Y., Hagen, G., and Guilfoyle, T. J. (1991) An auxin-responsive promoter is differentially induced by auxin gradient during tropisms. *Plant Cell* **3(11),** 1167–1175.
7. Lorberth, R., Damman, C., Ebneth, M., Amati, S., and Sanchez-Serrano, J. J. (1992) Promoter elements involved in environmental and developmental control of potato proteinase inhibitor II expression. *Plant J.* **2(4),** 477–486.
8. Luan, S. and Bogorad, L. (1992) A rice *cab* gene promoter contains separate *cis*-acting elements that regulate expression in dicot and monocot plants. *Plant Cell* **4(8),** 971–981.
9. Kertbundit, S., DeGreve, H., Debroeck, F., Van Montagu, M., and Hernalsteens, J. P. (1991) *In vivo* random β-glucuronidase gene fusions in *Arabidopsis thaliana. Proc. Natl. Acad. Sci. USA* **88,** 5212–5216.
10. Topping, J. F., Wei, W., and Lindsey, K. (1991) Functional tagging of regulatory elements in the plant genome. *Development* **112,** 1009–1019.
11. Lindsey, K., Wei, W., Clarke, M. C., McArdle, H., Rooke, L. M., and Topping, J. F. (1993) Tagging genomic sequences that direct transgene expression by activation of a promoter trap in plants. *Transgenic Res.* **2,** 33–47.
12. Gordon-Kamm, W. J., Spencer, M. T., Mangano, M. L., Adams, T. R., Daines, R. J., Start, W. G., O'Brien, J. V., Chambers, S. A., Adams, W. R., Willets, N. G., Rice, T. B., Mackey, C. J., Krueger, R. W., Krausch, A. P., and Lemaux, P. G. (1990) Transformation of maize cells and regeneration of transgenic plants. *Plant Cell* **2(7),** 603–618.
13. Omirulleh, S., Abraham, M., Golovkin, M., Stefanov, I., Karabaev, M. K., Mustardy, L., Morocz, S., and Dudits, D. (1993) Activity of a chimeric promoter with the doubled CaMV 35S enhancer element in protoplast-derived cells and transgenic plants in maize. *Plant Mol. Biol.* **21,** 415–428.
14. Mozo, T. and Hooykaas, P. J. J. (1992) Factors affecting the rate of T-DNA transfer from *Agrobacterium tumefaciens* to *Nicotiana glauca* plant cells. *Plant. Mol. Biol.* **19(6),** 1019–1030.
15. Vancanneyt, G., Schmidt, R., O'Conner-Sanchez, A., Willmitzer L., and Rocha-Sosa, M. (1990) Construction of an intron-containing marker gene: splicing of the intron in transgenic plants and its use in monitoring early events in *Agrobacterium* mediated plant transformation. *Mol. Gen. Genet.* **220,** 245–250.
16. Jefferson, R. A. and Wilson, K. J. (1991) The GUS gene fusion system. *Plant Mol. Biol. Man.* **B14,** 1–33.
17. Kavanagh, T. A., Jefferson, R. A., and Bevan, M. W. (1988) Targeting a foreign protein to chloroplasts using fusions to the transit peptide of a chlorophyll a/b protein. *Mol. Gen. Genet.* **215,** 38–45.
18. Schmitz, U. K and Lonsdale, D. M. (1989) A yeast mitochondrial presequence functions as a signal for targeting to plant mitochondria *in vivo. Plant Cell* **1(8),** 783–791.

19. Iturriaga, G., Jefferson, R. A., and Bevan, M. W. (1989) Endoplasmic reticulum targeting and glycosylation of hybrid proteins in transgenic tobacco. *Plant Cell* **1(3),** 381–390.
20. Restrepo, M. A., Freed, D. D., and Carrington, J. C. (1990) Nuclear transport of plant potyviral proteins. *Plant Cell* **2(10),** 987–998.
21. Twell, D. (1992) Use of nuclear-targeted β-glucuronidase fusion protein to demonstrate vegetative cell-specific gene expression in developing pollen. *Plant J.* **2,** 887–892.
22. Jefferson, R. A. (1988) Plant reporter genes: the *gus* gene fusion system, in *Gen. Eng.* **10,** 247–263 (Setlow, J. K., ed).
23. Bevan, M. W. (1984) Binary *Agrobacterium* vectors for plant transformation. *Nucleic Acids Res.* **12,** 8711–8721.
24. Jefferson, R. A. (1987) Assaying chimeric genes in plants: the GUS gene fusion system. *Plant Mol. Biol. Rep.* **5,** 387–405.
25. GUS protocols: Using the GUS Gene as a Reporter of Gene Expression (1992) (Gallagher, S. R., ed.), Academic, San Diego, CA, pp. 1–221.
26. Breyne, P., Loose, De M., Dedonder, A., Van Montagu, M., and Depicker, A. (1993) Quantitative kinetic analysis of β-glucuronidase activities using a computer-directed microtitre plate reader. *Plant Mol. Biol. Rep.* **11(1),** 21–31.
27. Rao Gururaj, A. and Flynn, P. (1990) A quantitative assay for β-D-glucuronidas (GUS) using microtitre plates. *BioFeedback* **8(1),** 38–41.
28. DeBlock, M. and Debrouwer, D. (1992) *In situ* enzyme histochemistry on plastic-embedded plant material. The development of an artefact-free beta-glucuroridase assay. *Plant J.* **2,** 261–266.
29. Plegt, L. and Bino, R. J. (1989) β-glucuronidase activity during development of the male gametophyte from transgenic and nontransgenic plants. *Mol. Gen. Genet.* **216,** 321–327.
30. Hodal, L., Bochardt, A., Nielsen, J. E., Mattsson, O., and Okkels, F. T. (1992) Detection, expression, and specific elimination of endogenous beta-glucuronidase activity in transgenic and non-transgenic plants. *Plant Sci.* **87,** 115–122.
31. Kosugi, S., Ohashi, Y., Nakajiama, K., and Arai, Y. (1990) An improved assay for beta-glucuronidase in transformed cells: methanol almost completely suppresses a putative endogenous beta-glucuronidase activity. *Plant Sci.* **70,** 133–140.
32. Alwen, A., Moreno Benito, R. M., Vicente, O., and Heberle-Bors, E. (1992) Plant endogenous β-glucuronidase activity: how to avoid interference with the use of the *E. coli* β-glucuronidase as a reporter gene in transgenic plants. *Transgen. Res.* **1,** 63–70.
33. Caissard, J. C. (1993) Reperage ultrastructural de la beta-glucuronidase sur des plantes transgeniques comportant le gene GUS: methodologies et études de cas; Thèse de doctorat, Université Pierre et Marie Curie, Paris VI.
34. Caissard, J. C., Rembur, J., and Chriqui, D. (1992) Electron microscopy and X-ray microanalysis as tools for fine localisation of the beta glucuronidase activity in transgenic plants harbouring the GUS reporter gene. *Protoplasma* **70,** 68–76.
35. Martin, T., Schmidt, R., Altmann, T., and Frommer, W. (1992) Nondestructive assay systems for detection of beta-glucuronidase activity in higher plants. *Plant Mol. Biol. Rep.* **10,** 37–46.

CHAPTER 11

Chloramphenicol Acetyl Transferase Assay

Michael R. Davey, Nigel W. Blackhall, and J. Brian Power

1. Introduction

The chloramphenicol acetyltransferase *(cat)* gene, isolated from transposon Tn9 of *Escherichia coli (1)*, is a convenient genetic marker for studies of transformation. (For a detailed description of the CAT gene, *see* Chapter 1.) In plants, the CAT system is generally used in transient, rather than in stable, gene expression studies. Gene products are detected by assaying crude protein extracts of putatively transformed cells, usually within a few hours of transformation.

The procedure normally employed utilizes the radiolabelled substrates [^{14}C]-acetylcoenzyme A or [^{14}C]-chloramphenicol. CAT inactivates chloramphenicol by acetylation, to produce three forms of radiolabeled acetylated chloramphenicol [1-acetyl chloramphenicol (1-AC), 3-acetyl chloramphenicol (3-AC), and 1,3-diacetyl chloramphenicol (1,3-AC)]. These products can be separated by thin layer chromatography and detected by autoradiography (Fig. 1). Recently, CAT assay kits, which utilize nonradioactive substrates, have become available commercially. An example is the FAST CAT kit (available from Cambridge Bioscience, Cambridge, UK), which contains chloramphenicol conjugated to the fluorescent molecule BODIPY. This procedure has been used to measure CAT activity in crude cellular extracts of transfected ovarian granulosa cells *(2)*. The K_m and V_{max} values for purified CAT enzyme are similar when employing the same concentration of enzyme

From: *Methods in Molecular Biology, Vol. 49: Plant Gene Transfer and Expression Protocols*
Edited by: H. Jones Humana Press Inc., Totowa, NJ

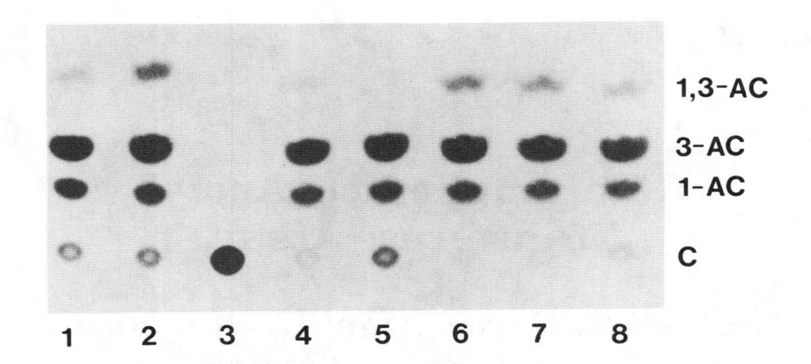

Fig. 1. Autoradiograph of a TLC plate of CAT reaction products. As an example, pCaMVI$_1$CN or pDW2 were introduced into protoplasts, isolated from barley cell suspensions, using 20% (w/v) PEG 1500. Protoplasts were cultured for 36 h prior to extraction. Lanes 1 and 2—1 and 5 units respectively of purified CAT (Sigma); lane 3—protoplasts mixed with 50 μg pCaMVI$_1$CN but not PEG treated; lane 4—protoplasts mixed with 50 μg pDW2, followed by PEG treatment; lanes 5–8—protoplasts mixed with 5, 25, 50, and 100 μg pCaMVI$_1$CN prior to PEG treatment. In lanes 3–8, 2.0–2.5 × 10^6 protoplasts were used per sample.

and equivalent concentrations of either BODIPY-chloramphenicol or ^{14}C-labeled chloramphenicol.

A high performance liquid chromatographic procedure, based on assays developed to detect CAT activity in mammalian cells *(3)* and bacteria *(4)*, has also been reported for plant protoplasts *(5)*. A comparison of the radioisotopic and HPLC methods has confirmed that these procedures have a similar detection limit of 10 mU CAT. However, HPLC requires a shorter assay time, once the chromatography equipment and the procedures are operational.

The procedure employing radiolabeled substrate described here is based on that of Gorman et al. *(6)* and is convenient for detecting transient CAT activity following direct DNA uptake into isolated plant protoplasts.

2. Materials

1. Plasmid DNA isolation kit: Wizard Megapreps DNA Purification System, catalog no. A7300 (Promega Corp., Madison, WI; Promega, Southampton, UK).
2. Extraction buffer: 200 m*M* Tris-HCl, pH 7.8.

3. Bio-Rad Protein Assay Kit I: Catalog no. 500-0001, Bio-Rad Laboratories Ltd., Hemel Hempstead, UK.
4. Purified chloramphenicol acetyltransferase: Sigma (Poole, UK); a 50% (v/v) glycerol solution containing 5 m*M* Tris-HCl, pH 7.8, and 0.5 m*M* 2-mercaptoethanol.
5. Acetylcoenzyme A, lithium salt (Sigma): 10 m*M* aqueous solution.
6. Chloramphenicol: Unlabeled reagent (Sigma); [^{14}C]-chloramphenicol, catalog no. CFA754, 1.85 GBq/mmol, 925 kBq/mL, Amersham International plc. (Buckinghamshire, UK).
7. Ethyl acetate: HPLC grade.
8. Thin layer chromatography plates: 20 cm × 20 cm C18 bonded silica gel on glass (Sigma).
9. TLC solvent: chloroform:methanol, 95:5, (v:v).
10. X-ray film: Hyperfilm-MP, catalog no. RPN34 (Amersham).

3. Methods

1. Isolate protoplasts of the plant species under investigation by enzymatic digestion of suitable source material (e.g., leaves of glasshouse-grown plants of *Petunia parodii* or cell suspensions of albino *Petunia hybrida [7]*).
2. Introduce a suitable plasmid carrying the CAT gene into freshly isolated protoplasts using a PEG uptake procedure or electroporation *(8,9)*. Suitable plasmids include pNOSCAT and pCaMVCAT *(10)* with nopaline synthase and cauliflower mosaic virus 35S promoters respectively, pDW2 (tandem 19S and 35S promoters from CaMV; this plasmid is derived from pDH51; *11*) and pCaMVI$_1$CN (35S promoter and *Adh* 1 intron; courtesy P. Holm, Carlsberg Research Laboratories, Denmark). All plasmids can be amplified in *E. coli* strain HB101 and harvested using a DNA isolation kit (e.g., Wizard Megapreps) following the manufacturer's instructions.
3. Maintain the protoplasts, post transformation, for 2–48 h in the appropriate liquid culture medium normally employed for the protoplast species under investigation.
4. Transfer 1-mL aliquots of the protoplast suspensions, at a density of 1.0 × 10^5/mL (including appropriate controls), to 1.5-mL Eppendorf tubes and centrifuge (5000*g*, 2 min). Discard the supernatants and add 200 µL of ice cold extraction buffer.
5. Lyse the protoplasts by passing six times through a 25G hypodermic needle. Incubate the lysed preparations at 65°C for 10 min (*see* Note 1). Centrifuge at 13,000*g* for 5 min and save the supernatants.
6. Determine the protein content of the supernatants using a commercial protein assay kit, in accordance with the manufacturer's instructions (*see* Note 2).

7. Dispense duplicate aliquots of crude extracts containing between 4.0 and 20 μg protein into 1.5-mL Eppendorf tubes and add extraction buffer to a total volume of 180 μL. Prepare suitable positive control samples with 1, 2, 5, and 10 units of purified CAT enzyme and extraction buffer in place of the protein extracts.
8. Add 0.5 μL of 1 m*M* [^{14}C]-chloramphenicol together with 9.50 μL of unlabeled 1 m*M* chloramphenicol to the Eppendorf tubes. Incubate at 37°C for 10 min; initiate the reaction by adding 10 μL of 10 m*M* acetylcoenzyme A.
9. Incubate the reaction tubes at 37°C for 1 h.
10. Add 300 μL of ice cold ethyl acetate and vortex each tube three times for 5 s.
11. Centrifuge (13,000*g*, 20 s) and transfer the upper aqueous phase from each sample tube to a clean Eppendorf tube.
12. Repeat steps 10 and 11 and pool the ethyl acetate extracts from individual samples.
13. Remove the ethyl acetate in the pooled fractions by evaporation under vacuum and redissolve the radiolabeled and acetylated chloramphenicol in 20 μL of ethyl acetate. Vortex for 5 s.
14. Spot 10 μL of the samples onto a thin layer chromatography plate, 1 cm from the edge of the plate and 1.5 cm apart. Dry using warm air.
15. Carry out ascending chromatography in a tank containing TLC solvent to a depth of 5 mm (*see* Note 3). Allow the solvent front to ascend 5–15 cm.
16. Remove the TLC plate from the tank and allow to dry. Wrap in cling film and use for autoradiography with X-ray film for 12–36 h (*see* Note 4).
17. Look for the presence of acetylated forms of chloramphenicol as in Fig 1. An approximate indication of CAT activity can be obtained by comparing the intensity of the autoradiograph spots with the positive controls that employ specific amounts of purified enzyme. Absence of ^{14}C-label indicates failure of acetylation that may be attributable to several factors (*see* Note 5).

4. Notes

1. Heating the protein extract inactivates plant substances that inhibit CAT activity.
2. A minimum concentration of 25 μg/mL is required. A fresh extract should be prepared if this concentration has not been obtained.
3. The TLC tank should be lined with filter paper and equilibrated with solvent prior to use.
4. Specific CAT activity for each sample can be determined by marking the position of the product bands on the TLC plate, transferring the silica layer to vials containing 5 mL of scintillation cocktail (Emulsifier Safe, Packard Instrument Co., Downers Grove, IL) and counting in a scintillation spectrophotometer. Specific CAT activity (mU CAT/μg protein) is determined as

$$[(N \cdot R)/(A \cdot M \cdot t)] \times 10^{12}$$

where

N: Number of disintegrations counted per 10 min;

R: Ratio of nonradioactive molecules of chloramphenicol in the reaction mixture at $t = 0$ to ^{14}C disintegrations counted in 10 min;

A: Avogadro's constant (6.02×10^{23});

M: Mass of crude protein used in the reaction in μg, corrected for the fraction of ethyl acetate extract spotted onto the TLC plate;

t: Time of incubation in hours.

5. Failure to acetylate choramphenicol can be caused by:
   a. An incorrect reaction mixture. If the positive control samples have been acetylated, the reaction mixture is correct.
   b. Inhibition of CAT activity by substances in the plant extracts. This can be tested by setting up a fresh reaction containing protein extracted from one of the putative negative samples, together with 5 U of purified CAT. Failure to acetylate by this second reaction mixture indicates problems with the protein extract.
   c. Failure to introduce, or lack of expression of, the *cat* gene in the plant cells.

References

1. Bolivar, F. (1978) Construction and characterisation of new cloning vehicles: III. Derivatives of plasmid pBR322 carrying unique EcoRI sites for selection of EcoRI generated recombinant DNA molecules. *Gene* **4,** 121–136.
2. Young, S. L., Barbera, L., Kaynard, A. H, Haugland R. P., Kang, H. C., Brinkley, M., and Melner, M. H. (1991) A nonradioactive assay for transfected chloramphenicol acetyltransferase activity using fluorescent substrates. *Anal. Biochem.* **197,** 401–407.
3. Burzio, L. O., Brito, M., Zarraga, A. M., and Siddiqui, M. A. Q. (1988) Assay of chloramphenicol acetyl transferase by high-performance liquid-chromatography. *Gene Anal. Tech.* **5,** 5–8.
4. Lovering, A. M., White, L. O., and Reeves, D. S. (1986) Identification of individual aminoglycoside-inactivating enzymes in a mixture by HPLC determination of reaction-products. *J. Antimicrob. Chemother.* **17,** 821–825.
5. Davis, A. S., Davey, M. R., Clothier, R. C., and Cocking, E. C. (1992) Quantification and comparison of chloramphenicol acetyltransferase activity in transformed plant protoplasts using high-performance liquid chromatography- and radioisotope-based assays. *Anal. Biochem.* **201,** 87–93.
6. Gorman, C. M., Moffat, L. F., and Howard, B. H. (1982) Recombinant genomes which express chloramphenicol acetyltransferase in mammalian cells. *Mol. Cell. Biol.* **2,** 1044–1051.
7. Power, J. B., Davey, M. R., McLellan, M., and Wilson, D. (1990) Isolation, culture and fusion of protoplasts—1. Isolation and culture of protoplasts. *Biotechnol. Educ.* **1,** 115–124.

8. Maas, C. and Werr, W. (1989) Mechanism and optimised conditions for PEG mediated DNA transfection into plant protoplasts. *Plant Cell Rep.* **8,** 148–151.
9. Golds, T. J., Davey, M. R., Rech, E. L., and Power, J. B. (1990) Methods of gene transfer and analysis in higher plants, in *Methods in Molecular Biology,* vol. 6, *Plant Cell and Tissue Culture* (Pollard, J. W. and Walker, J. M., eds.), Humana Press, Clifton, NJ, pp. 341–371.
10. Fromm, M., Taylor, L. P., and Walbot, V. (1985) Expression of genes transferred into monocot and dicot plant cells by electroporation. *Proc. Natl. Acad. Sci. USA* **82,** 5824–5828.
11. Pietrzak, M., Shillito, R. D., Hohn, T., and Potrykus, I. (1986) Expression in plants of two bacterial antibiotic resistance genes after protoplast transformation with a new plant expression vector. *Nucl. Acids Res.* **14,** 5857–5868.

CHAPTER 12

NPTII Assays for Measuring Gene Expression and Enzyme Activity in Transgenic Plants

Ian S. Curtis, J. Brian Power, and Michael R. Davey

1. Introduction

The stable introduction of genes into plants through genetic engineering normally necessitates the use of a selectable marker, especially when the transformation frequency is low (e.g., 1.0×10^{-3} to 10^{-6}). Marker genes enable quantification of both the transformation efficiency and gene expression *(1)*. The most commonly used selectable marker is the gene from Transposon 5 (Tn5) of *Escherichia coli* K12 *(aphA2) (2)*, which encodes aminoglycoside 3-phosphotransferase II (APH [3] II, Chemical Abstracts Registry number 58943-39-8) activity. This enzyme, also known as neomycin phosphotransferase II (NPTII), inactivates by phosphorylation the sugar-containing antibiotics, neomycin, kanamycin, geneticin (G418), and paromomycin. To date, the gene has been introduced into over 30 plant species. (For a detailed description of the *nptII* gene, *see* Chapter 1.)

The antibiotic, kanamycin, is a trisaccharide composed of one deoxystreptamine and two glucosamine units (2-deoxystreptamine-6-D-glucosamine) *(3)*. Neomycin is a tetrasaccharide *(3)*, whereas geneticin is a 3'-OH-containing gentamycin-derivative, both with chemical properties similar to kanamycin. Paromomycin is a broad spectrum aminoglycoside,

From: *Methods in Molecular Biology, Vol. 49: Plant Gene Transfer and Expression Protocols*
Edited by: H. Jones Humana Press Inc., Totowa, NJ

having a molecular structure similar to monomycin A. Kanamycin is bacteriocidal to both gram-positive and gram-negative bacteria, interacting with the 30S and 50S subunits of ribosomes *(4)* and inhibiting protein synthesis. In plant cells, kanamycin exerts it effect on mitochondria and chloroplasts by impairing protein synthesis, resulting in chlorosis *(5)*. Neomycin is about 10 times more toxic than kanamycin, as based on intraveinal and oral applications on mice.

The NPTII enzyme consists of 264 amino acids and has a high substrate specificity. It catalyses the ATP-dependent phosphorylation of the 3'-hydroxy moeity of the aminohexose ring *(3)*. Phosphorylation of these antibiotics prevents them from interacting with the 30S ribosomal subunit *(6)*, which inhibits protein synthesis. The use of the *npt*II gene in biotechnology increased rapidly when it was discovered that it could be expressed in plants when linked to eukaryotic regulatory sequences *(7)*.

The biosafety of the *npt*II gene and its protein has been reviewed extensively *(1,3,8,9)*. Current evidence suggests that the *npt*II gene and its protein product are safe.

The first measurements of NPTII activity were performed on bacterial extracts to determine the presence of R-factors responsible for antibiotic resistance. Enzyme activity was based on its ability to transfer gamma-labeled [^{32}P]-ATP cofactor to the antibiotic. The radiolabeled product, bound to phosphocellulose paper, was quantified using a scintillator *(10)*. Initial NPTII assays on plants involved the electroseparation of the enzyme and antibiotic phosphorylation in an agarose gel, followed by blotting with phosphocellulose paper *(11,12)*. However, these earlier methods were time-consuming (2–3 d), involved high levels of radiation (45 μCi of ^{32}P/sample), and restricted the number of samples that could be analyzed. Methodology diverged toward a modified dot blot assay, similar to the earlier techniques used on bacteria. McDonnell et al. *(13)* diluted gamma-labeled [^{32}P]-ATP with "cold" nucleotides and found this reduced the background signal often seen in gel assays. Reducing the amount of isotope lessens the occurrence of nonspecific protein kinase side reactions and more ATP in the reaction mixture increases the formation of more product. The inclusion of the phosphatase inhibitor, sodium fluoride, further prevents background problems, which can arise through the hydrolysis of product (kanamycin phosphate) and free ATP. This method can be employed to screen a large number of samples rapidly (2–3 h), without the need for high amounts of radiation (0.045 μCi/

sample). The assay can be applied to a range of plant species and tissues, with the same sensitivity as gel assays. Further improvements *(14)* have provided greater sensitivity for detecting NPTII activity. This method has become the procedure of choice for detecting NPTII activity *(15)*. Chromatographic separation of the reaction between NPTII with kanamycin and gamma-labeled [^{32}P]-ATP on polyethyleneimine cellulose (PEI) plates is also simple, rapid, sensitive, and safe *(16)*. The levels of radiation employed (0.18 µCi/sample) are lower compared to gel assays, but the radioactivity associated with the chromatographic procedure is still four times higher compared to the improved dot blot method.

Although quantitation of NPTII protein from crude cellular extracts can be performed using Western blotting, a safer, nonlabeling technique is now available. The NPTII enzyme-linked immunosorbant assay (ELISA) kit (5 Prime-3 Prime, Inc., Boulder, CO) is a sandwich immunoassay that is being used by several laboratories to detect and to quantify the enzyme in plant tissues *(17)*. Rabbit polyclonal antibody specific to the NPTII protein is fixed to a polystyrene microwell, blocked, and plant protein samples are added. During subsequent incubation, the NPTII enzyme binds to the antibody. After washing, a biotinylated secondary antibody is added that binds to the antibody–NPTII protein complex. Quantification of protein is measured colorimetrically by incubating the samples with streptavidin conjugated alkaline phosphatase and substrate.

This chapter describes the most popular methods of measuring NPTII activity *(14)* and the presence of NPTII protein in transgenic plants.

2. Materials

2.1. NPTII Dot Blot Assay

1. Whatman P81 Ion Exchange Chromatography Paper (Whatman Ltd., Maidstone, UK).
2. Soaking buffer: 20 m*M* adenosine 5'-triphosphate (ATP) sodium salt (Sigma, St. Louis, MO), 100 m*M* tetrasodium pyrophosphate (BDH). Dissolve tetrasodium pyrophosphate in aliquots of 10 mL of distilled water and autoclave. Add ATP prior to use.
3. Wash buffer: 10 m*M* sodium phosphate buffer, pH 7.50 *(18)*. Prepare separately, 1*M* stocks of Na_2HPO_4 and NaH_2PO_4 in distilled water. Add the required volumes of each solution to obtain the correct molarity and pH; make to volume with distilled water. Prepare the working solution before use.

4. Lysis buffer: 50 m*M* sodium phosphate, pH 7.00, 1 m*M* EDTA, 0.1% v/v Triton X-100, 10 m*M* 2-mercaptoethanol, 0.1% w/v sodium lauryl sarkosine, dissolved in water. Sterilize by filtration through a 0.2-μm membrane (Minisart, Epsom, UK) and store at 4°C.
5. 5X Reaction buffer: 335 m*M* Tris-base, 210 m*M* magnesium chloride, 2*M* ammonium chloride, dissolve in distilled water. Titrate to pH 7.10 with 1*M* maleic acid (dissolved in distilled water).
6. Liquid scintillation analyzer (Packard Instrument Co., Downers Grove, IL).

2.1.1. Assay Mixture

1. 10 m*M* ATP stock in distilled water, prepared immediately before use.
2. 22 m*M* neomycin sulfate (Sigma) in distilled water, freshly prepared prior to use.
3. 1*M* sodium fluoride (BDH) stock in distilled water.
4. 1X reaction buffer: Mix 5X reaction buffer with distilled water (1:4, v:v).
5. Assay mixture: 4.936 mL 1X reaction buffer, 5 μL 10 m*M* ATP solution, 7 μL 22 m*M* neomycin sulfate solution, 50 μL 1*M* sodium fluoride solution, 15 μCi [^{32}P]-ATP, 250 mg bovine serum albumin (BSA) (Sigma).

All solutions are sterilized by autoclaving at 121°C for 20 min, except for the lysis buffer, assay mixture, and solutions containing ATP or neomycin sulfate.

2.1.2. Standard Curve

1. Protein standards: Prepare serial dilutions of BSA fraction V (Sigma) in a range of 1–50 μg protein in 2-mL volumes, dissolved in distilled water.
2. Lysis buffer (*see* Section 2.1).
3. BioRad protein assay solution (Bio-Rad Laboratories, Hemel Hempstead, UK).
4. Blank sample: 5 μL lysis buffer, 200 μL Bio-Rad protein assay solution; make up to 1 mL with distilled water.
5. Test samples: 5 μL plant protein extract, 200 μL Bio-Rad protein assay solution, 795 μL distilled water.
6. Spectrophotometer for measuring absorbance at 595 nm.

2.2. NPTII ELISA

2.2.1. NPTII ELISA Kit

The NPTII ELISA Kit (5 Prime-3 Prime, Inc.) contains microwell strips, strip frame and container, NPTII coating antibody, biotinylated antibody to NPTII, NPTII enzyme standard concentrate, streptavidin conjugated alkaline phosphatase, coating buffer, 5 mg *p*-nitrophenyl phosphate tablets, 10X washing buffer, diethanolamine buffer, stop solution, 10X blocking/dilution buffer, and a 250 mL wash bottle. Instructions are also included.

1. Microwell strips (1 × 8 well format), strip frame, and container.
2. 1X Washing buffer: Prepare a 1:9 (v:v) dilution of 10X washing buffer with glass distilled or deionized water. Use only fresh solutions. Approximately 100 mL is required for each microwell strip. Store 10X stock solutions at room temperature.
3. 1X Phosphate buffered saline (PBS): Mix 10X PBS with distilled or deionized water (1:9, v:v). Prepare fresh before use; approx 12 mL is required for each microwell strip.
4. Coating buffer: store at 2–8°C.
5. 1X Blocking/dilution buffer: Mix 10X blocking/dilution buffer with glass distilled or deionized water (1:9, v:v). Warm 10X blocking/dilution buffer at 30–37°C to dissolve any crystals prior to diluting. Approximately 16 mL of 1X blocking/dilution buffer is required for the standard curve and 14 mL for processing each microwell strip. Prepare before use. Store 10X blocking/dilution buffer at 2–8°C.
6. Other reagents: Biotinylated antibody to NPTII, NPTII enzyme standard concentrate, and streptavidin conjugated alkaline phosphatase samples should be made in 1X blocking/dilution buffer prior to use. Pulse spin the tubes in a microfuge (13,500 rpm, 5 s) before opening to collect contents. Store stock solutions at –20°C.
7. NPTII coating antibody: Store stock solutions at 2–8°C and aliquots at –20°C.
8. *p*-Nitrophenyl phosphate tablets: Store at –20°C.
9. Diethanolamine buffer: Store at room temperature in an amber colored bottle.
10. Stop solution: Store at room temperature.
11. ELISA microwell reader or spectrophotometer with microcuvets for use at $A_{405\ nm}$.

2.2.2. Protein Extraction

1. Leaves of transformed and nontransformed plants.
2. Extraction buffer: 0.25*M* Tris-HCl, pH 7.8, 1 m*M* phenylmethylsulfonylfluoride (PMSF).
3. Microfuge.

2.2.3. Standard Curve

1. 1 µg NPTII/mL: A total volume of 1.3 mL is required to create a four point standard curve.
2. 1X Dilution buffer: This should be added to 1 µg NPTII/mL stock to prepare the correct concentration of protein for the standard curve.

3. Methods

3.1. NPTII Dot Blot Assay

1. Draw out a grid (2 cm × 2 cm/sample) in pencil on a sheet of Whatman P81 paper (*see* Note 1). Immerse the paper in soaking buffer (10 mL/20 cm × 20 cm) and blot (*see* Note 2). Hang the paper to drip dry (approx 1 h, room temperature).
2. Add 50μL aliquots of chilled (4°C) lysis buffer to 5–10 mg samples of plant material contained in Kontes "pellet-pestle" tubes (Scotlab, Strathclyde, Scotland) (*see* Note 3). Grind each sample with 20–30 strokes on ice. Centrifuge for 10 min (13,500 rpm, 4°C) to collect debris (*see* Note 4).
3. Set up a standard curve for quantifying protein concentration in the supernatant using a spectrophotometer ($A_{595\ nm}$) *(19)*. Add known amounts of protein standards (1–50 μg protein in solution) to 5 μL lysis buffer, 200 μL BioRad protein assay solution; make up to 1 mL with distilled water in plastic cuvets. Invert each cuvet several times to mix. Place into the spectrophotometer and record the optical density for each protein standard. Repeat 3–4 times until a reliable figure is obtained for each protein standard.
4. Quantify the protein concentration in each plant supernatant. Add 5 μL of each plant sample to a mixture of 200 mL BioRad protein assay solution and 795 μL distilled water contained in plastic cuvets. Invert and quantify as in Section 3.1.3. Extrapolate the protein concentration from the standard curve.
5. For each sample, add 20 μL supernatant containing 5 μg of plant protein to a 1.5-mL capacity Eppendorf tube on ice. Dilute the protein samples to the required concentration with distilled water.
6. Prepare assay mixture (*see* Section 2.1.1.4.), without the addition of isotope or BSA, in a 30 mL capacity plastic universal tube (Bibby Sterilin, Stone, UK). Add 1.5 μL of gamma-labeled [^{32}P]-ATP to the assay mixture, followed by the BSA. Gently agitate the mixture to dissolve (*see* Note 5).
7. Place the Eppendorf tubes containing the plant protein samples in perspex blocks, and add 20 μL of assay mixture to each sample. Pulse spin to collect the sample at the bottom of the tube and incubate in a water bath (30 min, 37°C).
8. Place samples on ice (2 min) to terminate the reaction. Pulse spin to collect the samples. Add 10 μL of each sample to its respective 2 cm × 2 cm area on the air-dried Whatman P81 paper. Repeat until 30 μL of each sample has been transferred, allowing the samples to air dry prior to adding more supernatant (*see* Note 6).
9. Wash the dried blot in 200 mL of preheated wash buffer (2 min, 80°C) with constant agitation (*see* Note 7). Remove the blot and immerse in 200 mL of wash buffer at room temperature on a horizontal rocking platform

(10 min). Repeat washes at room temperature two to three times until the level of the radioactivity of the plant samples is 5–10 cpm. Leave blots on a sheet of aluminum foil to air-dry in a fume cupboard (60–90 min).

10. Wrap the blot in Saran Wrap (Genetic Research Instruments, Cambridge, UK) and place into an autoradiographic cassette containing X-ray film (Hyperfilm, Amersham, UK). Expose for 2 d at –70°C (*see* Note 8).
11. Quantify the level of NPTII activity of each plant sample by cutting out each 2 cm × 2 cm square and measuring the radioactivity of each square in a liquid scintillation analyzer (*see* Note 9).

3.2. NPTII ELISA

1. Remove enough microwell strips from the frame without touching the bottom of the microwells and secure with the strip retainer (*see* Note 10).
2. Dilute the NPTII coating antibody with coating buffer. Add 200 µL of the antibody/well (approx 2 mL of antibody/microwell strip). Incubate at 37°C for 2 h (*see* Note 11).
3. Empty each well by inverting the plate (*see* Note 12). Add 400 µL of 1X blocking/dilution buffer to each well to block all additional protein binding sites. Incubate at room temperature for 30 min.
4. Finely mince 1–2 g of leaf material with a scapel blade, transfer to a homogenizer and add 3-4 mL of chilled (4°C) extraction buffer (*see* Note 3). Homogenize tissue at 4°C until the material is fully disrupted. Transfer the supernatant to a 1.5 mL capacity Eppendorf tube and centrifuge for 30 min (13,500 rpm, 4°C). Determine the protein concentration of the supernatant as in Section 3.1.4. Adjust the concentration of protein/sample to 400 µg/mL using 1X blocking/dilution buffer.
5. Prepare the NPTII enzyme dilutions for establishing the standard curve. Enough NPTII standard should be prepared to produce four standard curves for each assay as outlined herein:

NPTII Protein Standards Conc/200 µL	Preparation
150 pg	6.3 µL (1 µg NPTII/mL) + 8.4 mL of 1X dilution buffer (=A)
100 pg	1.6 mL (A) + 0.8 mL of 1X dilution buffer (=B)
50 pg	1.0 mL (B) + 1.0 mL of 1X dilution buffer (=C)
20 pg	0.4 mL (C) + 0.6 mL of 1X dilution buffer

These dilutions are enough to produce a standard curve spanning the linear range of the assay.

6. Empty the wells by inverting the plate. Wash each well with 1X washing buffer using a wash bottle; invert the plate as before. Repeat the washing procedure twice.
7. To the appropriate wells, add 200 μL of undiluted and the same volume of diluted crude plant protein extracted from a nontransformed control plant and from transformed plants to the appropriate wells. The dilutions should be 1:10, 1:25, 1:50, and 1:100 (v:v) using 1X blocking/dilution buffer (*see* Note 13). Incubate at room temperature for 2 h.
8. Dilute the NPTII Biotinylated antibody with 1X blocking/dilution buffer, 5–10 min before the end of the incubation described in Section 3.2.7. Each microwell requires approx 2 mL of diluted antibody.
9. Empty the wells and wash five times with 1X washing buffer as described in Section 3.2.6. Add 200 μL of the biotinylated anti-NPTII dilution to each well. Incubate for 1 h at room temperature.
10. Dilute the streptavidin conjugated alkaline phosphatase (SA/AP) with 1X blocking/dilution buffer, 5–10 min before the end of the incubation (*see* Section 3.2.9.). Approximately 2 mL of diluted SA/AP will be required for each microwell strip.
11. Empty the wells and wash five times with 1X washing buffer as described in Section 3.2.6. Add 200 μL of SA/AP per well and incubate for 30 min at room temperature.
12. Dissolve one 5 mg *p*-nitrophenyl phosphate tablet in 2.5 mL of diethanolamine buffer. Approximately 2.5 μL of the color development substrate solution will be required for each microwell strip. Prepare the solution 5 min before the end of the incubation (*see* Section 3.2.11.).
13. Empty the wells and wash five times with 1X washing buffer as in Section 3.2.6. Add 200 μL of substrate solution to each well. Incubate at room temperature for 30-40 min in the dark.
14. Read each well at 405 nm against the reagent blank in a microtiter well reader. The reaction must be terminated by adding 50 μL of stop solution to each well, if manual spectrophotometric recordings are made (*see* Note 14). If a 250 μL cuvet is not available, dilute 1:1 (v:v) with 0.4*N* NaOH and read samples in a 500 μL cuvet against the reagent blank.

4. Notes

1. Do not use ink to draw grids as it will smear.
2. This procedure eliminates nonspecific binding to the paper and background in the autoradiogram.
3. NPTII activity can be highly variable in different leaves even on the same plant, especially in chimeric plants. Several leaf samples should be assayed for each plant. Choose only young root tips when analyzing NPTII activity in roots, as old roots can give false positives.

4. Centrifugation prevents the supernatant becoming gelatinous; gelatinous supernatants can become labeled and stick nonspecifically to the blot, causing false positives.
5. Vigorous shaking to dissolve the BSA will cause foaming and may cause radioactive hot spots to occur.
6. Adding the reaction mixture to a wet blot may cause the dots to expand giving halos on the X-ray film when the blot is developed. The use of a hair dryer can assist drying of the blots.
7. 10 m*M* phosphate buffer helps to eliminate background counts. Most of the radioactivity will be removed in the first wash; extra care is needed at this time in disposal.
8. It is important not to wrap the blot when wet, as this can cause the formation of halos when the X-ray film is developed.
9. Alternatively, the blot can be analyzed by scanning with an AMBIS 2 D beta scanner (AMBIS Systems, Inc., San Diego, CA) in 2 h *(20)*.
10. Do not mix the components from NPTII ELISA kits purchased from different manufacturers. Strips to be used at a later date should be stored in a zip lock bag.
11. All working solutions should be freshly prepared. Storage and reuse of diluted solutions may give inaccurate results. Polystyrene tubes should be avoided, as they may bind proteins.
12. When emptying the wells, hold the frame and retainer firmly to secure the strips. Shake several times with a sharp downward motion to empty each well. Solutions containing sodium azide should not be dispensed down the sink, unless the drains are flushed with a large volume of water to prevent azide accumulation.
13. A serial dilution of the plant extracts is required to ensure that the concentration of NPTII falls within the linear range of the assay (20–150 pg NPTII/200 μL 1X blocking/dilution buffer).
14. The stop solution contains 2N NaOH, which is caustic; avoid contact with skin, eyes, or clothing.

References

1. Flavell, R. B., Dart, E., Fuchs, R. L., and Fraley, R. T. (1992) Selectable marker genes: safe for plants? *Bio/Technol.* **10,** 141–144.
2. Beck, E., Ludwig, G., Auerswald, E. A., Reiss, B., and Schaller, H. (1982) Nucleotide sequence and exact localization of the neomycin phosphotransferase gene from transposon Tn5. *Gene* **9,** 327–336.
3. Nap, J.-P., Bijvoet, J., and Stiekema, W. J. (1992) Biosafety of kanamycin-resistant transgenic plants. *Trans. Res.* **1,** 239–249.
4. Davis, B. D. (1988) The lethal action of aminoglycosides. *J. Antimicrob. Chemother.* **22,** 1–3.

5. Weide, R., Koornneef, M., and Zabel, P. (1989) A simple, nondestructive spraying assay for the detection of an active kanamycin resistance gene in transgenic tomato plants. *Theor. Appl. Genet.* **78,** 169–172.
6. Dickie, P., Bryan, L. E., and Pickard, M. A. (1978) Effect of enzymatic adenylation on dihydrostreptomycin accumulation in *Escherichia coli* carrying an R-factor: model explaining aminoglycoside resistance by inactivating mechanisms. *Antimicrob. Agents Chemother.* **14,** 569–580.
7. Bevan, M. W., Flavell, R. B., and Chilton, M.-D. (1983) A chimaeric antibiotic resistance marker gene as a selectable marker for plant cell transformation. *Nature* **304,** 184–187.
8. Fuchs, R. L., Ream, J. E., Hammond, B. G., Naylor, M. W., Leimgruber, R. M., and Berberich, S. A. (1993) Safety assessment of the neomycin phosphotransferase II (NPTII) protein. *Bio/Technol.* **11,** 1543–1547.
9. Miller, H. I., Huttner, S. L., and Beachy, R. (1993) Risk assessment experiments for "genetically modified" plants. *Bio/Technol.* **11,** 1323,1324.
10. Ozanne, B., Benveniste, R., Tipper, D., and Davies, J. (1969) Aminoglycoside antibiotics: inactivation by phosphorylation in *Escherichia coli* carrying R factors. *J. Bacteriol.* **100,** 1144–1146.
11. Reiss, B., Sprengel, R., Will, H., and Schaller, H. (1984) A new sensitive method for qualitative and quantitative assay of neomycin phosphotransferase in crude cell extracts. *Gene* **30,** 211–218.
12. Schreier, P. H., Seftor, E. A., Schell, J., and Bohnert, H. J. (1985) The use of nuclear-encoded sequences to direct the light-regulated synthesis and transport of a foreign protein into plant chloroplasts. *EMBO J.* **4,** 25–32.
13. McDonnell, R. E., Clark, R. D., Smith, W. A., and Hinchee, M. A. (1987) A simplified method for the detection of neomycin phosphotransferase II activity in transformed plant tissues. *Plant Mol. Biol. Rep.* **5,** 380–386.
14. Tomes, D. T., Ross, M., Higgens, P., Rao, A. G., Stabell, M., and Howard, J. (1990) Direct DNA transfer into intact plant cells and recovery of transgenic plants via microprojectile bombardment, in *Plant Molecular Biology Manual* A13 (Gelvin, S. B., Schilperoot, R. A., and Verma, D. P. S., eds.), Dordrecht, Kluwer, The Netherlands, pp. 1–22.
15. Curtis, I. S., Power, J. B., Blackhall, N. W., de Laat, A. M. M., and Davey, M. R. (1994) Genotype-independent transformation of lettuce using *Agrobacterium tumefaciens. J. Exp. Bot.* **45,** 1441–1449.
16. Cabanes-Bastos, E., Day, A. G., and Lichtenstein, C. P. (1989) A sensitive and simple assay for neomycin phosphotransferase II activity in transgenic tissue. *Gene* **77,** 169–176.
17. Franklin, C. I., Trieu, T. N., Cassidy, B. G., Dixon, R. A., and Nelson, R. S. (1993) Genetic transformation of green bean callus via *Agrobacterium* mediated DNA transfer. *Plant Cell Rep.* **12,** 74–79.
18. Sambrook, J., Fritsch, E. F., and Maniatis, T. (1989) *Molecular Cloning. A Laboratory Manual*, 2nd ed., vol. 3 (Ford, N., Nolan, C., and Ferguson, M., eds.), Cold Spring Harbor Laboratory, Cold Spring Harbor, NY.

19. Bradford, M. M. (1976) A rapid and sensitive method for the quantitation of microgram quantities of protein utilizing the principle of protein-dye binding. *Anal. Biochem.* **72,** 248–254.
20. Nye, L., Colclough, J. M., Johnson, B. J., and Harrison, R. M. (1988) Radioanalytical imaging: high speed radioisotope detection, imaging, and quantitation. *Am. Biotech. Lab.* **6,** 18–26.

PART IV

Study of Gene Organization by Southern Blotting and Inverse PCR

CHAPTER 13

Gene Characterization by Southern Analysis

Paolo A. Sabelli and Peter R. Shewry

1. Introduction

The study of how genes are organized within the genome is a major step in the characterization of cloned DNA sequences and is important in the analysis of plant transformation. In both cases, accurate information is required about the number of copies of a given DNA sequence and their location in the genome. The analysis may be complicated by the presence of between one and several hundred related sequences in the genome.

The most widely used method to study DNA sequence organization is Southern blotting *(1)*. In this method, DNA is extracted from cells, digested with appropriate restriction endonucleases, size-separated by agarose gel electrophoresis, denatured into single-stranded molecules, transferred (blotted) onto a solid support, and annealed (hybridized) to labeled single-stranded nucleic acid probes. Under suitable conditions related sequences anneal by hydrogen bonding between complementary bases. Labeling of the probe strand makes it possible to identify related sequences in the genomic DNA. In addition, the experimental conditions can be altered to vary the specificity of hybridization and, within certain limits, to discriminate between genomic sequences that are identical or less closely related, to the probe.

Although a range of protocols for Southern blotting is available, disappointing results are often obtained, especially when high sensitivity is

From: *Methods in Molecular Biology, Vol. 49: Plant Gene Transfer and Expression Protocols*
Edited by: H. Jones Humana Press Inc., Totowa, NJ

required. We describe here a method that is currently used in our laboratory, and has proved to be versatile, reliable in different hands, and highly sensitive *(2–7)*. We then give examples of its applications in various studies of plant molecular biology.

Although nonradioactive methods for labeling probes and signal detection have been developed, we feel that isotopic probes and detection procedures are still more sensitive. Sensitivity depends on the abundance of both target and probe sequences and on their similarity. High sensitivity is crucial when low-copy-number sequences within large genomes are targeted or the degree of complementarity between target sequences and probe is low or unknown (this is often the case when screening gene libraries using heterologous probes). However, nonradioactive methods *(8)* are safe for the operator, fast, can be used for simultaneous detection of different sequences *(9)*, and may well suit experiments in which high sensitivity is not required (e.g., analysis of small genomes, Southern hybridization of plasmids). The method described here does not include protocols for the preparation of probes as these are described in Chapter 18 of this volume.

2. Materials

Store all solutions at room temperature unless otherwise indicated.

2.1. Isolation of Genomic DNA

1. Homogenization buffer: 17% w/v sucrose, 50 m*M* Tris-HCl, pH 8.0, 20 m*M* EDTA, 25 m*M* KCl, 1% w/v *n*-lauroylsarcosine. Prepare fresh before use.
2. Ethidium bromide: 10 mg/mL. Store at 4°C. Light sensitive.
3. TE buffer: 10 m*M* Tris-HCl, pH 8.0, 1 m*M* EDTA, pH 8.0.
4. 50% w/v CsCl in TE buffer.
5. TE-saturated butan-1-ol.
6. Ethanol absolute.

2.2. Digestion and Electrophoresis

1. 10X Endonuclease restriction buffers and endonucleases supplied by manufacturers. Store at –20°C.
2. 100 m*M* Spermidine. Store at –20°C.
3. Ethidium bromide: 10 mg/mL. Store at 4°C. Light sensitive.
4. Sample loading buffer: 20% w/v Ficoll 400, 0.1*M* EDTA, pH 8.0, 0.25% w/v bromophenol blue, 0.25% w/v xylene cyanol.
5. 10X TBE buffer: 0.9*M* Tris-base, 0.9*M* boric acid, 25 m*M* EDTA, pH 8.3.

2.3. Blotting

1. Denaturing buffer: 0.5*M* NaOH, 1.5*M* NaCl.
2. Neutralizing buffer: 0.5*M* Tris-HCl, 1.5*M* NaCl, pH 7.5.
3. 20X SSC buffer: 3.0*M* NaCl, 0.3*M* trisodium citrate, pH 7.0.
4. Nylon transfer membrane (Hybond from Amersham [Arlington Heights, IL] or equivalent).

2.4. Hybridization

1. 20X SSPE: 3.6*M* NaCl, 0.2*M* NaH_2PO_4, 20 m*M* EDTA, pH 7.7.
2. Prehybridization and hybridization buffer: 1.5X SSPE (from stock 20X SSPE), 0.5% w/v commercial low-fat dried milk, 1% w/v sodium dodecyl sulfate (SDS), 6% w/v polyethylene glycol (PEG) 6000, pH 7.7. Filter through Whatman (Maidstone, UK) No. 1 filter paper and store at –20°C in aliquots. Do not autoclave.

2.5. Posthybridization

1. 20X SSC buffer: 3.0*M* NaCl, 0.3*M* trisodium citrate, pH 7.0.
2. Fuji RX or Kodak X-OMAT AR5 X-ray films.
3. Alkali wash solution: 0.4*M* NaOH.
4. Neutralization buffer: 0.2*M* Tris-HCl, 0.1X SSC, 0.1% w/v SDS, pH 7.5.

3. Methods

3.1. Isolation of Genomic DNA

1. Harvest plant material (4–10 g) and freeze in liquid N_2. Material previously frozen and stored at –80°C can also be used (*see* Notes 1 and 2).
2. Grind the material in an ice-cold mortar in the presence of liquid N_2 until a fine powder is obtained and the liquid N_2 is fully evaporated.
3. Prepare a solution of 26.5 mL of CsCl with 25 mL of ice cold homogenization buffer. Add to the powdered plant tissue and grind the mixture for a few minutes. Keep mixing to prevent a CsCl precipitate forming owing to the low temperature (any precipitate will redissolve). DNA is relatively stable in this solution and an increase in temperature (up to room temperature) does not affect the quality of the preparation.
4. Filter the homogenate through four layers of muslin, speeding up the procedure by gently squeezing the dense suspension through the muslin.
5. Collect the eluate into a corex tube and centrifuge at 10,000 rpm (r_{av} = 7840*g*) for 30 min at room temperature in a Beckman type JA20 rotor or equivalent.
6. Decant the supernatant into a 30-mL polycarbonate ultracentrifuge tube. Add 850 µL of ethidium bromide solution (10 mg/mL) (*see* Note 3), check the density of the solution (it should be 1.55–1.60 g/mL, if it is too low add

solid CsCl), and centrifuge at 38,000 rpm (r_{av} = 130,000*g*) for 16 h at 15°C in a Beckman (Fullerton, CA) type 50.2 Ti rotor or equivalent (*see* Note 4).

7. Visualize the band of high molecular weight DNA with a long-wave (e.g., 305 nm) UV lamp. Collect the DNA band by gentle aspiration with a Pasteur pipet, taking care not to contaminate the DNA with the impurities in the tube. Discard the RNA pellet at the bottom of the tube and the cushion of impurities floating on the top.
8. Transfer the DNA to another polycarbonate ultracentrifuge tube and fill with a solution of 50% w/w CsCl in TE buffer. Do not add any ethidium bromide. Centrifuge the sample as in Section 3.1., step 6.
9. Remove the DNA as in Section 3.1., step 7. and note the sample volume. Remove the ethidium bromide by extracting the sample four to six times with three to four volumes of TE-saturated butan-1-ol. This is conveniently carried out in disposable 15–50 mL plastic tubes (Falcons [Becton Dickinson, Cowley, Oxford, UK] or similar). Allow the two phases to separate each time and discard the upper phase with a Pasteur pipet. Note that the sample volume has increased.
10. Dilute the DNA with two volumes of TE buffer (relative to the volume of the sample removed from the CsCl gradient) and add two volumes (with respect to the diluted sample) of absolute ethanol to precipitate the DNA. Recover the gelatinous DNA with a glass rod (a flame-bent Pasteur pipet) and allow 2 min for the excess ethanol to evaporate. Transfer the DNA to a 1.5-mL Eppendorf tube containing 0.5–1.0 mL of TE buffer. If the DNA is allowed to dry completely, it takes a very long time to resuspend. Gently swirl the rod in the Eppendorf tube to release the DNA, and dissolve the DNA by leaving it at 4°C for several hours (preferably on a rotating rack).
11. Estimate the DNA concentration by absorbance measurement at 260 nm. One A_{260} unit corresponds to a concentration of 50 µg/mL of double stranded nucleic acids.
12. Carefully adjust the DNA concentration determined in the previous step by comparison with high molecular weight DNAs (e.g., bacteriophage λ DNA) of known concentration run on a 0.4% agarose gel. Check the integrity of the DNA: Only a high molecular weight band (more than 20 kb) should be obtained, with no smearing. *Caution*: overloading causes sample smearing (*see* Note 5).
13. The DNA is ready for further analyses or for storing at –20°C (*see* Note 6).

3.2. Digestion and Electrophoresis

1. Digest 2–10 µg of genomic DNA with appropriate endonuclease (6 U/µg DNA) overnight in 50-µL reaction volume containing 4 m*M* spermidine (*see* Notes 7 and 8).

2. Add 5 µL of sample loading buffer, mix thoroughly, and load onto a 0.8% w/v agarose gel containing 0.5 µg/mL ethidium bromide. Run the gel at 2–5 V/cm in 1X TBE buffer containing 0.5 µg/mL ethidium bromide until good separation is achieved as judged by visualizing the DNA samples and the size markers on a long-wave UV transilluminator (this may take from 6 to 20 h, depending on the gel size) (*see* Note 9).
3. Photograph the gel with a Polaroid MP4 camera using Polaroid 667 film and a long-wave UV transilluminator. Place a ruler at one edge of the gel (with zero set relative to the wells) to indicate the migration distance. Cut off the size-marker lane, the bottom of the gel if the length is more than 20 cm, and a corner of the gel as a useful reference point (*see* Notes 10 and 11).

3.3. Blotting

1. Denature the DNA by incubating the gel twice in denaturing buffer (approx 1.5 mL/cm^3 of gel volume) for 15 min each.
2. Rinse briefly in distilled water and neutralize by incubating the gel twice in neutralizing buffer (volume as in the prevous step) for 15 min each (*see* Note 12).
3. Cut a nylon transfer membrane (Hybond from Amersham or equivalent) a few mm larger than the gel and equilibrate it in 2X SSC buffer for 5–10 min (*see* Note 13). Also cut four sheets of Whatman 3 MM filter paper slightly larger than the membrane and soak in 2X SSC buffer.
4. Set up the blotting apparatus as follows: Place a glass plate across a glass or plastic tray. Place a sheet of Whatman 3 MM paper on the glass plate so that the ends hang down into the tray. Pour 20X SSC buffer (250–500 mL) into the tray and over the 3 MM paper and then place two of the prewetted sheets of 3 MM paper on top (*see* Note 14). Remove any air bubbles by gently rolling a glass rod between each layer of the blot (*see* Note 15). Slide the gel onto the paper sheets and mask off any exposed 3 MM paper by surrounding the gel with clingfilm. Place the nylon membrane on the gel, taking care to align the top with the wells of the gel, then put the remaning two sheets of 3 MM paper on top. Place a 5–10-cm stack of adsorbent paper towels on top of the 3 MM paper followed by a glass plate and a 0.5 kg weight. For a diagrammatic description of the blot set up, refer to Dyson *(10)* or Sambrook et al. *(11)*.
5. Leave overnight to allow the DNA to transfer by passive diffusion of the buffer, replacing the wet stack of paper towels twice (*see* Note 16).
6. Disassemble the apparatus and cut one corner of the nylon membrane to match the cut corner of the gel (*see* Section 3.2., step 3.).
7. Rinse the membrane briefly in 2X SSC, wrap in a single layer of clingfilm, and crosslink the DNA by irradiation on a long-wave UV transilluminator

for about 2–3 min (DNA-side down). Finally, bake the membrane at 80°C for 1 h. The membrane is now ready for hybridization. Once crosslinked to a membrane the DNA is stable for many months at room temp.
8. Stain the gel with ethidium bromide to check the blotting efficiency (optional).

3.4. Hybridization

1. Place the membrane into a glass hybridization bottle (Hybaid or similar) and incubate with preheated prehybridization buffer (0.075 mL/cm^2 of membrane), containing 30μg/mL of salmon sperm DNA (denatured by boiling for 5–10 min and cooled on ice). Avoid overlapping of the membrane by using thin nylon meshes. Gently invert the bottle to wet the membrane completely. Prehybridize for 4–12 h in a rotatingoven at 50–65°C (*see* Note 17).
2. Replace the prehybridization buffer with the hybridization buffer (*see* Notes 18 and19) containing the denatured labeled probe (obtained by boiling the probe for 5–10 min and immediately cooling on ice) and denatured salmon sperm DNA (30 μg/mL).The probe concentration should correspond to about 5×10^5 dpm/mL activity. Incubate in a rotating oven, at the same temperature used for prehybridization, for 8–24 h.

3.5. Posthybridization

1. Pour off the hybridization buffer, containing the nonbound probe, and rinse the membrane briefly, by inverting the bottle several times, with two washes of 2X SSC buffer at room temperature.
2. Transfer the membrane to a plastic box and wash at the required stringency (e.g., 1X SSC, 0.1% w/v SDS, at 65°C) in a shaking incubator for 15–30 min (*see* Note 20). Place the membrane on a sheet of clingfilm or aluminum foil and check the residual activity bound to the membrane with a portable Geiger counter. If the activity exceeds 10–50 cps, wash the membrane again.
3. Remove any remaining unbound probe by rinsing the membrane an additional 2–3 times with 2X SSC buffer at room temperature. Cover the membrane with clingfilm and place in an autoradiography cassette with two intensifying screens. Thoroughly dry the clingfilm surface with absorbent paper. Working in a darkroom, place one X-ray film in the cassette, seal and leave at –80°C overnight to expose the film (bend one corner of the X-ray film as a reference point).
4. Allow the autoradiography cassette to warm up at room temperature for 15 min. In a darkroom open the cassette and slowly remove the film. *Caution:* If this is done quickly, the developed film may become partly fogged by a discharge of static electricity corresponding to the areas of the film that

have become stuck to the clingfilm covering the membrane; this is a particularly common problem if the clingfilm has not been completely dried in Section 3.5., step 3. Develop the film manually or automatically in a film processing machine, and evaluate the hybridization signal. If the signal is too low, expose again for a longer period of time. If high background is detected, rewash the membrane as in Section 3.5., step 2, either for a longer time or under higher stringency conditions. *Caution:* Do not allow the membrane to dry during the washing/exposure procedure as this causes irreversible binding of the probe.

5. The membrane can be hybridized again with a new probe. In this case the probe is stripped off by incubating with about 0.4 mL/cm^2 of alkali wash solution at 45°C for 30 min with gentle shaking.
6. Neutralize by incubating with neutralization buffer as in the previous step. After a brief rinse in distilled H_2O, the membrane is ready for reuse. Up to six to eight rounds of hybridization can be carried out with the same membrane without significant loss of signal.

3.6. Applications

In this section we provide examples of how Southern blotting can be used to determine the copy number and chromosomal locations of sequences, and to analyze the structures of multigene families.

3.6.1. Gene Copy Number

The number of copies of a cloned sequence or similar sequences in a genome can be estimated by filter hybridization experiments. The genomic DNA is restricted with one endonuclease (or preferably several endonucleases in independent digestions) and is separated on a gel with a dilution series of the cloned DNA sequence under analysis (as used to prepare the probe). Comparison of the intensities of the hybridization signals of the genomic DNA sequences and the reference DNA allows a good approximation of the number of copies of the cloned DNA within the genome. For a faithful evaluation of gene copy numbers several points should be considered:

1. The number of copies of a given DNA sequence are conventionally based on the haploid complement of the genome (1C value).
2. Accurate determination of the concentrations of both genomic and reference DNAs is crucial (*see* Note 5). For example, to obtain a reference signal equivalent to one copy per haploid genome, the moles of reference DNA must be equal to the moles of genomic DNA used. The following is

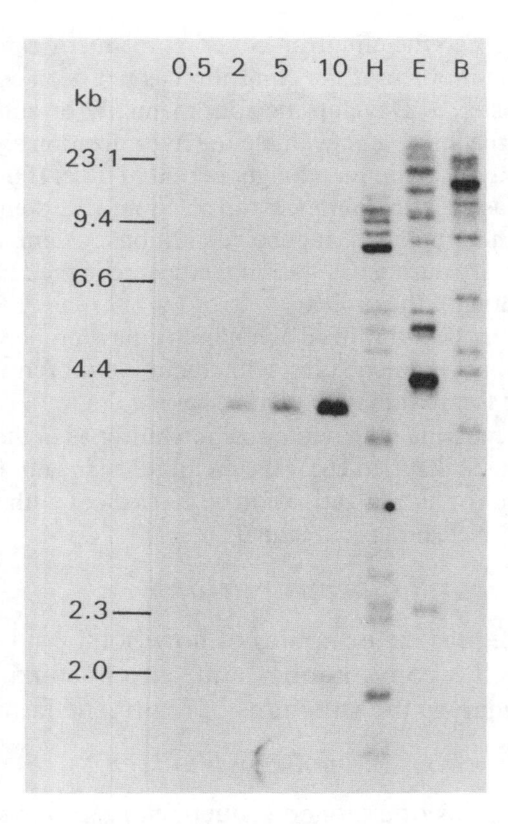

Fig. 1. Estimation of gene copy number by Southern analysis. Bread wheat (cv, Chinese Spring) genomic DNA is digested with *Hin*dIII (H), *Eco*RI (E), and *Bam*HI (B), and hybridized with pKAP1a, a γ-gliadin cDNA *(2)*. Reference amounts of linearized plasmid containing the insert probe corresponding to 0.5, 2, 5, and 10 copies/haploid genome are included, as indicated. Individual genomic fragments are estimated to correspond to between one and fifteen copies of related sequences. Hybridization was at 65°C. Washing was in 0.5X SSC, 0.1% w/v SDS at 65°C. Exposure was for 1 d.

an example of such a calculation for the pKAP1a clone and hexaploid bread wheat (cv, Chinese Spring) (*see* Note 7 and Fig. 1):

a. 1C DNA complement = 17.3 pg
b. 8 μg DNA/lane = 4.62×10^5 moles of haploid genome
c. 1 bp = 1.036×10^{-9} pg/mol
d. Plasmid size = 3.714×10^3 bp

   e. Amount of the plasmid corresponding to 1 copy/haploid genome = 4.62 $\times 10^5 \times 1.036 \times 10^{-9} \times 3.714 \times 10^3 = 1.77$ pg
   f. Large compilations of nuclear DNA contents of many higher plant species have been reported *(12,13)*.
3. The genomic DNA should be digested with enzymes that do not cut within the probe, otherwise more complex patterns that are difficult to interpret could be obtained.
4. This approach often does not discriminate between active and inactive genes (pseudogenes).
5. When applied to studies of gene families, variation in the degree of similarity between target sequence and probe, and the occurrence of repetitive motifs within and between individual genes, can significantly affect copy number estimations.
6. If linearized plasmids are used as reference DNAs, the possibility of cross-hybridization between probe and vector sequences should be ruled out by preliminary hybridization experiments, or by computer-aided sequence comparison analyses *(14)*. Alternatively, vector DNA without any insert may be run on the same gel and hybridized as a negative control.

3.6.2. Chromosomal Localization

Mapping a given sequence to one, or a set of, restriction fragments is the basis for the construction of genetic maps based on RFLPs, a major area of research that has developed based on Southern blotting techniques. In addition, probes can also be mapped to chromosomes if suitable genetic stocks are available. An example of this is shown in Fig. 2. In this the chromosomal locations of LMW glutenin gene sequences (which encode gluten protein components) are determined in hexaploid bread wheat by comparison of the hybridization pattern of *Eco*RI digests of euploid (CS) line and aneuploid nullisomic-tetrasomic lines *(15)*. The latter lines lack one pair of homoeologous group 1 chromosomes, but this is compensated for by the presence of four copies of another group 1 chromosome (e.g., nulli 1A, tetra 1B). This allows the identification of restriction fragments present on specific group 1 chromosomes. Similar comparisons with ditelosomic lines ($1A^s$, $1A^l$, $1B^s$, $1B^l$) lacking the short or long arms of chromosomes 1A or 1B allow hybridizing bands to be located on specific chromosomal arms. Figure 2 also shows the localization of hybridizing fragments to specific chromosomes in tetraploid durum wheat (cv, Langdon) by comparing the patterns of the euploid line with those of substitution lines in which chromosomes 1A or 1B of

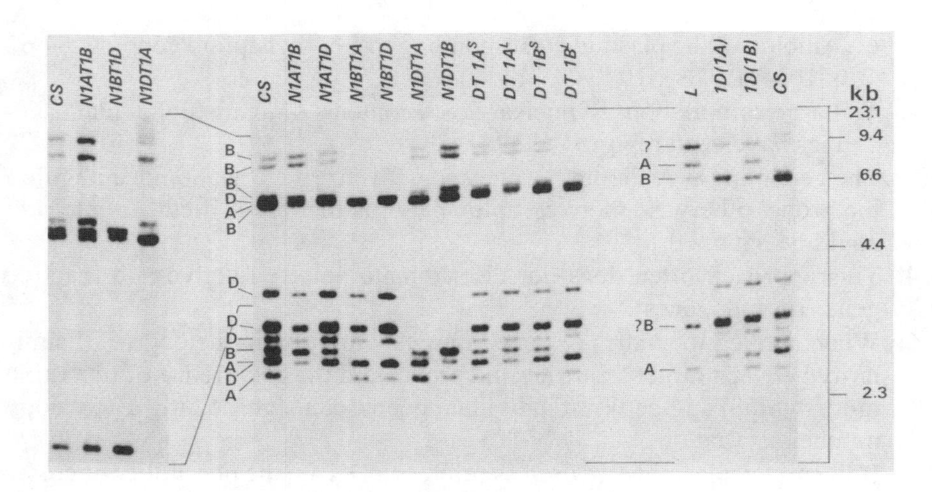

Fig. 2. Determination of the chromosomal localization of sequences. An LMW-glutenin probe is hybridized to DNAs from euploid bread wheat (cv, Chinese Spring, CS), and durum wheat (cv, Langdon, L), from nullisomic-tetrasomic and ditelosomic lines of Chinese Spring (center part), and from disomic substitution lines of Langdon that have chromosomes 1A or 1B substituted by 1D chromosomes from Chinese Spring (right part). Comparisons of the hybridization patterns allow all the bands present in Chinese Spring and most of the bands in Langdon to be assigned to group 1 chromosomes (and to chromosomal arms in the case of Chinese Spring). Question marks indicate bands whose chromosomal locations are uncertain. The left part of the figure shows that high resolution of bands is obtained after a long gel separation of about 50 h *(2)*. Hybridization was at 65°C. Washing was in 0.5X SSC, 0.1% w/v SDS at 65°C. Exposure was for 3 d.

Langdon has been substituted by a pair of 1D chromosomes from Chinese Spring *(16)*. The resolution of fragments can be improved by increasing the running time of the gel up to 50–60 h *(2)*, as shown in the left part of Fig. 2. This simple modification can significantly improve band resolution especially in the high molecular size range (*see also* Fig. 3).

3.6.3. Multigene Families

Multigene families usually contain genes with different degrees of sequence similarity *(17)*. It is therefore possible to identify subsets of sequences, which are more or less closely related to the probe available, by Southern or northern analyses, using different stringency conditions.

An example is given in Fig. 3. *Eco*RI-digested DNAs from euploid and aneuploid lines of bread wheat cv. Chinese Spring were hybridized to pKAP1a, a γ-gliadin cDNA whose repetitive 5'-end has sequence similarity to genes encoding ω-gliadins *(2)*. Under conditions of moderate stringency (m) a complex hybridization pattern is obtained, indicating a relatively large multigene family. However, under conditions of increased stringency, several bands either did not hybridize or hybridized only weakly (indicated by arrows in Fig. 3), indicating that their sequences are less closely related to that of pKAP1a. Further studies using family-specific probes and analysis of deletion mutants have shown that these bands may in fact contain ω-gliadin sequences *(2,3)*.

4. Notes

4.1. Endonuclease-Free Environment

1. DNA is subject to enzymatic degradation by nucleases that can be present on glassware, plasticware, and the operator's skin. Wherever possible, glassware should be treated with sulfuric/chromic acid, plasticware, and solutions autoclaved, and disposable gloves used.

4.2. Isolation of Genomic DNA

2. Plant material can alternatively be ground with a polytron or Waring blender. This increases the speed of the operation but can result in undesirable cross-contamination when several samples are to be processed. Although a low level of cross-contamination does not affect many applications, such as restriction fragment length polymorphism (RFLP) analysis, it could be a serious problem for more sensitive applications such as amplification by the polymerase chain reaction (PCR).
3. *Caution:* Ethidium bromide is a nucleic acid intercalating agent and mutagen; avoid contact with the skin.
4. The method described here is based on that of Kreis et al. *(18)* utilizing isopycnic centrifugation of nucleic acids on a CsCl gradient. Separation is based on the different densities of double-stranded DNA and single stranded RNA, that result, in part, from their different ethidium bromide binding capacities. By ultracentrifugation on a gradient of CsCl, it is possible to separate DNA (which will form a discrete band) from RNA (which will form a bright fluorescent pellet). Other impurities, mostly proteins, will remain on top of the gradient. The density of the sample is critical for a proper separation and should be checked, before centrifugation, by measuring both volume and weight. Values between 1.55–1.60 g/mL are ideal. If the density of the sample is too low, it can be adjusted by adding solid CsCl. (Note that this will also result in an increase in the volume.)

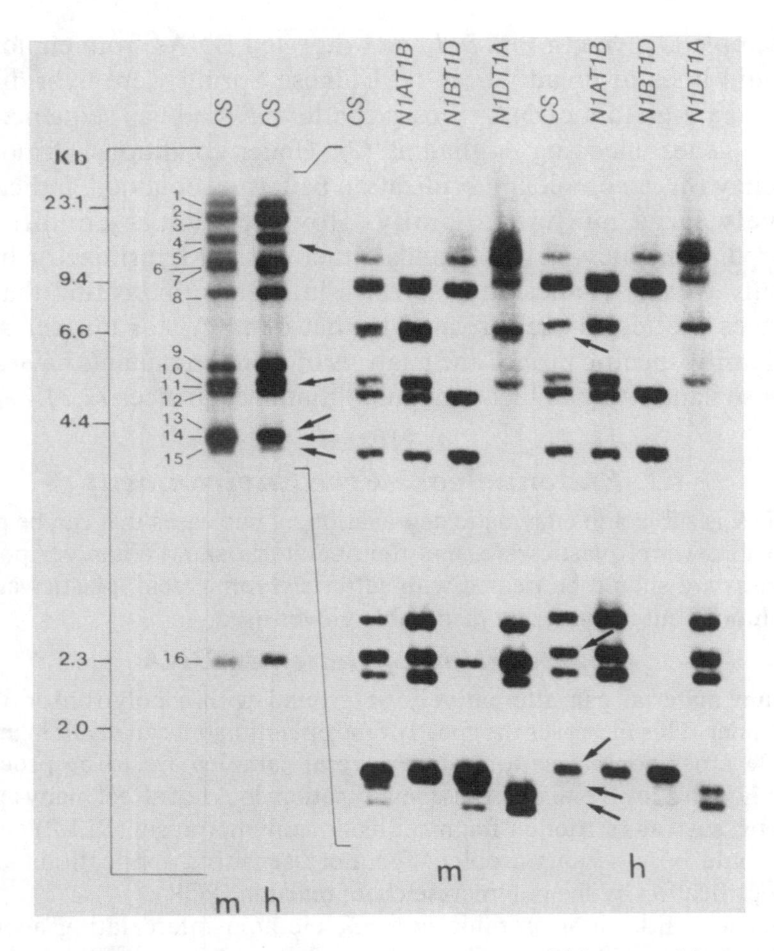

Fig. 3. Identification of a subset of sequences within a multigene family, characterized by a low degree of homology with the probe used. *Eco*RI digests of DNAs from euploid bread wheat, cv, Chinese Spring (CS) and group 1 nullisomic-tetrasomic lines are hybridized to pKAP1a, a γ-gliadin cDNA. Washing was carried out under moderate stringency conditions (m), and then under high stringency conditions (h). Arrows indicate restriction fragments with a low sequence similarity with pKAP1a that hybridize weakly or not at all under high stringency conditions. Further analyses with a γ-gliadin specific probe and analysis of mutants have shown that these fragments encode a subset of less-closely related gluten components (which probably correspond to ω-gliadins) *(2,3)*. Hybridization was at 65°C. Washing was in 0.5X SSC (m) or 0.2X SSC (h), 0.1% w/v SDS at 65°C. Exposure was for 2 d.

5. We have found that spectrophotometric readings often underestimate the concentrations of nucleic acids. The resulting errors in gel loading can in turn result in poor resolution, and inaccurate estimates of gene copy numbers. The preliminary estimates of DNA concentration based on absorbance measurements should therefore be adjusted by comparing aliquots of each sample with dilutions of high molecular weight marker DNA after brief electrophoresis on a low-concentration agarose gel. The estimates are most accurate when the quantity of DNA loaded is in the range of 50–100 ng/lane.
6. The DNA isolation procedure described here is suitable for relatively large preparations (final DNA yield about 0.2–1.5 mg). It can be scaled down to half quantities using smaller ultracentrifugation tubes (e.g., Beckman heat-sealed polyallomer tubes). In this case different rotors and centrifugation conditions should be used (e.g., Beckman type 80Ti rotor at 45K rpm for 15 h, or type 90VTi rotor at 80K rpm for 6 h). Disposable syringes and needles should also be used to recover the DNA from the CsCl gradient. We have found that DNA prepared by this method is stable at –20°C for several years. However, if a large number of samples must be processed, other protocols for small scale preparation might be more convenient and faster to use *(19,20)*.

4.3. Digestion and Electrophoresis

7. The amount of DNA to be digested depends on the size of the genome and the redundancy of the target sequence in the genome. For small genomes (e.g., *Arabidopsis thaliana* L. with a haploid genome of about 0.08 pg *[21]*) a small amount of DNA (i.e., 1–2 μg) will be sufficient for most purposes. Analyses of species with large genomes require larger amounts of DNA, as target sequences are more diluted. For example, 5–10 μg of genomic DNA should be used in order to detect a single-copy target sequence in hexaploid bread wheat (haploid genome size = 17.3 pg *[12]*) in a reasonably short exposure time (1–3 d).
8. We have found that the conditions recommended in Section 3.2., step 1 give reproducibly complete digestion of DNA. This is presumably owing to the high purity of the DNA prepared on a CsCl gradient, the relatively large reaction volume (in which any inhibitors are diluted out below a critical level), and the use of spermidine *(22)*, which binds and stabilizes the DNA double helix increasing the site specificity of the restriction enzyme. However, spermidine can also have an inhibitory role with some endonucleases, and its utilization should be evaluated for each enzyme.

9. Digesting the DNA in a relatively large volume (50 μL) requires large wells for loading the samples. For high resolution gels, we routinely use 20 cm × 25 cm × 0.5 cm gels with a 20-well comb (2–3 mm thick) (Gibco-BRL, Gaithersburg, MD). This allows the entire digestion reaction to be loaded without any spillage or overflowing. We do not recommend phenol/chloroform extraction followed by ethanol precipitation, to concentrate the sample after digestion, as this may give quantitatively heterogeneous results between the samples loaded. Sample concentration by evaporation in a speedvac centrifuge is also to be avoided because the concentration of salts present in the sample can cause severe migration artifacts.
10. It is important to keep a photographic record of the gel because the sizes of the hybridization bands are estimated by comparing their migration distances with those of the ethidium bromide stained size markers. A fluorescent ruler placed at one edge of the gel can be used to measure the migration distances of the nucleic acids. A photograph of the gel will also be useful to help explain abnormalities or quantitative differences in the hybridization patterns. Whether to transfer the DNA size markers is a matter of personal preference. We prefer not to blot them because probes sometimes hybridize to minor components (undetectable by staining) present in the size markers, which can result in errors in estimating the sizes of the hybridization bands *(23)*. Similarly, we prefer not to use ^{32}P or ^{35}S labeled size markers as they often require different exposure times from the hybridization products, and, because of their relatively fast radio-decay rates, are of little use in hybridization experiments performed several weeks or months after gel transfer.
11. Many sets of DNA size markers are commercially available, and additional markers can be prepared by digesting cloning vectors with suitable restriction enzymes. However, bacteriophage λ DNA digested with *Hin*dIII and a 1-kb ladder (Gibco), cover most requirements.

4.4. Blotting

12. The method recommended here does not include any depurination treatment (usually by incubating the gel with 0.25*M* HCl at room temperature for 15–30 min). This step is often recommended in order to nick the DNA to facilitate the transfer of large fragments. However, we have found it difficult to control and reproduce. Excessive DNA nicking can result in the production of short fragments and consequent loss of hybridization signal. On the other hand, up to 15–20% of the high molecular size DNA (larger than 15 kb) may remain in the gel if depurination is omitted. We prefer to avoid acid nicking and to compensate for this in evaluating autoradio-

graphs. However, depurination is necessary for Southern analysis of very large DNA fragments, e.g., following pulsed field gel electrophoresis (PFGE) *(24)*.

13. A range of commercial hybridization membranes is available, and the choice largely depends on personal preferences. However, nylon membranes are generally preferred for their superior physical strength, allowing several rounds of hybridization to be performed. Positively charged membranes have been claimed to give improved performances, but we have not tested them.
14. There is a direct relationship between salt concentration and the efficiency of transfer by capillarity, especially for large fragments *(1)*. It is advisable therefore, to use high salt concentrations (i.e., 20X SSC buffer) for blotting genomic DNA, whereas lower salt concentrations (typically 10X SSC buffer) should be used for short sequences (e.g., plasmids and PCR products).
15. Air bubbles can become trapped between the layers of the blotting apparatus and must be removed by rolling a glass rod over the surface. The bubbles interfere with the DNA transfer and are easily identifiable on the autoradiograph as a circular area lacking any hybridization signal, including background.
16. Although in this paper we describe the use of blotting procedures originally developed by Southern *(1)*, a number of variants have been reported. These include the use of alkaline blotting buffers *(25)*, positive or negative pressure *(26)*, electroblotting *(27,28)*, and downward capillarity blotting *(29,30)*. They are usually faster than the standard Southern blotting procedure and would be worth adopting for projects which require the analysis of many samples (e.g., RFLP mapping).

4.5. Hybridization and Posthybridization

17. The introduction of glass hybridization bottles and rotating hybridization ovens (Hybaid or similar) has considerably simplified the practical aspects of the procedure. Heat-sealed plastic bags, traditionally used in hybridization experiments, can still be useful, especially when processing large membranes that can give high background signal due to overlapping of the membrane when used in a bottle without separating meshes.
18. The conditions in the prehybridization and hybridization steps are designed to promote specific annealing of the probe to target sequences and, at the same time, to control nonspecific binding to both the membrane and non-target DNA. The buffer recommended in this method *(31)* includes dried milk *(32)* and denatured salmon sperm DNA as blocking agents, and PEG 6000 as a high molecular size polymer which occupies most of the reaction volume resulting in much higher probe concentrations *(33)*. This buffer

allows the use of relatively low nominal probe concentrations (resulting in saving of probe) in a relatively large reaction volume (which simplifies the practical aspects of the procedure). Furthermore, dried milk and PEG are both inexpensive.

19. T_m and stringency. A number of factors affect hybridization experiments and have been comprehensively reviewed elsewhere *(34–36)*. Among these, factors that affect the thermal stability of hybrid nucleic acid molecules play a major role. The stability of nucleic acid duplexes is given by the melting temperature (T_m), the temperature at which the hybrid is 50% denatured for a given set of conditions used. Equation (1) is used widely for calculating T_m values in hybridization experiments *(37)*:

$$T_m\ (^\circ C) = 81.5 + 16.6\ (\log_{10}M) + 0.41\ (\%\ G + C) - 0.72\ (\%\ f) - 500/n \quad (1)$$

where M is the molarity of monovalent cations (typically Na^+), (% G + C) is the percentage of guanosine and cytosine nucleotides (% f) is the percentage of formamide (if used), and n is the length of the hybrid (in bases).

Although the above equation derives from analyses of hybridization in solution, it gives a good approximation also for hybridization on solid supports. It shows clearly that, for 100% complementary nucleic acid hybrids, the thermal stability of nucleic acid duplexes is directly related to the salt concentration, the proportion of triple hydrogen bond-forming nucleotides (i.e., G + C), and the length of the hybrid. It also depends inversely on the concentration of double helix-destabilizing agents (typically, formamide). Although formamide allows the use of lower hybridization temperatures, this does not represent in our experience a practical advantage, and we prefer not to use it.

The base composition does not significantly affect the stability of most hybrids, unless it is dramatically unbalanced. However, when short probes are used (i.e., oligonucteotides < 50 nucleotides in length) it can play a major role and the T_m is more accurately calculated using Eq. (2):

$$T_m\ (^\circ C) = 4\ (G + C) + 2\ (A + T) \quad (2)$$

for a standard 0.9*M* NaCl concentration. A, C, G, and T are the numbers of the corresponding nucleotides in the oligonucleotide. *See* Chapter 18, this volume, for specific considerations regarding hybridizations using oligonucleotide probes.

Although Eqs. (1 and 2) are valid for perfectly matching hybrids, mismatches play a major role and have been estimated on average to decrease the T_m by approx 1°C for each 1% of mismatch *(38)*. However, mismatch distribution may also be a major factor in determining the stability of mismatched duplexes.

The criterion value *(C)* is an inverse measure of the stringency used in the experiment. It is given by Eq. (3):

$$C\ (^{\circ}C) = T_m - T_i \quad (3)$$

where T_i is the experimental temperature. When T_i is low, C is high and the stringency is low, and vice versa. As a rule of thumb, closely related sequences can be detected by using high stringency conditions (e.g., $5 < C < 10$), and low stringency conditions can be used to detect less complementary sequences.

20. As a general rule, it is advisable initially to carry out hybridizations under relatively low stringency conditions and progressively to increase the stringency in the subsequent washing steps. This allows the probe to hybridize to relatively distantly related sequences (if any) that can be detected initially by autoradiography and subsequently removed by high stringency washes. This can provide very interesting information, as described in Section 3.6.3. We hybridize initially under relatively low stringency conditions: about T_m - 30 for an average probe at 65°C. However, for specific applications, such as gene library screening, the stringency can be reduced further by lowering the temperature. The stringency is then increased simply by decreasing the salt concentration in the washings (Fig. 3).

We give herein a practical example of the calculation of T_m and C values, using the γ-gliadin cDNA pKAP1a *(2,39)*, which was used in the experiments shown in Figs. 1 and 3:

a. pKAP1a size = 1049 bp
b. G + C content = 49%
c. T_i = 65°C
d. Na^+ concentrations:
   i. Hybridization conditions (1.5X SSPE, 1% w/v SDS) = 0.32*M*
   ii. Moderate-stringency washing conditions (0.5X SSC, 0.1% w/v SDS) = 0.1*M*
   iii. High-stringency washing conditions (0.2X SSC, 0.1% w/v SDS) = 0.04*M*
e. Hybridization conditions:

$$T_m = 81.5 + 16.6\ (-0.49) + 0.41\ (49) - 500/1{,}049 = 94^{\circ}C \quad (4)$$

$$C = T_m - T_i = 94^{\circ}C - 65^{\circ}C = 29^{\circ}C \quad (5)$$

f. Moderate-stringency washing conditions: T_m = 85°C; C = 20°C
g. High-stringency washing conditions: T_m = 79°C; C = 14°C

Acknowledgments

The authors thank Shirley Burgess and John Lenton for comments on the manuscript and Bob Harvey and Ken Williams for photographic assistance.

References

1. Southern, E. M. (1975) Detection of specific sequences among DNA fragments separated by gel electrophoresis. *J. Mol. Biol.* **98,** 503–517.
2. Sabelli, P. A. and Shewry, P. R. (1991) Characterization and organization of gene families at the *Gli-1* loci of bread and durum wheats by restriction fragment analysis. *Theor. Appl. Genet.* **83,** 209–216.
3. Sabelli, P. A., Lafiandra, D., and Shewry, P. R. (1992) Restriction fragment analysis of 'null' forms at the *Gli-1* loci of bread and durum wheats. *Theor. Appl. Genet.* **83,** 428–434.
4. Sabelli, P. A., Burgess, S. R., Carbajosa, J. V., Parker, J. S., Halford, N. G., Shewry, P. R., and Barlow, P. R. (1993) Molecular characterization of cell populations in the maize root apex, in *Molecular and Cell Biology of the Plant Cell Cycle* (Ormrod, J. C. and Francis, D., eds.), Kluwer, Dordrecht, pp. 97–109.
5. Alderson, A., Sabelli, P. A., Dickinson, J. R., Cole, D., Richardson, M., Kreis, M., Shewry, P. R., and Halford, N. G. (1991) Complementation of *snf1,* a mutation affecting global regulation of carbon metabolism in yeast, by a plant protein kinase cDNA. *Proc. Natl. Acad. Sci. USA* **88,** 8602–8605.
6. Hull, G., Sabelli, P. A., and Shewry, P. R. (1992) Restriction fragment analysis of the secalin loci of rye. *Biochem. Genet.* **30,** 85–97.
7. Halford, N. G., Vicente-Carbajosa, J., Sabelli, P. A., Shewry, P. R., Hannappel, U., and Kreis, M. (1992) Molecular analyses of a barley multigene family homologous to the yeast protein kinase gene *SNF1. Plant J.* **2,** 791–797.
8. Mundy, C. R., Cunningham, M. W., and Read, C. A. (1991) Nucleic acid labeling and detection, in *Essential Molecular Biology. A Practical Approach*, vol. II (Brown, T. A., ed.), IRL Press, Oxford, pp. 57–110.
9. Hoeltke, H. J., Ettl, I., Finken, M., West, S., and Kunz, W. (1992) Multiple nucleic acid labeling and rainbow detection. *Anal. Biochem.* **207,** 24–31.
10. Dyson, N. J. (1991) Immobilization of nucleic acids and hybridization analysis, in *Essential Molecular Biology. A Practical Approach,* vol. II (Brown, T. A., ed.), IRL Press, Oxford, pp. 111–156.
11. Sambrook, J., Fritsch, E. F., and Maniatis, T. (1989) *Molecular Cloning. A Laboratory Manual.* Cold Spring Harbor Laboratory Press, Cold Spring Harbor, NY.
12. Bennett, M. D. and Smith, J. B. (1976) Nuclear DNA amounts in angiosperms. *Proc. R. Soc. Lond. Ser. B.* **274,** 227–274.
13. Bennett, M. D., Smith, J. B., and Heslop-Harrison, J. S. (1982) Nuclear DNA amounts in angiosperms. *Proc. R. Soc. Lond. Ser. B.* **216,** 179–199.
14. Staden, R. (1982) An interactive graphics program for comparing and aligning nucleic acid and amino acid sequences. *Nucl. Acids Res.* **10,** 2951–2961.
15. Sears, E. R. (1954) *The Aneuploids of Common Wheat.* University of Missouri College of Agriculture Agricultural Experiment Station, Columbia, MO, pp. 1–59.

16. Joppa, L. R. and Williams, N. D. (1988) Langdon durum disomic substitution lines and aneuploid analysis in tetraploid wheat. *Genome* **30,** 222–228.
17. Beltz, G. A., Jacobs, K. A., Eickbush, T. H., Cherbas, P. T., and Kafatos, F. C. (1983) Isolation of multigene families and determination of homologies by filter hybridization methods. *Meth. Enzymol.* **100,** 266–285.
18. Kreis, M., Shewry, P. R., Forde, B. G., Rahman, S., and Miflin, B. J. (1983) Molecular analysis of a mutation conferring the high-lysine phenotype on the grain of barley (*Hordeum vulgare*). *Cell* **34,** 161–167.
19. Dellaporta, S. L., Wood, J, and Hicks, J. B. (1983) A plant DNA minipreparation: version 2. *Plant Mol. Biol. Rep.* **1,** 19–22.
20. Tai, T. and Tanskley, S. (1991) A rapid and inexpensive method for isolation of total DNA from dehydrated plant tissue. *Plant Mol. Biol. Rep.* **8,** 297–303.
21. Meyerowitz, E. M. (1992) Introduction to the *Arabidopsis* genome, in *Methods in* Arabidopsis *Research* (Koncz, C., Chua, N.-H., and Schell, J., eds.), World Scientific Publishing, Singapore, pp. 100–118.
22. Pingoud, A. (1985) Spermidine increases the accuracy of type II restriction endonucleases. *Eur. J. Biochem.* **147,** 105–109.
23. Sabelli, P. A. and Shewry, P. R. (1993) Nucleic acids blotting and hybridisation, in *Methods in Plant Biochemistry* (Bryant, J. A., ed.), Academic, London, pp. 79–100.
24. Sørensen, M. B. (1989) Mapping of the *Hor2* locus in barley by pulsed field gel electrophoresis. *Carlsberg Res. Commun.* **54,** 109–120.
25. Reed, K. C. and Mann, D. A. (1985) Rapid transfer of DNA from agarose gels to nylon membranes. *Nucl. Acids Res.* **13,** 7207–7221.
26. Olszewska, E. and Jones, K. (1988) Vacuum blotting enhances nucleic-acid transfer. *Trends Genet.* **4,** 92–94.
27. Smith, M. R., Devine, C. S., Cohn, S. M., and Lieberman, M. W. (1984) Quantitative electrophoretic transfer of DNA from polyacrylamide or agarose gels to nitrocellulose. *Anal. Biochem.* **137,** 120–124.
28. Ishihara, H. and Shikita, M. (1990) Electroblotting of double-stranded DNA for hybridization experiments: DNA transfer is complete within 10 minutes after pulsed-field gel electrophoresis. *Anal. Biochem.* **184,** 207–212.
29. Lichtenstein, A. V., Moiseev, V. L., and Zaboikin, M. M. (1990) A procedure for DNA and RNA transfer to membrane filters avoiding weight-induced gel flattening. *Anal. Biochem.* **191,** 187–191.
30. Chomczynski, P. (1992) One-hour downward alkaline capillary transfer for blotting of DNA or RNA. *Anal. Biochem.* **201,** 134–139.
31. Boulnois, G. J. (1987) *University of Leicester Gene Cloning and Analysis: A Laboratory Guide.* Blackwell Scientific, Oxford, pp. 47–60.
32. Johnson, D. A., Gautsch, J. W., Sportsman, J. R., and Elder, J. H. (1984) Improved technique utilizing nonfat dry milk for analysis of proteins and nucleic acids transferred to nitrocellulose. *Gene Anal. Techn.* **1,** 3–8.
33. Amasino, R. M. (1986) Acceleration of nucleic acid hybridization rate by polyethylene glycol. *Anal. Biochem.* **152,** 304–307.
34. Hames, B. D. and Higgins, S. J. (1985) *Nucleic Acid Hybridisation. A Practical Approach.* IRL Press, Oxford.

35. Meinkoth, J. and Wahl, G. (1984) Hybridization of nucleic acids immobilized on solid supports. *Anal. Biochem.* **138,** 267–284.
36. Wetmur, J. G. (1991) DNA probes: applications of the principles of nucleic acid hybridization. *Crit. Rev. Biochem. Mol. Biol.* **26,** 227–259.
37. Schildkraut, C. and Lifson, S. (1965) Dependence of the melting temperature of DNA on salt concentration. *Biopolymers* **3,** 195–208.
38. Bonner, T. I., Brenner, D. J., Neufield, B. R., and Britten, R. J. (1973) Reduction in the rate of DNA reassociation by sequence divergence. *J. Mol. Biol.* **81,** 123–135.
39. Bartels, D., Altosaar, I., Harberd, N. P., Baker, R. F., and Thompson, R. D. (1986) Molecular analysis of γ-gliadin gene families at the complex *Gli-1* locus of bread wheat *(T. aestivum). Theor. Appl. Genet.* **72,** 845–853.

CHAPTER 14

Isolation and Characterization of Plant Genomic DNA Sequences via (Inverse) PCR Amplification

Remko Offringa and Frederique van der Lee

1. Introduction

Today the isolation and characterization of a gene of interest from any organism has become a standard procedure. One of the milestones that have facilitated this procedure in particular, has been the development of in vitro techniques for amplification of specific RNA or DNA sequences. By far the most important amplification technique, the polymerase chain reaction (PCR), was proposed more than 20 years ago *(1)*, but only reached its practical form in 1985 *(2)*. The basic principle is that following heat denaturing of the DNA sample two oligodeoxynucleotide primers are allowed to anneal to their target DNA sequences located at the opposite DNA strand. In the presence of deoxynucleotide triphosphates, DNA polymerase synthesizes new strands from the 3' hydroxyl ends of the primers, doubling the number of copies of the target DNA segment. Repeated cycles of denaturation, primer annealing and extension eventually result in accumulation of the DNA segment defined by the primers. In 1985 Saiki et al. *(2)* used the Klenow fragment of *Escherichia coli* DNA polymerase I, which because of its heat-sensitivity had to be added to the reaction mixture after each denaturation step. The discovery and application of the heat stable DNA polymerase from the thermophilic bacterium *Thermus aquaticus* for PCR and the automation of the procedure *(3)* laid the foundation for the development of numerous and diverse PCR methods *(4)*.

From: *Methods in Molecular Biology, Vol. 49: Plant Gene Transfer and Expression Protocols*
Edited by: H. Jones Humana Press Inc., Totowa, NJ

The introduction of PCR technology in plant molecular genetics has provided a new set of molecular markers that are used in plant breeding to identify quantative trait loci (QTL) and in fundamental scientific research to map mutations. Moreover, PCR has proven to be a valuable tool in the more recently developed plant transformation technology. The aim of this chapter is to provide a short overview on several of the PCR based methods that are presently used by plant molecular geneticists, with the emphasis on the inverse PCR method and its use in the different research projects in our laboratories. Ready-to-use protocols are described for PCR- and inverse PCR amplification on genomic DNA of *Nicotiana tabacum* and *Arabidopsis thaliana*.

PCR can be used to detect allelic polymorphisms such as deletions, insertions, and basepair substitutions in a known DNA sequence. The detection of larger inserts (such as transposons or transgenes) or deletions is very straightforward, leading to new PCR fragments or PCR fragments of different size. Of the small changes in a specific sequence (including basepair mutations), only those that lead to the introduction or removal of a restriction site can be detected by restriction fragment length polymorphism (RFLP) analysis. The traditional RFLP analysis includes time-consuming Southern blot analysis on a larger quantity of digested genomic DNA, whereas the PCR based RFLP method allows detection of the presence or absence of a restriction site by gel electrophoresis of the digested PCR fragments amplified from a relatively small amount of genomic DNA *(5)*. These cleaved amplified polymorphic sequences (CAPS) are especially useful for plant species that do not allow Southern blot analysis because of their large genome size. In the more abundant cases where small changes do not lead to an RFLP, detection has become possible via denaturing gradient gel electrophoresis (DGGE, *6*) or single-stranded conformation polymorphism (SSCP) analysis *(7,8)* of PCR fragments covering the mutated DNA sequence. The presence of a small change in the sequences of a PCR fragment results in a mobility shift caused by a difference in melting properties of double-stranded PCR products on a denaturing gel or a conformational change of denatured (single-stranded) PCR products on a nondenaturing gel, respectively.

A PCR-based approach for which the target DNA sequences are not predetermined, involves the use of a single 9 or 10 bp PCR primer of arbitrary sequence. PCR with such a primer results in random amplified polymorphic DNA (RAPD) fragments, that can be genotype specific depending on the combination of primer and annealing temperature *(9)*. A drawback on this technique is that the pattern very much depends on the PCR conditions and may vary per experiment and per thermal cycler.

As mentioned above the more recently developed plant transformation technology requires specific PCR techniques for the analysis of transgenic plant lines. Most transformation protocols for dicotyledonous plant species rely on the capacity of the soil bacterium *Agrobacterium* to transfer a defined part of an extrachromosomal plasmid, the transferred- or T-DNA, to the nucleus of the plant cell, where it is inserted in the plant genome via a process of nonhomologous recombination (*see* ref. *10* for review). The borders of the T-DNA on the plasmid are defined by 24 bp imperfect repeats, which are the targets for the transfer machinery of the bacterium. A nice property of *Agrobacterium*-mediated DNA transfer is that a large proportion of the transgenic plant lines contains only one or a few intact copies of the T-DNA inserted at one location. Especially the right border end of the T-DNA is conserved in many inserts. *Agrobacterium* T-DNA has proven to be a valuable tool for insertional mutagenesis of the *Arabidopsis thaliana* genome *(11,12)*. Moreover, transformations with a T-DNA construct carrying the promoterless gene encoding β-glucuronidase (*gusA*), a histochemical marker, with its translational start adjacent to the right border repeat, resulted in gene activity in a high percentage of the transgenic lines *(13)*. This indicated that T-DNA is preferentially inserted in regions of the plant genome that are transcriptionally active. Such T-DNA tagged lines are now used in our laboratories to isolate tissue specific promoters *(14)*.

One of the projects in our laboratories, for which several PCR methods were developed and applied, concerns the study of site-directed integration of *Agrobacterium* T-DNA via homologous recombination in the plant genome (gene targeting). Transgenic (target) lines of *Nicotiana tabacum* were obtained with a T-DNA construct that carries a defective kanamycin resistance (*nptII*) gene with a deletion in its 3' end. Protoplasts of one selected target line were retransformed with a repair T-DNA construct carrying a defective *nptII* gene with a nonoverlapping deletion in its 5' end. Calli were selected for resistance to kanamycin and the resistant calli were screened for the presence of an intact *nptII* gene by a PCR approach described by Kim and Smithies *(15)*, using opposite primers annealing in the regions that are deleted in the defective *nptII* genes. Amplification of a fragment of expected size was observed in only a few of the resistant clones. These calli indeed contained an intact *nptII* gene *(16,17)*. Sometimes PCR resulted in amplification of unspecific fragments, but this could be solved by increasing the annealing temperature. No recombination products were formed due to annealing of incomplete elongation products during PCR, a phenomenon reported by Meyerhans et al. *(18)* and Marton et al. *(19)*. In more recent experiments, it appeared to be possible to detect one recombinant clone in a pool of 25–50 resistant calli *(19a)*.

The different tobacco lines transgenic for the target T-DNA were screened for single copy T-DNA inserts. With inverse PCR, a PCR-based technique with which regions flanking known DNA sequences can be amplified *(4,18,20)*, a quick and reliable estimation could be obtained of the T-DNA copy number in these transgenic lines *(21)*. The basic methodology of inverse PCR, as depicted in Fig. 1A, includes four steps, i.e.

1. A first digestion with restriction enzyme X;
2. Recircularization by ligation;
3. A second digestion with restriction enzyme Y; and
4. PCR amplification.

Fig. 1. *(opposite page)* **(A)** Outline of the inverse PCR protocol. Genomic DNA is digested with restriction enzyme X and the linear border fragment between T-DNA insert (striped box) and the unknown flanking chromosomal sequence (thick line) is circularized by ligation under diluted conditions. Incubation with restriction enzyme Y again results in a linear fragment, from which the sequence flanking the T-DNA can be amplified in a PCR with primers 1 and 2 (arrows). **(B)** Overview of different inverse PCR strategies. Sequences of the different oligonucleotide primers (arrows) are presented in the material section. **1.** Does et al. (21): X = *Taq*I (T), Y = *Ssp*I (S1), *Sac*II (S2) or *Mst*II (M2); **2.** Offringa et al. (17): X = *Hpa*II (H2), Y = *Sac*II (S2); **3.** Goddijn et al. (14): X = *Nsp*I (N), *Eco*RV (E) or *Msc*I (M), Y = *Sna*BI (Sn). In strategy 3 the restriction enzyme recognition sites in the flanking chromosomal sequence are indicated at positions according to their chance of occurrence. *Nsp*I recognizes a 5-bp sequence whereas *Eco*RV and *Msc*I both have a 6-bp recognition sequence. General legends: RB: indicates the nick site at the right T-DNA border repeat; NPT: open reading frame encoding neomycin phosphotransferase II; P-NOS: promoter of the nopaline synthase gene; GUS: open reading frame encoding β-glucuronidase; ATG: start codon of GUS; Intron: second intron of the potato ST-LS1 gene *(23)*.

A

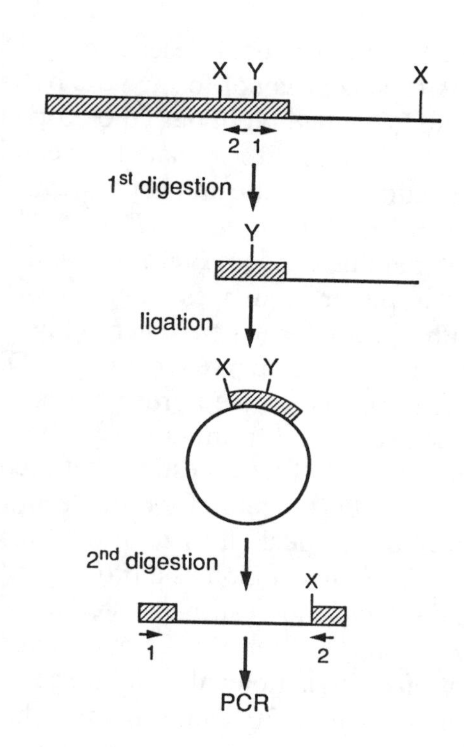
X
Y
X
2
1
1st digestion
Y
ligation
X
Y
2nd digestion
X
1
2
PCR

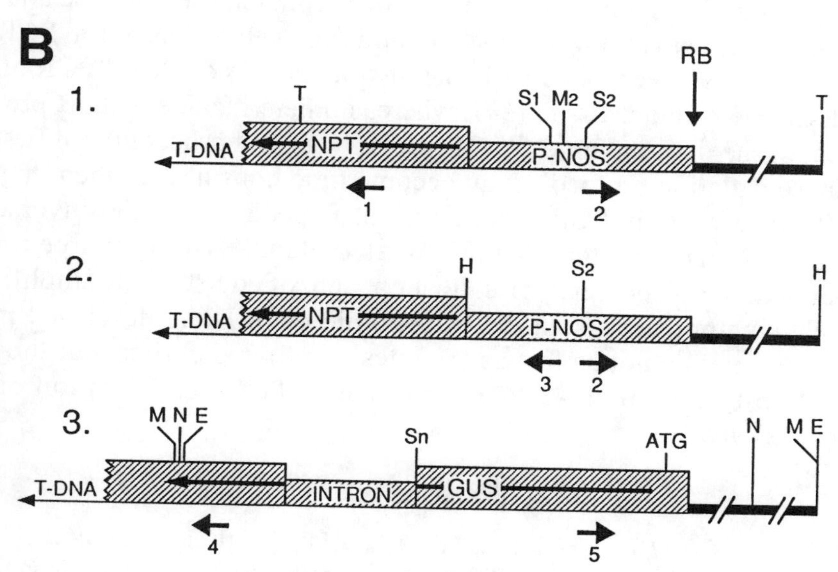
B
1.
T
S1
M2
S2
RB
T
T-DNA
NPT
P-NOS
1
2
2.
H
S2
H
T-DNA
NPT
P-NOS
3
2
3.
M N E
Sn
ATG
N
M E
T-DNA
INTRON
GUS
4
5

The enzymes used in the protocol described by Does et al. *(21)* are indicated in Fig. 1B.1. This protocol formed the basis for several applications of the inverse PCR method in our laboratories. Inverse PCR has proven to be a valuable tool to discriminate between independent transgenic lines of potato cultivars Bintje and Escort *(22)*, for detailed molecular analysis of sequences flanking the *nptII* gene that was corrected via homologous recombinant in tobacco (Fig. 1B.2) and for the isolation and characterization of promoter sequences from *Arabidopsis* identified by T-DNA tagging with a promoterless *gusA* gene (Fig. 1B.3). The protocol described in this chapter can be adapted to every T-DNA construct, provided that the DNA sequence of the border region is known. The best way is to look in the sequence for unique restriction sites. A risk when using restriction enzymes for first digestion that recognize 6 pb sites is that the resulting fragment is too large for reliable amplification by PCR. It is advisable, therefore, to check the size of the fragments by Southern analysis using the right border region as a probe.

An important and maybe underestimated step in PCR amplification is the isolation of DNA. In Chapter 13 of this volume methods are presented for the isolation of high molecular weight genomic plant DNA for Southern analysis. For standard PCR amplification, however, these methods are too time-consuming, since many different clones have to be analyzed. Recently, several very fast methods to isolate DNA for PCR amplification have been described, one of which relies on direct PCR on a small sample of plant tissue *(24)*, whereas others include a short pretreatment of the tissue sample *(25,26)*. Being very fast procedures at first sight, these isolation protocols may become time-consuming when they do not lead to reliable PCR amplification. Especially for the inverse PCR protocol it is important that the isolated plant DNA is suitable for digestion with restriction enzymes, ligation, and, of course, PCR amplification. The methods for isolation of genomic plant DNA described in this chapter may be not as fast as the ones mentioned earlier, but they have proven to give reliable PCR and inverse PCR amplification on many occasions.

2. Materials

1. TE buffer: 10 m*M* Tris-HCl, 1 m*M* EDTA, pH 8.0; high salt TE buffer: 1*M* NaCl, 10 m*M* Tris-HCl, 1 m*M* EDTA, pH 8.0 (both sterilized by autoclaving).
2. 1X TBE: 90 m*M* Tris-base, 90 m*M* boric acid, 2 m*M* EDTA, pH 8.0.

3. Extraction buffer: 0.14*M* sorbitol, 0.22*M* Tris-HCl, 22 m*M* EDTA, 0.8*M* NaCl, 0.8% (w/v) hexadecyltrimethylammonium bromide (CTAB), 1% *N*-laurylsarcosine, pH 8.0 (sterilized by autoclaving).
4. Chloroform: Chloroform (Merck, Darmstadt, Germany) and isoamylalcohol (Merck) are mixed in a ratio of 24:1 and equilibrated with TE buffer.
5. Phenol/chloroform: Phenol (Merck) containing 0.1% (w/v) 8-hydroxyquinoline (Sigma, St. Louis, MO) is equilibrated as follows: first with 1 vol 0.1*M* Tris and subsequently two times with 1 vol TE buffer. One vol of phenol is mixed with one vol of chloroform *(see earlier)* and 1/10 vol TE buffer is added.
6. Heat stable DNA polymerases were purchased from different manufacturers: *Taq* DNA polymerases from Promega (Madison, WI) (3 U/reaction), Perkin-Elmer (Norwalk, CT) (1 U/reaction), or HT Biotechnology (Cambridge, MA) (0.1 U/reaction), and *Pfu* DNA polymerase from Stratagene (La Jolla, CA) (2.5 U/reaction) (*see* Note 10).
7. Restriction enzymes, T4 DNA ligase, and appropriate digestion buffers as recommended by the manufacturer.
8. RNase A (Boehringer [Indianapolis, IN] #109-142) at 10 mg/mL, heated for 20 min at 100°C.
9. Sequences of the oligonucleotide primers indicated in Fig. 1 are:
   1. 5'-TTG.TCA.AGA.CCG.ACC.TGT.CC-3';
   2. 5'-CGT.TGC.GGT.TCT.GTC.AGT.TCC-3';
   3. 5'-CGC.CTA.AGG.TCA.CTA.TCA.GC-3';
   4. 5'-TAA.TGC.TCT.ACA.CCA.CGC.CG-3';
   5. 5'-CTT.TCC.CAC.CAA.CGC.TGA.TC-3'.
10. Nucleotides: Ultrapure dNTP set, Pharmacia (Upsala, Sweden) #27-2035-01.
11. Lambda DNA: Pharmacia #27-4111-01.
12. T-vector: Self-made (*see* Section 3.4., steps 4–7) or purchased from Promega (pGEM-T vector, #A3600).
13. Thermocycler: Perkin-Elmer/Cetus (Norwalk, CT) DNA thermal cycler 480.
14. Gel purification of PCR fragments: The Jet Sorb kit (GENOMED, Bad Orynhausen, Germany) or the Prep-A-Gene DNA purification kit (Bio-Rad, Richmond, CA) are both suitable for this purpose.

3. Methods

3.1. Isolation of Genomic DNA for (Inverse) PCR Amplification

1. A miniprep method for isolation of plant genomic DNA is adapted from Lassner et al. *(27)*. In our hands, this method yields DNA suitable for PCR

from callus and leaf tissue of *Nicotiana tabacum* as well as *Arabidopsis thaliana*. The *Nicotiana tabacum* DNA can very well be used in the inverse PCR protocol (*see* Note 1).

2. Put 50–100 mg callus tissue or 30–50 mg leaf tissue (in vitro as well as green house material) in an Eppendorf tube.
3. Add 100 µL extraction buffer of 65°C. Mix well before use, since the extraction buffer will form two layers at this temperature.
4. Homogenize the tissue for maximal 30 s with an Eppendorf potter (moderate rotation speed).
5. Add 20 µL chloroform, mix, and incubate at 65°C for at least 5 min.
6. Do 12 samples in a row (*see* Section 3.1., steps 3–5).
7. When the last samples have been incubated at 65°C for 5 min, centrifugate at full speed for 5 min.
8. Transfer the supernatant to a fresh tube and add 1 vol ice-cold 2-propanol (Merck).
9. Invert and centrifugate at full speed for 5 min.
10. Wash with ice-cold 70% ethanol and dry the pellet.
11. Dissolve the pellet in 35 µL TE buffer.
12. Determine the DNA concentration by electrophoresis on a 0.7% agarose, 1X TBE gel next to a lambda DNA standard.

3.2. PCR Amplification

1. Add approx 0.5 µg of tobacco genomic DNA (for *Arabidopsis* approx 25 ng genomic DNA is sufficient; *see* Note 2) to a small Eppendorf tube. Important for efficient amplification is not to use too much template DNA.
2. Make up the reaction mix in a separate tube and distribute over the tubes containing the DNA samples (do not forget to include a water control; *see* Note 3). The reaction mix should contain the appropriate (1X) reaction buffer (manufacturer), 200 µ*M* of the four dNTPs, 25 pmol of each primer, *Taq* DNA polymerase (*see* Materials) and the final vol should be 50 µL.
3. Overlay the reaction mixture with two drops of mineral oil (Perkin-Elmer, No. 186-2302) using a P1000 Gilson (Paris, France) pipet (blue tip).
4. Close the tubes and put them into the thermocycler without mixing.
5. For standard PCR amplification 30 cycles of 1 min 95°C denaturation, 1 min 57°C annealing, and 2 min 72°C elongation can be used. The first denaturation step is for 2 min and the elongation time in the last cycle is 10 min. An annealing temperature of 57°C works fine for 20-bp primers with a GC-contents of 55%. For primers that do not fall into this class the annealing temperature (T_{an}) can be estimated using the formula: $T_m s = 4 \times G/C + 2 \times A/T$ (in °C) and $T_{an} = T_m s - 5°C$ (*see* ref. *4*).

3.3. Inverse PCR Amplification

1. Digest 0.6–1 µg DNA for 6 h or overnight at the appropriate temperature in a 50 µL reaction mixture containing 1X digestion buffer, 10 U of the desired enzyme, and 0.4 µg/µL RNase A.
2. Add TE buffer to a final volume of 100 µL and inactivate the restriction enzymes by extraction with 1 vol phenol/chloroform and subsequently with 1 vol chloroform.
3. Transfer the upper phase to a fresh tube, add 1/10 vol 3*M* sodium acetate pH 5.2, mix, and add 2 vol ice-cold 96% ethanol. At this point the mixture can be stored at –20°C until further use.
4. If continued, incubate for 15 min at –80°C and centrifugate at full speed for at least 10 min.
5. Remove the supernatant and wash the pellet with 200 µL ice-cold 70% ethanol. Centrifugate 5 min at full speed.
6. Remove the supernatant and dry the pellet.
7. Dissolve the pellet in sterile water and add ligation buffer and 3 U ligase to a final vol of 250 µL (*see* Note 4).
8. Ligation is performed overnight at 14°C.
9. Preferentially heat-inactivate the ligase by incubation for 20 min at 65°C and use one-third of the ligation mixture for the second digestion (*see* Note 5). Adapt the salt/buffer concentrations to the digestion buffers recommended by the manufacturer by the addition of extra Tris, NaCl, $MgCl_2$, and/or DTT.
10. Alternatively, the enzyme can be inactivated by phenol/chloroform extraction followed by ethanol precipitation (*see* Section 3.3., steps 2–7). In this case the reaction buffer supplied by the manufacturer can be used.
11. Perform the second digestion using 10–20 U of enzyme for 2 h at the appropriate temperature (*see* Note 5).
12. Inactivate the enzyme by phenol/chloroform extraction and recover the DNA by ethanol precipitation (*see* Section 3.3., steps 2–7).
13. Dissolve the pellet in 10 µL TE buffer (*see* Note 6).
14. Use 2 µL for PCR amplification. PCR is performed as described earlier, except that 35 instead of 30 cycles are used. The yield of an inverse PCR product can be improved either by serial amplification or by reamplification using 1/100 of the original product as template.
15. Analyze 15 µL of the reactions on a 1–2% agarose, 1X TBE gel. Which percentage agarose to use, depends on the size of the fragments.
16. Several protocols have been described for direct sequencing of PCR products *(4,30)*. However, in our hands more reliable sequence data are obtained by first cloning the fragments as described later.

3.4. Cloning (Inverse) PCR Fragments

1. Gel purify PCR fragments and dissolve in 15 µL TE buffer. Fragments can be cloned either in a blunt-ended vector or a T-vector (*see* Note 7).
2. For this purpose digest 3 µg of a *lacZ* containing multicopy plasmid with *Eco*RV or *Sma*I in 40 µL (check 5 µL on gel whether digestion is complete).
3. Do a phenol/chloroform extraction and ethanol precipitation (*see* Section 3.3., steps 2–7). Dissolve the pellet in 50 µL TE buffer. Use 1 µL for ligation (*see* Section 3.4., step 8).
4. For a self-made T-vector (*see* Note 7) make up a 100 µL reaction mixture containing 25 µL of the linearized plasmid, the appropriate reaction buffer, 2 m*M* dTTP, and a fourfold amount of *Taq* DNA polymerase (in comparison with a standard PCR).
5. Incubate for 2 h at 72°C.
6. Extract with phenol/chloroform and ethanol precipitate (*see* Section 3.3., steps 2–7). Dissolve the pellet in 30 µL TE buffer and use 1 µL for ligation.
7. Ligate overnight at 14°C in a 20 µL reaction mixture containing 50 ng vector, a tenfold molar excess of fragment, 1 U T4 DNA ligase, and ligation buffer.
8. Half of the ligation mixture is used for transformation of *Escherichia coli*. In our laboratories the DH5α strain is commonly used for standard cloning procedures *(32)* and the boiling lysis procedure for extraction of DNA *(32,33)*.

3.5. Sequence Analysis

1. Sequence analysis can be performed on double-stranded DNA obtained either by the boiling lysis procedure *(32,33)* or by the alkaline lysis procedure *(34)*. Phenol extraction steps are not necessary (*see* Notes 8 and 9).
2. Add in one Eppendorf tube: 3 µg plasmid DNA, 1 µL primer (25–50 ng), 1 µL 0.5*M* sodium hydroxide, and TE buffer to a final vol of 7 µL.
3. Mix and heat at 80°C for 5 min.
4. Let the primers anneal by placing the tubes in a small beaker filled with water cooling down from 80°C to 30°C in 10–15 min.
5. Centrifugate for 5 s and add 1 µL 0.5*M* hydrochloric acid and 2 µL sequenase buffer.
6. Perform sequence analysis as described in the Sequenase protocol (United States Biochemicals [Cleveland, OH] Version 2.0) (*see* Note 10).

4. Notes

1. To isolate genomic DNA of *Arabidopsis* (approx 1 g rosettes) for inverse PCR or Southern blotting we use the CTAB based protocol as described by Kilby and Furner *(28)*, but we leave out the cesium chloride purification step. The pellet obtained after the addition of CTAB III is dissolved in

0.5 mL high salt TE buffer, the solution is transferred to a fresh Eppendorf tube and 2 vol ice cold 96% ethanol are added. Following centrifugation for 15 min, the pellet is washed with ice cold 80% ethanol, dried and dissolved in 200 μL 1/10 TE buffer. The concentration is estimated by gel electrophoresis and a minimum of 1 μg DNA is used per restriction enzyme digestion. For Southern analysis on *Nicotiana tabacum* genomic DNA, we prefer to include a cesium chloride centrifugation step to purify DNA obtained via the method of Mettler *(29)*. The DNA concentration is determined by OD_{260} measurement and a minimum of 10 μg DNA is used per digestion. The Hybond N^+ membrane (Amersham Life Science [Arlington Heights, IL] #RPN303B) is used for DNA transfer and hybridization according to the protocol of the manufacturers.

2. The *Arabidopsis* genome is 20-fold smaller in size than the tobacco genome. The purification methods described above will always give mixture of nuclear and plastid genomic DNA. One plant cell may contain many plastid genome copies. Thus, plastid specific bands will preferably be amplified in a PCR.
3. A water control is a PCR in which TE buffer is added instead of DNA, to check chemicals, enzymes, solutions, plastics, and pipets for contamination with (plasmid) DNA from which a fragment can be amplified.
4. Ligation in a large volume favors circularization.
5. In our hands the efficiency of PCR amplification is much increased by a second digestion. In case the second enzyme cuts within the fragment to be amplified, either choose another restriction enzyme (*see* Fig. 1B.1) or just try amplification without second digestion. Direct PCR on the circularized DNA does lead to fragment amplification, albeit at lower efficiency.
6. After this step the DNA must be used immediately for PCR analysis or stored at –20°C during precipitation.
7. The use of a T-vector is based on the principle that *Taq* DNA polymerase has a template independent terminal transferase activity, by which preferentially an extra adenosine is added to the 3' ends of a PCR product. One may try ligation of the fragment with both a T-vector *(31)* and a blunt-end vector to see what works best. Alternatively, restriction sites may be added to the 5' end of the oligonucleotide primers, so that cleavage of the PCR product allows directed cloning.
8. In case the protocol described in Section 3.5. does not work, we advise the reader to try the following procedure. Extract plasmid DNA from 6 × 1.5 mL overnight culture and dissolve the pooled isolates in 500 μL TE buffer. Incubate for 10 min with 200 μg/mL RNase A at 37°C, extract two times with 1 vol phenol/chloroform, one time with 1 vol chloroform and ethanol precipitate (*see* Section 3.3., steps 2–7). Wash with ice-cold 70% ethanol,

dry, and dissolve pellet in 150 µL TE buffer. Test 2 µL on gel to estimate the concentration. Take ±5 µg double-stranded circular plasmid DNA and add H_2O to a final vol of 48 µL. Denature the DNA by addition of 2 µL 5*M* NaOH and 30 min incubation at 60°C. Add 5 µL 3*M* sodium acetate pH 5.2 and 120 µL ice-cold ethanol. Invert and incubate for 15 min at –80°C, centrifugate at full speed, wash with ice-cold 70% ethanol, and dry. The pellet is dissolved in 7 µL H_2O. Immediately add 1 µL primer (25 ng/µL for a 17-mer) and 2 µL 5X sequenase buffer and incubate for 2 min at 65°C. Let the primers slowly anneal by incubation in a small beaker with water cooling down from 65°C to below 30°C. Use this annealing mix according to the Sequenase protocol.

9. The CircumVent™ thermal cycle dideoxy sequencing protocol (New England Biolabs, Beverly, MA) was used as well, and seems to provide a quick and reliable alternative for the Sequenase protocol. Only 1 µg double-stranded plasmid DNA is needed.
10. Because *Taq* DNA polymerase lacks proofreading 3–5' exonuclease activity, misincorporations can occur during PCR. These errors have been found to occur at a frequency of 2.1×10^{-4} per base per cycle and to result predominantly in A.T to G.C transitions *(35)*. The problem of basepair substitutions may be overcome during sequence analysis by pooling DNA of 10 different clones or analyzing three or four different clones separately. Alternatively, one may determine the sequence of the ends of the flanking sequence in one clone, design primers to these ends, and reamplify the fragment from genomic DNA using a thermostable DNA polymerase with proofreading activity, such as the *Thermococcus litoralis* DNA polymerase (VENT™, New England Biolabs; *36,37*) or the *Pyrococcus furiosus* DNA polymerase (Strategene, *38*). The latter has been used successfully in our laboratories. Reamplification also provides a verification for the presence of the DNA sequence as identified by inverse PCR in the genomic DNA.

References

1. Kleppe, K., Ohtsuka, E., and Kleppe, R. (1971) Studies on polynucleotides XCVI. Repair replications of short synthetic DNAs as catalyzed by DNA polymerases. *J. Mol. Biol.* **56,** 341–361.
2. Saiki, R. K., Scharf, S., Faloona, F., Mullis, K. B., Horn, G. T., Erlich, H. A., and Arnheim, N. (1985) Enzymatic amplification of β-Globin genomic sequences and restriction site analysis for diagnosis of sickle cell anemia. *Science* **230,** 1350–1354.
3. Saiki, R. K., Gelfand, D. H., Stoffel, S., Scharf, S. J., Higuchi, R., Horn, G. T., Mullis, K. B., and Erlich, H. A. (1988) Primer-directed enzymatic amplification of DNA with a thermostable DNA polymerase. *Science* **239,** 487–491.

4. Innes, M. A., Gelfland, D. H., Shinsky, J. J., and White, T. J. (1990) *PCR Protocols. A Guide to Methods and Applications.* Academic, San Diego.
5. Williams, M. N. V., Pandé, N., Nair, S., Mohan, M., and Bennett, J. (1991) Restriction fragment length polymorphism analysis of polymerase chain reaction products amplified from mapped loci of rice (*Oryza sativa* L.) genomic DNA. *Theor. Appl. Genet.* **82,** 489–498.
6. Meyer, R. M., Maniatis, T., and Lerman, L. S. (1987) Detection and localization of single-base-pair changes by denaturing gradient gel electrophoresis. *Meth. Enzymol.* **155,** 501–527.
7. Hayashi, H. (1991) PCR-SSCP: a simple and sensitive method for detection of mutations in the genomic DNA. *PCR Meth. Appl.* **1,** 34–38.
8. To, K.-Y., Liu, C.-I., Liu, S. -T., and Chang, Y.-S. (1993) Detection of point mutations in the chloroplast genome by single-stranded conformation polymorphism analysis. *Plant J.* **3,** 183–186.
9. Williams, J. G. K., Kubelik, A. R., Livak, K. J., Rafalski, J. A., and Tingey, S. V. (1990) DNA polymorphisms amplified by arbitrary primers are useful as genetic markers. *Nucleic Acids Res.* **18,** 6531–6535.
10. Hooykaas, P. J. J. and Schilperoort, R. A. (1992) *Agrobacterium* and plant genetic engineering. *Plant Mol. Biol.* **19,** 15–38.
11. Feldmann, K. A. (1991) T-DNA insertion mutagenesis in *Arabidopsis*: mutational spectrum. *Plant J.* **1,** 71–82.
12. Koncz, C., Németh, K., Rédei, G. P., and Schell, J. (1992) T-DNA insertional mutagenesis in *Arabidopsis*. *Plant Mol. Biol.* **20,** 963–976.
13. Koncz, C., Martini, N., Mayerhofer, R., Koncz-Kalman, Z., Korber, H., Redei, G. P., and Schell, J. (1989) High-frequency T-DNA-mediated gene tagging in plants. *Proc. Natl. Acad. Sci. USA* **86,** 8467–8471.
14. Goddijn, O. J. M., Lindsey, K., van der Lee, F. M., Klap, J. C., and Sijmons, P. C. (1993) Differential gene expression in nematode-induced feeding structures of transgenic plants harbouring promoter-*gusA* fusion constructs. *Plant J.* **4,** 863–873.
15. Kim, H.-S., and Smithies, O. (1988) Recombination fragment assay for gene targeting based on the polymerase chain reaction. *Nucleic Acids Res.* **16,** 8887–8903.
16. Offringa, R., de Groot, M. J. A., Haagsman, H. J., Does, M. P., van den Elzen, P. J. M., and Hooykaas, P. J. J. (1990) Extrachromosomal homologous recombination and gene targeting in plant cells after *Agrobacterium*-mediated transformation. *EMBO J.* **9,** 3077–3084.
17. Offringa, R., Franke-van Dijk, M. E. I., de Groot, M. J. A., van den Elzen, P. J. M., and Hooykaas, P. J. J. (1993) Nonreciprocal homologous recombination between *Agrobacterium* transferred DNA and a plant chromosomal locus. *Proc. Natl. Acad. Sci. USA* **90,** 7346–7350.
18. Meyerhans, A., Vartanian, J.-P., and Wain-Hobson, S. (1990) DNA recombination during PCR. *Nucleic Acids Res.* **18,** 1687–1691.
19. Marton, A., Delbecchi, L., and Bourgaux, P. (1991) DNA nicking favors PCR recombination. *Nucleic Acids Res.* **19,** 2423–2426.

19a. Risseeuw, E., Offringa, R., Franke-van Dÿk, and Hooykaas, P. J. J. (1995) Targeted recombination in plants using *Agrobacterium* coincides with additional rearrangements at the target locus. *Plant J.* **7,** 109–119.
20. Ochman, H., Gerber, A. S., and Hartl, D. L. (1988) Genetic applications of an inverse polymerase chain reaction. *Genetics* **120,** 621–623.
21. Does, M. P., Dekker, B. M. M., de Groot, M. J. A., and Offringa, R. (1991) A quick method to estimate the T-DNA copy number in transgenic plants at an early stage after transformation, using inverse PCR. *Plant Mol. Biol.* **17,** 151–153.
22. Jongedijk, E., de Schutter, A. A. J. M., Stolte, T., van den Elzen, P. J. M., and Cornelissen, B. J. C. (1992) Increased resistance to potato virus X and preservation of cultivar properties in transgenic potato under field conditions. *Bio/Technol.* **10,** 422–429.
23. Vancanneyt, G., Schmidt, R., O'Conner-Sanchez, A., Willmitzer, L., and Rocha-Sosa, M. (1990) Construction of an intron-containing marker gene: Splicing of the intron in transgenic plants and its use in monitoring early events in *Argobacterium*-mediated plant transformation. *Mol. Gen. Genet.* **220,** 245–250.
24. Berthomieu, P. and Meyer, C. (1991) Direct amplification of plant genomic DNA from leaf and root pieces using PCR. *Plant Mol. Biol.* **17,** 555–557.
25. Langridge, U., Schall, M., and Langridge, P. (1991) Squashes of plant tissue as substrate for PCR. *Nucleic Acids Res.* **19,** 6954.
26. Klimyuk, V. I., Carroll, B. J., Thomas, C. M., and Jones, J. D. G. (1993) Alkali treatment for rapid preparation of plant material for reliable PCR analysis. *Plant J.* **3,** 493–494.
27. Lassner, M. W., Peterson, P., and Yoder, J. I. (1989) Simultaneous amplification of multiple DNA fragments by polymerase chain reaction in the analysis of transgenic plants and their progeny. *Plant Mol. Biol. Rept.* **7,** 116–128.
28. Kilby, N. J. and Furner, I. J. (1993) Another CTAB protocol: Isolation of high molecular weight DNA from small quantities of *Arabidopsis* tissue, in *AAtDB Research Companion for Arabidopsis Information; WWW: http: weeds. mgh. harvard. edu; Arabidopsis: Compleat Guide* (Dean, C. and Flanders, D., eds.), Massachusetts General Hospital, Boston.
29. Mettler, I. J. (1987) A simple and rapid method for minipreparation of DNA from tissue cultured plant cells. *Plant Mol. Biol. Rept.* **5,** 346–349.
30. Rao, V. B. (1994) Direct sequencing of polymerase chain reaction-amplified DNA. *Anal. Biochem.* **216,** 1–14.
31. Marchuk, D., Drumm, M., Saulino, A., and Collins, F. S. (1991) Construction of T-vectors, a rapid and general system for direct cloning of unmodified PCR products. *Nucleic Acids Res.* **19,** 1154.
32. Sambrook, J., Fritsch, E. F., and Maniatis, T. (1989) *Molecular Cloning: A Laboratory Manual.* Cold Spring Harbor Laboratory Press, Cold Spring Harbor, NY.
33. Holmes, D. S. and Quigley, M. (1981) A rapid boiling method for the preparation of bacterial plasmids. *Anal. Biochem.* **114,** 193–197.
34. Birnboim, H. C. and Doly, J. (1979) A rapid alkaline extraction procedure for screening recombinant plasmid DNA. *Nucleic Acids Res.* **7,** 1513.

35. Keohavong, P. and Thilly, W. G. (1989) Fidelity of DNA polymerase in DNA amplification. *Proc. Natl. Acad. Sci. USA* **86,** 9253–9257.
36. Cariello, N. F., Swenberg, J. A., and Skopek, T. R. (1991) Fidelity of *Thermococcus litoralis* DNA Polymerase (Vent™) in PCR determined by denaturing gradient gel electrophoresis. *Nucleic Acids Res.* **19,** 4193–4198.
37. Mattila, P., Korpela, J., Tenkanen, T., and Pitkänen, K. (1991) Fidelity of DNA synthesis by the *Thermococcus litoralis* DNA polymerase—an extremely heat stable enzyme with proofreading activity. *Nucleic Acids Res.* **19,** 4967–4973.
38. Lundberg, K. S., Shoemaker, D. D., Adams, M. W. W., Short, J. M., Sorge, J. A., and Mathur, E. J. (1991) High-fidelity amplification using a thermostable DNA polymerase isolated from *Pyrococcus furiosus*. *Gene* **108,** 1–6.

PART V

RNA Techniques for Studying Gene Expression

CHAPTER 15

Isolation of Whole Cell (Total) RNA

Rod J. Scott

1. Introduction

Efficient extraction of high quality RNA from a variety of plant tissues is an important first step in many procedures, such as analysis of gene expression, cDNA library construction, and in vitro translation. The procedure described here, which is essentially the same as described in Draper et al. *(1)*, involves grinding and phenol extraction of plant material followed by differential precipitation of RNA with sodium acetate. The protocol has been successful with leaf material, various floral organs, and cultured cells from a large number of species. However, slight adjustments may be necessary to optimize extraction from other tissues.

The method of Hall et al. *(2)* may be appropriate where RNA is required from particularly intractable material, such as seeds. In cases where only a small amount of material is available, or the intention is to analyze a large number of samples, the micromethod of Verwoerd et al. *(3)* is highly recommended. For a comprehensive list of various other protocols that may be better suited to particular needs, the reader is referred to Speirs and Longhurst *(4)*.

Successful RNA preparation depends on the inhibition of both endogenous RNase activity liberated on cell lysis and contamination of preparations by exogenous RNases. In the method described here treatment of equipment and solutions with diethyl pyrocarbonate (DEPC), a strong inhibitor of RNases, is used to prevent degradation of samples. In addition working quickly and keeping preparations on ice whenever possible will help to minimize problems with RNase activity.

From: *Methods in Molecular Biology, Vol. 49: Plant Gene Transfer and Expression Protocols*
Edited by: H. Jones Humana Press Inc., Totowa, NJ

2. Materials

As in any procedure involving the handling of RNA it is vital to ensure that all equipment and solutions are treated to remove ribonuclease activity and that gloves are worn at all times to prevent contamination of the preparations with ribonucleases present in perspiration. Glassware should be siliconized, soaked in 0.2% DEPC for 1 h, and then baked at 180°C for 2 h. Solutions, with the exception of Tris-HCl solutions and organic solvents, should be DEPC-treated prior to autoclaving (*see* Note 1). Tris-HCl solutions should be prepared with DEPC-treated and autoclaved water. All chemicals used should be AnalaR grade.

2.1. Phenol Method—Macroscale

1. Alumina Type A5 (Sigma, St. Louis, MO) (*see* Note 2).
2. Grinding buffer: 6% 4-aminosalicylate, 1% triisopropyl napthalene sulfonate (*see* Note 3), 6% phenol (phenol [as supplied by Fisons plc] contains 0.1% 8-hydroxyquinoline and is equilibrated against 100 m*M* Tris-HCl, pH 7.5. Phenol is highly toxic; *see* Note 4 for precautions when handling.), 50 m*M* Tris-HCl, pH 8.4. This solution should be freshly prepared prior to use and kept on ice in the dark.
3. Phenol resistant plastic centrifuge tubes.
4. Phenol/chloroform: 49.95% Phenol, 0.05% 8-hydroxyquinoline (equilibrated against Tris-HCl, pH 7.5 as above), 50% chloroform/isoamyl alcohol (24:1). The phenol/chloroform should be mixed in a fume-hood and left for a few hours until the mixture clears.
5. 4*M* Sodium acetate adjusted to pH 6.0 with acetic acid.
6. Absolute ethanol stored at –20°C.
7. 80% Ethanol stored at –20°C.
8. Sterile DEPC-treated water: To DEPC treat solutions, add 0.1% diethyl pyrocarbonate (DEPC), shake to disperse, and incubate at 37°C for 2 h; the solution should then be autoclaved to destroy the DEPC (*see* Note 1).

2.2. Hot Phenol—Microscale

1. Extraction buffer: 0.1*M* LiCl, 100 m*M* Tris-HCl, pH8.0, 10 m*M* EDTA, 1% sodium dodecyl sulfate (SDS) mixed 1:1 with phenol. Phenol as supplied by Fisons contains 6% phenol, 0.1% 8-hydroxyquinoline and is equilibrated against 100 m*M* Tris-HCl, pH 7.5. Phenol is highly toxic; *see* Note 4 for precautions when handling. Solution maintained at 80°C in a water bath.
2. Chloroform:isoamyl alcohol (24:1).
3. 4*M* LiCl.

4. 3*M* Sodium acetate adjusted to pH 5.2 with acetic acid.
5. 70% ethanol stored at –20°C.
6. Sterile DEPC-treated water.

3. Methods

3.1. Phenol Method—Macroscale

As discussed earlier, RNA preparations are vulnerable to degradation by RNases. For this reason, it is important to work quickly and as far as possible, keeping preparations on ice at all times.

1. Harvest and weigh fresh plant material and freeze in liquid nitrogen until required. The material should be frozen as soon as possible after collection and should not be damaged or allowed to wilt.
2. Precool a pestle and mortar on ice. Grind frozen plant tissue in the presence of alumina (approx 0.5 g alumina/5 g plant material) (*see* Note 2). It is vital that the tissue is kept frozen during grinding by the addition of liquid nitrogen when required.
3. Half-fill the mortar with liquid nitrogen and before it evaporates drop in grinding buffer (2 mL/g of tissue) (*see* Note 4). Continue grinding until material and buffer are ground to a fine powder; then transfer to a 50 mL plastic centrifuge tube. Place on ice.
4. Add 1 vol phenol/chloroform and thaw on ice occasionally shaking the tube gently to homogenize contents (*see* Note 4).
5. Mix by inversion and centrifuge at 4°C in a fixed angle rotor at 3000 for 10 min.
6. Remove the aqueous (top) layer, being careful to avoid the protein at the interface and transfer to a fresh tube. Repeat the phenol/chloroform extraction three times or until no protein interface is visible.
7. Transfer to Corex tubes, add 0.05 vol 4*M* sodium acetate, pH 6.0 and 2.5 vol absolute ethanol that has been stored at –20°C. Mix thoroughly and spin (4°C) at 12,000*g* for 10 min to pellet nucleic acids.
8. Remove the supernatant carefully and wash the pellet with 80% ethanol, dry very briefly, and then dissolve pellet in sterile distilled water (dissolve in as small a volume as possible). It may be necessary to leave the pellet dissolving on ice for up to 1 h with occasional mixing. If the preparation still contains insoluble material, centrifuge at 12,000*g* for 5 min and discard the pellet before continuing. Depending on the volume, it may be possible to transfer the dissolved nucleic acid to a sterile microcentrifuge tube at this stage.

9. Add 3 vol 4*M* sodium acetate, pH 6.0, mix, and leave on ice for at least 3 h to precipitate the RNA. The preparation can be left overnight at this stage if required.
10. Centrifuge (4°C) at 12,000*g* for 20 min to pellet the RNA. Discard the supernatant and dissolve pellet in 100–500 µL of sterile water. Reprecipitate by adding 0.05 vol 4*M* sodium acetate and 2.5 vol ethanol (–20°C), incubate at –20°C for 30 min, and spin at 12,000*g* for 20 min. Wash the pellet with 80% ethanol, dry, and redissolve in a minimum volume of sterile water (approx 50 µL/g tissue). For long-term storage RNA should be stored as aliquots at –80°C.
11. RNA purity and concentration can be assessed by scanning spectrophotometry. Take a 1:100 dilution of RNA in sterile water and scan between 200 and 300 nm in 1-cm quartz cuvets. A clear peak of absorbance should be visible at 260 nm. An A_{260} of 1 corresponds to an RNA concentration of approx 37 mg/mL. Pure RNA has an A_{260}/A_{280} ratio of 2.0. Values significantly less than this, i.e., below 1.8, indicate contamination of the preparation, most usually with protein or phenol (*see* Notes 5 and 6). RNA quality, can be checked by running on a gel (*see* Chapter 18 of this volume, and Note 7).
12. Poly A+ RNA can be prepared from total RNA samples as described in Chapter 16.

3.2. Hot Phenol—Microscale

This method is designed for the simple and rapid isolation of total RNA from small amounts of starting material. The protocol lends itself to the preparation of large numbers of individual samples, as might be required in analyzing a set of transgenic plants for transgene expression. The RNA quality is sufficiently good for RNA gel and dot blots. The original authors report yields of between 25 and 50 µg total RNA for tobacco and tomato leaf tissue. In our experience, the method also works well for tobacco and brassica anthers and other floral organs.

1. Collect fresh leaf material in 1.5-mL microcentrifuge tubes (*see* Note 8). Store on ice until all samples have been collected.
2. Quick freeze samples in liquid nitrogen. Store at –80°C until ready for further processing.
3. Grind to a fine powder with a metal rod precooled in liquid nitrogen.
4. Add 500 µL hot extraction buffer (80°C)—ensure that the organic and aqueous phases of the extraction buffer are mixed together by shaking thoroughly immediately prior to dispensing. Vortex for 30 s.
5. Add 250 µL chloroform:isoamyl alcohol (24:1) and vortex for 30 s.

6. Centrifuge in a minifuge for 5 min at 4°C.
7. Transfer upper aqueous phase to a new 1.5-mL microcentrifuge tube and add 1 vol 4*M* LiCl.
8. Precipitate at 4°C for a minimum of 2 h or leave overnight.
9. Centrifuge in a minifuge for 5 min at 4°C to pellet RNA.
10. Discard supernatant and redissolve pellet in 250 μL sterile distilled water.
11. Reprecipitate RNA by adding 0.1 vol of 3*M* sodium acetate, pH 5.2 and 2 vol of ethanol (–20°C). Incubate at –20°C for 5–10 min.
12. Centrifuge in a minifuge for 5 min at 4°C to pellet RNA. Wash pellet in 70% ethanol (–20°C) and dry under vacuum.

4. Notes

1. DEPC is highly flammable and must be handled in a fume hood; it is also thought to be a carcinogen and so it should be handled with care.
2. A face mask should be worn while weighing out alumina and grinding tissue to avoid accidental inhalation of alumina or frozen grinding buffer.
3. Both 4-aminosalicylate and triisopropyl napthalene sulfonate are toxic. A face mask should be worn while weighing them out.
4. Phenol is highly toxic and should be handled with great care and where possible confined to a fume hood. Protective clothing, i.e., lab coat, disposable gloves, and safety glasses should be worn at all times. If phenol, grinding buffer, or phenol/chloroform come into contact with the skin, the area should be soaked in 70% PEG 300, 30% IMS, and medical advice should be sought.
5. If protein contamination is a problem, further phenol/chloroform extractions (steps 4–6) followed by ethanol precipitation (steps 7–8) should be carried out.
6. Phenol contamination of the preparation can be addressed by repeated ethanol precipitation (steps 7–8) as phenol is very soluble in ethanol.
7. RNA samples when run on a gel should show no degradation. The two major ribosomal RNA species (28S and 18S) should be clearly visible, well defined bands. If the RNA appears to be degraded it indicates that exogenous and/or endogenous RNases were not sufficiently inhibited during the procedure and greater care must be taken. In practice, if the precautions set out in the protocol are adhered to, no problems with RNA degradation should be experienced.
8. For ease of collection and to avoid cross-contamination, leaf samples can be taken by simply closing the microcentrifuge lid onto the leaf lamina. In many cases, a single leaf piece (100 mg or less) provides sufficient RNA for 2–3 Northern gel tracks.

References

1. Draper, J., Scott, R. J., Armitage, P., and Walden, R. (1988) *Plant Genetic Transformation and Gene Expression: A Laboratory Manual.* Blackwell Scientific, Oxford, pp. 226–230.
2. Hall, T. C., Ma, Y., Buchbinder, B. U., Pyne, J. W., Sun, S. M., and Bliss, F. A. (1978) Messenger RNA for G1 protein of French bean seeds: Cell-free translation and product characterization. *Proc. Natl. Acad. Sci. USA* **75,** 3196–3200.
3. Verwoerd, T. C., Dekker, B. M. M., and Hoekema, A. (1989) A small-scale procedure for the rapid isolation of plant RNAs. *Nucleic Acids Res.* **17,** 2362.
4. Speirs, J. (1993) RNA extraction and fractionation. *Methods in Plant Biochemistry*, vol. 10 (Bryant, J., ed.), Academic, London, pp. 1–32.

CHAPTER 16

Poly(A)$^+$ RNA Isolation

Rod J. Scott

1. Introduction

Eukaryotic messenger RNA (mRNA) can be separated from the other RNA species in a total RNA preparation by affinity chromatography by virtue of the presence of a polyadenylic acid "tail," 20–25 bases long, at the 3' end of the molecule *(1)*. Oligo(dT)-cellulose is used routinely for homemade affinity columns, but a range of ready-made products is now available from a number of suppliers (e.g., Hybond-mAp, Amersham, Arlington Heights, IL; PolyATract™, Promega, Madison, WI). Oligo(dT)-cellulose consists of a polymer of 10–20 T-residues, covalently linked to a cellulose matrix, which will hybridize to and bind poly(A)$^+$ containing RNA providing the tail is at least 15–20 bases long. Ribosomal and transfer RNA do not bind efficiently to oligo(dT) and can be washed through the column. Bound poly(A)$^+$ RNA can then be eluted with low salt buffers. The protocol given below is a general purpose method designed to give high yields of cDNA synthesis-/in vitro translation-quality mRNA from relatively large amounts of starting material (1–2 mg total RNA).

2. Materials

1. Sterile, siliconized Pasteur pipets. Polypropylene wool; washed and autoclaved. Clamp and stand. 65°C water bath. Silica microcuvets (300 µL volume). Store microcuvets under concentrated HCl and rinse thoroughly in sterile distilled water before use.
2. Oligo(dT)-cellulose (e.g., Sigma, St. Louis, MO).
3. Absolute ethanol. 80% Ethanol.

From: *Methods in Molecular Biology, Vol. 49: Plant Gene Transfer and Expression Protocols*
Edited by: H. Jones Humana Press Inc., Totowa, NJ

4. 2X Binding buffer: 40 m*M* Tris-HCl, pH 7.6, 1.0*M* LiCl, 2 m*M* EDTA, 0.2% sodium dodecyl sulfate (SDS). Autoclave and store a room temperature.
5. 1X Binding buffer: 20 m*M* Tris-HCl, pH 7.6, 0.5*M* LiCl, 1 m*M* EDTA, 0.1% SDS. Autoclave and store a room temperature.
6. Elution buffer: 10 m*M* Tris-HCl, pH 7.6, 1 m*M* EDTA, 0.05% SDS.
7. 4*M* sodium acetate, pH 6.0 (pH with acetic acid).
8. DEPC-treated water. To DEPC treat solutions, add diethyl pyrocarbonate (DEPC), to 0.1%, shake to disperse and incubate at 37°C for 2 h. Autoclave solution to destroy residual DEPC (*see* Note 1).

3. Methods

Where carbohydrate contamination is likely to interfere with the oligo(dT)-cellulose chromatography, this contaminant can be removed by a prior purification step (*see* Note 2).

1. Prepare the column by plugging a sterile siliconized Pasteur pipet within the capillary with a small amount of sterile polypropylene wool (*see* Note 3). Support the column in a clamp and stand.
2. Suspend an appropriate amount of oligo(dT)-cellulose (5–15 mg) in 2 mL of sterile 1X binding buffer in a sterile Corex™ tube (*see* Note 4). Allow the oligo(dT)-cellulose to settle out and remove the supernatant. Repeat with sterile 1X binding buffer until the supernatant remains absolutely clear. Load the slurry into the plugged Pasteur pipet and measure the flow rate while washing the column with 1–2 vol of 1X binding buffer (*see* Note 3). At this or any other stage, flow through the column can be stopped by sealing the top of the pipet with Nescofilm. Flow through should be reinitiated by first pricking a hole in the film and then removing it (*see* Note 5).
3. Dissolve a pellet of total RNA (e.g., 1 mg) in 200 μL DEPC-treated water, add an equal volume of 2X binding buffer and 600 μL of 1X binding buffer, in a 1.5 mL micocentrifuge tube. Incubate at 65°C for 7 min and then place on ice for 5–10 min.
4. Drain the column until the buffer meniscus is about to reach the bed surface and gently load the RNA sample. Collect the eluate in the original microcentrifuge tube and reload onto the column six times.
5. Wash the column with 10–50 mL 1X binding buffer to remove tRNA and rRNA (*see* Note 6).
6. Heat the elution buffer to 45°C. Allow the last batch of 1X binding buffer to run through the column. Elute the mRNA in five separate aliquots by loading 5X 300 μL aliquots of elution buffer onto the column and collecting the eluant in five numbered microcentrifuge tubes (*see* Note 7).

7. Measure the concentration of poly(A)$^+$ mRNA in each tube and calculate the yield (*see* Note 8).
8. Add 4*M* sodium acetate pH 6.0 to obtain a 0.2*M* concentration, mix, and then add 2.5 vol of absolute ethanol to precipitate the mRNA. Store under ethanol at –80°C until required.
9. When required, pellet RNA by centrifugation (5 min in a microfuge), remove ethanol, wash pellet in 80% ethanol, and briefly vacuum dry. Resuspend poly(A)$^+$ RNA in sterile distilled water to an appropriate concentration.

4. Notes

1. DEPC is highly inflammable and should be handled in a fume hood. DEPC is also a suspected carcinogen and should be treated accordingly.
2. To remove contaminating carbohydrates, the RNA is passed through a cellulose column before oligo(dT)-cellulose chromatography *(2)*. Treat cellulose powder (e.g., Sigmacell type 500) with 0.25*M* NaOH, autoclave, and then wash repeatedly with DEPC-treated water until all trace of dark coloration is removed. Reautoclave the cellulose in DEPC-treated water and pack in a 1 mL bed volume column (*see* steps 1 and 2 of the Methods section). Equilibrate with several column volumes of 1X binding buffer. Dissolve a pellet of total RNA (e.g., 1 mg) in 200 µL DEPC-treated water, add an equal volume of 2X binding buffer, and 600 µL of 1X binding buffer. Incubate at 65°C for 7 min and then place on ice for 5–10 min. Pass total RNA through the cellulose column. Collect eluant and pass through the column twice more. *Save the eluant.* Wash the RNA from the column with 2 × 1 mL of 1X binding buffer and combine with the eluate fractions. Incubate RNA at 65°C for 7 min and then place on ice for 5–10 min. Load RNA onto the oligo(dT) column (*see* step 4 of main Method).
3. It is important to achieve a moderate flow rate through the column. A fast flow rate will result in poor binding, and a slow rate will increase the time taken for the procedure, perhaps to several hours. Therefore, care should be taken when packing the polypropylene wool and the flow rate should be checked by passing 1X binding buffer through the column before packing with oligo(dT)-cellulose. Aim for 1–2 drops/s at this stage and 1 drop/5 s when packed with oligo(dT)-cellulose.
4. This size column will accommodate up to 5 mg of total RNA.
5. Simply pulling the film from the pipet will disrupt the oligo(dT)-cellulose bed.
6. A 10-mL wash is sufficient for the preparation of poly(A)$^+$ mRNA for Northern blotting analysis. For cDNA synthesis, washing should be more thorough, using up to 50 mL of 1X binding buffer, and in most cases a second round of purification should be performed to yield poly(A)$^{++}$ mRNA.

7. The elution volume should be related to the volume of the microcuvet used for quantification.
8. RNA concentration can be assessed by scanning spectrophotometry. A clear peak of absorbance should be visible at 260 nm. An A_{260} of 1 corresponds to an RNA concentration of approx 37 mg/mL. Most mRNA elutes in fractions 2–3. The yield will depend on the content of poly(A)$^+$ mRNA within the starting tissue, but recoveries average between 1–2% of total RNA.

References

1. Aviv, H. and Leder, P. (1972) Purification of biologically active globin messenger RNA by chromatography on oligothymidylic acid-cellulose. *Proc. Natl. Acad. Sci. USA* **69,** 1408–1412.
2. Mozer, T. J. (1980) Partial purification and characterisation of the mRNA for a α-amylase from barley aleurone layers. *Plant Physiol.* **65,** 834–837.

CHAPTER 17

In Vitro Translation

Rod J. Scott

1. Introduction

In vitro translation remains a powerful technique for the identification, characterization, and quantification of mRNAs from a wide range of sources. The technique is also useful for assessing the quality of mRNA prior to cDNA synthesis. There are numerous in vitro translation systems, but among the most useful are derived from wheat germ *(1,2)* and rabbit reticulocytes *(3,4)*. The rabbit reticulocyte system is probably the more efficient of the two, particularly in translating larger mRNAs. However, the wheat germ system does offer a genuine "animal-friendly" alternative. Both systems are commercially available as well defined and highly active preparations, and are therefore the most convenient way of carrying out in vitro translations. However, for those with sufficient incentive, the wheat germ system is relatively easy to prepare and the reader is referred to Speirs *(5)*.

2. Materials

Typical commercial preparations, e.g., Promega (Madison, WI) Wheat Germ Extract and Rabbit Reticulocyte Lysate System, are supplied complete with an energy generating system (phosphocreatine kinase and phosphocreatine), spermidine (to reduce premature termination), magnesium acetate, and an amino acid mixture lacking the desired labeled amino acid. The user provides only the mRNA, a ribonuclease inhibitor (optional), RNase-free water, and an isotopically labeled amino acid. The products of in vitro translation are typically analyzed by dena-

From: *Methods in Molecular Biology, Vol. 49: Plant Gene Transfer and Expression Protocols*
Edited by: H. Jones Humana Press Inc., Totowa, NJ

turing polyacrylamide gel electrophoresis (PAGE) *(6)* followed by fluorography *(7)*.

2.1. In Vitro Translation

1. Commercial in vitro translation system (e.g., Promega Wheat Germ Extract or Rabbit Reticulocyte Lysate System). This should be kept as 40–100-µL aliquots stored over liquid nitrogen (or at –80°C for short periods).
2. DEPC-treated distilled water.
3. Ribonuclease inhibitor (e.g., Promega RNasin^R). This component is optional.
4. [^{35}S]methionine (1200 Ci/mmol) at 10 mCi/mL (e.g., Amersham International, Arlington Heights, IL).

2.2. Quantification of Radioactive Amino Acid Incorporation

1. Whatman (Maidstone, UK) GF/A glass fiber filters.
2. 1*M* NaOH, 2% H_2O_2.
3. Ice cold 25% trichloroacetic acid (TCA), 2% casamino acids (Difco [E. Molesley, Surrey, UK] vitamin assay grade).
4. 5% TCA acetone.
5. Scintillation cocktail (add 3 g 2,5-diphenyloxazole (PPO) and 0.5 g 1,4-bis-[4-methyl-5-phenyl-2-oxazolyl]benzene (POPOP) to 1 L of toluene.

2.3. SDS Polyacrylamide Gel Electrophoresis (SDS-PAGE)

1. Sturdier-type vertical slab gel apparatus.
2. 10X Running buffer (Tris-glycine pH 8.3). For 1 L: 30 g Tris-base, 144 g glycine, 100 mL 10% SDS, distilled water to volume. This solution is at the correct pH without titration. Store at room temperature.
3. Acrylamide/*bis*-acrylamide (30%:0.8%) solution. For 100 mL: 30 g Acrylamide, 0.8 g *bis*-acrylamide, distilled water to volume. Store at 4°C.
4. 1*M* Tris-HCl pH 8.8: Dissolve 121.1 g of Tris-base in 800 mL of distilled water and titrate to correct pH with concentrated HCl. Make to 1 L with distilled water. Store at room temperature.
5. 1*M* Tris-HCl pH 6.8: Dissolve 121.1 g of Tris-base in 800 mL of distilled water and titrate to correct pH with concentrated HCl. Make to 1 L with distilled water. Store at room temperature.
6. 10% Sodium dodecyl sulfate (SDS).
7. Tetramethylethylene diamine (TEMED). Store in the dark at 4°C.
8. 1.5% Ammonium persulfate. Make immediately before use.
9. Propanol/distilled water (1:1).

2.4. Fluorography

1. Gel drying apparatus.
2. Liquid fluor (e.g., Amersham's Amplify™ or New England Nuclear's [Boston, MA] Enhance™).
3. X-ray film (e.g., Kodak's X-Omat AR or Amersham's Hyperfilm).

3. Methods

3.1. In Vitro Translation

1. Carry out in vitro translation reactions according to the manufacturer's instructions.
2. Analyze the results of translation by performing an incorporation assay (*see* Section 3.2.) and SDS-PAGE (*see* Section 3.3.).

3.2. Quantification of Radioactive Amino Acid Incorporation

1. Following completion of the translation reaction, remove 2 μL from the reaction and add it to 98 μL of 1*M* NaOH, 2% H_2O_2.
2. Vortex briefly and incubate at 37°C for 10 min.
3. Precipitate the translation products by adding 900 μL ice cold 25% TCA, 2% casamino acids. Incubate on ice for 30 min.
4. Collect the precipitate onto a Whatman GF/A glass fiber filter by vacuum filtering 250 μL of the TCA/reaction mix. Rinse filter 3X with 2 mL ice-cold 5% TCA and 1X with 2 mL acetone. Allow filter to dry completely either at room temperature or under a heat lamp.
5. Put filter in 2 mL of scintillation cocktail, invert to mix, and count in a liquid scintillation counter on the appropriate setting for the chosen isotope. This gives the *incorporated counts.*
6. To determine the *input counts* present in the reaction, spot 5 μL of the TCA/reaction mix directly onto a Whatman GF/A glass fiber filter. Dry filter as in step 4, and count as in step 5. This gives the *input counts* in 5 μL of reaction mix.
7. Calculate percentage incorporation (*see* Note 1):

$$\text{cpm of washed filter/cpm of unwashed filter} \times 50$$

3.3. SDS Polyacrylamide Gel Electrophoresis (SDS-PAGE)

1. Clean glass plates, first with detergent (e.g., 0.1% Decon), then with ethanol/acetone mix (1:1). Allow to air dry.
2. Assemble the electrophoresis apparatus.

3. Prepare the resolving gel: Mix, in a 250-mL Buchner flask, the following *in order*, adding TEMED and persulfate after degassing. This mixture is sufficient to pour a 10% resolving gel (*see* Note 2) with a volume of about 37 mL.

Acrylamide/*bis*-acrylamide (30:0.8%)	13.4 mL
1*M* Tris-HCl, pH 8.8	15.0 mL
Distilled water	7.0 mL
Degas for 1–5 min	
10% SDS	370 µL
TEMED	2.0 µL
1.5% ammonium persulfate	2.0 mL

4. Pour the solution between the glass plates, leaving enough space for the well former and 2–3 cm of stacking gel.
5. Carefully overlay with 1–2 mL of propanol/water, and allow to polymerize at room temperature for 30–60 min.
6. Mix the stacking gel:

Acrylamide/*bis*-acrylamide (30:0.8%)	2.5 mL
1*M* Tris-HCl, pH 6.8	2.5 mL
Distilled water	13.8 mL
Degas for 1–5 min	
SDS	200 µL
TEMED	5.0 µL
1.5% ammonium persulfate	1.0 mL

7. Tip off the propanol/water and rinse the gel surface with distilled water. Remove any remaining liquid with filter paper—but do not touch the gel surface.
8. Pour stacking gel solution onto the resolving gel and immediately insert the well former. Polymerize gel at room temperature for 15–30 min.
9. Carefully remove the well former and rinse wells with 1X running buffer.
10. Place completed gel into the electrophoresis apparatus and charge with 1X running buffer. Do not prerun before loading samples.
11. Prepare samples: Add a 5-µL aliquot of the completed reaction (or intermediate timepoint) to 20 µL SDS sample buffer. Store remainder of reaction at –20°C.
12. Cap tube and incubate at 100°C for 2 min to denature the proteins.
13. Load a 10 µL sample onto the gel and electrophorese: 15 mA in the stacking gel and 30 mA in the separating gel. Run until the bromophenol dye front is about to enter the bottom buffer tank (*see* Note 3).
14. Remove the gel from the apparatus and carry-out fluorography.

3.4. Fluorography

Fluorography dramatically increases the sensitivity of ^{35}S detection and is therefore recommended for the analysis of in vitro translation products.

1. Impregnate gel with fluorography reagent according to the manufacturer's instructions.
2. Dry gel under vacuum.
3. Expose the gel to X-ray film for 1–6 h at –80°C.

4. Notes

1. Values of 2–10% should be expected.
2. This percentage resolving gel gives good separation of peptide mixtures between 20,000 and 100,000 Daltons, with peptides between 55,000 and 60,000 Daltons migrating halfway down the gel. Other percentages or a gradient (e.g., 15–20%) can be used as appropriate.
3. Since the dye front also contains the free labeled amino acids, disposal of unincorporated label is facilitated if the electrophoresis is stopped while the dye front remains within the gel.

References

1. Roberts, B. E. and Paterson, B. M. (1973) Efficient translation of tobacco mosaic virus RNA and rabbit globin 9S RNA in a cell-free system from commercial wheat germ. *Proc. Natl. Acad. Sci. USA* **70,** 2330–2334.
2. Marcu, K. and Dudock, B. (1974) Characterisation of a highly efficient protein synthesizing system derived from commercial wheat germ. *Nucleic Acids Res.* **1,** 1385–1397.
3. Schimke, R. T., Rhoads, R. E., and McKnight, S. (1974) Assay of ovalbumin mRNA in reticulocyte lysate. *Meth. Enzymol.* **30,** 694–701.
4. Pelham, H. R. B. and Jackson, R. J. (1976) An efficient mRNA-dependent translation system from reticulocyte lysates. *Eur. J. Biochem.* **67,** 247–256.
5. Speirs, J. (1993) *In vitro* translation of plant messenger RNA. *Methods in Plant Biochemistry*, vol. 10 (Bryant, J., ed.), Academic, London, pp. 33–56.
6. Laemmli, U. K. (1970) Cleavage of structural proteins during assembly of the head of Bacteriophage T4. *Nature* **227,** 680–685.
7. Laskey, R. A. and Mills, A. D. (1975) Quantitative film detection of ^{3}H and ^{14}C in polyacrylamide gels by fluorography. *Eur. J. Biochem.* **56,** 335–341.

CHAPTER 18

Northern Analysis and Nucleic Acid Probes

Paolo A. Sabelli and Peter R. Shewry

1. Introduction

Soon after the introduction of the Southern blotting procedure *(1)*, an analogous technique, known as northern blotting, was developed for the analysis of RNA sequences *(2)*. Both procedures rely on the annealing (hybridization) of complementary single-stranded nucleic acid molecules under suitable conditions *(3)*. Nucleic acids separated by size using agarose gel electrophoresis are transferred onto a solid support (membrane) and hybridized to complementary radiolabeled nucleic acid sequences. Whereas in Southern blotting experiments, DNA is transferred from a gel to a membrane and hybridized (*see* Chapter 13, this volume), in northern blotting it is RNA that is fractionated, immobilized on a membrane, and hybridized to a probe.

Since the RNA present in a given cell or tissue type represents the portion of the genome that is expressed in that cell or tissue, it is not surprising that northern analysis is regarded as a powerful tool for studying gene expression. In plants, this technique has been extensively used for the characterization of cloned cDNAs and genes, in order to determine the sizes of transcripts, to analyze the tissue/organ specificity of gene expression, and to monitor gene activity following the introduction of foreign DNA into transgenic plants. Northern blotting, therefore, allows the expression (or lack of expression) of specific DNA sequences to be related to the physiological and morphological properties of living organisms. Although more sensitive techniques have been developed for

From: *Methods in Molecular Biology, Vol. 49: Plant Gene Transfer and Expression Protocols*
Edited by: H. Jones Humana Press Inc., Totowa, NJ

the analysis of rare transcripts (RNase protection assay [*see* Chapter 20], RNA-PCR [*see* Chapters 22 and 24]), northern hybridization still remains a widely used procedure.

We have described a general method for the hybridization of DNA (*see* Chapter 13, this volume), which is also suitable for the analysis of RNA. We therefore describe here only the modifications that are required for northern hybridization experiments. In addition, we also discuss several methods for preparing radiolabeled nucleic acid probes.

2. Materials

All solutions should be treated with the RNase inhibitor diethylpyrocarbonate (DEPC) *(4)*, which is removed whenever possible by autoclaving (*see* Note 1). Unless otherwise indicated, store all solutions at room temperature.

2.1. Gel Electrophoresis

1. RNase-free, electrophoresis-grade agarose.
2. 10X MOPS/EDTA buffer: 0.5*M* 3-(*N*-morpholino)propanesulfonic acid (MOPS), 10 m*M* EDTA, pH 7.0. Store at 4°C. Light sensitive.
3. 12.3*M* Formaldehyde. This concentration is usually supplied by the manufacturer (*see* Note 2).
4. Reagent-grade formamide, preferably deionized with Dowex XG-8 resin. Store at –20°C.
5. Sample-loading buffer: 50% w/v glycerol, 1 m*M* EDTA, pH 8.0, 0.25% w/v xylene cyanol FF, 0.25% w/v bromophenol blue.
6. Running buffer: 1X MOPS/EDTA buffer, pH 7.0. Prepare fresh before use.
7. Ethidium bromide stock solution: 10 mg/mL. Prepare with DEPC-treated H_2O. Store at 4°C. Light sensitive.
8. Staining solution: 5µg/mL ethidium bromide (from ethidium bromide stock). Prepare fresh before use with DEPC-treated H_2O (*see* Note 3).

2.2. Blotting and Hybridization

1. 20X SSC buffer: 3.0*M* NaCl, 0.3*M* triNa-citrate, pH 7.0.
2. 10X SSC buffer: Prepare from 20X SSC stock solution.
3. Stripping buffer: 5 m*M* Tris-HCl, pH 8.0, 2 m*M* EDTA, pH 8.0.

2.3. Probe Preparation

1. 10X Random priming buffer: 0.5*M* Tris-HCl, pH 8.0, 0.05*M* $MgCl_2$, 0.1*M* 2-mercaptoethanol. Store in aliquots at –20°C.
2. 10X dNTPs mix: 0.2 m*M* each of dCTP, dGTP, and dTTP. Store in aliquots at –20°C.

3. 10X Primers: Random hexadeoxyribonucleotides (Pharmacia, Uppsala, Sweden), resuspended at 0.4 A_{260}/mL. Store in aliquots at –20°C.
4. BSA: 5 mg/mL. Store in aliquots at –20°C.
5. [α-^{32}P]dATP (~3000 Ci/mmol). Store at –20°C (*see* Note 4).
6. Klenow fragment of *E. coli* DNA polymerase I.
7. TE buffer: 10 m*M* Tris-HCl, pH 8.0, 1.0 m*M* EDTA, pH 8.0.
8. Sephadex G-50: Prepare a slurry by washing the matrix several times with TE buffer. Store at 4°C.
9. Terminal dimethyl transferase (TdT). Store at –20°C.
10. 5X TdT buffer. From the same manufacturer as the enzyme. Store at –20°C.
11. [α-^{32}P]dATP (~6000 Ci/mmol). Store at –20°C (*see* Note 4).
12. Sephadex G-25: Prepare as for solution 8. Store at 4°C.
13. NH_4-acetate: 7.5*M* NH_4-acetate, pH 7.5.
14. Ethanol, absolute.
15. Yeast tRNA: 10 mg/mL. Store in aliquots at –20°C.

3. Methods

3.1. Gel Electrophoresis

1. Prepare a 1.5% w/v gel by dissolving agarose in H_2O using a microwave oven. Cool to 70°C and add 10X MOPS/EDTA buffer, and formaldehyde. Mix gently and pour the gel in a fumehood. For example, for 100 mL volume of gel use 1.5 g agarose, 72 mL H_2O, 10 mL MOPS/EDTA buffer and 18 mL formaldehyde (*see* Note 5).
2. Prepare the RNA samples by mixing the following in an Eppendorf tube (*see* Notes 6 and 7):
   a. RNA (0.5–20 μg), 5.5 μL
   b. 10X MOPS/EDTA, 1.0 μL
   c. Formaldeyde, 3.5 μL
   d. Formamide, 10.0 μL

   Heat at 70°C for 10 min and cool on ice. Add 2 μL of sample-loading buffer and mix.
3. Prerun the gel in running buffer for 5–10 min at 2–5 V/cm. Load the samples and the RNA size-markers and run at the same voltage until the bromophenol blue has migrated approx 2/3 of the length of the gel. The buffer from the anode and cathode reservoirs is mixed and returned to the apparatus halfway through the run.
4. Excise the lane containing the RNA size-markers and incubate in the staining solution for 2 min with gentle shaking (*see* Note 8). *Caution:* Staining for longer than 3 min may result in excessive gel staining.

5. Destain the gel in H_2O for about 1 h. Wash the remainder of the gel containing the sample RNAs in H_2O (with two or three changes) in order to remove formaldehyde.
6. Photograph the RNA size-markers on a UV transilluminator as described in Chapter 13, this volume.

3.2. Blotting and Hybridization

RNAs are single-stranded, relatively short sequences. No depurination, denaturation, or neutralization steps are required before blotting. Use 10X SSC buffer to transfer the RNA by capillarity. Refer to Chapter 13 of this volume for the prehybridization and hybridization procedures and discussion of hybrid stability and stringency. If multiple hybridizations are to be performed on the same membrane, do not strip off the probe by alkali treatment as this hydrolyzes the RNA. Instead, incubate the membrane with the stripping buffer described in Section 2.2. at 65°C for 1 h.

3.3. Probe Preparation

A range of labeling methods are available. These include nick translation *(5)*, random priming *(6, 7)*, primer extension *(8–10)*, and labeling of oligonucleotides *(11–13)*. The choice of method depends on the particular application or research goal. Here we give protocols for random priming and oligonucleotide labeling.

3.3.1. Random Priming (6,7)

1. Purify DNA by fractionation on low melting point agarose for use as a template.
2. Denature the template DNA by boiling for 5–10 min, then cool quickly on ice.
3. Set up the following labeling reaction in an Eppendorf tube kept on ice:
   a. Template DNA (25 ng/μL), 2 μL
   b. 10X random priming buffer, 2 μL
   c. 10X dNTPs mix, 2 μL
   d. 10X primers, 2 μL
   e. BSA (5 mg/mL), 1 μL
   f. [α-^{32}P] dATP (~3000 Ci/mmol), 50 μCi
   g. Klenow fragment (5 U/μL), 1 μL
   h. Sterile H_2O to a total of, 20 μL

   Incubate the reaction at room temperature for 2-5 h.
4. Dilute the reaction mixture to 100 μL with TE buffer and load onto a Sephadex G-50 (medium) minicolumn prepared in a Pasteur pipet. Elute with 150 μL aliquots of TE buffer to separate the probe from unincorporated nucleotides. Collect about 15 fractions (4 drops/fraction) in Eppen-

dorf tubes. Before starting the chromatography, neutralize static electricity on the pipet and Eppendorf tubes with a zerostat antistatic instrument (Sigma, St. Louis, MO).

5. Count 2 µL of each fraction in a Beckman LS 1800 scintillation counter (or equivalent). Two peaks of radioactivity should be observed. The first peak contains the labeled probe (usually included between fractions 5–10), whereas the second peak contains the unincorporated ^{32}P nucleotide. Pool the fractions corresponding to the first peak and count 2 µL of the pool. Estimate the specific activity of the probe as follows:
   a. Template DNA = 50 ng
   b. Labeled probe ~100 ng
   c. Conversion factor for 1 µg of probe = 10
   d. Activity of the 2 µL aliquot = 4×10^5 dpm (2×10^5dpm/µL)
   e. Volume of pooled fractions = 5×10^2 µL
   f. Specific activity: $2 \times 10^5 \times 5 \times 10^2 \times 10 = 10^9$ dpm/µg

3.3.2. Labeling Oligonucleotides by TdT (13,14)

1. Dry the oligonucleotide (~150 ng) in a speedvac and resuspend in 2 µL of TE buffer.
2. Set up the following labeling reaction:
   a. Oligonucleotide (75 ng/µL), 2 µL
   b. 5X TdT buffer (Gibco, Gaithersburg, MD), 4 µL
   c. BSA (5 mg/mL), 0.5 µL
   d. [α-^{32}P] dATP (6000 Ci/mmol), 125 µCi
   e. TdT (5 U/µL) (Gibco), 3 µL
   f. Sterile H_2O to a total of, 20 µL

   Incubate at 37°C for 1 h.
3. Separate the labeled probe from unincorporated nucleotides by gel filtration on Sephadex G-25 (medium) as described in Section 3.3.1.
4. Precipitate with 0.5 vol of ammonium acetate and 3 vol of absolute ethanol. Addition of 50 µg of carrier yeast tRNA can significantly increase the recovery of the probe after centrifugation. Count an aliquot into a scintillation counter and estimate the probe activity as described in Section 3.3.1. For further information regarding probe preparation and other hybridization procedures, *see* Notes 9–15.

4. Notes

4.1. RNase-Free Environment

1. RNA is extremely susceptible to degradation by RNases. All solutions should be treated with the RNase inhibitor DEPC *(4)*. DEPC is a suspected carcinogenic agent and should always be handled in a fumehood. In addi-

tion, disposable gloves should be worn at all times. The prehybridization and hybridization buffers should be treated with DEPC as dried milk is a rich source of RNases *(15)*. However, since this buffer does not withstand autoclaving it should be handled with care (preferably in a fumehood).

2. Formaldehyde is toxic and gels should be cast and run in a fumehood.
3. Ethidium bromide is a powerful mutagen; avoid contact with the skin.
4. Special care should be taken when working with radioactive sources. ^{32}P should be handled behind a 1-cm thick perspex screen. When the hybridization procedure is completed (after the washing steps), the amount of radioactivity remaining on the membrane is usually low and does not require the use of screens. ^{35}S has a much lower penetration power but, because of the longer half-life, contamination can be regarded as equally dangerous as from ^{32}P sources.

4.2. Gel Electrophoresis

5. The method described here is based on RNA fractionation on a denaturing formaldeyde-agarose gel *(2)*. Although other methods are available this is perhaps the most commonly used. In order to estimate the sizes of RNA molecules by their migration distances it is necessary to eliminate both intramolecular secondary structures and intermolecular interactions. This is effectively carried out by heat-denaturation and gel electrophoresis in the presence of formaldehyde.
6. The amount of RNA to be analyzed depends on the abundance of the target sequence within the sample. Usually, about 80% of the cellular RNA in eukaryotes is ribosomal RNA (rRNA), whereas messenger RNA (mRNA) represents 2–5% of the total. The remaining 15–18% is small RNA species (i.e., transfer RNA [tRNA], small cytoplasmic RNA [scRNA], and small nuclear RNA [snRNA]). If the experimental goal is the detection of abundant RNAs (considered to be ≥0.1% of the mRNA population), 10–20 μg of total cellular RNA should be used. For the analysis of rare mRNAs (<0.1% of the total mRNAs), 1–3 μg of purified mRNA should be used. Finally, the analysis of 1–2 μg of total cellular RNA should be sufficient for the detection of very abundant RNA species such as rRNA.
7. RNA size-markers should be used because, on formaldehyde gels, DNA migrates more slowly than RNA of equivalent size *(16)*. Often, the 28S and 18S rRNA bands (as visualized after staining with ethidium bromide) are used as size-markers. However, a migration curve created on the basis of only two reference points does not give accurate size-estimates for the hybridization bands. For more accurate estimates we prefer to use commercial RNA ladders (e.g., 0.24–9.5 kb ladder from Gibco) (*see* Fig. 1).

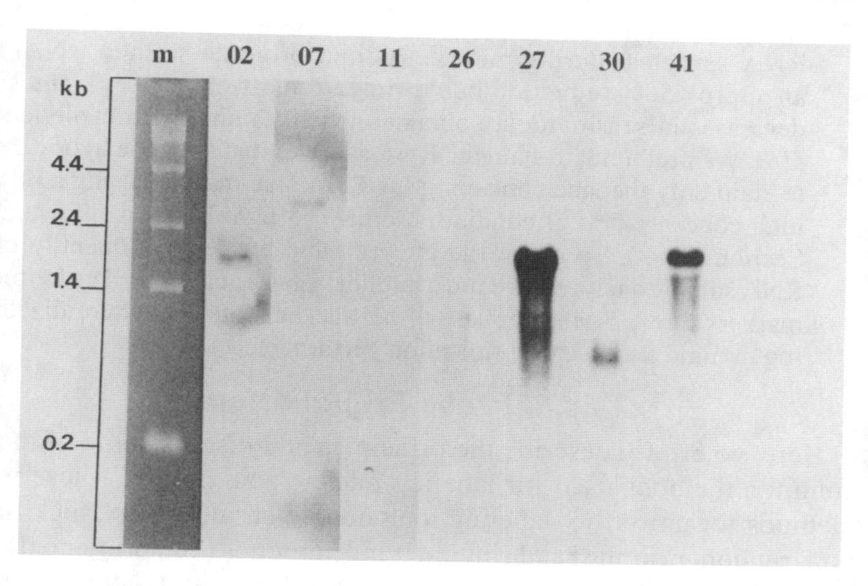

Fig. 1. Example of Northern hybridization. Total cellular RNA from maize root tips (10 µg/lane) was hybridized with several cDNA probes differentially expressed within the root apex *(32)*. The different probes, labeled by random-priming, are indicated by numbers at the top. The RNA size-marker is m (0.24–9.5 kb ladder from Gibco), stained with ethidium bromide. Note the different abundancies of the hybridizing transcripts in the RNA population. Interestingly, probe 30 detects two transcripts having different sizes. The membranes hybridized to probes 02, 07, and 11 were washed with 1 x SSC, 0.1 w/v SDS at 65°C and exposed for 3 d. The other membranes were washed with 0.5 w/v SSC, 0.1% w/v SDS at 65°C and exposed overnight.

8. Several methods can be used to visualize size-separated RNAs. The whole gel can be stained with a low concentration of ethidium bromide solution (0.5 µg/mL) for 30–45 min *(12)*, ethidium bromide can be added to the samples before loading the gel *(17)*, or RNA can be stained on the membrane after transfer using methylene blue *(18)*. It should be noted that ethidium bromide staining of single-stranded nucleic acids is far less efficient than that for double-stranded nucleic acids. For this reason either higher concentrations of dye or longer staining periods should be used. Excessive staining of the gel after electrophoresis is a common problem and should be avoided because subsequent visualization on a UV transilluminator may result in intense flourescence that makes it very difficult to visualize the

RNA samples. Furthermore, ethidium bromide-stained RNA shows an approx 50% reduction in hybridization efficiency *(19)*, and similar decreases in sensitivity are obtained when staining with methylene blue *(18)*. We prefer not to stain the RNA samples that are to be hybridized, but to stain only the lane containing the RNA size-markers using a relatively high concentration of ethidium bromide (5 µg/mL) for a short time (*see* Section 3.1.4.). To check the integrity and the relative quantity of each RNA sample analyzed, we run a duplicate gel that is stained as for the size-markers above. Such a replica gel can also be blotted and hybridized, bearing in mind its lower hybridization performance.

4.3. Probe Preparation

Here we briefly describe the principles and discuss some of the most common methods used for labeling nucleic acid probes. These include methods for uniformly labeling long double-stranded (i.e., nick translation, random priming) and single-stranded probes (i.e., primer extension, riboprobes), and for labeling of short (<50 nt) single-stranded probes (oligonucleotide probes end-labeled with T4 kinase or homo-tailed with terminal dimethyl transferase). Detailed protocols for these and other methods have been reviewed elsewhere *(12)*.

4.3.1. Nick-Translation (5)

9. The DNA probe is labeled by the incorporation of a α-^{32}PdNTP (usually dATP or dCTP) by *E. coli* DNA polymerase I to the 3'-hydroxylterminus of double-stranded DNA that has been nicked by DNase I. As *E. coli* DNA polymerase I has 5' → 3' exonucleotide activity, the nick "moves" from 5' to 3' and the nucleotides in the molecule are replaced by the nucleotides from the pool that includes the radiolabeled precursor. The resulting probe is of variable length (average size of about 500 bp), double-stranded and of relatively high specific activity (~10^8 dpm/µg).

4.3.2. Random Priming (6,7)

10. This method is based on the incorporation of radiolabeled dNTPs by primer extension using the Klenow fragment of *E. coli* DNA polymerase I. The DNA template is denatured (by boiling and rapidly cooling on ice), annealed to short deoxynucleotides (6–9 nt) of random sequence and the radiolabel incorporated into the DNA synthesized by the Klenow fragment. The resulting probe is double stranded with a high specific activity (≥10^9 dpm/µg). The size of the probe depends on the size of the template DNA and on the concentration of the primers (being larger for low concentrations as a result of a smaller number of priming sites *[20]*). Random prim-

ing is often the method of choice for labeling DNA for filter hybridization experiments, as it gives probes of high specific activity and does not require a high work input. Commercial kits are also available from several manufacturers (e.g., Stratagene [La Jolla, CA], Pharmacia, and Amersham [Arlington Heights, IL]). The latest versions use T7 DNA polymerase or Klenow fragment lacking exonucleotide activity that shortens the labeling step to a matter of minutes.

4.3.3. Single-Stranded Probes

11. Both nick translation and random priming work best with purified templates and give double-stranded probes that must be denatured before use in hybridization experiments. However, when the target molecule consists of only one nucleic acid strand, as in a northern blot, only one strand of the probe is required for hybridization. The other strand is not involved in any specific hybridization and can give increased background. In addition, the efficiency of the hybridization to the target sequence is reduced by reannealing of the probe strands. Although in most cases this does not represent a problem and random priming is widely used, methods to generate only labeled single-strand probes may be more appropriate for certain applications (e.g., S1-nuclease mapping, and *in situ* hybridization).

    Labeling of single-stranded probes is based on the *de novo* synthesis of DNA or RNA following primer extension. Both DNA (from template sequences cloned into a phagemid or bacteriophage M13 vectors *[21]*) or RNA (from template sequences cloned into vectors carrying a promoter for a bacteriophage DNA-dependent RNA polymerase) probes can be obtained. In the case of DNA probes, the probe is separated from vector sequences by digestion with endonucleases and denatured by purification on denaturing polyacrylamide gel. In the case of RNA probes *(9)*, the DNA template is digested downstream from the promoter, the synthesized RNA labeled, and the DNA template removed with RNase-free DNase I. In both cases, probes of fixed length and of very high specific activity (>10^9 dpm/μg) are obtained.

4.3.4. Oligonucleotide Probes

12. Two widely used methods for labeling oligonucleotides use T4 polynucleotide kinase *(11)* or terminal dimethyl transferase *(13)*. T4 poynucleotide kinase is used to transfer the γ-^{32}P from [γ-^{32}P]ATP to the 5'-termini of oligonucleotides (synthetic oligonucleotides lack a phosphate group at the 5'-termini). Although virtually every molecule of oligonucleotide can become labeled using this method, only one nucleotide/molecule is labeled; the resulting probes generally have a low specific activity.

However, with the addition of a homopolymeric tail of α-^{32}PdATP to the oligonucleotide 3'-end by TdT, specific activities in excess of 10^9 dpm/µg can be obtained *(13)*. Since each probe molecule has a tail of about 4–10 labeled nucleotides, care must be taken to specifically block hybridization to T-rich sequences of the immobilized nucleic acid. This can be done by using polyadenylic acid (about 200 µg/mL) in the prehybridization and hybridization buffers.

4.3.5. General Considerations

13. Some of the important parameters that influence the choice of labeling procedures are summarized in Table 1.
    a. Template. Although not always necessary, best results are obtained with purified DNA. Vector sequences are sometimes labeled in addition to the cloned inserts in order to obtain a stronger signal as a result of the binding of a network of probes to the target sequence. However this is not necessary, except in exceptional cases, and may result in nonspecific hybridization of the vector probe sequence.
    b. The percentage of incorporation can be measured by determining the proportion of radioactivity in TCA-precipitable nucleic acids or, for oligonucleotides, by thin-layer chromatography on polyethyleneimine *(12)*.
    c. Probes should be separated from unincorporated nucleotides by gel filtration on Sephadex G-50 or G-25 (for oligonucleotides), or by polyacrylamide gel electrophoresis. A range of spin- or push-columns are available from manufacturers and can be used to speed up the probe purification.
    d. ^{32}P-labeled probes should be used for filter hybridization experiments.
    e. We recommend counting the radioactivity of each probe synthesized. This allows the amount of probe used to be controlled for maximum sensitivity while limiting the background. With the hybridization buffer described in Chapter 13 (this volume), the best results are obtained when the concentration of the probe in the hybridization buffer corresponds to approx 5×10^5 dpm/mL. The counting procedure described in Section 3.3.1. is simple and gives reproducible results. More accurate figures can be obtained using other methods *(22)*.
    f. We have found that random priming can be used for short templates (100–200 bp) (e.g., PCR fragments), and usually it gives a discrete band (Fig. 2). However, in our experience, the specific activity is about tenfold lower than that for larger templates. Alternative methods that use Taq DNA polymerase have recently been shown to give higher specific activity using short templates (<100 bp) *(23)*.

Table 1
Requirements and Properties of Probes Obtained with Different Labeling Methods

Labeling method	Template	Probe-size	Specific activity	Strand specificity	Hybridization specificity	Hybrid stability	Work input
Nick translation	Double-stranded DNA	Variable	Medium	No	Good	High	Low
Random priming	Single-stranded DNA	Variable	High	No	Good	High	Low
Primer-extension	Single-stranded DNA	Fixed	High	Yes	Excellent	High	High
Riboprobes	Single-stranded DNA	Fixed	High	Yes	Excellent	High	High
Oligonucleotides:							
by TdT[a]	Single-stranded DNA	Variable	High	Yes	Good	Low	Low
by T4 K.[b]	Single-stranded DNA	Fixed	Low	Yes	Excellent	Low	Low

[a]Indicates oligonucleotide homopolymeric tailing by terminal dymethyl transferase (TdT).
[b]Indicates oligonucleotide end-labeling by T4 polynucleotide kinase.

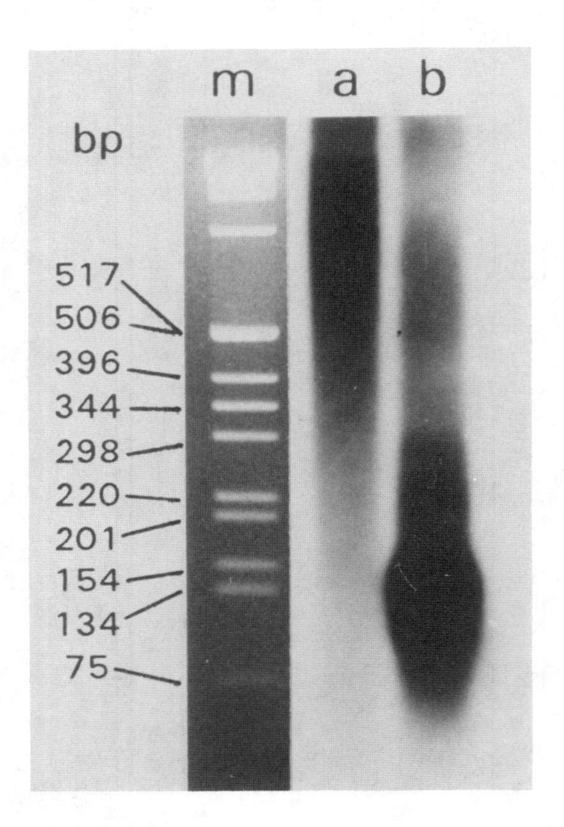

Fig. 2. Gel electrophoresis of two probes obtained by random priming. A 1.6 kb **(a)** and a 148 bp **(b)** DNA fragments were labeled using the Prime II kit (Stratagene) folowing the manufacturer's instructions, then fractionated on a 2% w/v agarose gel. m is 1 kb ladder DNA (Gibco) stained with ethidium bromide. The gel was vacuum dried and exposed to an X-ray film. Note the different average sizes of the two probes (>500 bp in [a], ~130 bp in [b]).

4.4. Hybridization with Oligonucleotide Probes

14. Oligonucleotide probes have the advantage of high hybridization specificity, and sequences that differ by as little as a single nucleotide can be distinguished *(24,25)*. However, short hybrid molecules are highly unstable. Hybridization is usually carried out under high stringency conditions ($Tm - 10 < T_i < Tm - 5$, *see* Chapter 13, this volume), but under these conditions the hybridization reaction can be considered reversible. This does not affect the hybridization step because a relatively high concentration of probe is present, but care should be taken during the washing steps, which should

be kept as short as possible (generally, 2–5 min). However, perhaps surprisingly, low-stringency conditions have been also successfully used with oligonucleotides shorter than 10 nucleotides *(26)*. The stability of oligonucleotide duplexes is greatly affected by mismatches, especially those occurring in the middle rather than at the ends of the duplex.

It is often necessary to use degenerate oligonucleotide probes, making it difficult to estimate a consensus T_m. Specific conditions to avoid this problem have been described, including:

a. The use of conditions equivalent to T_m - 2 for the most (A + T)-rich member of the pool *(12)*.
b. The use of hybridization buffers that contain tetramethylammonium chloride (TMACl) instead of NaCl. In the presence of this salt the T_m of duplexes seems to depend only on their length *(27)*.
c. The design of oligonucleotide probes that contain deoxyinosine at each position of degeneracy, as neutral bases such as inosine bind to all four purine and pyrimidine bases with about the same efficiency *(28)*. However, it should be emphasized that time-consuming optimization of conditions is required to obtain efficient hybridization of oligonucleotide probes to large genomes or complex RNA populations.

4.5. Dot Blot Hybridization

15. In Southern and northern blotting analyses, important information is obtained about the sizes of the hybridization products. This in turn can be used to study a range of problems of molecular biology. However, dot blot hybridization can be used for studies in which a large number of samples must be processed, or when information about the sizes of hybridizing bands is not required *(29–31)*. In addition, it is also possible to quantify the relative abundances of target molecules in different samples. For dot blotting, the nucleic acid concentrations of samples are determined carefully, and the samples are then blotted directly onto a transfer membrane using a vacuum manifold. The immobilized nucleic acid is single stranded as in standard Southern and northern blotting techniques. For more accurate estimates of abundance, the hybridization signals on the autoradiographs can be scanned using a densitometer and compared with standard samples. For a correct evaluation of quantitative data several points should be considered:
    a. The manufacturer's instructions for the blotting must be followed carefully as these may vary with the apparatus used.
    b. It is important to use control samples and probes with known hybridization properties. It helps also to use serial dilutions of each sample.
    c. Dot blot hybridization does not discriminate between signal and background. Some samples may contain a high level of impurities that increases the background.

d. Dot blot hybridization is often used to quantify differences in gene expression in samples that differ in developmental stages, positions, treatments applied, and so on. As with comparisons of hybridization signals on northern blots, care should be taken to normalize data on a per cell basis and not per mass unit of RNA. Indeed differences in signals may be owing to different numbers of cells in the samples, or different levels of transcriptional or metabolic activities. The use of constitutively expressed transcripts as probes may help to minimize artifacts.

Acknowledgments

The authors thank Shirley Burgess and Peter Holloway for reviewing the manuscript and Bob Harvey and Ken Williams for their photographic assistance.

References

1. Southern, E. M. (1975) Detection of specific sequences among DNA fragments separated by gel electrophoresis. *J. Mol. Biol.* **98,** 503–517.
2. Alwine, J. C., Kemp, D. J., and Stark, G. R. (1977) Method for detection of specific RNAs in agarose gels by transfer to diazobenzyloxymethyl-paper and hybridization with DNA probes. *Proc. Natl. Acad. Sci. USA* **74,** 5350–5354.
3. Denhardt, D. (1966) A membrane filter technique for the detection of complementary DNA. *Biochem. Biophys. Res. Commun.* **23,** 641–646.
4. Kumar, A. and Lindberg, U. (1972) Characterization of messenger ribonucleoprotein and messenger RNA from KB cells. *Proc. Natl. Acad. Sci. USA* **69,** 681–685.
5. Rigby, P. W. J., Dieckmann, M., Rhodes, C., and Berg, P. (1977) Labeling deoxyribonucleic acid to high specific activity *in vitro* by nick translation with DNA polymerase I. *J. Mol. Biol.* **113,** 237–251.
6. Feinberg, A. P. and Vogelstein, B. (1983) A technique for radiolabeling DNA restriction endonuclease fragments to high specific activity. *Anal. Biochem.* **132,** 6–13.
7. Feinberg, A. P. and Vogelstein, B. (1984) Addendum: a technique for radiolabeling DNA restriction endonuclease fragments to high specific activity. *Anal. Biochem.* **137,** 266,267.
8. Ley, T. J., Anagnou, N. P., Pepe, G., and Nienhuis, A. W. (1982) RNA processing errors in patients with β-thalassemia. *Proc. Natl. Acad. Sci. USA* **79,** 4775–4779.
9. Melton, D. A., Krieg, P. A., Rebagliati, M. R., Maniatis, T., Zinn, K., and Green, M. R. (1984) Efficient *in vitro* synthesis of biologically active RNA and RNA hybridization probes from plasmids containing a bacteriophage SP6 promoter. *Nucleic Acids Res.* **12,** 7035–7056.
10. Srivastava, R. A. K. and Schonfeld, G. (1991) Use of riboprobes for northern blotting analysis. *Biotechniques* **11,** 584–587.
11. Richardson, C. C. (1971) Polynucleotide kinase from *Escherichia coli* infected with bacteriophage T4, in *Procedures in Nucleic Acid Research*, vol. 2 (Cantoni, G. L. and Davies, D. R., eds.), Harper and Row, New York, pp. 815–828.

12. Sambrook, J., Fritsch, E. F., and Maniatis, T. (1989) *Molecular Cloning: A Laboratory Manual.* Cold Spring Harbor Laboratory Press, Cold Spring Harbor, NY.
13. Collins, M. L. and Hunsaker, W. R. (1985) Improved hybridization assays employing tailed oligonucleotide probes: a direct comparison with 5'-end-labeled oligonucleotide probes and nick-translated plasmid probes. *Anal. Biochem.* **151,** 211–224.
14. Henderson, G. S., Conary, J. T., Davidson, J. M., Stewart, S. J., House, F. S., and McCurley, T. L. (1991) A reliable method for northern blot analysis using synthetic oligonucleotide probes. *Biotechniques* **10,** 190–197.
15. Siegel, L. I. and Brosnick, E. (1986) Northern hybridization analysis of RNA using diethylpyrocarbonate-treated nonfat milk. *Anal. Biochem.* **159,** 82–87.
16. Wicks, R. J. (1986) RNA molecular weight determination by agarose gel electrophoresis using formaldehyde as denaturant: comparison of RNA and DNA molecular weigth markers. *Int. J. Biochem.* **18,** 277,278.
17. Fourney, R. M., Miyakoshi, J., Day III, R. S., and Paterson, M. C. (1988) Northern blotting: efficient RNA staining and transfer. *Bethesda Res. Lab. Focus* **10 (1),** 5,6.
18. Wilkinson, M., Doskow, J., and Lindsey, S. (1991) RNA blots: staining procedures and optimization of conditions. *Nucleic Acids Res.* **19,** 679.
19. Sabelli, P. A. and Shewry, P. R. (1993) Nucleic acids blotting and hybridization, in *Methods in Plant Biochemistry* (Bryant, J., ed.), Academic, London, pp. 79–101.
20. Hodgson, C. P. and Fisk, R. Z. (1987) Hybridization probe size control:optimized "oligolabelling." *Nucleic Acids Res.* **15,** 6295.
21. Messing, J. (1983) New M13 vectors for cloning. *Methods Enzymol.* **101,** 20–78.
22. Mundy, C. R., Cunningham, M. W., and Read, C. A. (1991) Nucleic acid labeling and detection, in *Essential Molecular Biology: A Practical Approach*, vol. II (Brown, T. A., ed.), IRL Press, Oxford, pp. 57–110.
23. Mizobuchi, M. and Frohman, L. A. (1992) A rapid and simple method for labeling short DNA fragments using *Taq* polymerase. *Biotechniques* **12,** 350–353.
24. Wallace, R. B., Shaffer, J., Murphy, R. F., Bonner, J., Hirose, T., and Itakura, K. (1979) Hybridization of synthetic oligodeoxyribonucleotides to $\phi\chi$ 174 DNA, the effect of single base pair mismatch. *Nucleic Acids Res.* **6,** 3543–3556.
25. Conner, B. J., Reyes, A. A., Morin, C., Itakura, K., Teplitz, R. L., and Wallace, R. B. (1983) Detection of sickle cell β-globin allele by hybridization with synthetic oligonucleotides. *Proc. Natl. Acad. Sci. USA* **80,** 278–282.
26. Drmanac, R., Strezoska, Z., Labat, I., Drmanac, S., and Crkvenjakov, R. (1990) Reliable hybridization of oligonucleotides as short as six nucleotides. *DNA Cell Biol.* **9,** 527–534.
27. Wood, W. I., Gitschier, J., Lasky, L. A., and Lawn, R. M. (1985) Base composition-independent hybridization in tetramethylammonium chloride: a method for oligonucleotide screening of highly complex gene libraries. *Proc Natl. Acad. Sci. USA* **82,** 1585–1588.
28. Ohtsuka, E., Matsuki, S., Ikehara, M., Takahashi, Y, and Matsubara, M. (1984) An alternative approach to deoxyoligonucleotides as hybridization probes by insertion of deoxyinosine at ambiguous codon positions. *J. Biol. Chem.* **260,** 2605–2608.
29. Kafatos, F. C., Weldon Jones, C., and Efstratiadis, A. (1979) Determination of nucleic acid sequence homologies and relative concentrations by a dot hybridization procedure. *Nucleic Acids Res.* **7,** 1541–1552.

30. White, B. A. and Bancroft, F. C. (1982) Cytoplasmic dot hybridization. *J. Biol. Chem.* **257,** 8569–8572.
31. Yu, S.-M. and Gorovsky, M. A. (1986) *In situ* dot blots: quantitation of mRNA in intact cells. *Nucleic Acids Res.* **14,** 7597–7615.
32. Sabelli, P. A., Burgess, S. R., Carbajosa, J. V., Parker, J. S., Halford, N. G., Shewry, P. R., and Barlow, P. W. (1993) Molecular characterization of cell populations in the maize root apex, in *Molecular and Cell Biology of the Plant Cell Cycle* (Ormrod, J. C. and Francis, D., eds.), Kluwer, Dordrecht, pp. 97–109.

CHAPTER 19

Nuclear "Run-On" Transcription Assays

John Gatehouse and Andrew J. Thompson

1. Introduction

Nuclear "run-on" (or "run-off") transcription assays have been used to obtain quantitative information about the relative rates of transcription of different genes in nuclei isolated from a particular tissue or organ. This information can then be used in analyzing the factors that control the expression of specific genes. These assays are complimentary to measurements of the levels of mRNA species present in particular tissues or organs by hybridization to specific nucleic acid probes (*see* Chapter 18). Hybridization assays give a measure of how much mRNA is present; this is termed a "steady-state" level, since it reflects the balance between mRNA synthesis and degradation (although, of course, the "steady-state" level itself can change with time), whereas transcription assays give a direct measure of the activity of genes.

The overall process of gene expression can be considered to include all the factors that influence how much of a particular gene product is present at a particular time in an organism's development, under the influence of endogenous and exogenous effects. The process includes many steps, including signal transduction, transcriptional level controls, posttranscriptional effects on mRNA level, translational controls, and protein turnover; however, it is gene transcription that is (almost invariably) the crucial step in the whole process, without which no gene expression is possible. It is perhaps unfortunate that direct measurements of the tran-

From: *Methods in Molecular Biology, Vol. 49: Plant Gene Transfer and Expression Protocols*
Edited by: H. Jones Humana Press Inc., Totowa, NJ

scription rates of specific genes in higher organisms are difficult to make, and, as a result, measurements of steady-state levels of mRNA species have often been considered to be equivalent to measuring "gene expression." However, the accumulation of an mRNA species is itself subject to many controls, its synthesis being determined not only by transcription, but also by posttranscriptional processing, and the steady-state level reflects both the rates of synthesis and degradation. If methods for measuring the rates of the different processes are available, comparisons of the rates of mRNA synthesis and degradation, and measurements of the steady-state levels of mRNAs in nuclear and cytoplasmic or total RNA allow conclusions to be drawn about the points of regulation.

1.1. In Vivo Assays for Transcriptional Activity of Specific Genes

A simple approach to measuring the rates of synthesis and degradation of a specific mRNA is by measuring the time course of changing mRNA levels, after a stimulus is used to perturb the system and cause a change in the steady-state levels of the mRNA. The kinetics of the approach to a new steady-state level are then followed by hybridization assay. The rate of change of mRNA level, dC/dt (molecules mRNA per cell/min) is given by:

$$dC/dt = k_s - k_d C_t$$

where k_s is the zero-order rate constant of synthesis (molecules RNA per cell/min), k_d is the first-order rate constant of degradation (min^{-1}) and C_t is the molecules of RNA per cell at time t. If dC/dt is plotted against C_t both k_s (y-intercept) and k_d (slope) can be measured *(1,2)*.

Where the steady-state levels of the mRNA are constant, or changing very slowly, and are not susceptible to perturbation, k_s and k_d can be measured by determining the kinetics of radioactive labeling of mRNA. In this case, to make measurements for specific mRNA species, they must be purified by hybridization techniques (e.g., hybridization to specific nonlabeled probes followed by separation from single-stranded nucleic acids by hydroxylapatite chromatography). In an example using sea urchin embryos, organisms were incubated continuously with [^{3}H]-guanosine; the specific activity of the GTP precursor pool, S_t (dpm/mol nucleotide), and the incorporation of radioactivity into a specific mRNA species, ${}^rC_t$ (dpm specific mRNA/cell) were measured throughout the labeling period *(3)*. k_s and k_d can be estimated from the equation:

$$d^{r}C/dt = k_{s}S_{t} - k_{d}{}^{r}C_{t}$$

Such experiments are technically difficult, since the amount of label incorporated into a specific mRNA species will be very small unless its encoding gene is very actively transcribed.

If some assumptions about k_d, the rate of degradation of a specific mRNA species, can be made, then experimental protocols that are less technically demanding (and prone to error) can be adopted. By administering a pulse of RNA precursor to cells, the rate of transcription of a specific gene can be measured directly by measuring incorporation of label into a specific mRNA species, measured by hybridization to a specific DNA probe. The assumption is made that k_d is sufficiently low that no significant degradation of the primary transcripts occurs, and thus a single measurement gives k_s. This method has been employed with moderate success in mammalian cell cultures, using [^{3}H]-uridine as the labeled precursor; for example, enough label was taken up in a 5-min pulse to allow measurement of labeled globin nuclear RNA *(4)*, and this time scale is known to be much faster than that normally required to see changes in globin mRNA levels.

Although these in vivo labeling techniques allow measurement of transcription rates under conditions approaching ideal, i.e., where the tissue, organ, or cell culture is in a state as close to normal as possible, they are often not possible to apply. Suitable cell or organ culture systems may not be available. Even where they are, the incorporation of label is usually so low with higher plants that only very highly transcribed genes can be assayed. Since comparatively few genes (and by no means the most interesting) are highly transcribed, an alternative system is necessary.

1.2. Assays for Transcriptional Activity of Specific Genes Using Isolated Nuclei

The use of isolated nuclei to generate labeled transcripts *(5)* has a significant advantage over the use of whole cells in that it is able to generate highly labeled RNA by the use of α[^{32}P]-labeled nucleoside triphosphates (usually UTP) as labeled precursors. These transcripts can then be hybridized to specific (unlabeled) DNA probe sequences, allowing detection of transcripts from most expressed genes. It is generally accepted from studies on animal cell nuclei that little or no initiation of transcription occurs in vitro *(6)*, and thus the transcripts detected represent preinitiated transcripts on a given gene, which are elongated in vitro and incorporate the labeled nucleotide. By comparing the relative amounts of label incorporated into different mRNA molecules in the same labeled RNA preparation, as detected by hybridization, the relative amounts of transcripts produced by different genes can be measured, and a good estimate of the relative transcription rates of the genes in vivo can be obtained. It is generally accepted that transcription in isolated nuclei gives a good estimation of the relative transcription rates of different genes in vivo *(7)*, although measured rates for the same gene in vivo and in vitro may differ *(5)*.

This type of transcription assay, first used with plant nuclei by Gallagher and Ellis *(8)*, is preferred for plants owing to severe technical difficulties encountered when trying to label transcripts in vivo, and many authors have subsequently used the method to study transcription of plant genes.

2. Materials

1. 125-μm and 53-μm nylon mesh; sterilize by boiling for 10 min.
2. Homogenizer: A glass or Teflon-glass Quick-fit homogenizer is good for small amounts (<5 g) of material. For larger amounts, use a conventional food mixer (Osterizer) or, better, a Ystral homogenizer (Ystral Gmbh, Dottingen, Germany). Gentle homogenization is essential; avoid homogenizers that generate ultrasound like the Polytron.
3. Percoll (Pharmacia, Uppsala, Sweden).
4. Low speed swinging bucket centrifuge.
5. Buffers for isolation of nuclei. Autoclave individual components where appropriate (make up concentrated stocks), use sterile (autoclaved) distilled deionized water. Buffers should be made up fresh as required.
   a. Buffer A: 0.7*M* mannitol, 10 m*M* MES, 5 m*M* EDTA, 0.1% BSA (bovine serum albumin), 0.2 m*M* PMSF (phenylmethyl sulfonyl fluoride). Filter sterilize.
   b. Buffer B: Buffer A supplemented with 0.1 mg/mL cellulase and 0.05 mg/mL pectinase. In order to dissolve the enzymes the pH is raised to 11 with 2*M* NaOH for 3 min, and then returned to pH 5.8 with acetic acid. Add PMSF after dissolving enzymes. Filter sterilize.
   c. Buffer C: 0.25*M* sucrose (grade I, Sigma, St. Louis, MO), 10 m*M* NaCl, 5 m*M* EDTA, 0.15 m*M* spermine, 0.5 m*M* spermidine, 20 m*M* 2-mercaptoethanol, 0.2 m*M* PMSF, 0.6% Nonidet P-40, 0.1% BSA. Filter sterilize.
   d. Buffer D: 30 g 5X buffer C (except Nonidet P-40 should be kept at 0.6%) added to 116 g Percoll. Adjust to pH 6.0 with 1*N* HCl.

e. Buffer E: 12 g 5X buffer C (except Nonidet P-40 should be kept at 0.6%) added to 90 g Percoll. Adjust to pH 6.0 with 1*N* HCl.

6. Buffers for transcription assays. *See notes* for *(5)*.
   a. Buffer I: 20 m*M* Tris-HCl, pH 7.4, 5 m*M* $MgCl_2$, 0.1 m*M* EDTA, 0.5% BSA; the buffer also contains glycerol at concentrations from 0–50% (qv).
   b. Buffer II: buffer I supplemented with 240 m*M* $(NH_4)_2SO_4$.
   c. Buffer V: buffer II supplemented with 2 m*M* ATP, 2 m*M* GTP, 2 m*M* CTP. SET buffer: 10 m*M* Tris-HCl, pH 7.5, 5 m*M* EDTA, 1% SDS.
   d. DNase I buffer: 50 m*M* Tris-HCl, pH 7.5.
   e. Column buffer: 0.3*M* NaCl, 0.1% SDS, 100 m*M* Tris-HCI, pH 7.5, 1 m*M* EDTA.
7. [α^{32}P]UTP; 410 Ci/mmole. Higher specific activities can be used if necessary.
8. Human placental ribonuclease inhibitor (RNasin).
9. Other enzymes, reagents as specified in procedures. Normal laboratory glass and plastic ware. All glassware used for procedures handling "naked" RNA (i.e., the transcription assay) should be treated with diethylpyrocarbonate and autoclaved to remove RNase contamination. A scintillation counter or laser densitometer is necessary to enable hybridization to be quantified, and a standard laboratory microscope with fluorescence excitation is also useful for examining nuclei.

3. Methods

3.1. Isolation of Nuclei

The method is based on that of Willmitzer and Wagner *(9)*, with some modifications *(10)* (*see* Note 1). All manipulations are carried out on ice unless otherwise stated, and all centrifugation steps are carried out at 4°C using swing-out rotors with no braking. Glass and plastic ware should be sterile.

1. Harvest plant tissue (*see* Note 2) into a suitable flat dish in an ice bucket, cover with buffer A, and chop into small piece with a razor blade. Filter through 125-µm nylon mesh, wash in buffer A. Incubate in buffer B for 4 h, then filter through 125-µm nylon mesh, wash in buffer C, and resuspend in buffer C at 20 mL/g of tissue.
2. Homogenize tissue gently using a hand or powered homogenizer *(see the preceding)*. Filter the homogenate through a 125-µm nylon mesh, and centrifuge the filtrate at 2000*g* for 10 min. Resuspend the pellet in buffer C at 2 g buffer per g tissue. Weigh the resulting suspension, and add 1.5 times this weight of buffer D. At this final Percoll concentration of 47.6%, nuclei and starch sediment, and the majority of cell wall material remains buoyant.

3. Filter the suspension through 53-μm nylon mesh, and centrifuge the filtrate at 1000*g* for 5 min. Remove the supernatant, mix thoroughly, and repeat the centrifugation. This step should be repeated an additional 2–3 times, and then all the pellets are combined by resuspending in 5–10 mL buffer E. Centrifuge the resulting suspension at 5000*g* for 15 min. Nuclei band as a floating layer, which is removed and diluted with 5–10 vol of buffer C, and centrifuged for 10 min at 1000*g*. Wash the pellet twice with buffer C.
4. Resuspend the pellet in 0.5–1.0 mL of transcription buffer I, containing 50% glycerol. Nuclei may be examined by fluorescence microscopy after dilution with buffer C and staining with 20 μg/mL DAPI (4,6-diamidino-2-phenylindole); stained nuclei can be counted using a hemocytometer slide (*see* Note 3). Nuclear preparations should contain little or no cell debris not staining with the DAPI agent, and nuclei should appear rounded in shape and intact, although most preparations will also contain some nuclei that are obviously damaged or fragmented. The presence of damaged nuclei does not seem to significantly affect the results of subsequent transcription assays. Nuclear suspensions can be stored for up to 6 mo at –80°C.

3.2. Transcription Reactions Using Isolated Nuclei

A method adapted from that of McKnight and Palmiter *(11)* is used *(12)*.

1. Nuclei stored in 50% glycerol are diluted to 15% glycerol with glycerol-free buffer I. Centrifuge for 5 min at 1000*g* in a 15-mL Corex tube. The pellet is washed twice with buffer I containing 15% glycerol, and then resuspended in 50 μL buffer I (containing 15% glycerol). Add 20 μL buffer II (*see* Note 5) and incubate the suspension on ice for 15 min.
2. Lyophilize an aliquot of [α-^{32}P]UTP and redissolve in 20 μL buffer V containing 140 U of RNasin.
3. Add the suspension of nuclei to the solution of labeled UTP (final reaction volume 140 μL) (*see* Note 4). Incubate at 26°C for 10 min.
4. Stop the transcription by digestion with DNase I added to 10 μg/mL for 5 min at 26°C. Deproteinize the reaction by treatment with 100 μg/mL proteinase K in SET buffer (make to appropriate concentrations by adding 10X stock) at 45°C for 30 min in the presence of 25 μg of purified yeast tRNA (from 5 mg/mL stock).
5. Phenol/chloroform extract the mixture; back extract the organic phase with 50 μL 10 m*M* Tris-HCl, pH 7.5. Combine the aqueous phases and extract with chloroform. Precipitate overnight at –20°C by adding NaCl to 0.1*M* and 2.5 vol of ethanol.

6. Centrifuge at 12,000*g* for 15 min and wash the precipitate with 40 m*M* NaCl, 66% ethanol. Dissolve the nucleic acid in 100 µL DNase I buffer and incubate for 1 h at 26°C with 10 µg/mL DNase I and 140 U RNasin. Deproteinize the reaction with proteinase K at 100 µg/mL for 30 min at 45°C after adjusting the buffer to 0.2X SET. Phenol/chloroform and chloroform extract as before.
7. Separate the RNA from unincorporated labeled nucleotide by gel filtration on a column (5–10 mL) of Sephadex G-50 in column buffer. Precipitate fractions from the first peak of radioactivity overnight at –20°C by adding 2.5 vol of ethanol; wash the precipitate with 70% ethanol, and redissolve in water (200 µL). Take an aliquot (1 µL) for scintillation counting. Make the solution up to 0.1% SDS, and, if necessary, store at –20°C before hybridization; storage should be for the minimum time possible, and no longer than a few hours.

3.3. Hybridization of Labeled Transcripts to DNA Blots

DNA blots, either in the form of Southern blots of agarose gel separations of probe sequences, or more conveniently, as dot-blots of probe sequences, are prepared by normal techniques (*see* Note 6). The maximum amount of DNA convenient should be loaded on gels or dots, for maximum hybridization efficiency; in practice, this is approx 5 µg DNA per band or dot. A selection of different probe DNA sequences, corresponding to the different genes to be assayed, is loaded onto each blot (Fig. 1). The protocol given *(13)* maximizes hybridization efficiency.

1. Prehybridize blots for 4 h at 41°C in 50% deionized formamide, 0.5*M* NaCl, 40 m*M* PIPES-NaOH pH 6.5, 0.4% SDS, 1 m*M* EDTA, 100 µg/mL polyadenylic acid and 100 µg/mL *E. coli* tRNA. Cut blots down to minimum size before prehybridization.
2. Add labeled transcripts in a minimum volume (1 mL or less for dot-blots, 2 mL or less for gel blots) of the same buffer and hybridize for 72 h at 41°C.
3. Wash the blots four times in 1.4X SSC, 0.1% SDS at 60°C, two times in 0.1X SSC, 0.1% SDS at 60°C, once in 0.3*M* NaCl, 10 µg/mL RNase A at room temperature and once in 0.3*M* NaCl at room temperature. All washes are for 30 min. Transcripts may be saved from the hybridization solution and hybridized to a second prehybridized blot to estimate the extent of hybridization; this should be at least 90%.
4. Autoradiograph blots, then cut out radioactive dots or bands and dissolve each in 5 mL scintillation fluid containing 60% toluene, 40% methoxyethanol, 1.1 g/L Omnifluor for scintillation counting. Blots can be cali-

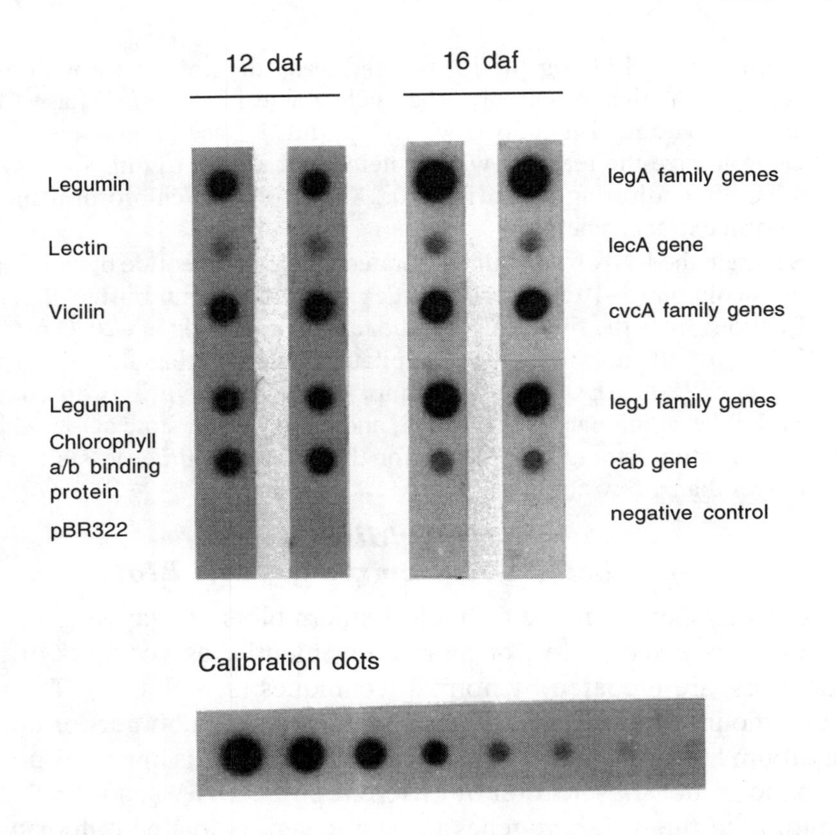

Fig. 1. Example of transcription assays using isolated nuclei *(13)*. Autoradiographs of DNA dot-blots (5 μg DNA per dot) hybridized to transcripts synthesized in nuclei isolated from cotyledons of developing pea seeds, harvested at 12 and 16 d after flowering (daf). Dots are made with DNA of plasmids containing gene fragments to detect transcripts from seed storage protein genes (legumin, vicilin) and other genes as designated. 3×10^8 nuclei were used for each transcription reaction, with 0.75 and 1.0 mCi [α^{32}P]UTP as labeled nucleotide for 12 and 16 daf cotyledons, respectively. Each transcription reaction was hybridized to three dot-blots, of which only two are shown; each dot-blot received 7.8×10^7 cpm (12 daf) or 1.7×10^8 cpm (16 daf) of incorporated counts. Legumin (legA) transcripts make up 3.5×10^{-6} of total transcripts at 12 daf, and 6.1×10^{-6} of total transcripts at 16 daf. Blots were autoradiographed for 24 h.

brated by adding dots containing, in serial dilution, known amounts of end-labeled DNA. Quantitative estimation of transcripts that have hybridized can be made by scintillation counting, or by densitometry of autoradiographs using a laser densitometer (LKB Ultroscan XL).

4. Notes

1. Protocols for nuclei isolation and run-off transcription are also given by Cox and Goldberg *(14)*. The methods are broadly similar, but differ in detail.
2. The method given for isolation of nuclei has been used successfully with leaves and seedling tissue from several dicot plant species, and with developing legume seeds. However, other sources of tissues (other plant species, or other organs/tissues) may not give viable nuclear preparations using this technique. Some plant tissues cannot be used as a source of nuclei; woody or fibrous tissue will not work!
3. Yields of nuclei obtained are typically 10^6 per g of developing seed tissue, or 10^7 per g of leaf tissue. (The yield will depend on the age of the tissue, i.e., the cell number and size of cells.)
4. A typical transcription reaction will use 10^7 nuclei, and add 250 µCi of [α-^{32}P]UTP to label the transcripts produced. These quantities can be scaled up (or down) according to the performance of the preparation of nuclei used. Blots should be hybridized with at least 10^8 cpm of incorporated radioactivity in the transcripts to give clean results (*see* Fig. 1).
5. The concentration of $(NH_4)_2SO_4$ in the transcription reaction may need to be adjusted to give optimal incorporation of label (i.e., maximal transcription).
6. Interpretation of results, and experimental design. The "run-on" transcription assay gives a comparison of the transcriptional activities of different genes in the same nuclear preparation. It is not suitable for comparing the activity of the same gene in different nuclear preparations (e.g., at different developmental stages) unless some kind of internal standard (e.g., a gene that is expressed at a constant level) can be devised. However, the relative levels of expression of two or more genes in different nuclear preparations can be measured, so that it is quite easy to show that the expression of one gene relative to another changes. Experimental design must take the shortcomings of this assay into account. The possibility of cross-hybridization of the probe sequences used must also be considered; often, the transcription of a subfamily or family of genes, rather than a single gene, is measured. There is no simple way to deal with this problem, since using shorter, potentially more specific probe sequences adversely affects the efficiency of hybridization.

References

1. Kafatos, F. C. (1972) mRNA stability and cellular differentiation. *Karolinska Symposia on Research.Methods in Reproductive Endocrinology vol. 5: Gene Transcription in Reproductive Tissue.* Karolinska Institute, Stockholm, pp. 319–345.
2. Guyette, W. A., Matusik, R. J., and Rosen, J. M. (1979) Prolactin-mediated transcriptional and posttranscription control of casein gene expression. *Cell* **17,** 1013–1023.

3. Cabrera, C. V., Lee, J. J., Ellison, J. W., Britten, R. J., and Davidson, E. H. (1984) Regulation of cytoplasmic mRNA prevalence in sea urchin embryos: rates of appearance and turnover for specific sequences. *J. Mol. Biol.* **174,** 85–111.
4. Ganguly, S. and Skoultchi, A. I. (1985) Absolute rates of globin gene transcription and mRNA formation during differentiation of cultured mouse erythroleukaemia cells. *J. Biol. Chem.* **260,** 12,167–12,173.
5. Cox, R. F. (1976) Quantification of elongating form A and B RNA polymerases in chick oviduct nuclei and effects of estradiol. *Cell* **7,** 455–465.
6. Weber, J., Jelinek, W., and Darnell, J. E. (1977) The definition of a large viral transcription unit latr in Ad2 infection of HeLa cells: mapping of nascent RNA molecules labelled in isolated nuclei. *Cell* **10,** 611–616.
7. Darnell, J. E. (1982) Variety in the level of gene control in eukaryotic cells. *Nature* **297,** 365–371.
8. Gallagher, T. F. and Ellis, R. J. (1982) Light-stimulated transcription of genes for two chloroplast polypeptides in isolated pea leaf nuclei. *EMBO J.* **1,** 1493–1498.
9. Willmitzer, L. and Wagner, K. G. (1981) The isolation of nuclei from tissue cultured plant cells. *Exp. Cell Res.* **135,** 69.
10. Evans, I. M., Gatehouse. J. A., Croy, R. R. D., and Boulter, D. (1984) Regulation of the transcription of storage-protein mRNA in nuclei isolated from developing pea (*Pisum sativum* L.) cotyledons. *Planta* **160,** 559–568.
11. McKnight, G. S. and Palmter, R. D. (1979) Transcriptional regulation of the ovalbumin and conalbumin genes by steroidal hormones in chick oviduct. *J. Biol. Chem.* **254,** 9050–9058.
12. Thompson, A. J.. Evans, I. M., Boulter, D., Croy, R. R. D., and Gatehouse, J. A. (1989) Transcriptional and posttranscriptional regulation of seed storage protein gene expression in pea (*Pisum sativum* L.). *Planta* **179,** 279–287.
13. Thompson, A. J. (1989) Regulation of gene expression in developing pea seeds. Ph.D. thesis, University of Durham.
14. Cox, K. H. and Goldberg, R. B. (1988) Analysis of plant gene expression, in *Plant Molecular Biology—A Practical Approach* (Shaw., C. H., ed.), IRL Press, Oxford, UK, pp. 1–35.

CHAPTER 20

RNase A/T_1 Protection Assay

Craig G. Simpson and John W. S. Brown

1. Introduction

RNase A/T_1 mapping provides a sensitive and quantitative method of expression analysis of known gene sequences. It differs from primer-derived analyses such as primer extension and reverse transcriptase-PCR by the use of a probe that is colinear with the transcript under study. This protocol has been applied to the analysis of plant RNA transcripts, for example, in the expression of endogenous plant genes *(1)*, introduced genes *(2)*, mutated or tagged genes *(3)*, antisense gene copies *(4)*, and in the analysis of plant nuclear pre-mRNA splicing *(5)*. There are four steps involved in RNase A/T_1 analysis. First, the transcription of a labeled RNA probe complementary to the transcript under study. Second, hybridization of the labeled probe to the target RNA transcript. Third, digestion of the noncomplementary regions of the probe with ribonucleases. Finally, analysis of the digested products on denaturing polyacrylamide gels and autoradiography.

In vitro preparation of a single-stranded RNA probe (riboprobe), labeled to high specific activity, involves a number of steps. First, the sequence under study needs to be cloned into a plasmid vector that contains RNA polymerase promoter sequences (Fig. 1A). There are a large number of commercially available vectors from various sources for RNA transcription, but virtually all are based on bacteriophage SP6, T7, and T3 RNA polymerases. A number of vectors are available that contain two promoter sequences in opposite orientations and either side of the multiple cloning site, permitting transcription of either sense or antisense

From: *Methods in Molecular Biology, Vol. 49: Plant Gene Transfer and Expression Protocols*
Edited by: H. Jones Humana Press Inc., Totowa, NJ

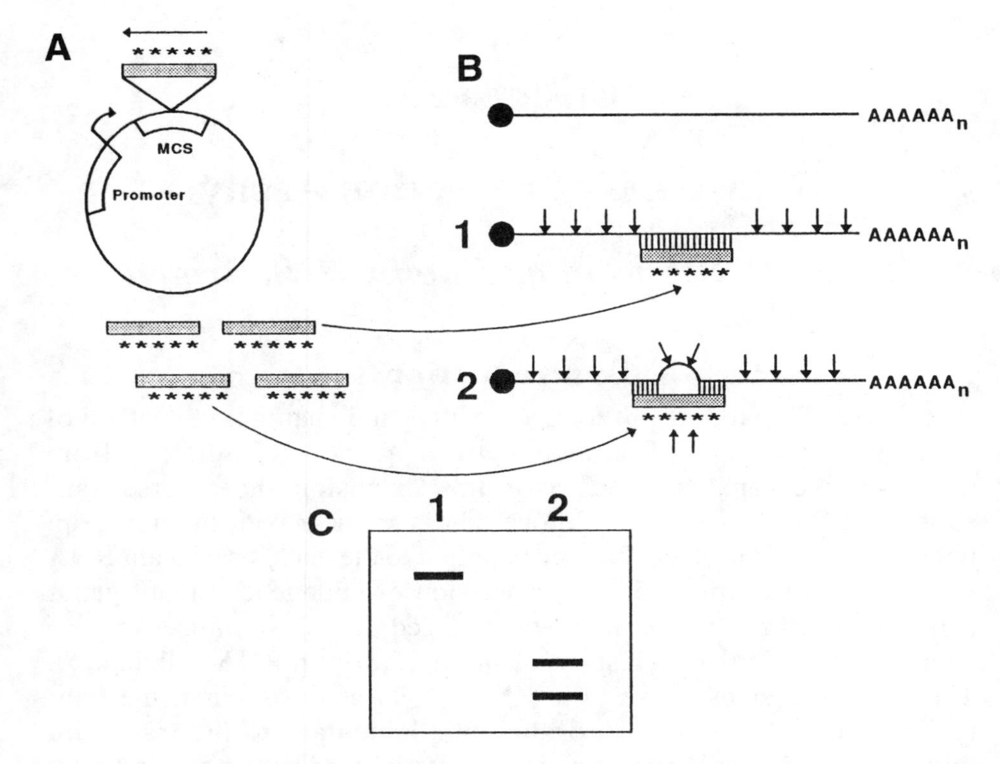

Fig. 1. RNase A/T_1 Analysis. **A.** Antisense RNA transcripts (shaded boxes) are efficiency synthesized by transcription from cloned DNA fragments (inserted into the multiple cloning site, MCS) of an in vitro transcription vector. Transcription is carried out from linearized vector by the RNA polymerases from T3, T7, or SP6 bacteriophage promoter and radiolabeled nucleotides (*) are incorporated into the transcripts. **B.** In vitro synthesized antisense RNA transcripts are hybridized to total plant RNA. Perfect hybridization with no mismatches results in full length protection of the labeled transcript from RNase A/T_1 *(1)*. Any mismatch between the target RNA and the labeled transcript will result in digestion by RNase A/T_1 (arrows) that results in shorter labeled digestion products *(2)*. **C.** Full length protection products (lane 1) and products of partial digestion (lane 2) can be identified by separation on polyacrylamide gels.

transcripts from the same construct. Second, the vector needs to be linearized downstream of the insert to allow production of discrete RNA transcripts that represent the antisense sequence of the RNA under study. Third, the riboprobe is transcribed in vitro by an RNA polymerase from the vector transcription start site initiated at the phage promoter (Fig. 1A). The RNA polymerase incorporates labeled nucleotides into the growing polymer before running off the end of the linearized template. Finally, the riboprobe is gel purified to remove the DNA template and to allow isolation of the full length riboprobe eliminating premature stops in the RNA transcription reaction.

The [^{32}P]-labeled riboprobe is hybridized in solution to total RNA extracted from plant material of the desired tissue or developmental stage (Fig. 1B). Treatment of the RNA–RNA hybrids with RNase A/T_1 results in hydrolysis of unprotected single-stranded [^{32}P]RNA probe and RNA target (as well as the nonhybridized total plant RNA) (Fig. 1B). RNase A digests RNA after pyrimidine nucleotides, whereas RNase T_1 digests RNA specifically after guanosine nucleotides. Double-stranded [^{32}P]RNA–RNA duplexes remain resistant to the ribonucleases. The analysis is completed by fractionating the digestion products on denaturing polyacrylamide gels (Fig. 1C).

Until the development of RT-PCR, RNase A/T_1 was the most sensitive method of RNA analysis available. However, it is especially useful when studying expression of multigene families, where a high level of sequence conservation exists between gene members and their RNA transcripts, but where variation at a number of base positions is observed. The colinear RNase A/T_1 assay allows a direct analysis of the specific RNA transcript under study, whereas the variant transcripts are digested into smaller fragments at the regions where base pairing is absent. Indeed, in some cases, RNase A/T_1 may allow mapping of the variable sequences. Such direct analysis is more difficult by primer methods, although single-strand conformational polymorphism (SSCP) does allow the identification of related DNA or RNA transcripts that vary in sequence. Here, we describe a widely used RNase A/T_1 protocol *(6)*.

2. Materials

All chemicals used were molecular biology grade and, where available, were RNase-free.

2.1. Preparation of Labeled Probe

1. Restriction enzymes. Store at –20°C.
2. Phenol/chloroform: Melt 250 g phenol with 1*M* Tris-HCl, pH 8.0, at 40°C for 1 h. Add an equal volume of chloroform and 0.1% (w/v) 8-hydroxyquinoline. Store at 4°C.

3. Chloroform/isoamyl alcohol: Mix chloroform and isoamyl alcohol at a ratio of 24:1. Store at 4°C.
4. 3*M* Sodium acetate, pH 6. Store at room temperature.
5. Ethanol. Store at –20°C.
6. 70% Ethanol: 70% ethanol in water. Store at –20°C.
7. NTP mix: 5 m*M* ATP, 5 m*M* CTP, 5 m*M* GTP, 2 m*M* UTP. Store at –20°C.
8. RNase inhibitor (30 U/mL). Store at –20°C.
9. 0.1*M* Dithiothreitol (DIT). Store at –20°C.
10. Borine serum albumin (BSA) (10 mg/mL). Store at –20°C.
11. α-[^{32}P]UTP (800 Ci/mmol). Store at –20°C.
12. 5X Transcription buffer: 200 m*M* Tris-HCl, pH 7.5, 150 m*M* $MgCl_2$, 50 m*M* spermidine. Store at –20°C.
13. SP6 RNA polymerase (10 U/mL). Store at –20°C.
14. 7.5*M* Ammonium acetate. Store at 4°C.
15. 80% Ethanol: 80% ethanol in water. Store at –20°C.
16. Urea (Analar).
17. 40% Acrylamide, bisacrylamide (19:1). Store at 4°C.
18. 10X TBE: 89 m*M* Tris-base, 89 m*M* boric acid, 2 m*M* EDTA. Store at room temperature.
19. 10% (w/v) ammonium persulfate. Store at 4°C.
20. TEMED. Store at 4°C.
21. Load buffer: 75% (v/v) formamide, 4 m*M* NaOH, 1.6 m*M* EDTA, 0.01% (w/v) xylene cyanol FF, 0.01% (w/v) Bromophenol blue. Store at –20°C.
22. RNA elution buffer: 0.3*M* sodium acetate, 1 m*M* EDTA. Store at 4°C.

2.2. RNA: RNA Annealing and RNase Digestion

1. Deionized formamide: Deionize by stirring 200 mL of formamide with 5 g BioRad (Richmond, CA) AG 501-X8 mixed bed resin for 1 h and filter through Whatman (Maidstone, UK) 1 MM paper. Store at –70°C.
2. 5X PEX: 2*M* NaCl, 50 m*M* Pipes pH 6.4. Store at –20°C.
3. 1*M* Tris-HCl, pH 7.5. Store at 4°C.
4. 0.4*M* EDTA pH 8. Store at 4°C.
5. 4*M* LiCl. Store at 4°C.
6. 1*M* NaCl. Store at 4°C.
7. RNase A (2 μg/mL). Store at –20°C.
8. RNase T_1, (100 U/μL). Store at 4°C.
9. 10% (w/v) SDS. Store at room temperature.
10. Proteinase K (10 mg/mL). Store at –20°C.
11. 2.5 mg/mL transfer RNA. Store at –20°C.

2.3. DNase Treatment

1. TE: 10 m*M* Tris-HCl, pH 8, 1 m*M* EDTA. Store at 4°C.
2. 1*M* $MgCl_2$. Store at 4°C.
3. RQ1 DNase (Promega, Madison, WI) (1 U/μL). Store at –20°C.

3. Methods

All centrifugation steps were performed at 4°C in a Sigma (St. Louis, MO) 2K-15 centrifuge or at room temperature in a Sorvall microfuge 24 (DuPont, Newtown, CT). Precautions are necessary when working with RNA (*see* Note 1) and successful RNase A/T_1 mapping requires good quality total plant RNA (*see* Note 2).

3.1. Preparation of Labeled Probe

1. Digest 20 μg of template vector with the appropriate restriction enzyme (downstream from insert) (*see* Note 3). Ensure complete digestion by using 5 U of enzyme/μg of DNA and digest overnight or for at least 5 h. Complete digestion can be determined by agarose gel electrophoresis.
2. Extract the linearized vector with a single phenol/chloroform extraction. After centrifugation at room temperature for 5 min, the aqueous phase is precipitated by the addition of 0.1 vol sodium acetate, pH 6, and 2.5 vol of ethanol. Precipitate at –20°C for 30 min followed by collection of the precipitate by centrifugation at full speed for 5 min. Wash the pellet with 100 μL of 70% ethanol and centrifuge for a further 2 min. Dry the pellet in a vacuum desiccator and suspend in 40 μL of water (0.5 μg/μL).
3. Prepare the in vitro transcription reaction by adding the following in this order at room temperature: 6 μL H_2O, 2 μL NTP mix, 1 μL RNase inhibitor (30 U), 2 μL 0.1*M* DTT, 1 μL BSA (1 μg), 1 μL α[^{32}P]UTP (10 μCi), 4 μL 5X transcription buffer, 2 μL linearized plasmid DNA (1 μg), 1 μL SP6 RNA polymerase (10 U). Incubate at 40°C for 75 min. (*See* Note 4.)
4. Add 28 μL 7.5*M* NH_4 acetate and 152 μL of H_2O for easier extraction of labeled RNA with a single phenol/chloroform extraction, followed by a single chloroform/isoamyl alcohol extraction and ethanol precipitation. After precipitation at –20°C for 30 min, collect the precipitate by centrifugation and wash pellet in 80% ethanol before drying and dissolving the pellet in 2 μL of loading buffer. The riboprobe is purified by gel electrophoresis, although RNase treatment can be used to remove plasmid template (*see* Note 5).
5. Prepare a 6% polyacrylamide-urea denaturing gel 20 × 40 cm using 0.25-mm spacers: 21 g urea, 7.5 mL 40% stock acrylamide, bisacrylamide (19:1), 10 mL 10X TBE. Water to 50 mL, filter, and degas. Polymerize with the addition of 150 μL 10% ammonium persulfate and 75 μL TEMED.

6. Prerun the gel at 1200 V for 20 min prior to loading the whole in vitro transcription reaction onto the gel. Run the gel at 1750 V until the Bromophenol blue has run the full length of the gel.
7. With due consideration for radioactive safety procedures, separate the gel plates and cover the gel with Saran wrap. In a dark room, tape a piece of X-ray film to the gel and stick two pieces of tape along each side of the film. Using a pencil, mark the film and the surrounding tape to allow orientation of the film after the film is developed. Expose the film for 1–2 min.
8. Develop the film and then orientate the pencil marks on the film with the marks on the tape. Cut out the most intense band, ignoring the lower molecular weight products owing to premature termination and in some cases higher molecular weight products (residual circular DNA template) (*see* Note 6).
9. Separate the gel piece from the Saran wrap and place the gel in an Eppendorf containing 400 µL RNA elution buffer and 2 µL of phenol/chloroform. This is incubated at 4°C for 16 h.
10. Remove the 400 µL of eluted RNA transcript to a fresh tube and remove 10 µL to count using a radioisotope counter. With the remainder perform a single phenol/chloroform extraction, followed by a single chloroform/isoamyl alcohol extraction and ethanol precipitation. Precipitate at –20°C for 30 min. Centrifuge to collect precipitate at full speed at 4°C followed by a wash with 100 µL 80% ethanol before further centrifugation for 5 min. Partly dry the pellet in a vacuum desiccator.
11. Calculate the total amount of activity in disintegrations per minute (dpm) from the 10 µL sample and dissolve the labeled RNA pellet in water at 10,000 dpm/µL.

3.2. RNA: RNA Annealing, RNase Digestion, and Gel Fractionation

1. Suspend 8 µg total plant RNA (*see* Note 2) in a volume of 20 µL. Add 15 µL of labeled riboprobe (150,000 dpm) to the RNA. Precipitate the RNA and labeled RNA together by the addition of 0.1 vol sodium acetate pH 6.0 and 2.5 vol of ethanol. Precipitate at –20°C for 30 min before centrifugation, washing pellet and drying as described.
2. Resuspend the pellet in 8 µL deionized formamide and 2 µL 5X PEX. Incubate at 90°C for 2 min before bringing the temperature down to 45°C and incubate at this temperature for 16 h (*see* Notes 7–9).
3. Prepare the RNase mix as follows: 15 µL 1*M* Tris HCl pH 7.5, 19 µL 0.4*M* EDTA, 75 µL 4*M* LiCl, and 150 µL 1*M* NaCl. Water to 1500 µL. To this add 6 µL RNase A (12 µg) and 2 µL RNase T_1 (200 U) (*see* Notes 7 and 10).

4. To each RNA hybridization add 100 μL of RNase mix. Mix the reaction, spin briefly, and incubate at 25°C for 40 min.
5. Add 2 μL 10% SDS and 3 μL proteinase K (10 mg/mL) to the RNase A/ T_1 reaction. Mix and spin briefly before incubating for a further 10 min at 37°C.
6. Add water to make the reaction up to 200 μL and extract by phenol/chloroform and chloroform/isoamyl alcohol treatment (*see* Note 11). Add 10 μg carrier tRNA and 0.1 vol sodium acetate, 2.5 vol of ethanol and precipitate for 30 min at –20°C. Centrifuge at 4°C at full speed to collect precipitate, wash the pellet with 80% ethanol, and dry. Suspend the pellet in 2 μL of load buffer.
7. Prepare a 6, 8, or 10% polyacrylamide-urea denaturing gel, depending on the size and separation of the expected products. For a 6% gel, mix 21 g urea, 7.5 mL 40% stock acrylmide, bisacrylamide (19:1) and 10 mL 10X TBE. Water to 50 mL, filter, and degas. Polymerize with the addition of 150 μL ammonium persulfate and 75 μL TEMED. For an 8% gel use 10 mL and for a 10% gel use 12.5 mL of the 40% acrylamide, bisacrylamide (19:1) stock in the above gel solutions.
8. Prerun the 4 gel at 1200 V for 20 min prior to loading samples. Incubate the samples at 95°C for 2 min prior to loading on the gel. It is recommended that a sequencing reaction be included as a marker and/or an end-labeled size ladder to facilitate the sizing of the labeled products. Run the gel at 1750 V until the Bromophenol blue dye is at the bottom of the gel or longer, if desired, and autoradiograph.

4. Notes

4.1. Working with RNA

1. Precautions must be taken when working with RNA because of the presence of active RNases on dust, fingers, and so on. However, we have found that good laboratory practice, the use of autoclaved plasticware, such as tips and Eppendorf tubes, heat-sterilized glassware, autoclaved water and solutions, and wearing gloves, is sufficient to obtain and preserve high quality RNA. We have not found the need to add diethyl pyrocarbonate (DEPC) to our water or solutions as is widely reported.
2. An increasing number of procedures are available for the extraction of high quality RNA. We use a protocol based on the single step method of RNA isolation by acid guanidinium thiocyanate-phenol-chloroform extraction *(7)*. We find this a quick and relatively simple procedure to perform and it has allowed the successful extraction of RNA from a number of different tissues. This procedure consistently provides high quality RNA and appears to facilitate the stripping of ribonucleoproteins from the RNA

allowing access to antisense riboprobes or primers. The amount of RNA used in the analysis can be varied. We generally use 8 μg of total RNA but have used between 5 and 20 μg of RNA depending on the abundance of the transcript under study and the availability of material.

4.2. Riboprobe

3. In our hands transcript lengths of between 200 and 800 nt have proved successful in RNase A/T_1 analysis based on efficiency of transcription, gel recovery, and gel analysis.
4. The method described above is for use of SP6 RNA polymerase to generate in vitro RNA transcripts. We find that the same buffers can be used for T3 and T7 derived RNA transcription, but manufacturer's instructions can be consulted if required.
5. Although it is recommended that the plasmid template and the full length RNA transcript is separated by gel purification, vector template can be removed by DNase treatment and the probe used directly. Add 36.5 μL TE, 1 μL 1*M* $MgCl_2$, 3 μL RNase inhibitor (90 U), and 10 μL RQ1 DNase (10 U) to one in vitro RNA transcription reaction and bring to a 100 μL vol with water. After incubation at 37°C for 20 min add 2 μL of 0.5*M* EDTA and 2 μL of 10% SDS followed by extraction of the RNA with a single phenol/chloroform extraction and ethanol precipitation by the normal procedure. The collected RNA pellet was resuspended in water at 10,000 dpm/μL.
6. If premature termination occurs from a particular template the insert can be recloned into another in vitro transcription plasmid vector to allow another promoter/RNA polymerase system to be used.

4.3. RNase A/T_1 Analysis

7. To obtain optimal signal strength while keeping background to a minimum, it may be necessary to vary the hybridization temperature and/or the concentration of RNases. In the laboratory we have only found the need to alter hybridization temperature between 45 and 55°C.
8. It is useful to include a nonspecific RNA (e.g., tRNA) or no RNA control to determine nonspecific RNase resistant fragments.
9. Quantitation of mRNA by RNase A/T_1 digestion can be achieved by comparing the intensity of the band owing to mRNA to that produced from a known amount of in vitro synthesized sense RNA (designed to give a product of detectably different size), hybridized to the labeled antisense probe. The internal control RNA can be added to the same RNase A/T_1 reaction.
10. It should be noted that, because RNases A and T_1 cut RNA after pyrimidines and guanosine, respectively, single nucleotide mismatches involving an adenosine in the riboprobe will not be cleaved.

11. An RNase A/T_1 (Ambion, Austin, TX) kit is available that has the advantage of eliminating the many phenol/chloroform steps.

References

1. Brown, J. W. S. and Waugh, R. (1989) Maize U2 snRNAs: gene sequence and expression. *Nucleic Acids Res.* **17,** 8991–9001.
2. Vankan, P., Edoh, D., and Filipowicz, W. (1988) Structure and expression of the U5 snRNA gene of *Arabidopsis thaliana.* Conserved upstream sequence elements in plant U-RNA genes. *Nucleic Acids Res.* **16,** 10,425–10,440.
3. Vaux, P., Guerineau, F., Waugh, R., and Brown, J. W. S. (1992) Characterisation and expression of U1snRNA genes from potato. *Plant Mol. Biol.* **19,** 959–971.
4. Cannon, M., Platz, J., O'Leary, M., Sookdeo, C., and Cannon, F. (1990) Organ specific modulation of gene expression in transgenic plants using antisense RNA. *Plant Mol. Biol.* **15,** 39–47.
5. Goodall, G. J. and Filipowicz, W. (1989) The AU-rich sequences present in the introns of plant nuclear pre-mRNAs are required for splicing. *Cell* **58,** 473–483.
6. Goodall, G. J., Wiebauer, K., and Filipowicz, W. (1990) Analysis of pre-mRNA processing in transfected plant protoplasts. *Methods Enzymol.* **181,** 148–161.
7. Chomczynski, P. and Sacchi, N. (1987) Single-step method of RNA isolation by acid guanidinium thiocyanate-phenol-chloroform extraction. *Anal. Biochem.* **162,** 156–159.

CHAPTER 21

Primer Extension Assay

Craig G. Simpson and John W. S. Brown

1. Introduction

A number of protocols are available for the analysis of RNA transcripts: Northern analysis (*see* Chapter 18), S1 mapping, primer extension, RNase AT_1 mapping (*see* Chapter 20), and the more recent reverse transcriptase/PCR amplification (*see* Chapters 22 and 24). These protocols have been applied to the analysis of plant RNAs and all vary in the degree of sensitivity of detection and the information generated. Primer extension assays have been used mainly in mapping of 5' termini of RNA transcripts, but have also been of value, in, for example, analysis of nuclear pre-mRNA splicing *(1–3)*, and in the detection of low abundance RNA species *(2)*. There are four steps involved in performing a primer extension assay. First, selection and preparation of a labeled primer complementary to the RNA transcript of interest. Second, hybridization of the primer complementary to a region of the RNA under study. Third, extension from the primer that is catalyzed by an RNA-dependent DNA polymerase (reverse transcriptase) using RNA as the template to synthesize a cDNA strand. Fourth, analysis of the extended cDNA products on denaturing polyacrylaminde gels and autoradiography.

The development of primer technology has allowed the ready availability of oligonucleotides for use in primer extension techniques, such that the use of restriction fragments as primers in primer extension assays is no longer required. For best results it is preferable to design the primer such that small extended products (<100 nt) are synthesized. Smaller

From: *Methods in Molecular Biology, Vol. 49: Plant Gene Transfer and Expression Protocols*
Edited by: H. Jones Humana Press Inc., Totowa, NJ

extended products are preferred because premature termination of transcription can occur using RNA templates owing to, for example, secondary structures, RNA degradation and pausing by the reverse transcriptase. Primers are usually designed to be about 20 nt long with GC content of about 50%. In addition, it is recommended to have G or C nucleotides at the termini of the oligonucleotide to provide a good "anchor" for the primer and permit strong initiation of transcription. Primers are efficiently labeled by polynucleotide kinase, but it is important to remove aberrantly short oligonculeotides by gel purification, formed during synthesis of the primer, as they may cause distracting background signals.

Successful primer extension depends first on the efficient and specific hybridization of the primer to the transcript of interest. This is generally derived empirically and variation in the annealing temperature may be required. Second, primer extensions can depend on the reverse transcriptase used. There are two commercially available reverse transcriptases, Avian Myeloblast Virus (AMV) reverse transcriptase and Moloney Murine Leukemia Virus (M-MLV) reverse transcriptase. An RNase H^- M-MLV reverse transcriptase is available that can increase the yield of cDNA copies. Necessary controls include primer extension without reverse transcriptase, without RNA, and, if possible, the use of RNA isolated from the same plant species under study, under conditions, or from tissue, in which the gene of interest is not expressed.

Besides the standard primer extension assay (Fig. 1, 1) to map the 5' ends of transcripts, the procedure has also been modified to produce double-stranded products to allow detailed examination of closely related transcripts (for example, alternatively spliced mRNAs). Double primer extension analysis (Fig. 1, 2) *(2)* involves preparing a cDNA transcript in a similar way to a standard primer extension reaction except that the primer is not labeled. A second primer, complementary to the 3' end of the cDNA, is end-labeled and used as a primer for second strand synthesis with a DNA-dependent DNA polymerase. The advantage of double primer extensions over standard primer extension reactions is, first, the increased specificity that a second primer provides, and second, only full length products between the primers are visualized because premature terminations are not observed. In addition, these labeled double-stranded DNA products can be directly digested by restriction enzymes for discrimination of closely related products. This procedure has been largely superseded by RT-PCR (*see* Chapters 22 and 24).

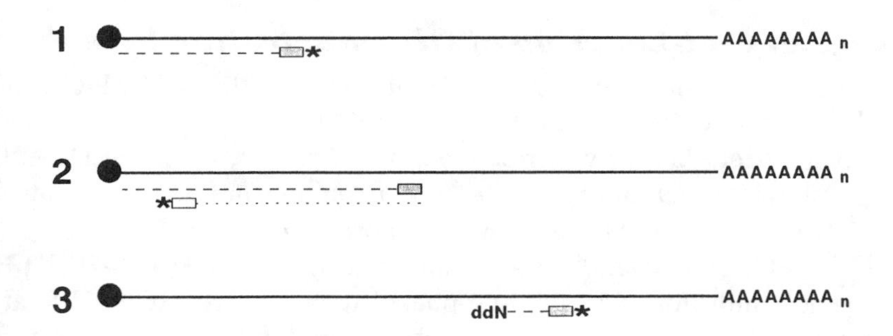

Fig. 1. Primer extension reactions. 1. Standard primer extension reactions involve the annealing of an end-labeled primer complementary to the RNA transcript (white box with *) followed by the synthesis of cDNA by reverse transcriptase (dashed line). Synthesis continues until the reverse transcriptase drops off at the 5' end of the transcript. 2. Double primer extension involves annealing of a cold primer complementary to the RNA transcript (white box), which is followed by primer extension with reverse transcriptase (dashed line). Second-strand synthesis to make a double-stranded cDNA product is performed by annealing an end-labeled primer complementary to the 3' end of the cDNA transcript (white box with *). Synthesis of the second strand is performed by a DNA dependent DNA polymerase such as T7 DNA polymerase (dotted line) and continues until the polymerase drops off at the 5' end of the initial cDNA transcript. 3. Chain termination primer extension involves annealing of an end-labeled primer complementary to the RNA transcript (white box with *) followed by synthesis of cDNA by reverse transcriptase (dashed line). Synthesis continues until the addition of a dideoxy nucleotide (ddN) terminates synthesis.

Finally, where primer extension is used to map internal RNA sites (for example, intron branchpoints or RNA secondary structures), a necessary control is primer extension of debranched or denatured RNA. To terminate cDNA synthesis before the 5' terminus of the RNA, a dideoxy nucleotide triphosphate is incorporated into the reaction. The ddNTP is chosen such that the termination event occurs just upstream of the internal site being mapped and this will usually require a primer lying in close proximity to this site. An extension of this process is direct sequencing of RNA transcripts using reverse transcriptase *(4)*.

2. Materials

All chemicals used were molecular biology grade and, where available, were RNase-free.

2.1. End-Labeling of Oligonucleotide Primer

1. 10X PNK buffer: 500 m*M* Tris-HCl, pH 7.6, 100 m*M* $MgCl_2$, 50 m*M* dithiothreitol. Store at –20°C.
2. Load buffer: 75% (v/v) formamide; 4 m*M* NaOH, 1.6 m*M* EDTA, 0.01% (w/v) xylene cyanol FF, 0.01% (w/v) Bromophenol blue. Store at –20°C.
3. 10X TBE: 89 m*M* Tris-base, 89 m*M* boric acid, 2 m*M* EDTA.
4. Oligo elution buffer: 0.3*M* ammonium acetate, 10 m*M* Tris-HCl, pH 7.5.
5. Phenol/chloroform: Melt 250 g phenol with 1*M* Tris-HCl, pH 8, at 40°C for 1 h. Add an equal volume of chloroform and 0.1% (w/v) 8-hydroxy-quinoline. Mix and store at 4°C.
6. Chloroform/isoamyl alcohol: Mix chloroform and isoamyl alcohol at a ratio of 24:1.
7. 10 m*M* Spermidine. Store at –20°C.
8. 5 m*M* EGTA. Store at –20°C.
9. γ-[^{32}P]ATP (5000 Ci/mmol). Store at –20°C.
10. T4 polynucleotide kinase. Store at –20°C.
11. 2.5*M* Ammonium acetate.
12. Ethanol. Store at –20°C.
13. 2.5 mg/mL Transfer RNA. Store at –20°C.
14. 40% Acrylamide, bisacrylamide (19:1). Store at 4°C.
15. 10% (w/v) Ammonium persulfate. Store at 4°C.
16. TEMED. Store at 4°C.
17. Urea (Analar).

2.2. Primer Extension

1. 5X PEX: 2*M* NaCl, 50 m*M* Pipes, pH 6.4. Store at –20°C.
2. 1*M* Tris-HCl, pH 8.3.
3. 1*M* KCl.
4. 0.1*M* $MgCl_2$.
5. 20 m*M* of 4 each dNTP. Store at –20°C.
6. 1*M* DTT. Store at –20°C.
7. M–MLV reverse transcriptase (200 U/μL). Store at –20°C.
8. 0.3*M* NaOH.
9. 3*M* Sodium acetate, pH 6.
10. 1*M* Tris-HCl, pH 7.5.

2.3. Double Primer Extension

1. Annealing buffer: 40 m*M* Tris-HCl, pH 7.5, 50 m*M* NaCl, 20 m*M* $MgCl_2$. Store at –20°C.
2. 10X TNM: 100 m*M* Tris-HCl, pH 7.5, 125 m*M* NaCl, 50 m*M* $MgCl_2$. Store at –20°C.

3. 5 m*M* dNTP mix. Store at –20°C.
4. 0.1*M* DTT. Store at –20°C.
5. Sequenase ver. 2.0 T7 DNA polymerase. Store at –20°C.

2.4. Chain Termination Primer Extension

0.5 m*M* ddNTP. Store at –20°C.

3. Methods

All centrifugation steps were performed at 4°C in a Sigma (St. Louis, MO) 2K-15 centrifuge or at room temperature in a Sorvall microfuge 24 (DuPont, Newtown, CT).

3.1. End-Labeling of Primer

1. Prepare 2X: PNK buffer containing: 4 µL 10X PNK, 4 µL 10 m*M* spermidine, 2 µL 5 m*M* EGTA, and 10 µL H_2O.
2. Add the components of the kinase reaction in the following order at room temperature: 10 µL 2X PNK, 0.5 µL oligonucleotide (0.5 µg), 5.5 µL H_2O, 3 µL γ-[^{32}P]ATP (30 µCi), and 1 µL T4 polynucleotide kinase. Mix the components and incubate at 37°C for 60 min.
3. To stop the reaction and precipitate the labeled primer add: 80 µL 2.5*M* ammonium acetate, 300 µL ethanol, and 2 µL tRNA (5 µg). Store at –20°C for 30 min and then centrifuge at full speed at 4°C for 15 min.
4. After centrifugation, remove supernatant and dry under vacuum. Suspend the primer in 1.5 µL of H_2O and 1.5 µL load buffer and incubate at 95°C for 2 min prior to gel purification.
5. Prepare a 16% polyacrylamide/urea denaturing gel 20 × 40 cm using 0.25-mm spacers as for a sequencing gel: 21 g urea, 20 mL 40% stock acrylamide, bisacrylamide (19:1), 10 mL 10X TBE. Water to 50 mL, filter, and degas. Polymerize with the addition of 150 µL 10% ammonium persulfate and 75 µL TEMED.
6. Prerun the gel at 1200 V for 20 min prior to loading the whole end-labeling reaction onto the gel. Run the gel at 1750 V until the Bromophenol blue is 3/4 of the way down the gel.
7. Adopting radioactivity safety procedures to minimize exposure, separate the gel plates and cover the gel with Saran wrap. In a dark room, tape a piece of X-ray film to the gel and apply two pieces of tape (about 2 × 1 cm) to each of the edges of the film. Using a pencil, mark the film and the surrounding tape to allow orientation of the developed film. Expose the film for 1–2 min.
8. Develop the film and orientate the pencil marks on the film with those on the tape. Labeled primers sometimes show the major primer with addi-

tional smaller oligonucleotides. Cut out the highest molecular weight (and usually most intense) band by cutting through the Saran wrap and the gel. Separate the gel piece from the Saran wrap and place the gel in an Eppendorf containing 400 µL of oligo elution buffer and leave overnight at 4°C.

9. Remove the 400 µL of eluted primer to a fresh tube and remove 10 µL to count using a radioisotope counter. To the remainder perform a single phenol/chloroform extraction, followed by a single chloroform/isoamyl alcohol extraction, and ethanol precipitation with 5 µg of carrier tRNA.
10. Calculate the total amount of activity in disintegrations per min and dissolve in H_2O at 50,000 dpm/µL.

3.2. Primer Extension Reactions

1. Precipitate and dissolve 10 µg of total plant RNA in 6 µL of H_2O and add 2 µL of end-labeled primer (100,000 dpm). To anneal the primer incubate at 90°C for 2 min and cool on ice. Add 2 µL 5X PEX and incubate for a further 20 min at 50°C followed by 15 min at room temperature.
2. Prepare the primer extension mix: 5 µL 1*M* Tris-HCl, pH 8.3, 4 µL 1*M* KCl, 1 µL 1*M* DTT, 3 µL 0.1*M* $MgCl_2$, 2.5 µL 20 m*M* dATP, 2.5 µL 20 m*M* dCTP, 2.5 µL 20 m*M* dGTP, 2.5 µL 20 m*M* dTTP, and 62 µL H_2O. This is a reaction mix for a single primer extension reaction (*see* Note 1).
3. Add 85 µL F of the primer extension mix to the annealed sample, followed by the addition of 5 µL (1000 U) M-MLV reverse transcriptase and incubate for 75 min at 42°C.
4. Perform a single phenol/chloroform extraction followed by a chloroform/isoamyl alcohol extraction and precipitate the primer extension by the addition of 250 µL of ethanol. Precipitate at –20°C for 30 min. Spin the precipitate down at full speed at 4°C for 15 min. Remove the supernatant and dry the pellet under vacuum.
5. To ensure easy loading of the sample onto a gel, degrade the RNA by suspending the pellet in 50 µL of 0.3*M* NaOH and incubate for 30 min at 65°C. Add 60 µL 1*M* Tris-HCl, pH 7.5, to neutralize, 2 µL tRNA (5 µg), 12 µL 3*M* sodium acetate, pH 6.0, and 250 µL ethanol, and precipitate at –20°C for 30 min (*see* Note 2).
6. Centrifuge at full speed at 4°C for 15 min to collect the precipitate. Remove the supernatant and wash the pellet with 70% ethanol followed by a 5 min spin at full speed at 4°C. Remove the supernatant, dry the pellet under vacuum, and suspend in 2 µL of load buffer.
7. Prepare a 6, 8, or 10% polyacrylamide/urea denaturing gel, depending on the size and separation of the expected products. For a 6% gel: 21 g urea, 7.5 mL 40% stock acrylamide, bisacrylamide (19:1), 10 mL 10X TBE. Water to 50 mL, filter, and degas. Polymerize with the addition of 150 µL

ammonium persulfate and 75 μL TEMED. For an 8% gel use 10 mL and for a 10% gel use 12.5 mL of the 40% acrylamide bisacrylamide (19:1) stock.

8. Prerun the gel at 1200 V for 20 min prior to loading samples. Incubate the primer extension products in load buffer at 95°C for 2 min prior to loading on the gel. It is recommended that a sequencing reaction is run alongside the primer extension reaction to facilitate the sizing of the labeled cDNA products. Run the gel at 1750 V until the Bromophenol blue dye is at the bottom of the gel or if greater separation is required and the sizes of the products are large, the Bromophenol blue can be run off the bottom of the gel.

3.3. Double Primer Extension

1. End label the primer complementary to the 3' end of the cDNA transcript as described previously (*see* Section 3.1.).
2. Perform the reverse transcription as described previously (*see* Section 3.2.) using 10–40 μg of total RNA and 500 ng of the primer complementary to the 3' end of the region of RNA under study.
3. After precipitation of the cDNA products, suspend the pellet in 36 μL of annealing buffer and 4 μL of the end-labeled primer (50,000 dpm/μL). This was incubated for 5 min at 95°C.
4. Prepare second strand run off reaction buffer: 2 μL 10X TNM, 5 μL 5 m*M* dNTP mix, 4 μL 1*M* DTT, and 8 μL H_2O. This is for a single second-strand run off reaction.
5. Dissolve the pellet in second-strand run off reaction buffer and add 1 μL of a 1 in 4 dilution of Sequenase (approx 3U) to begin second-strand synthesis and incubate for 15 min at 37°C.
6. After precipitation, the pellet is treated with NaOH and loaded on gels in the same way as previously described (*see* Section 3.2.).

3.4. Chain Termination Primer Extension

1. Prepare end-labeled primer as described previously and suspend at 30,000 dpm/μL.
2. Precipitate 10 μg of RNA and suspend in 4 μL of water. Add 1 μL primer (30,000 dpm) and incubate at 95°C for 2 min followed by cooling on ice.
3. Add 5X PEX and anneal as before, followed by the addition of the primer extension mix that contains 1.5 μL of 0.5 m*M* of a specific dideoxy nucleotide. Alternatively, a single deoxynucleotide is excluded from the primer extension mix.
4. Extend with reverse transcriptase, followed by RNA degradation and precipitation before suspending in gel load buffer as described for standard primer extension reactions (*see* Section 3 2.).
5. Reactions are loaded onto 8 or 10% acrylamide gels as described in the standard primer extension protocol.

6. It is recommended that end-labeled oligos are included as size markers. These can be prepared in exactly the same way as the oligos used for primer extensions (*see* Section 3.1.). A total of 1000 dpm marker oligo loaded on a gel will give a strong signal after overnight radiography.

4. Notes

1. Modifying enzymes, such as reverse transcriptase, are often bought with their own reaction buffer. In this case replace the reaction buffer described here with the reaction buffer supplied with the enzyme.
2. As an alternative to RNA degradation by NaOH, RNA can be degraded by the addition of RNase. If RNase is used, the cDNA/RNA pellet can be resuspended in 2.5 μL of RNase A (1 μg/μL) and incubated for 20 min at 37°C. Add an equal volume of load buffer, incubate at 95°C for 2 min and load directly on the gel.

References

1. Brown, J. W. S., Feix, G., and Frendeway, D. (1986) Accurate *in vitro* splicing of two pre-mRNA plant introns in a HeLa cell nuclear extract. *EMBO J.* **5,** 2749–2758.
2. Hershberger, R. P. and Culp, L. A. (1990) Cell-type-specific expression of alternatively spliced human fibronectin IIICS mRNAs. *Mol. Cell. Biol.* **10,** 662–671.
3. Reich, C. I., Vanhoy, R. W., Porter, G. L., and Wise, J. A. (1992) Mutations at the 3' splice site can be suppressed by compensatory base changes in U1snRNA in fission yeast. *Cell* **69,** 1159–1169.
4. Hahn, C. S. Strauss, E. G., and Strauss, J. M. (1989) Dideoxy sequencing of RNA using reverse transcriptase. *Methods Enzymol.* **180,** 121–130.

CHAPTER 22

Applications of RT-PCR

Craig G. Simpson and John W. S. Brown

1. Introduction

Reverse transcriptase-polymerase chain reaction (RT-PCR) is a highly sensitive method that permits the detection and quantification of RNA transcripts. It is particularly useful where transcripts are in very low abundance and where the amount of starting RNA is fairly limited, such as from a protoplast transfection. The procedure is based on production of a cDNA copy of the transcript by primer extension (*see* Chapter 21) and PCR amplification of the cDNA transcript by *Taq* DNA polymerase. The ability to amplify cDNA transcripts by *Taq* polymerase gives the flexibility and sensitivity, found with standard DNA PCR amplification, in the analysis of RNAs.

A detailed description of the RT-PCR reaction (Fig. 1A) and analysis is given in Chapter 24. In addition, there is now a large body of literature dealing with RT-PCR methods and approaches that are too numerous to extensively review in this chapter. The aim of this chapter is to describe our standard RT-PCR reaction using an end-labeled primer to produce radioactive RT-PCR products *(1)*, and how we have used this to adopt more novel approaches to study gene expression in plants. We shall cover the method of quantifying PCR products that we use routinely, and describe briefly, analyses of expression in a plant multigene family, and detection of antisense and sense RNAs in transgenic plants.

Quantification of RT-PCR products based simply on band intensities is not possible because PCR amplification is an exponential process and any variation in the efficiency at which DNA is amplified can alter the

From: *Methods in Molecular Biology, Vol. 49: Plant Gene Transfer and Expression Protocols*
Edited by: H. Jones Humana Press Inc., Totowa, NJ

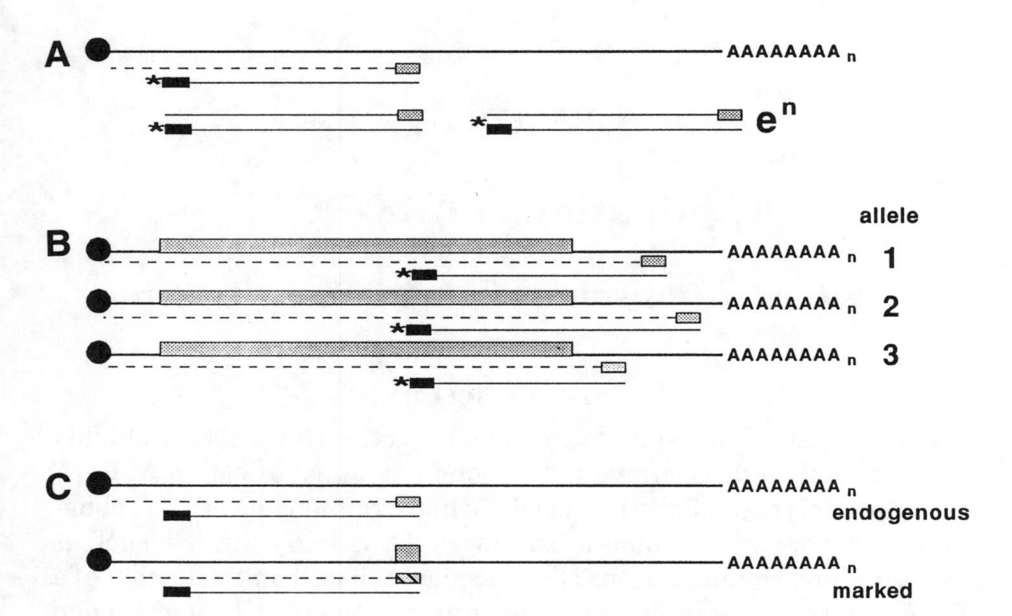

Fig. 1. End-labeled RT-PCR Reactions. **(A)** The standard RT-PCR reaction involves the annealing of a cold 3' primer (shaded box) complementary to the RNA transcript followed by primer extension with reverse transcriptase (dashed line). Second strand synthesis and PCR amplification of cDNA is performed by the addition of the same cold 3' primer, an end-labeled 5' primer (dark box and *), deoxynucleotides, and *Tag* DNA polymerase. DNA is exponentially amplified (e) over a number of cycles (n) of denaturation, primer annealing, and DNA synthesis (e^n). Amplified cDNA products are fractionated on polyacrylamide gels. **(B)** Expression analysis of multigene families involves annealing of a cold 3' primer (shaded box) complementary to each allele in the 3' untranslated region. After reverse transcription, PCR amplification is performed using the same allele-specific 3' primers and a single labeled 5' primer (dark box and *) common to all alleles and usually in the coding sequence (long shaded box). Alleles can be separated on the basis of RT-PCR product size. **(C)** Expression analysis of "marked" genes in transfected protoplasts or transgenics. If a gene can be "marked" by insertion of a novel sequence (short shaded box) a primer complementary to this sequence (cross hatched box) can be used to specifically amplify this transcript. The relative level of the marked gene transcript can be directly compared to that from endogenous genes by using a gene-specific 3' primer (shaded box).

product yield. Such variation is primarily owing to the efficiency of *Taq* DNA polymerase that is affected by the base composition and length of the DNA being amplified. Second, the rate of amplification decreases in later cycles owing to depletion of reaction components, accumulation of PCR products, and loss of enzyme activity. To quantify the amount of cDNA product after the original reverse transcription reaction we determine the efficiency of the *Taq* DNA polymerase by measuring the incorporation of end-labeled primer into the PCR product as described previously by Golde et al. *(2)*. This is done by taking aliquots of the reaction at specific cycles and measuring the amount of incorporated primer. The efficiency of the *Taq* DNA polymerase and the amount of cDNA at the start of the PCR reaction can then be calculated *(2)*.

To compare different RNA samples it is necessary to include an internal standard against which amounts of products can be compared. In some cases the strategy provides such a standard. For example, in our analysis of pre-mRNA splicing the levels of unspliced and spliced products can be directly compared *(3)*, or when comparing transcripts from different alleles, allele-specific RT-PCR products can be compared *(4)*. Where a single transcript is being analyzed, for example, an endogenous transcript or transcripts from introduced genes (sense or antisense) *(5)* reference signals can be produced in two ways. First, RT-PCR of an unrelated endogenous transcript can be carried out by including specific primers for this target in the RT-PCR reaction. We have used primers to amplify transcripts of the small subunit of Rubisco genes as standards. Second, a specific amount of an in vitro transcribed RNA can be added to each RT-PCR reaction along with specific primers to amplify part of this sequence.

The analysis of expression of genes from multigene families can be readily addressed by RT-PCR using at least one primer that is allele-specific. For quantitation, the primers should be carefully designed to give products of slightly different lengths, and ensure that the sequence to be amplified for each allele is very similar (to minimize the effects of differential *Taq* DNA polymerase efficiency for different DNA sequences). The strategy for such an analysis is shown in Fig. 1B, where one primer is designed to a region conserved in all alleles (usually in the coding sequence) and the allele-specific primers are designed to variable regions in the different alleles (coding or flanking sequences) to give products of slightly different sizes. The RT-PCR reactions are carried out in one tube with the common primer end-labeled. The allele-specific RT-PCR products can be compared and their relative amounts quantified.

We have also used this RT-PCR method to analyze expression of genes introduced into protoplasts or transgenic plants. In the first case, where a gene has been "marked" by introduction of an oliognucleotide, specific RT-PCR amplification of the transcript of the introduced gene is possible with no amplification of endogenous transcripts *(6)*. Inclusion of a primer to allow amplification of endogenous transcripts will allow quantification of the relative expression of the introduced "marked" gene compared to endogenous genes (Fig. 1C). Second, RT-PCR can be used to detect endogenous sense and introduced antisense transcripts in antisense transgenic plants *(5)*. The same two primers can be utilized and specificity of sense and antisense transcripts determined by which primer is used in the reverse transcription reaction.

Finally, another approach to analysis of expression of multigene families is to combine RT-PCR with single-strand conformational polymorphism (SSCP) (Fig. 2) *(7)*. RT-PCR products that vary in their sequence content by as little as a single nucleotide can be separated by denaturing the double-stranded DNA fragments and then loading on a neutral polyacrylamide gel. Semistable conformations are generated by heating and cooling of the sample and mobility shifts are found where the single-stranded DNA molecules adopt different conformations dependent on their sequence. This method is particularly useful where alleles are too similar in sequence to design allele-specific primers to generate length differences in the RT-PCR products. It gives information on sequence diversity, even when the sequences of all alleles are not known and is useful in comparing expression of genes in different plant tissues/organs *(7)*.

2. Materials

All chemicals used were molecular biology grade and, where available, were RNase-free.

2.1. End-Labeling

All materials are as listed in Chapter 21.

2.2. RT-PCR

2.2.1. Reverse Transcription

1. 3' Primers for reverse transcription are diluted to 1 µg/mL. Store at –20°C.
2. 5 m*M* dNTPs: Dilute 20 m*M* stocks of dATP, dCTP, dGTP, and dTTP at a ratio of 1:1:1:1. Store at –20°C.

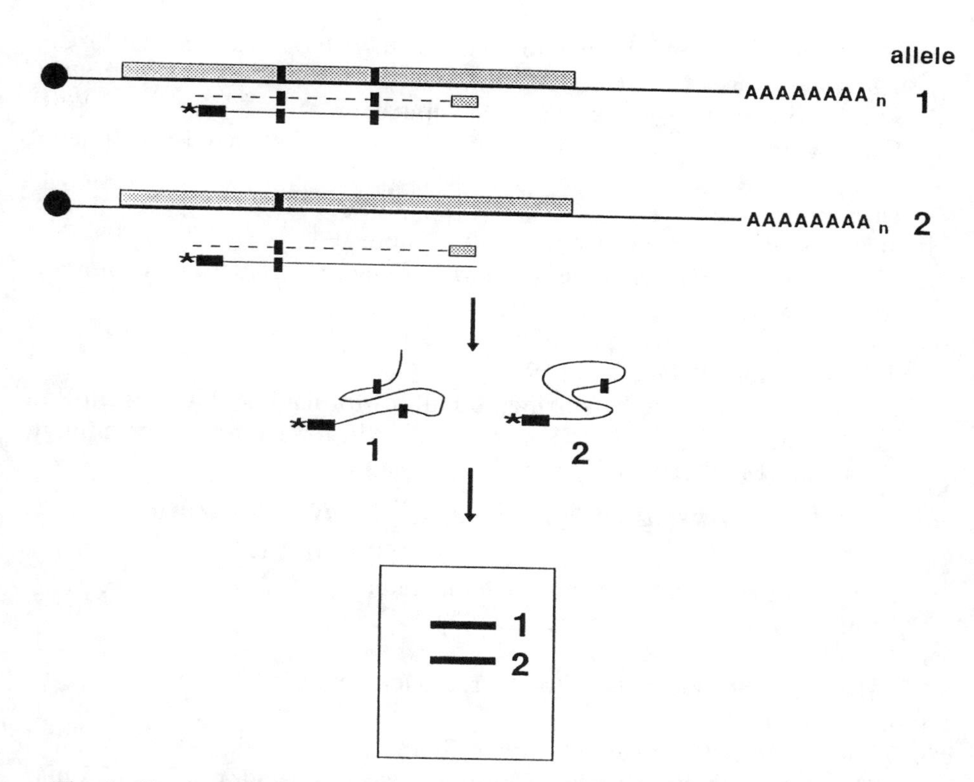

Fig. 2. Single-strand conformational polymorphism. Two allels that vary only in specific bases (black boxes) can be amplified in a standard end-labeled RT-PCR reaction using 3' specific primer (shaded box) and 5' specific primer (black box and *). RT-PCR products from both alleles are identical in length, but after denaturation by heating and cooling to generate semistable conformations that vary between alleles, mobility shifts between the allelic conformations can by observed on neutral polyacrylamide gels. Only the end-labeled DNA strands are observed.

3. 5X Reverse transcriptase buffer: 250 m*M* Tris-HCl, pH 8.3, 375 m*M* KCl, and 15 m*M* $MgCl_2$. Store at –20°C.
4. 0.1*M* DTT. Store at –20°C.
5. M-MLV reverse transcriptase (200 U/µL). Store at –20°C.

2.2.2. PCR Reaction

1. 10X PCR buffer: 100 m*M* Tris-HCl, pH 8.3, 500 m*M* KCl, 15 m*M* $MgCl_2$. Store at 4°C.
2. 1.25 m*M* dNTP: Add 10 µl of each 20 m*M* stock of, dATP, dCTP, dGTP, and dTTP to 120 µL water. Store at –20°C.

3. 20 μ*M* Primer: Primer diluted to 20 μ*M* with water and stored at –20°C.
4. *Taq* polymerase (4 U/μL). Store at –20°C.
5. Sterile mineral oil. Store at room temperature.
6. Loading buffer: 95% formamide, 20 m*M* EDTA, 0.05% w/v Bromophenol blue, 0.05% xylene cyanol. Store at –20°C.
7. Urea.
8. 40% Acrylamide, bisacrylamide (19:1). Store at 4°C.
9. 10X TBE: 89 m*M* Tris-base, 89 m*M* boric acid, 2 m*M* EDTA. Store at room temperature.
10. 10% (w/v) Ammonium persulfate. Store at 4°C.
11. TEMED. Store at 4°C.
12. Formamide: Deionize by stirring 200 mL of formamide with 5 g BioRad (Richmond, CA) AG 501 X8 mixed bed resin for 1 h and filter through Whatman (Maidstone, UK) 1 MM paper. Store at –70°C.

2.3. Cloning of Specific RT-PCR Products

1. Stop buffer: 4*M* urea, 50% sucrose, 50 m*M* EDTA, pH 8, 0.1% Bromophenol blue. Store at room temperature.
2. Ethidium bromide: 10 mg/mL. Store at 4°C.
3. Chloroform.
4. 3*M* Sodium acetate, pH 6. Store at room temperature.
5. Ethanol. Store at –20°C.
6. 70% Ethanol: 70% ethanol in water. Store at –20°C.
7. 10X Klenow reaction buffer: 330 m*M* Tris-acetic acid, pH 7.8, 625 m*M* potassium acetate, 100 m*M* magnesium acetate, 40 m*M* spermidine, 5 m*M* DTT. Store at –20°C.
8. 20 m*M* dATP. Store at –20°C.
9. 20 m*M* dCTP. Store at –20°C.
10. 20 m*M* dGTP. Store at –20°C.
11. 20 m*M* dTTP. Store at –20°C.
12. Klenow enzyme (2 U/μL). Store at –20°C.
13. Phenol/chloroform: Melt 250 g phenol with 1*M* Tris-HCl, pH 8, at 40°C for 1 h. Add an equal volume of chloroform and 0.1% (w/v) 8-hydroxy-quinoline. Mix and store at 4°C.
14. Oligoelution buffer: 0.3*M* ammonium acetate, 10 m*M* Tris-HCl, pH 7.5. Store at room temperature.

2.4. Single-Strand Conformational Polymorphism (SSCP)

1. Mutation detection enhancement gel 2X concentrate. Purchase from AT Biochem Inc. (Malvern, PA).
2. Glycerol.

2.5. DNase Treatment

1. TE: 10 m*M* Tris-HCl, pH 8, 1 m*M* EDTA. Store at 4°C.
2. 1*M* $MgCl_2$. Store at 4°C.
3. RQ1 DNase (Promega, Madison, WI) (1 U/µL). Store at –20°C.
4. RNase inhibitor: RNase inhibitor (30 U/µL). Store at –20°C.
5. 0.5*M* EDTA, pH 8. Store at room temperature.
6. 10% (w/v) SDS. Store at room temperature.

2.6. Antisense Analysis

RNase A (2.5 mg/mL). Store at –20°C.

3. Methods

5' Primer denotes the upstream sense primer to the transcript under study, whereas 3' primer denotes the downstream antisense primer throughout. Centrifugation was performed at room temperature in a Sorvall microfuge 24 (DuPont, Newtown, CT).

3.1. End-Labeling of Primer

The 5' primer was usually selected for labeling. Primers were end-labeled exactly as described for primer extension reactions (*see* Chapter 21) and primers were dissolved in H_2O at 300,000 dpm/µL (*see also* Note 1).

3.2. RT-PCR

3.2.1. Reverse Transcription Reaction

1. Suspend 1 µg RNA (DNA free) (*see* Notes 2–5) in an 11 µL vol and add 0.5 µL of 3' primer. Incubate at 92–95°C for 2 min followed by cooling on ice.
2. To the primed sample add 5 µL 5 m*M* dNTPs, 5 µL 5X reverse transcriptase buffer, 2.5 µL 0.1*M* DTT, and 1 µL M-MLV reverse transcriptase (use 1 µL of water instead of reverse transcriptase for RT minus controls). Incubate the reverse transcription reaction for 75 min at 42°C.

3.2.2. PCR Reaction

1. To the 25 µL reverse transcriptase reaction add 10 µL 10X PCR buffer, 16 µL 1.25 m*M* dNTP, 5 µL 5' primer (1,500,000 dpm), 5 µL 20 µ*M* 3' primer, and 0.5 µL *Taq* polymerase (2 U). Add water to a final volume of 100 µL. Briefly vortex and centrifuge to collect all the reagents. Layer 100 µL of mineral oil on to the reaction prior to temperature cycling.
2. Temperature and cycling times are 94°C for 1 min, 55°C for 2 min, and 72°C for 3 min for up to 26 cycles. After completion of cycling, samples are heated to 72°C for 5 min and then the temperature is held at 15°C.

3. The RT-PCR reaction is removed, avoiding mineral oil, to a fresh Eppendorf tube. 2 µL of the reaction is added to 4 µL of loading buffer, incubated at 92°C for 2 min and 1–3 µL loaded on a gel.
4. Prepare a 6 or 8% polyacrylamide-urea denaturing gel, depending on the size and separation of the expected products. For a 6% gel: 21 g urea; 7.5 mL 40% stock acrylamide, bisacrylamide (19:1), and 10 mL 10X TBE. Water to 50 mL, filter, and degas. Polymerize with the addition of 150 µL 10% ammonium persulfate and 75 µL TEMED. For an 8% gel use 10 mL of the 40% acrylamide, bisacrylamide stock. An increase in denaturant may be required on some occasions (*see* Note 6). Where required, add 10 mL of formamide to the components of the gel solution before making up to 50 mL with water.
5. Prerun the gel at 1200 V for 20 min prior to loading samples. It is recommended that a sequencing reaction be included as a marker and/or an end-labeled size ladder to facilitate the sizing of the labeled products. Run the gel at 1750 V until the Bromophenol blue is at the bottom of the gel, or longer if greater separation is required and the size of the products are >250 bps.

3.3. Cloning of Specific RT-PCR Products

For RT-PCR products greater than 250 bp, purification can be performed by standard agarose gel separation, excision of bands from the gel, and elution of DNA fragments. However, we often use primers that give products shorter than 250 bp in order to achieve accurate sizing on polyacrylamide sequencing gels. To purify and clone small RT-PCR products, we separate part of the RT-PCR reaction on a nondenaturing polyacrylamide slab gel.

3.3.1. Purification of RT-PCR Products

1. To 40 µL of RT-PCR reaction, add 20 µL of stop buffer.
2. Prepare a 6% nondenaturing gel 15X 18 cm using 1.5 mm spacers: 7.5 mL 40% stock acrylamide, bisacrylamide (19:1); 10 mL 10X TBE. Water to 50 mL, filter, and degas. Polymerize with the addition of 150 µL 10% ammonium persulfate and 75 µL TEMED.
3. Load samples and run the gel at 250 V until Bromophenol blue is at the bottom of the gel.
4. Separate the gel from the plate and stain the gel in a bath of ethidium bromide (1 µg/mL).
5. Cut out appropriate band(s) from the gel and elute for 16 h in oligoelution buffer.
6. DNA fragments can be reamplified by PCR. Add 5 µL of elutant to 10 µL 10X PCR buffer, 16 µL 1.25 m*M* dNTP, 5 µL 20 µ*M* 5' primer, 5 µL 20 µ*M*

3' primer, 0.5 µL *Taq* polymerase (2 U) and make to 100 µL with water. Vortex briefly and spin to collect. Add 100 µL mineral oil. Use a temperature cycle course of 1 min 92°C, 1 min 55°C, 1 min 72°C for 30 cycles. Complete PCR reaction with 5 min at 72°C and reduce temperature to 15°C.

7. Remove RT-PCR reaction sample, avoiding mineral oil, to a fresh tube. Add 100 µL of chloroform and mix to remove remaining mineral oil. Remove aqueous phase to a fresh Eppendorf tube and precipitate DNA by the addition of 0.1 vol 3*M* sodium acetate pH 6 and 2.5 vol ethanol at –20°C for 30 min.
8. Centrifuge at full speed for 15 min to collect DNA, wash the pellet with 70% ethanol followed by a further spin at full speed for 5 min. Remove ethanol and dry pellet in a desiccator before resuspending the pellet in 20 µL of water.

3.3.2. Cloning of RT-PCR Products

1. The Klenow fragment of DNA polymerases I, is used to fill in the overhangs that are often left after *Taq* polymerase amplification.
2. Take 17.5 µL of purified sample and add 2.5 µL 10X Klenow reaction buffer, 1 µL 20 m*M* dATP, 1 µL 20 m*M* dCTP, 1 µL 20 m*M* dGTP, 1 µL dTTP, and 1 µL Klenow enzyme (2 U). Incubate at 37°C for 60 min.
3. Add 75 µL of water to the reaction and extract the DNA with the addition of 100 µL of phenol/chloroform. Mix and then centrifuge at full speed for 5 min. Remove the aqueous layer and precipitate with 0.1 vol sodium acetate pH 6 and 2.5 vol ethanol at –20°C for 30 min.
4. Centrifuge at full speed for 15 min to pellet DNA, wash the pellet with 70% ethanol, and centrifuge for a further 5 min. Dry pellet and resuspend in 20 µL of water.
5. Ligate the fragment into a suitable plasmid vector digested with, for example, *Sma*I or *Hinc*II to generate blunt ends, by standard ligation procedures followed by transformation into *E. coli.*
6. Alternatively, RT-PCR products before Klenow treatment can be directly cloned into specialized vectors containing T-overhangs. These and the methodology can be purchased from a number of manufacturers.

3.4. Quantitative PCR (QPCR)

1. Throughout QPCR care should be taken to minimize variation between samples. Accurate pipeting and preparing mixes of the reaction components helps to reduce tube-to-tube variation.
2. Prepare a 100 µL RT-PCR reaction with end-labeled primer as described previously (*see* Sections 3.1. and 3.2.). During the PCR reaction, remove aliquots of 10, 15, or 20 µL depending on the number of cycle points to be

analyzed (*see* Note 7). A range of cycle points between 16 and 26 cycles is usually used (*see* Note 8). Complete each cycle aliquot by incubating the sample at 72°C for 5 min.

3. Take 2 µL of each RT-PCR reaction cycle point and add 4 µL of stop buffer. Load 2 µL onto a 6% nondenaturing gel 15 × 18 cm using 1.5 mm spacers as described previously (*see* Section 3.3.1.).
4. After running and staining the gel as before (*see* Section 3.3.1.) the bands are isolated from the gel and placed in an Eppendorf tube for counting. Counts are determined by Cerenkov counting and log counts plotted against cycle point (Fig. 3). The slope of this plot determines the efficiency (R) of the enzyme in the reaction. The amount of cDNA before PCR amplification ($cDNA_o$) can then be calculated from the equation:

$$\log cDNA_n = \log cDNA_o + n\log(1 + R)$$

where: $\log cDNA_n$ is the amount of incorporated primer at cycle number n; $\log cDNA_o$ is the amount of incorporated primer at the first cycle; n is the cycle number; R is the efficiency of *Taq* polymerase; and log (1 + R) is the slope of the plot.

5. We routinely carry out QPCR measurements in three separate experiments to allow calculation of a standard error (*see* Notes 9 and 10).

3.5. Single-Strand Conformational Polymorphism (SSCP)

1. Prepare a 100 µL RT-PCR reaction with end-labeled primer as described previously (*see* Sections 3 1. and 3 2.). Take 2 µL of reaction and add to 2 µL of load buffer. Heat the sample to 94°C for 2 min and then place directly on ice for 5 min before loading on a gel.
2. Prepare a nondenaturing mutation detection enhancement (MDE) gel 20 × 40 cm using 0.25-nm spacers. The gel is prepared to the manufacturer's specification. For 50 mL: 12.5 mL MDE gel and 3.5 mL 10X TBE. Polymerize with the addition of 200 µL 10% ammonium persulfate and 20 µL TEMED. The mobility shift can sometimes be enhanced by the addition of glycerol. Gels are run at low power (2W for 16 h) to keep temperature of the gel low (*see also* Note 11).

4. Notes

4.1. RT-PCR

1. As an alternative to using an end-labeled primer, radiolabeled dCTP can be included in the PCR reaction. Remove dCTP from the 1.25 m*M* dNTP solution and add 50 µCi of α-[^{32}P]dCTP.
2. It is important to remove any contaminating DNA prior to RT-PCR. We routinely perform two DNase treatments to the extracted RNA to remove

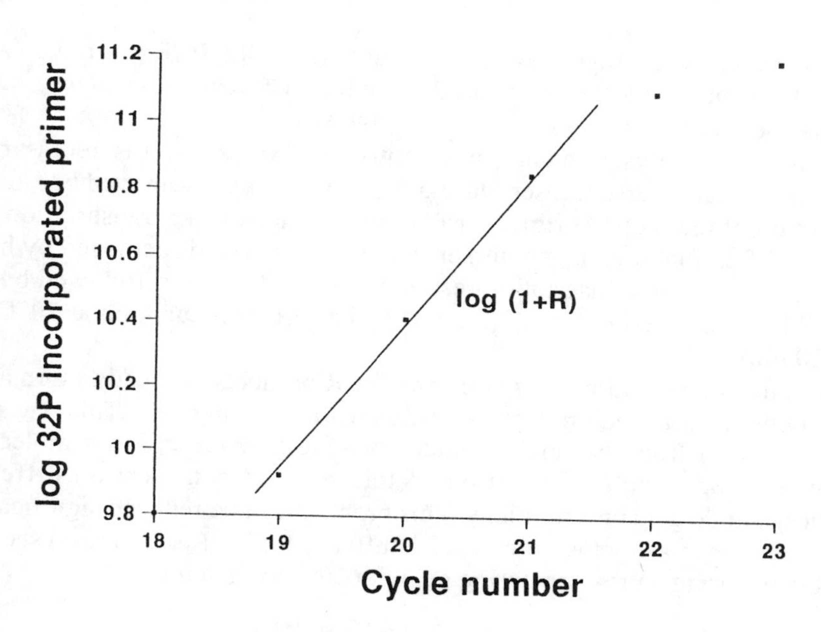

Fig. 3. Plot to determine *Taq* polymerase efficiency. Excised RT-PCR products were counted by Cerenkov counting to measure incorporated [^{32}P]-labeled primer. Log [^{32}P] incorporated primer (dpm) was plotted against cycle number. The plot indicates that the *Taq* DNA polymerase efficiency remained constant over 21 cycles, the efficiency of amplification becoming less thereafter. The best line through the linear portion of the plot and the slope of the line (log [1 + R]) was calculated using Genstat V.

any trace of DNA. DNase treatments can be performed as follows: To 50 µL of RNA (1 µg/µL) add 36 µL TE; 1 µL $MgCl_2$; 3 µL RNase inhibitor (90 U) and 10 µL RQ1 DNase (10 U). Incubate at 37°C for 20 min and terminate the reaction by the addition of 2 µL of 0.5*M* EDTA and 2 µL of 10% SDS. Extract the RNA with a single phenol/chloroform extraction and ethanol precipitation by the normal procedure. Collect the RNA by centrifugation, dry the pellet, and resuspend in 50 µL water. Repeat the DNase treatment and suspend the RNA at about 1 µg/µL. Determine absolute RNA concentration spectrophotometrically.

3. Even with two DNase treatments, occasionally contaminating DNA persists and it is, therefore, always important to carry out an RT-PCR reaction without reverse transcriptase as a control.
4. RNA is extracted by a protocol based on the single step method of RNA isolation by acid, guanidinium thiocyanate-phenol-chloroform extraction

(8) that provides high quality RNA suitable for RT-PCR reactions. Variable amounts of RNA can be used in the RT-PCR reactions and we usually use between 0.5 and 5 μg.

5. Antisense analysis. In analysis of antisense expression it is necessary to ensure that reverse transcriptase or template RNA is removed before addition of the second primer for PCR to prevent reverse transcription and amplification of endogenous sense transcripts. This is achieved by heating the reverse transcription reaction to 92°C for 4 min followed by the addition of 1 μL RNase A (2.5 μg) and a further incubation at 70°C for 10 min.
6. Larger concentrations of synthesized PCR products can lead to formation of double-stranded products, even in urea-denaturing gels. This is visible as a smear from the single-stranded band to an intense, lower molecular weight product. The position of this product can vary on different acrylamide gel concentrations. This problem can be reduced significantly by first increasing the ratio of load buffer to RT-PCR sample and second, by preparing a urea denaturing gel with 20% formamide.

4.2. Quantitative PCR

7. The number of cycle points can be increased by reducing the amount of sample removed at each cycle point.
8. A cycle course between 16 and 26 cycles is initially used. But depending on the abundance of the RNA under study, cycle courses can be increased or decreased.
9. To determine the quantitative accuracy of quantitative RT-PCR serial dilutions of the starting RNA can be made followed by quantitative RT-PCR. The $cDNA_0$ values should reflect the RNA dilution.
10. This protocol describes kinetic analysis of incorporation of an end-labeled primer and linear regression analysis to determine initial concentrations of RNA. A number of other quantitative methods are available using internal and external standard controls. A review of these methods can be found in: Quantitative RT-PCR, Methods and Applications book 3, from Clontech Laboratories, Inc. (Palo Alto, CA).

4.3. Single-Strand Conformational Polymorphism

11. Mobility shifts are caused by differences in sequence of single-stranded DNA fragments resulting in conformational variations of the molecules. This can be affected by the temperature of the gel, buffer concentration, and the presence of denaturing agents. The behavior of SSCPs can, therefore, be unpredictable and optimal electrophoretic conditions or specific reactions may need to be determined empirically.

References

1. Simpson, C. G., Sinibaldi, R., and Brown, J. W. S. (1992) Rapid analysis of plant gene expression by a novel reverse transcriptase-PCR method. *Plant J.* **2,** 835–836.
2. Golde, T. E., Estus, S., Usiak, M., Younkin, L. M., and Younkin, S. G. (1991) Expression of β-amyloid protein precursor mRNAs: recognition of a novel alternatively spliced form and quantitation in Alzheimer's disease using PCR. *Neuron* **4,** 253–267.
3. Brown, J. W. S., Simpson, C. G., Simpson, G. G., Turnbull-Ross, A. D., and Clark, G. P. (1993) Plant pre-RNA splicing and splicing components. *Phil. Trans. R. Soc. Lond. B.* **342,** 217–224.
4. Simpson C. G., Sawbridge, T. I., Jenkins, G. I., and Brown, J. W. S. (1992) Expression analysis of multigene families by RT-PCR. *Nucleic Acids Res.* **20,** 5861–5862.
5. Brown, J. W. S., Simpson, G. G., Clark, G., Lyon, J., Kumar, A., and Simpson, C. G. (1993) Detection of antisense transcripts in transgenic plants by RT-PCR. *Plant J.* **4,** 833–885.
6. Vaux, P., Guerineau, F., Waugh, R., and Brown, J. W. S. (1992) Characterization and expression of U1snRNA genes from potato. *Plant Mol. Biol.* **19,** 959–971.
7. Hedley, P. E., Machray, G. C., Davies, H. V., Burch, L., and Waugh, R. (1994) Potato *(Solanum tuberosum)* invertase-encoding cDNAs and their differential expression. *Gene* **145,** 211–214.
8. Chomczynski, P. and Sacchi, N. (1987) Single step method of RNA isolation by acid guanidinium thiocyanate-phenol-chloroform extraction. *Anal. Biochem.* **162,** 156–159.

References

CHAPTER 23

In Vitro Transcription of Class II Promoters in Higher Plants

Richard Cooke and Paul Penon

1. Introduction

Class II gene in vitro transcription systems from animal cells *(1–3)* or fungi *(4)* have been essential for elucidating the mechanisms regulating gene expression. Crude protein extracts from Hela cell nuclei that allow accurate transcription of class II promoters in vitro have led to the investigation of DNA sequences required for initiation of transcription (*cis*-acting-factors) and protein factors that bind to these regulatory elements of promoters (*trans*-acting-factors).

Transcription systems for plant genes have been unavailable until recently. Transient and transgenic in vivo expression systems have been employed to analyze the regulation of these genes *(5,6)*. These methods have given reliable results but are time-consuming, so research on in vitro transcription systems in plants has been pursued.

Wheat germ extracts were first used *(7)* but contain inhibitors that need to be removed in order to improve activity. Fractions of wheat germ extract are capable of replacing Hela factors in transcription of the adenovirus major late promoter. However, the system prepared from wheat germ alone is inefficient, and presents a block in the elongation of the RNA at the 20–30 nt level.

Other groups have also used wheat germ *(8)*. Their system was efficient with the TC7 promoter located on a linear template. Specificity of transcription was tested by the runoff assay. The measure of the size of

From: *Methods in Molecular Biology, Vol. 49: Plant Gene Transfer and Expression Protocols*
Edited by: H. Jones Humana Press Inc., Totowa, NJ

discrete length RNA transcribed by RNA polymerase II allows the localization of the transcription start site. Discrete length RNAs are generated by cleavage of a DNA template by a restriction enzyme downstream from a putative promoter. The accuracy of the runoff assay is limited by side reactions producing nonspecific incorporation of XTPs that induces background labeling and hinders the identification of the products of the template-dependent reaction. As this wheat germ system allows the transcription of circular templates, a 35S CaMV promoter linked to a fragment lacking guanosine residues on the sense strand, and directing the synthesis of the corresponding RNA alone in a medium without GTP has been used *(9)*. This allows the reduction of nonspecific background.

As wheat germ is a dormant material, it may be inactivated with regard to transcription factors and this could explain differences in the results reported with this material *(7,8)*. This suggests that the choice of the plant tissues used as starting material to prepare extracts is very important. Tobacco cells in suspension culture were used to prepare a whole cell extract *(10)* that is purified according to a procedure developed for the separation of specific transcription factors in Hela cell extracts *(11)*. These extracts were used to transcribe templates containing the 19S and 35S promoters of CaMV. Although it has been suggested that the 19S promoter is weaker than the more widely used 35S promoter *(12)*, preliminary experiments indicate that the 19S promoter is efficiently used in the tobacco system, giving a signal which is 5–10 times weaker than that observed with the major late promoter (the strongest promoter in the Hela system). The accuracy of transcription for the two CaMV templates, in this system, has been compared by using the runoff assay and SP6 mapping.

More recently *(13)*, cells from tobacco and rice converted to protoplasts were used to prepare a nuclear extract. Nuclei isolated and purified on a Percoll gradient were extracted with ammonium sulfate. The template was a gliadin cloned promoter that has a 41 nt poly-(dA) upstream from a *Sma*I site, allowing selection of the transcripts using oligo-(dT) chromatography, which retains only the gliadin RNA containing poly(A). This contributes to the specific recovery of this subset of transcripts and reduces the nonspecific background.

A major improvement has recently been introduced *(14)*. Soybean nuclear and whole cell extracts were prepared from cell suspension cultures. The templates that contain the start site of transcription for two chalcone synthase genes, for the adenovirus 2 major late promoter and for a chitinase gene, were coupled to avidin-agarose beads after biotinylation. RNA transcription complexes were formed with the bead linked templates. The immobilized transcription complexes were then washed and XTPs added. The transcripts were eluted and electrophoresed. Bean, tobacco, and rice suspension cultures have been also used to prepare active whole cell extracts. We describe here our method of preparation of a cell-free extract from tobacco cells in suspension culture.

2. Materials

2.1. Preparation of the Whole Tobacco Cell Extract

1. 3.8*M* saturated ammonium sulfate, pH 8, filtered, and stored at 4°C.
2. 0.2*M* dithiothreitol (DTT), in water, stored at –20°C.
3. 0.4*M* EGTA, stock solution in water pH 8, stored at –20°C.
4. Leupeptin, 1 mg/mL in water, stored at –20°C.
5. Stock solution of phenylmethylsulfonyl fluoride (PMSF) and benzamide, respectively 0.1 and 0.05*M* in absolute ethanol, stored at –20°C.
6. Pepstatin 2 mg/mL in methanol, stored at –20°C.
7. 3X extraction buffer (buffer A): 0.35*M* Tris-acetate, pH 7.9, 0.17*M* potassium acetate, 36 m*M* $MgSO_4$, 66% (v/v) glycerol. This buffer is prepared 1 d before extraction and stored at 4°C. From 5–15 min before grinding, the following reagents are added to buffer A, to give a final concentration of 39 m*M* β-mercaptoethanol, 18 m*M* dithiothreitol, 3.6 m*M* PMSF, 1.8 m*M* benzamide, 7 μ*M* pepstatin, 1.5 μ*M* leupeptin.
8. Dialysis buffer (buffer B): 20 m*M* HEPES-KOH, pH 7.9, 10 m*M* $MgSO_4$, 10 m*M* EGTA, 20% glycerol (v/v), 0.1*M* KCl, prepared 1 d before use and stored at 4°C. Fifteen minutes before use, the following reagents are added to give a final concentration of 5 m*M* DTT, 1 m*M* PMSF and 0.5 m*M* benzamide, 2 μ*M* pepstatin, 0.5 μ*M* leupeptin.
9. Miracloth (Calbiochem, La Jolla, CA).
10. Protein concentrations are determined by the method of Bradford *(15)* using a kit obtained from BioRad (Richmond, CA).

2.2. Preparation of Columns for Chromatographic Purification of the Extracts

1. Heparin-Sepharose chromatography: Heparin-Sepharose is obtained from Pharmacia (Uppsala, Sweden). 0.4 g gives 1 mL of gel. The dry powder (10 g) is suspended twice in 1 L of water, then in 1 L of buffer B containing 0.1*M* KCl, overnight. The slurry is poured into a glass column (1.7 × 10 cm) giving a bed volume of 25 mL.

2. A suspension of DEAE-sepharose-CL6B obtained from Pharmacia is washed twice with 1 L of distilled water then equilibrated overnight against 1 L of buffer B containing 0.1*M* KCl. The slurry is poured into a glass column (2 × 16 cm) giving a bed volume of 50 mL. Flow-through fractions of heparin-sepharose eluted at 0.1*M* KCl are separated on this column. Fractions eluted at 0.6*M* KCl are separated on a smaller DEAE-sepharose-CL6B (15 mL; 1.7 × 6 cm).

2.3. Buffers for Template Preparation and RNA Synthesis

1. Restriction endonucleases and other enzymes are obtained from Boehringer Mannheim (Mannheim, Germany).
2. Radioactive products are from Amersham (les Ulis, France).
   a. [^{3}H]UTP (40–60 Ci/mMol): 1 μCi/μL.
   b. [α-^{32} P] UTP (400 Ci/mMol): 10 μCi/μL.
3. XTPs 5 m*M* of each ATP, CTP, GTP, and UTP adjusted to pH 8 and stored separately at –20°C.
4. Nonspecific RNA polymerase II activity is measured in 50 μL of 11.2 m*M* Hepes-KOH, pH 7.9, 5.6 m*M* $MgSO_4$, 1.1 m*M* $MnCl_2$, 5.6 m*M* EGTA, 5.6 m*M* β-mercaptoethanol, 2.8 m*M* dithiothreitol, 50 m*M* KCl, 25 U/test RNAsin (Amersham), 11.2% (v/v) glycerol, 250 μg/μL calf thymus DNA, 0.4 m*M* each ATP, CTP, and GTP, 10 m*M* UTP, and 1.5 μCi [^{3}H]-UTP (40–60 Ci/mMol) test.
5. Specific transcription is measured in 10 m*M* HEPES-KOH, 10 m*M* $MgCl_2$, 50 m*M* KCl, 5 m*M* DTT, 10% glycerol, 125 ng poly(I):poly(C), 400 ng template and extract.
6. SP6 probes are synthesized in 10 μL of 40 m*M* Tris-HCl, pH 7.5, 6 m*M* $MgCl_2$, 2 m*M* spermidine, 10 m*M* DTT, 100 μg/mL BSA, XTPs (A, G, C) 330 μ*M* each, 10 μ*M* UTP, and 10 μCi of [α-^{32}P]UTP (400 Ci/mMol), 1 μL (5 U/μL) of RNAsin, 1 μL of SP6 RNA polymerase (3–10 U/μL), in the presence of 200 ng *Eco*RV-linearized pCa 19.3. Template DNA is removed by addition of 1 U RNase-free DNase (Boehringer Mannheim) and digestion for 5 min.

2.4. Gel Electrophoresis

1. 5% acetic acid, 20% ethanol (v/v).
2. Gels for electrophoresis of specific transcripts: 4% acrylamide/N-N-methylene bisacrylamide (19:1 w/w), in 89 m*M* Tris-borate, pH 8.3, 2 m*M* EDTA, 7*M* urea, 0.5μL/mL 10% ammonium persulfate, 5 μL/mL TEMED.
3. Ammonium persulfate: 10% (w/v).
4. Loading buffer: 90% formamide, 1X TBE, 0.1% Bromophenol blue, 0.1% xylene cyanol.

3. Methods

3.1. Suspension Culture

Established suspension cultures of tobacco are grown at 25°C as previously described *(17)*. They are subcultured every week by 10 times dilution. Four-day-old cultures containing rapidly dividing cells are harvested by filtration, washed with distilled water, frozen in liquid nitrogen, and stored at –80°C until extraction. All steps after thawing are carried out at 0–4°C.

3.2. Preparation of Runoff Templates

1. Prepare plasmids by the alkaline lysis method *(15)* and purify by caesium chloride gradient centrifugation. Digest with an appropriate restriction enzyme to isolate a fragment containing the transcription initiation site and several hundred basepairs downstream. For our purposes, the 19S template (Fig. 1A) is prepared by *Eco*RI digestion of plasmid pCa 19.2, which contains a 459bp *Eco*RI fragment of CaMV DNA overlapping the 19S promoter (coordinates 5646–6105, all coordinates given are from the CM1841 isolate) cloned into pUC8. In vivo transcription is initiated at 5762bp *(18)*. The 35S template (Fig. 1C) is recovered from plasmid pCa35.1, which contains a 719-bp *Sau*3AI fragment overlapping the 35S promoter (coordinates 6932–7651 on the CaMV genome) cloned into the *Bam*HI site of pUC8. This plasmid is digested by *Hind*II and the 643 bp fragment extending from the *Hind*II site at 7013 bp on the CaMV genome to the *Hind*II site in the vector recovered.
2. After digestion, add EDTA to 10 m*M*, heat the solution to 65°C for 10 min, cool, and load on a 10–40% sucrose gradient in TE buffer. Centrifuge at 39,000 rpm in the Beckman (Fullerton, CA) SW41 rotor (261,000*g*) for 22 h.
3. Recover 700-µL fractions and test for the presence of the insert fragment by agarose gel electrophoresis of a 20-µL aliquot.
4. Pool fractions containing insert DNA, precipitate with 95% ethanol, wash in 70% ethanol, and dissolve in TE buffer at a concentration of 200 µg/mL.

3.3. Preparation of the SP6 Mapping Template

1. Prepare a linear template for SP6 mapping by restriction enzyme digestion of the plasmid used to prepare the cell-free transcription template (*see* Note 11). We use a 202 bp *Eco*RI-*Hind*III fragment of CaMV DNA (5646–5848 bp) subcloned into pSP64 (plasmid pSC19.3, Fig. 1B) to map the initiation site of 19S transcript.
2. After phenol-chloroform (1/1) extraction, precipitate the DNA with ethanol, wash with 70% ethanol, and redissolve in TE buffer.

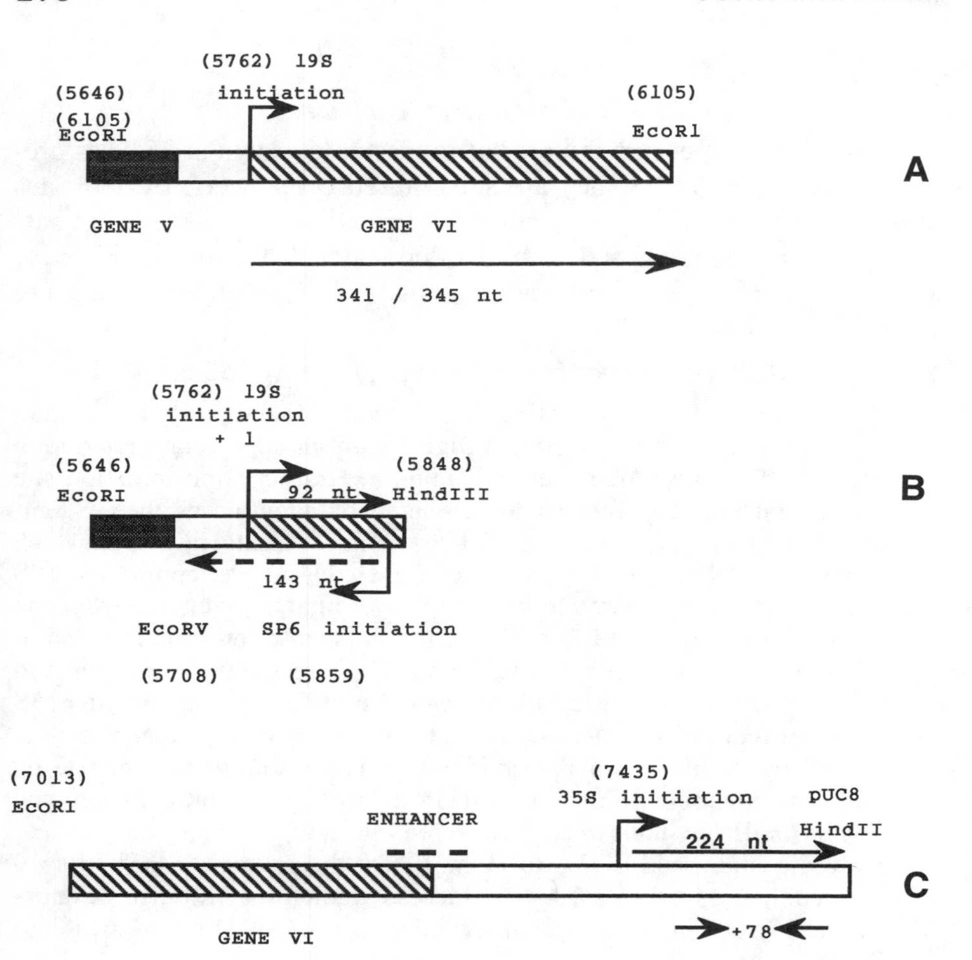

Fig. 1. The 19S promoter template used for runoff transcription. **(A)** The region of the CaMV genome contained on the 459 bp EcoRI fragment of pCa19.1. The extremities of genes V and VI are shown. The transcription initiation site is localized at 5762 bp *(18)*. The arrow shows the runoff RNA expected if transcription is initiated accurately. Its length, 341–345 nt, depends on the termination site on the staggered *Eco*RI end. The transcribed region is indicated by a thick arrow. **(B)** The template used for SP6 mapping. The limits of the 202 bp *Eco*RI-*Hind*III fragment which was subcloned into pSP 64 are shown. The transcription initiation sites for RNA polymerase II (19S) and the SP6 RNA polymerase are indicated with the transcribed region of each strands (solid and dotted line, respectively). The *Eco*RV site used for runoff transcription is also shown. Numbers correspond to distances from the 19S transcription

(continued)

3.4. Preparation of the Whole Cell Extract

1. Mix about 500 g of frozen cells with 180 mL of buffer A for 5 min, at low speed in a Waring blender (*see* Note 1). The temperature increases to 0°C, then after five pulses of 20s each at high speed, the slurry is homogenized in aliquots of 25 mL with a motor driven Teflon™ glass homogenizer and filtered through two layers of Miracloth. We estimate that 90% of the cell population is broken by this method.
2. Slowly add ammonium sulfate (3.8*M*) to the homogenate to give a final concentration of 0.38*M*. Leave the suspension for about 1 hour at 4°C with magnetic stirring.
3. Centrifuge the mixture at 40,000 rpm for 1.5 h in a Beckman 70 Ti rotor (164,000*g*).
4. Decant the supernatant into a sterile beaker and adjust to 75% saturation with solid ammonium sulfate (0.44 g/mL). Add 1*M* KOH (10 µL/g of ammonium sulfate) dropwise to obtain a balanced pH.
5. After stirring at 4°C for 45 min collect the precipitate by centrifugation at 11,500 rpm for 45 min in a Beckman JS-13 rotor (21,000*g*) and resuspend at a concentration between 15 and 30 mg/mL in buffer B containing 0.1*M* KCl, 2 µ*M* pepstatin, and 0.5 µ*M* leupeptin. Dialyze the whole cell extract overnight in 1.5 L of buffer B with two changes of buffer until the conductivity of a 1/1500 dilution of the extract reaches 10 µS. Centrifuge the dialyzed extract at 11,500 rpm in a Beckman JS-13 rotor (21,000*g*) for 5 min to remove insoluble proteins and quickly freeze to –80°C (*see* Notes 2–4).

3.5. Chromatographic Purification of the Whole Cell Extract

1. The method of Moncollin et al. *(11)* is used. The first step is to separate the extract on a heparin-sepharose column. Load the dialyzed extract (25 mL, 27 mg/mL) onto a 25 mL column of heparin-sepharose equilibrated in buffer B, 0.1*M* KCl. Wash the column with 50 mL of the same buffer, 25 mL of buffer B containing 0.24*M* KCl and elute with the same buffer containing 0.6*M* KCl. In extractions with a lower amount of material (250 g of cells), smaller columns (15 mL; 1.7 × 7 cm) can be used.

(Fig. 1 *continued*) initiation site. **(C)** The 35S promoter template showing the region of the CaMV genome contained on the 643 bp fragment of pCa 35.1. Gene VI is indicated and the putative start of transcription, at 7435 bp, is localized. Coordinates are those of the CaMV CM1841 isolate. The presumed initiation site is shown (+1). The arrow shows the runoff RNA expected when the transcription is initiated correctly. A potential cruciform centered at +78 is indicated.

2. Pool the fractions of the flow-through, which contain 90% of the proteins, and load onto a DEAE-Sepharose-CL6B column (50 mL). After loading and washing with two column volumes of buffer B containing 0.1*M* KCl, elute with 1.5 vol of buffer B containing 0.35*M* KCl, and collect 2.7-mL fractions. Pool peak fractions (2–3, generally containing 15–17 mg/mL) and dialyze against buffer B containing 0.1*M* KCl. These fractions constitute the flow-through (FT0.35) containing essential factor(s).
3. The other essential factors elute with about one-third of nonspecific RNA polymerase II activity in the leading edge of the 0.6*M* KCl peak. A key step in the characterization of the specific activity is to test the fractions eluted at 0.6*M* KCl to determine their ability to transcribe the CaMV 19S template in the presence of the FT0.35 fraction (*see* Section 3.6.2.). The nonspecific activity of RNA polymerase elutes before the bulk of eluted protein (Fig. 2A). The first polymerase containing fractions are capable of synthesizing 5–6 major RNA species, one of which corresponds to the expected 345 nt runoff (arrowed in Fig. 2B) (*see* Notes 3 and 4).
4. Pool the fractions active in transcription of the 19S template from large scale preparations (1000 g of cells), dialyze against buffer B 0.1*M* KCl, and precipitate by addition of solid ammonium sulfate (0.28 g/mL and 1 vol of saturated ammonium sulfate). Collect the precipitated material by centrifugation in a Beckman JS-13 rotor at 11,500 rpm (21,000*g*) for 45 min, resuspend in buffer B (5 mL) containing 0.05*M* KCl, and dialyze for 8 h against the same buffer. This constitutes the HEP 0.6 fraction.
5. Centrifuge the dialyzate for 5 min in the JS-13 rotor and apply to a DEAE-sepharose-CL6B column (15 mL; 1.7 × 6 cm). After washing with two column volumes of dialysis buffer, elute the proteins with 1.5 vol of buffer B containing successively: 0.15*M*, 0.20*M*, 0.25*M,* and 1*M* KCl (Fig. 3). For in vitro transcription, the fractions are pooled, precipitated by addition of solid ammonium sulfate, then 1 vol of saturated ammonium sulfate, as previously described, and dialyzed against buffer B containing 0.1*M* KCl. Only one fraction, eluted by 0.25*M* KCl, is capable of synthesizing the 345 nt runoff (*see* Notes 5 and 6).

3.6. In Vitro Transcription

3.6.1. Assay for Nonspecific Transcription

The extracts contain a large amount of RNA polymerase II, which is unable to perform specific transcription. It is important to localize this nonspecific activity, as specific activity elutes within its leading edge.

1. Incubate 5 µL of extract at 30°C in 50 µL of nonspecific transcription medium for 30 min.

2. Collect the solution on Whatman (Maidstone, UK) GFC filters and wash eight times with 10% TCA and twice with 95% ethanol before scintillation counting.
3. One unit of enzyme is defined as the amount of RNA polymerase catalyzing the incorporation of 1 pmol of [^{3}H]UTP into TCA insoluble material (6200 dpm in the test conditions).

3.6.2. Assay for Specific Transcription of the 19S CaMV Template

1. Prepare the preincubation mix (16 or 26 μL), containing 10 m*M* HEPES-KOH, pH 8.0, 10 m*M* $MgCl_2$, 50 m*M* KCl, 125 ng poly(I):poly(C), and required volumes of the FT0.35, HEP0.6, or DEAE column fractions. Incubate 5 min at 30°C.
2. Add 400 ng template DNA and incubate a further 10 min.
3. Initiate transcription by adding 4 μL of XTP mix to give a final concentration of 330 μ*M* ATP, CTP, GTP, and 16 μ*M* UTP containing 5 μCi [α-32P]-UTP (400 Ci/mmol) per reaction.
4. After 50 min at 30°C, stop the reactions by addition of 9 vol of 0.3*M* ammonium acetate, 0.2% SDS, 25 μg/mL tRNA. Extract once with phenol-chloroform (1/1) and precipitate with ethanol.
5. Carry out in vitro transcription with a Hela cell extract (BRL) as described by the suppliers and extract RNA as described earlier.
6. Dissolve samples in 6 μL of sequencing gel loading buffer, denature for 2 min at 90°C, and cool in an ice-methanol bath.
7. Load immediately on a 0.5–mm thick, 4% polyacrylamide-urea gel (32 × 18 cm). Migrate at 400 V with TBE as the electrode buffer, until the Bromophenol blue reaches the end of the gel.
8. Fix the gel with two washes of 15 min in 5% acetic acid, 20% ethanol, dry at 85°C for 75 min, and expose to Kodak X-Omat film using DuPont Cronex Quanta III intensifying screens at –80°C (*see* Notes 7–10).

3.6.3. SP6 Mapping

In order to confirm that the cell-free extract accurately transcribes the template, transcripts are mapped by SP6 mapping *(19)*.

1. Synthesize the complementary RNA probe in 10 μL, with the transcription buffer for SP6.
2. Synthesize unlabeled RNA in the cell-free system as described above but with 330 μ*M* unlabeled UTP, no radioactive tracer, and treat the RNA produced with RNase free DNase.

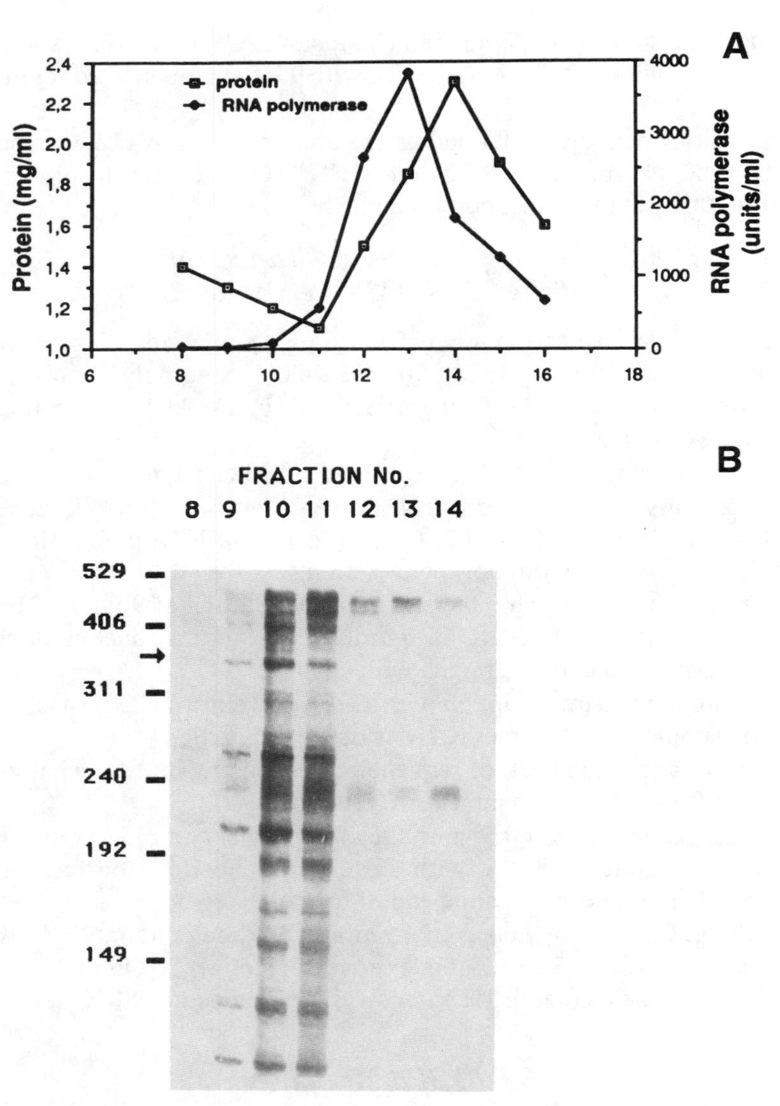

Fig. 2. Analysis of the runoff transcription by fractions from the heparin-sepharose column. **(A)** After loading tobacco cell extract (24 mL; 14.8 mg/mL) on a heparin-sepharose column (16 mL; 1.7 × 7 cm) and washing with buffer B containing 0.1*M* KCl (30 mL) and 0.24*M* KCl (16 mL), elution was performed with 0.6*M* KCl. Protein content and nonspecific RNA polymerase activity are indicated for the 0.6*M* eluate. **(B)** Fractions corresponding to the RNA polymerase peak were analyzed for transcription of the 19S template. Assays

(continued)

3. Hybridize 10^5 cpm or 2×10^5 cpm of labeled SP6 probe with unlabeled in vitro synthesized RNA from the Hela or tobacco systems in 100 µL 50% formamide, 0.4*M* NaCl, 40 m*M* PIPES pH 6.7, 1 m*M* EDTA at 37°C. These conditions take into account the low T_m of the expected hybrids under our conditions.
4. Digest the hybrids with RNase A (40 µg/mL) and RNase T1 (2 µg/mL) at 30°C for 90 min in 10 m*M* Tris-HCl, pH 7.5, 5 m*M* EDTA, and 0.3*M* NaCl.
5. Terminate RNase digestion by addition of 20 µL of 10% SDS and 5 µL of 10 mg/mL proteinase K and continuing incubation at 30°C for 10 min.
6. Analyze RNase-resistant molecules by electrophoresis at 650 V for 3 h on 0.5 mm thick, 8% polyacrylamide-urea gels (32 × 18cm) prepared with 1X TBE pH 8.3, 7*M* urea (*see* Notes 11 and 12).

4. Notes

4.1. Precautions in Preparing the Whole Cell Extract

1. Plant tissues are very rich in proteases and a cocktail of inhibitors is used during the first steps of grinding, ammonium sulfate precipitation, and dialysis. The half-life of PMSF in aqueous solution is particularly short (3 h at pH 8 at 4°C) so we add fresh PMSF at every purification step.
2. The dialyzed crude extract is inactive in nonspecific and specific transcription. Alternatively, in some preparations, the extracts are desalted by molecular sieving on a column of Sephadex G75. The gel, swollen overnight in buffer B containing 10% glycerol, is poured into a column (145 mL; 2.02 × 47 cm). The extracts prepared according to this procedure are capable of performing nonspecific transcription. This results from the removal of low-mol-wt inhibitors retained on the G75 column. The same effect is observed when insoluble polyvinyl pyrrolidone (1.5% w/v), which inactivates polyphenols, is added during grinding.
3. Initiation of transcription by RNA polymerase II occurs preferentially at the termini of DNA fragments or at internal nicks. Moreover, the enzyme promotes end-labeling of the template molecules. These reactions are

(Fig. 2 *continued*) (20 µL) contained 5 µL of 0.6*M* heparin-sepharose fractions: 8 (6.5 mg/mL protein, 0.2 U RNA polymerase), 9 (6 mg, 0.4 U), 10 (5.5 mg, 2.8 U), 11 (7.5 mg, 13.3 U), 12 (9 mg, 19.3 U), 13 (11.5 mg, 9.0 U), 14 (9.5 mg, 6.3 U), with 2 mL (35 mg/mL protein) of the FT0.35 fraction. RNAs are electrophoresed on 4% polyacrylamide/urea gels. The size of molecular marker fragments prepared by *Hpa*II digestion of pCa29 are indicated. The arrow indicates the expected 345 nt RNA.

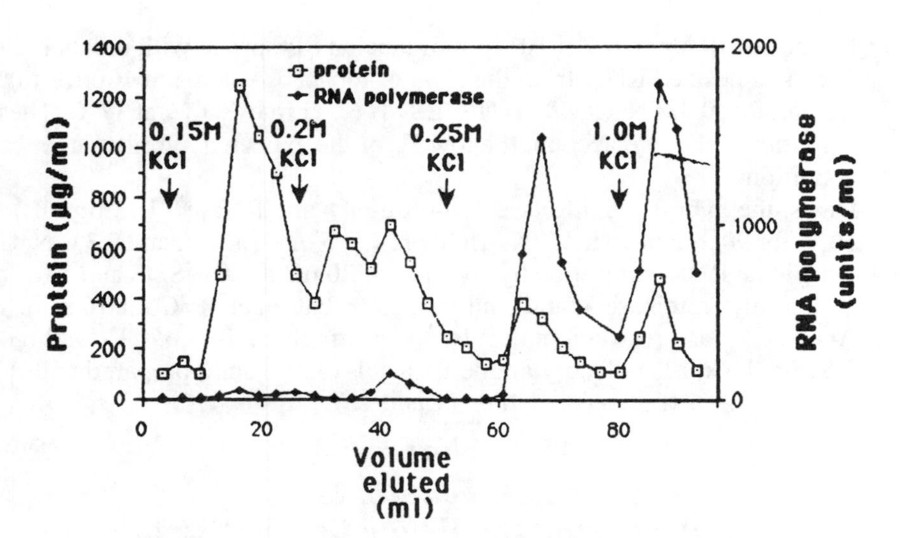

Fig. 3. A DEAE-sepharose CL6B (25 mL; 1.7 × 11 cm) was loaded with the pooled and dialyzed 0.6*M* KCl heparin-sepharose fractions corresponding to 1000 g of tobacco cells. The column was equilibrated in buffer B, 0.05*M* KCl. After washing with 30 mL of the same buffer, the column was eluted stepwise with 1.5 columns volumes of buffer B at successive concentrations of 0.15, 0.20, 0.25, and 1*M* KCl. Fractions of 1.6 mL were collected and tested for nonspecific RNA polymerase activity. 0.25*M* fractions were precipitated with ammonium sulfate, dialyzed, and used in specific transcription reactions.

inhibited by 1 µg/mL α-amanitin. Transcription by other polymerases may also interfere and can be identified by the sensitivity of these reactions to α-amanitin, insensitivity of transcription by RNA polymerase I, and low sensitivity (200 µg/mL) for RNA polymerase III.

4. As our whole cell extract possesses only a very low nonspecific activity, a chromatographic step *(11)* is needed to remove inhibitors. The extracts are fractionated on heparin-sepharose-CL6B that separates a tobacco cell extract into two fractions. The fractions corresponding to the majority of RNA polymerase activity in the nonspecific test (Fig. 2a) synthesize low amounts of RNA (fractions 12–14, Fig. 2B). This may indicate that inhibitory activities are present, as reported in the Hela system. The active fractions contain about one-third of the total nonspecific RNA polymerase activity. It should be noted that fraction 10, which shows the highest level of transcription of the CaMV template, contains only 5% of the polymerase activity measured on calf thymus DNA.

4.2. Transcription with the DEAE-CL6B Fractions

5. To obtain a higher level of specific transcription the amount of starting material is increased and the pooled active 0.6*M* KCl fractions from two heparin-sepharose columns are separated on a DEAE-CL6B column. Figure 3 shows the elution profile obtained. It is important to dialyze the pooled fractions to remove excess KCl that, in the presence of the large amount of ammonium sulfate used (more than 85% of saturation), induces a salt precipitation. Five fractions are recovered and tested for their transcriptional activity (Fig. 3). RNA polymerase II is abundant in the 1*M* KCl fraction, but this activity is highly unstable and is frequently lost during dialysis. The 0.25*M* KCl fraction contains about 40% of the polymerase loaded on the DEAE-CL6B column. The 0.20*M* KCl fraction tested on calf thymus DNA contains an activity resistant to α-amanitin, corresponding to RNA polymerase I. In contrast to the Hela system where the 0.15 and 0.25*M* KCl fractions contain essential factors, the tobacco 0.25*M* fraction alone is capable of synthesizing the 345 nt runoff RNA on the 19S template in the presence of FT0.35.
6. The effect of various parameters on the transcription reaction have been tested. The most striking effect is salt concentration. Figure 4 shows that increasing KCl concentration (from 25 to 50 m*M*) first leads to a general increase in the level of transcription products, the 345 nt RNA being the major transcribed RNA, the smaller RNAs being no longer observed. The other major band corresponds to end-to-end transcription of the template. When salt concentration is increased to 100 m*M*, end-to-end transcription is reduced and the 345 nt band constitutes more than half of the total transcription.

4.3. Transcription of the CaMV 35S Promoter

7. The number of promoters transcribed accurately by plant in vitro systems is rather low. However, successful and accurate transcription has been reported for the TC7 promoter of the Ti plasmid DNA *(8)*, the gliadin promoter of rice *(13)*, the 19S promoter of CaMV *(10)*, in some conditions the 35S promoter of CaMV, a strong viral promoter that is expressed at high level in many tissues *(9,13)*, the chalcone synthase promoter, and a chitinase promoter *(14)*. The tobacco cell system described here was used to transcribe, in vitro, the 35S promoter of CaMV *(10)*. The template used for the runoff transcription (Fig. 1C) is characterized by an initiation site localized at 7435 bp. The runoff transcript is an RNA of 224 nt. An as-1 element responsive to TGA-1a, a tobacco factor that stimulates transcription by increasing the number of preinitiation complexes, is found upstream from the TATA box.

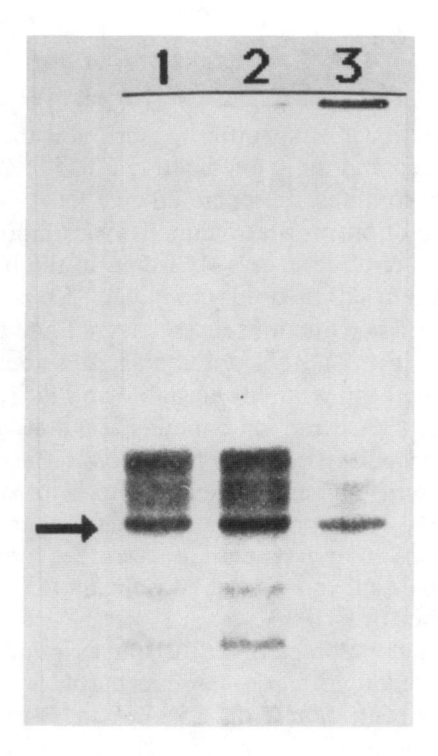

Fig. 4. Effect of ionic strength on transcription. Reactions were carried out under standard conditions using 4 µL of DEAE 0.25*M* fraction and 5 µL of FT0.35, except that the KCl concentration was 25 m*M* (lane 1), 50 m*M* (lane 2), or 100 m*M* (lane 3). The 345 nt band is arrowed.

8. With the wheat germ extract *(9)*, a modified 35S promoter allows the initiation of RNA synthesis, in the absence of GTP, on a G-free sequence. The construct used contains two copies of the as-1 sequence upstream from the TATA box and a fragment corresponding to the -343 to -91 region of the 35S promoter, devoid of TGA-1a binding sites when inserted in the reverse orientation, upstream from the two as-1. Its transcription is stimulated in the presence of purified TGA-1a.
9. Figure 5 shows the results obtained with the pCa 35.1 template and tobacco and Hela extracts. Lane 2 shows that a Hela system is capable of efficient transcription at this promoter. Three contiguous major bands (arrow) corresponding to the initiation sites are detected. With tobacco extracts no bands are observed comigrating with the bands present in the Hela system (lane 1). However, accumulation of low molecular weight components

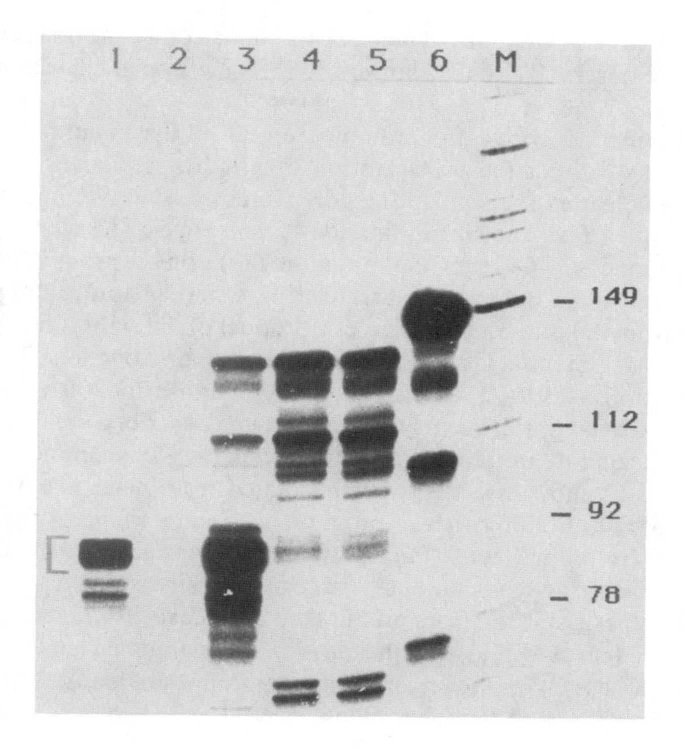

Fig. 5. Runoff transcription from the 35S promoter. Transcription reactions were carried out by the standard method for lanes 1 (tobacco extract) and 2 (Hela extract). For the reaction in lanes 3–5, after the preincubation of DNA and extract fractions, ATP and GTP were added to 330 μ*M* and CTP to 5 μ*M* with 5 μCi [α-32P]-UTP, without unlabeled UTP. After a further 10 min, CTP was added to 330 μ*M* and UTP to 50 μ*M* (lane 3), 100 μ*M* (lane 4), or 200 μ*M* (lane 5). The mixtures were incubated for 50 min. Reaction products were analyzed on a 5% polyacrylamide-urea gel. The size of the molecular weight markers are shown in the middle M lane.

(bracketed, lane 1) and larger RNAs including end-to-end products can be seen. The small RNAs could correspond to accurately initiated products whose synthesis prematurely terminated at an inverted repeat sequence centered at +78 bp of the initiation site, termination being induced by the limited amount of UTP in our test conditions. This hypothesis was verified by allowing the synthesis of short highly radioactive RNA chains with limiting amount of UTP and CTP. The concentration of CTP was then brought to 330 μ*M* and UTP was added to 50, 100, and 200 μ*M* (lanes 3, 4, and 5) to allow elongation. In lane 5, two bands comigrating with the Hela system are observed and the shorter products have been chased into larger molecules.

10. Another construct with a 35S promoter region similar to our promoter, has been used to direct the transcription of gliadin sequences in rice and tobacco nuclear extracts *(13)*. The 35S promoter is active, but the size of the template affects the size of the runoff transcripts. The whole promoter (with sequences –464 bp of the initiation site) gives a product that is 200 bases longer than the expected transcript, whereas a minimal promoter (–91 bp from the initiation site) gives a product of the right size. The whole promoter has potential CAAT-like/TATA like pairs; one located at nucleotide positions –236/–219 could account for a transcript 200 nt longer than the *in vivo* transcript. It seems that the systems described lack factors necessary for accurate initiation from the full length 35S promoter.

The results show that the systems isolated from plant material are not yet capable of initiating transcription on as broad a range of templates as extracts from animal cells. Purification of the extracts *(10)*, use of templates synthesizing RNA products that can be easily isolated *(9,13)*, initiation of synthesis starting from preinitiated complexes formed on templates immobilized on avidin-agarose beads *(14)*, are major steps in the construction of a tool useful for the understanding of important features of the regulation of transcription in higher plants.

4.4. SP6 Mapping of the Initiation Site

11. Confirmation of the localization of the transcription site at the 19S promoter is obtained by SP6 mapping *(17)*. As premature termination is often observed with the SP6 RNA polymerase, it may be necessary to subclone a shorter restriction fragment to prepare the mapping template. The template we use for synthesis of the RNA probe consists of a 202 bp *Hind*III-*Eco*RI fragment overlapping the 19S initiation site cloned in the vector pSP64 (Fig. 1B). The plasmid is linearized at the unique *Eco*RV site. The runoff transcript initiates at the phage promoter and terminates at the end of the linear template giving a 149 nt RNA. 143 nt of this RNA is identical to the transcribed strand of the CaMV insert. RNA correctly initiated at the 19S site should protect a 92 nt fragment. Figure 6 shows the result of mapping experiments carried out with unlabeled RNA synthesized by the tobacco and Hela cell extracts. One of the two major fragments observed for the Hela system corresponds to the expected 92 nt fragment (lane 1). The same fragments are observed when RNA synthesized in the tobacco extract is used in hybridization (lanes 4 and 5), whereas when the same amount of yeast RNA is used, the probe is completely degraded (lane 2). Longer pro-

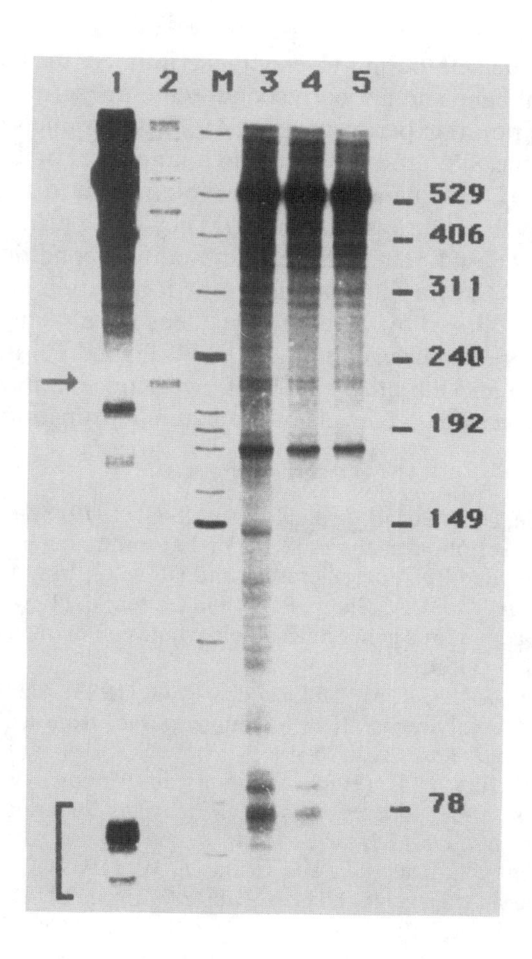

Fig. 6. SP6 mapping of transcription initiation sites for Hela and tobacco extracts on the 19S promoter. Hybridization reactions contained 10^5 cpm (lane 4) or 2×10^5 cpm (lanes 1, 2, 3, and 5) labeled probe and RNAs synthesized in the Hela extract (lanes 1 and 3; lane 1 is a shorter exposure of lane 3) or tobacco extract (lanes 4 and 5). In the latter case, five transcription reactions, using the 0.25*M* and the FT0.35 fractions were pooled for each hybridization reaction. RNase-resistant fragments were analyzed on an 8% polyacrylamide-urea gel. Lane 6 contains 10^4 cpm probe treated under hybridization conditions without RNase digestion. Lane 2 shows the complete degradation of the probe when hybridized with yeast RNA. The lengths of molecular markers are shown on the right hand side, lane labeled M.

tected fragments are present at the top of lanes 2, 3, and 4. One corresponds to products of the end-to-end transcription of the template (143 nt). The additional bands do not correspond to the transcription products that can be seen in runoff experiments (Fig. 4). These products are long enough to protect all the SP6 probes, leading to a high level of 143 nt fragments. The majority of these bands are also visible in hybridizations with RNA synthesized by the Hela system (lane 3). They could result from variations in the transcriptional pattern under the modified conditions used for the preparation of RNA for SP6 mapping or to degradation of the in vitro synthesized RNA induced by the additional steps in extraction.

12. SP6 probes were synthesized with 10 μ*M* UTP and 10 μCi [α-32P]-UTP (400 Ci/mmol) and the others XTPs. Lower concentrations of unlabeled UTP lead to unacceptable levels of premature termination of the probe.

References

1. Weil, P. A., Luse, D. S., Segall, J., and Roeder, R. G. (1979) Selective and accurate initiation of transcription at the Ad2 major late promoter in a soluble system dependent on purified RNA polymerase II and DNA. *Cell* **18,** 469–484.
2. Manley, J., Fire, A., Cano, A., Sharp, P. A., and Gefter, M. (1980) DNA-dependent transcription of adenovirus genes in a soluble whole cell extract. *Proc. Natl. Acad. Sci. USA* **77,** 3855–3859.
3. Dignam, J. D., Lebovitz, R. M., and Roeder, R. G. (1983) Accurate transcription initiation by RNA polymerase II in a soluble extract from isolated mammalian nuclei. *Nucleic Acids Res.* **11,** 1475–1489.
4. Tyler, B. M. and Giles, N. H. (1985) Accurate transcription of cloned Neurospora RNA polymerase II-dependent genes *in vitro* by homologous soluble extracts. *Proc. Natl. Acad. Sci. USA* **82,** 5450–5454.
5. Leisy, D. J., Hinlo, J., Zhao, Y., and Okita, T. W. (1989) Expression of a rice glutelin promoter in transgenic tobacco. *Plant Mol. Biol.* **14,** 41–50.
6. Lam, E., Benfey, P. N., Gilmartin, P. M., Fang, R. X., and Chua, N. H. (1989) Site specific mutations alter *in vitro* factor binding and change promoter expression pattern in transgenic plants. *Proc. Natl. Acad. Sci. USA* **86,** 7890–7894.
7. Flynn, P. A., Davis, E. A., and Ackerman, S. (1987) Partial purification of plant transcription factors. II. An *in vitro* transcription system is inefficient. *Plant Mol. Biol.* **9,** 159–169.
8. Yamazaki, K. and Imamoto, F. (1987) Selective and accurate initiation of transcription at the T-DNA promoter in a soluble chromatin extract from wheat germ. *Mol. Gen. Genet.* **209,** 445–452.
9. Yamazaki, K., Katagiri, F., Imaseki, H., and Chua, N. H. (1990) TGA1a, a tobacco DNA-binding protein, increases the rate of initiation in a plant *in vitro* transcription system. *Proc. Natl. Acad. Sci. USA* **87,** 7035–7039.
10. Cooke, R. and Penon, P. (1990) In vitro transcription from cauliflower mosaic virus promoters by a cell-free extract from tobacco cells. *Plant Mol. Biol.* **14,** 391–405.

11. Moncollin, V., Miyamoto, N. G., Zheng, X. M., and Egly, J. M. (1986) Purification of factor specific for the upstream element of the adenovirus 2 major late promoter. *EMBO J.* **5,** 2577–2584.
12. Guilley, H., Dudley, R. K., Jonard, J., Balazs, J. E., and Richard, K. E. (1982) Transcription of cauliflower mosaic virus DNA: detection of promoter sequences and characterisation of transcript. *Cell* **30,** 763–773.
13. Roberts, M. W. and Okita, T. W. (1991) Accurate *in vitro* transcription of plant promoters with nuclear extracts prepared from cultured plant cells. *Plant Mol. Biol.* **16,** 771–786.
14. Arias, A. A., Dixon, R. A., and Lamb, C. J. (1993) Dissection of the functional architecture of a plant defense gene promoter using a homologous in vitro transcription initiation system. *Plant Cell* **5,** 485–496.
15. Bradford, M. (1976) A rapid and sensitive method for the quantitation of microgram quantities of protein utilizing the principle of preotein-dye binding. *Anal. Biochem.* **72,** 248–254.
16. Jouanneau, J. P. and Tandeau de Marsac, N. (1973) Stepwise effects of cytokinin activity and DNA synthesis upon mitotic cycle events in partially synchronized tobacco cells. *Exp. Cell. Res.* **77,** 167–174.
17. Birboim, H. C. and Doly, J. (1979) A rapid alkaline extraction procedure for screening recombinant plasmid DNA. *Nucleic Acids Res.* **7,** 1513–1523.
18. Dudley, R. K., Odell, J. T., and Howell, S. H. (1982) Structure and 5' termini of the large en 19S RNA transcripts encoded by the cauliflower mosaic virus genome. Virology **117,** 19–28.
19. Melton, D. A., Krieg, P. A., Rebagliati, M. R., Maniatis, T., Zinn, K., and Green, M. R. (1984) Efficient *in vitro* synthesis of biologically active RNA and RNA hybridisation probes from plasmids containing a bacteriophage SP6 promoter. *Nucleic Acids Res.* **12,** 7035–7056.

CHAPTER 24

Analysis of Plant Gene Expression by Reverse Transcription-PCR

Mark Bell

1. Introduction

As described in previous chapters (Chapters 15–23), techniques for the measurement of gene expression at the RNA level include RNA gel blot analysis, S1 mapping, primer extension, RNase protection, and, more recently, reverse transcription-polymerase chain reaction (RT-PCR) *(1)*. This involves the reverse transcription of total RNA (or mRNA) from which cDNA is synthesized. Primers designed to the target cDNA are then used in a PCR reaction to amplify the desired product. This is then identified by the predicted size of the cDNA on agarose gels with size markers. The identity can also be confirmed by DNA gel blot analysis with a homologous probe or by the use of an end-labeled primer in the PCR reaction from which autoradiographs of gels run with end-labeled size markers that give both a sensitive and accurate determination of product *(2)*. This is necessitated when PCR conditions are suboptimal and result in low concentrations of product nonvisible in ethidium bromide stained gels. However, as PCR is a commonly used technique, the need to optimize should not act as a deterrent to the use of RT-PCR. The advantages of RT-PCR include its simplicity in comparison to techniques such as RNase protection and the ability to detect rare transcripts or,

From: *Methods in Molecular Biology, Vol. 49: Plant Gene Transfer and Expression Protocols*
Edited by: H. Jones Humana Press Inc., Totowa, NJ

similarly, mRNA from small tissue samples. Problems associated with RT-PCR, when the detection of a transcript from transgenic plants is complicated by cDNA copies in genomic DNA contamination of RNA, are discussed in Note 3.

PCR is a powerful amplification tool for small amounts of DNA; however, in most instances the method only gives qualitative results. This arises from reaction conditions that interfere with the exponential nature of the cycling reaction, thereby limiting the linear nature of PCR *(3)*. The essence of RT-PCR is to determine, by sampling at time points during PCR, the linear-product window from which quantitative results can be found.

The application of RT-PCR to an expression analysis must include the determination of the assay limits in terms of precision and reproducibility once PCR optimization has been achieved. The literature available is extensive and the experimenter should first decide what information is required, after all Northern blots often provide only limited information about the relative abundance of a transcript.

The following technique is not quantitative, but concentrates on the steps necessary for the successful development of RT-PCR toward a comparative measurement of mRNA. The technique is also very quick and allows for the screening of many samples to determine transcript presence. Oligo-dT is used as a primer so that more than one mRNA can be studied. In parallel with the cDNA synthesis, a small sample of reverse transcribed RNA is labeled with [^{32}P] dCTP, from which the size and quantity of cDNA is determined using alkaline gel electrophoresis. Similarly, the progress of PCR amplification is monitored by sampling reactions at four-cycle intervals to determine the linear amplification window, from which comparative results are taken. Ultimately, to check that RT-PCR produces a comparative result, tenfold dilutions of RNA prior to reverse transcription can be used. By comparing PCR products from tenfold dilutions of the same RNA, there should be a clear correlation between the dilution of transcript and amount of final PCR product, as seen from successive four-cycle samples. Similarly, comparing different samples in this way is a measure of reliability. Product identification is then achieved by agarose gel electrophoresis followed by DNA gel blot analysis.

It is important to read Note 2 before deciding on a pilot experiment, which would not include a DNase step prior to RT-PCR.

2. Materials

2.1. DNase Treatment of RNA and Reverse Transcription

Store items at –20°C unless otherwise indicated.

1. DNase 1-RNase free (1 μg/μL).
2. 10X DNase buffer: 100 m*M* Tris-HCl, pH 7.5, 100 m*M* $MgCl_2$. Store at room temperature (rt).
3. Tris-saturated phenol/chloroform (1:1 v/v). Store at 4°C.
4. Chloroform. Store at rt.
5. Ethanol.
6. RNase inhibitor (5 U/μL).
7. Diethlpyrocarbonate (DEPC)-treated dH_2O. Store at rt.
8. Moloney murine leukemia virus reverse transcriptase (MMLV-RT) 200 U/μL.
9. 5X RT buffer: 0.25*M* Tris-HCl, pH 8.3, 0.375*M* KCl, 15 m*M* $MgCl_2$.
10. 100 m*M* dithiothreitol (DTT). Items 1–3 are often supplied together.
11. Oligo-dT (12–18) template primer (1 mg/mL).
12. 10 m*M* dNTPs.
13. α-[^{32}P]dCTP.

2.2. Assessing cDNA Synthesis

Store items at room temperature unless otherwise indicated. Alkaline gel buffers should be made fresh as required.

1. Salmon sperm DNA 100 μg/mL in 20 m*M* EDTA; store at –20°C.
2. 2*M* NaOH.
3. 1*M* HCl.
4. 1*M* Tris-HCl, pH 8.3.
5. Tris-saturated phenol/chloroform (1:1, v/v). Store at 4°C.
6. Chloroform.
7. 4*M* and 2*M* ammonium acetate.
8. Ethanol: Store at –20°C.
9. Agarose gel buffer 50 m*M* NaCl, 1 m*M* EDTA.
10. 1X running buffer: 30 m*M* NaOH, 10 m*M* EDTA.
11. Alkali loading buffer: 50 m*M* NaOH, 1 m*M* EDTA, 2.5% Ficoll 400, 0.025% Bromophenol blue.
12. End-labeled 1 kb or *Hin*dIII markers.
13. 7% Trichloroacetic acid (TCA).

2.3. PCR

It is important to note that the use of *Taq* DNA polymerase is governed by US Patent no. 4,889,818 and therefore the experimenter should be aware that the use of *Taq* polymerase for PCR is under copyright.

1. Filtered pipet tips.
2. *Taq* DNA polymerase (5 U/µL).
3. 10X PCR buffer: 500 m*M* KCl, 100 m*M* Tris-HCl, pH 8.3, 0.01% (w/v) gelatin, 15 m*M* $MgCl_2$, 10 m*M* dNTPs.
4. 5' and 3' primers (50 µ*M*).
5. DEPC-treated dH_2O.
6. cDNA from Section 2.1.

3. Methods

3.1. DNase Treatment

1. For DNase digestion of the RNA, add the following to 1 µg of RNA resuspended in 10 µL of DEPC dH_2O: 4 µL RNase inhibitor (5 U/µL), 48 µL DEPC dH_2O, 7 µL 10X DNase buffer, 1 µL RNase-free DNase (5 mg/mL).
2. Incubate at 37°C for 10 min. Extract using an equal volume of phenol/chloroform, vortex, and centrifuge at full speed for 5 min in a microfuge. Remove the supernatant and extract using an equal volume of chloroform. To the supernatant add an equal volume of 4*M* ammonium acetate and 400 µL of cold ethanol. Precipitate either overnight at –20°C or at –70°C for 15 min. Spin for 15 min in a microfuge cooled to 4°C. Remove the supernatant and wash the pellet by adding 100 µL of 70% ethanol followed by a further 2 min spin at full speed. Remove the 70% ethanol and lyophilize the RNA pellet.

3.2. Reverse Transcription

1. Dilute 1 µg of total RNA (*see* Note 1) in a total volume of 19 µL DEPC-treated dH_2O and add 4 µL of oligo-dT in a microfuge tube. If RNA is plentiful, then 1 µg of RNA is ideally used in the control, with the addition of 5 µCi of label. If not, then a 5-µL aliquot of the nonlabeled sample from step 3 is added to 5 µCi of label in a separate microfuge tube as described in step 4.
2. The tube(s) are then incubated at 70°C for 5 min to denature the RNA and then placed on ice for 3 min to anneal the oligo-dT.
3. A reagent premix is used to standardize separate reactions. Mix the reagents in the following order: 0.5 µL RNase inhibitor (5 U/µL), 0.4 µL 100 m*M* DTT, 4 µL 10 m*M* dNTPs, 2 µL RT-MMLV, 8 µL RT buffer.

4. If RNA supply is limiting then a 5-μL aliquot from the above reaction mix is added to a microfuge tube containing 5 μCi of dried down α-[^{32}P]dCTP. Mix by gentle vortexing followed by centrifugation to collect reagents in the base of the tube. Then incubate both non- and label-added tubes at 37°C for 1 h for reverse transcription followed by 70°C for 10 min to activate the RT.
5. To the nonradiolabeled tube add 60 μL of dH_2O and store at –20°C. To the radiolabeled-tube add 35 μL of dH_2O.

3.3. Assessing cDNA Synthesis

1. To the radiolabeled reaction mix add 20 μL of salmon sperm DNA (carrier DNA) and 14 μL of 2*M* NaOH followed by incubation at 46°C for 20 min to hydrolyze the RNA.
2. Then add 14 μL 1*M* HCl, 14 μL 1*M* Tris-HCl, pH 8.3, and 100 μL of phenol/chloroform. Mix by vortexing and then microcentrifuge for 2 min. Re-extract the resulting supernatant with an equal volume of chloroform and repeat the centrifugation.
3. Add an equal volume of 4*M* ammonium acetate and 2 vol of –20°C ethanol and leave to precipitate either overnight at –20°C or in a –70°C freezer for 30 min.
4. Leave the tube to warm to room temperature to dissolve precipitated dNTPs. Spin down the pellet in a microcentrifuge for 10 min at room temperature. The warming step prior to centrifugation is important in separating unincorporated label from the cDNA pellet in the control. The supernatant is then carefully removed and the pellet washed to further reduce contamination. Add 50 μL 2*M* ammonium acetate and 200 μL –20°C ethanol and repeat the centrifugation. The wash step is then repeated minus the ammonium acetate and the supernatant removed.
5. Prepare a 1.4% agarose gel in gel buffer, microwave, and cool to 50°C in a waterbath before pouring.
6. Equilibriate the gel in 1X running buffer for at least 30 min. Do not use excess running buffer, but add just enough to cover the wells, thereby preventing the gel from floating.
7. Resuspend the pellet from step 4 in alkali loading buffer to the required well volume using reciprocation through a filtered micropipet tip.
8. Load end-labeled molecular weight markers and electrophorese until the dye front has migrated two-thirds distance. Run at no greater than 4 V/cm to reduce gel heating.
9. Soak the gel in 7% TCA twice for 30 min. Then gently transfer onto Whatman (Maidstone, UK) 3 MM paper and cover with clingfilm.
10. Dry the gel overnight at 60°C in a gel dryer and expose to X-ray film in a cassette with intensifier screens. Allow up to 24 h for exposure.

3.4. PCR

1. The order of reaction component addition is important as not to contaminate buffers with cDNA. To this end the use of PCR-only pipets is recommended combined with the use of filtered tips. Primers should also be aliquoted and mixed at the correct dilution and stored at –20°C. A fresh aliquot is used for each experiment.
2. Adequate PCR premix, 1X PCR buffer and primers diluted to 50 μM in DEPC-treated dH_2O should be prepared before handling cDNA.
3. To 90 μL of premix add a 10 μL aliquot of diluted cDNA. Add the *Taq* polymerase at 0.1 U/100 μL. Mix by gentle tapping of the tube-base followed by a brief spin in a microfuge. Finally, apply 80 μL of oil to the top of a 0.5 mL tube.
4. To sample PCR products from different cycle numbers two strategies are available. The first is the use of replicates of each reaction tube that can then be removed at four-cycle intervals from 16 cycles onward. In this case, aliquot the 100 μL mix (after *Taq* addition) into five separate tubes in 20-μL vol reducing the oil layer to 40 μL. The second strategy uses the full 100-μL reaction mix from which 10-μL samples are taken. At the end of the extension phase, minimal mixing of oil and aqueous is desired and is achieved by inserting the pipet tip down the side of the tube through the oil layer followed by expelling 10 μL of air to clear the tip head from oil, followed by sample aspiration.
5. Samples are then electrophoresed on an agarose gel and stained with ethidium bromide, blotted, and probed.

4. Notes

1. The success of reverse transcription is dependent on the preparation of high quality, intact RNA. Although a wide range of RNA isolation procedures is available, for most plant tissue of low starch content, a guanidinium hydrochloride method is recommended *(8)*. The integrity of the RNA is maintained by a range of procedures that prevent exposure to RNase *(9)*. Before cDNA synthesis, the integrity and quantity of total RNA can be checked by comparing $A_{260/280}$ ratios. Also electrophoresis through formaldehyde gels containing ethidium bromide is an essential step to ensure RNA integrity. These should show ribosomal RNA subunits intact against a faint background of mRNA.
2. A pilot RT-PCR should be performed to familiarize the worker with the technique. Therefore, to simplify the technique ascertain whether the DNase step is necessary. Avoiding a DNase step reduces RNA handling, and is one less variable. The second step is to optimize the PCR reaction using the primers to the cDNA of interest before attempting RT-PCR. It is

also useful to miss out the assessment of cDNA synthesis initially, thereby increasing the speed of pilot results. As the reverse transcription and PCR steps are simple to perform, it is therefore important to determine whether qualitatively there is any PCR product by agarose gel electrophoresis and DNA gel blot analysis before proceeding to compare transcript levels more thoroughly. Therefore, to simplify the PCR further, no multiple samples should be taken from each tube an a maximum of cycles employed.

PCR should include both no cDNA and RNA, only controls to determine whether products arise from cDNA contamination or a genomic source, respectively. The former involves the DNase step followed by the comparison of PCR products from the same sample before and after reverse transcription (Fig. 1A). The problem of priming from a genomic source can also be overcome by designing primers that flank the intron of the endogenous gene. When amplified, the PCR product would be of a larger size than that predicted for the amplified cDNA. However, if more than one product is obtained in the PCR reaction, then any quantitative analysis is negated because of the competition between products. This does not apply with the expression of a chimeric transgene, the cDNA of which contains no intron. Even if no apparent product is available on ethidium bromide stained agarose gels, the gel should be blotted and probed. Even if a result is not obtained by this more sensitive approach, the probe can be reused. An example of a pilot experiment in which the expression of the fission yeast *cdc*25 gene in tobacco was determined is given in Fig. 1B.

It is important to stress the need for caution when working with PCR as contamination is potentially everywhere, thereby the inclusion of controls and more expensive filtered tips is a necessity. These tips also prevent contamination of pipets with radiolabel.

3. The use of RT-PCR as a quantitative technique relies on the optimization and standardization of both the reverse transcription and amplification steps *(1)*. Accordingly, efforts to quantitate RNA have included internal competitive or noncompetitive templates to control for differences in amplification. Competitive internal standard protocols rely on modification of the internal standard to distinguish it from the target PCR product by inclusion of restriction sites or a different product size. The amount of mRNA is then quantitated by extrapolating against the standard curve generated with the internal standard *(4)*. The use of an internal RNA standard provides a control not only for the amplification but also for reverse transcription efficiency.

   The essence of RT-PCR quantitation is in determining between which cycles the amplification is linear. Several different approaches have been used to maximize this linear-product cycling span. These have included

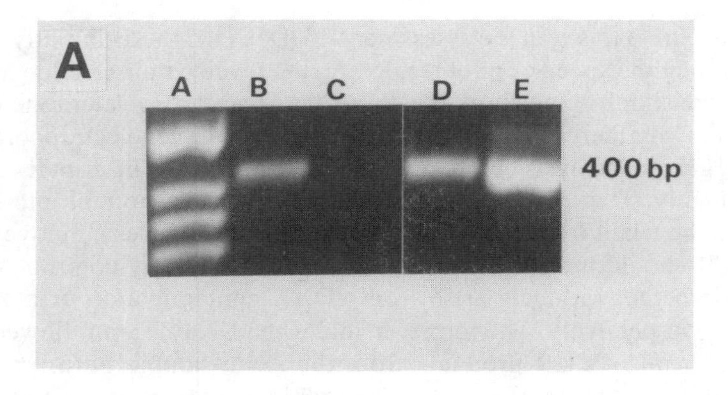

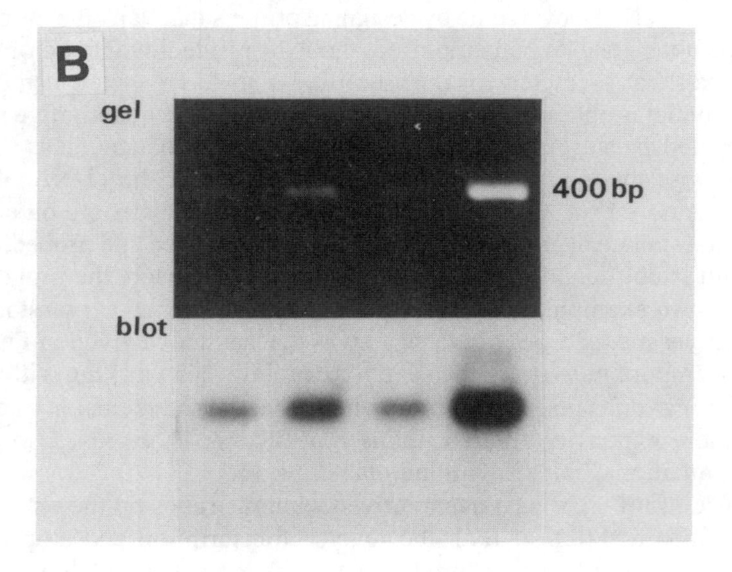

Fig. 1. A single 400-bp product is obtained from PCR using primers designed to fission yeast *cdc*25 transformed into tobacco. Primers: 5'GCCCTGCCTTACCGACTCC3' and 5'TGGTGGACAAGCTACCCGCC3' were annealed at 55°C. **(A)** The controls used to determine that the RT-PCR product is primed by the synthesized cDNA: A, 1 kb ladder; B, PCR from total RNA; C, PCR of RNA after DNase to remove genomic DNA contamination; D, RT-PCR of RNA from C; E, positive control (PCR of *cdc*25 cDNA). **(B)** The RT-PCR of 1 μg total RNA from three separate tobacco transformants with *cdc*25 after 30 cycles. 20 μL PCR product was electrophoresed on a 1.5% agarose gel and blotted to nitrocellulose and probed with the 400-bp ^{32}P-labeled positive control (far right).

the input of small quantities of RNA to ensure substrate excess combined with rapid cycling times to maximize the efficiency of *Taq* polymerase; however, this approach requires specialized equipment to achieve a total PCR cycle of 24-s duration *(6)*. The principles of optimizing PCR are beyond the scope of this work but further reading should extend to the use of hotstarts, primer design, and cycling parameters and, most important, minimal handling, such that the reverse transcription and subsequent PCR take place in the same tube *(5)*.

References

1. Ferre, I. (1992) Quantitative or semi-quantitative PCR: reality versus myth. *PCR Meth. Appl.* **2,** 1–9.
2. Simpson, C. G., Sinibaldi, R., and Brown, J. W. S. (1992) Rapid analysis of plant gene expression by a novel reverse transcription-PCR method. *Plant J.* **2(5),** 835,836.
3. Mullis, K. B. (1991) The polymerase chain reaction in an anemic mode: How to avoid cold oligodeoxyribonuclear fusion. *PCR Meth. Appl.* **1,** 1–4.
4. Wang, A. M., Doyle, M. V., and Mark, D. F. (1989) Quantitation of mRNA by the polymerase chain reaction. *Proc. Natl. Acad. Sci. USA* **86,** 9717–9721.
5. Robinson, M. O. and Simon, M. I. (1991) Determining transcript number using the polymerase chain reaction: Pgk-2, mP2, and PGK-2 transgene mRNA levels during spermatogenesis. *Nucleic Acids Res.* **19(7),** 1557–1562.
6. Tan, S. T. and Weis, J. H. (1992) Development of a sensitive reverse transcriptase PCR assay, RT-RPCR, utilizing rapid cycle times. *PCR Meth. Appl.* **2,** 137–143.
7. Rappolee, D. A., Mark, D., Banda, M. J., and Werb, J. (1988) Wound macrophages express TGF-α and other growth factors in vivo: analysis by mRNA phenotyping. *Science* **241,** 708–712.
8. Gurr, S. J. and MacPherson, M. J. (1991) In *Molecular Plant Pathology: A Practical Approach* (Gurr, S. J., MacPherson, M. J., and Bowles, D. J., eds.), IRL, Oxford.
9. Finer, M. H. (1991) In *Methods in Molecular Biology* vol. 7, *Gene Transfer and Expression Protocols* (Murray, E., ed.), Humana, Clifton, NJ, pp. 283–296.

Chapter 25

In Situ Hybridization to Plant Tissue Sections

Shirley R. Burgess

1. Introduction

An important step in the analysis of gene action is to determine the spatial and temporal regulation of gene expression. The patterns of gene expression can provide clues to the function of genes and facilitate the mapping of qualitative changes in steady state mRNA levels. *In situ* hybridization is a cytochemical procedure first used to map sequences on chromosomes *(1,2)*. It is also used to identify viral RNA in and around infection sites *(3,4)* and is commonly used for locating specific mRNA transcripts in tissue sections *(5–9)*. The technique can also be adapted for ultrastructural resolution particularly with biotin labeled probes *(10,11)*.

The positional information that can be obtained with *in situ* hybridization is not possible with conventional filter hybridization experiments. Combining immunocytochemical techniques *(12,13)* that identify the product of the mRNA transcripts, with *in situ* localization, can reveal important information on the sites of transcription and the targeting of proteins.

As with all nucleic acid hybridizations, *in situ* hybridization relies on the annealing of complementary single-stranded sequences under appropriate conditions. For light microscopy, the fixed tissue sections are attached to glass supports and hybridized to a nucleic acid probe, which may be RNA or DNA. After washing, the slides are coated with either an X-ray emulsion film for radioactively labeled probes, or developed with a substrate if using a nonradioactive probe, and finally viewed under a light microscope.

From: *Methods in Molecular Biology, Vol. 49: Plant Gene Transfer and Expression Protocols*
Edited by: H. Jones Humana Press Inc., Totowa, NJ

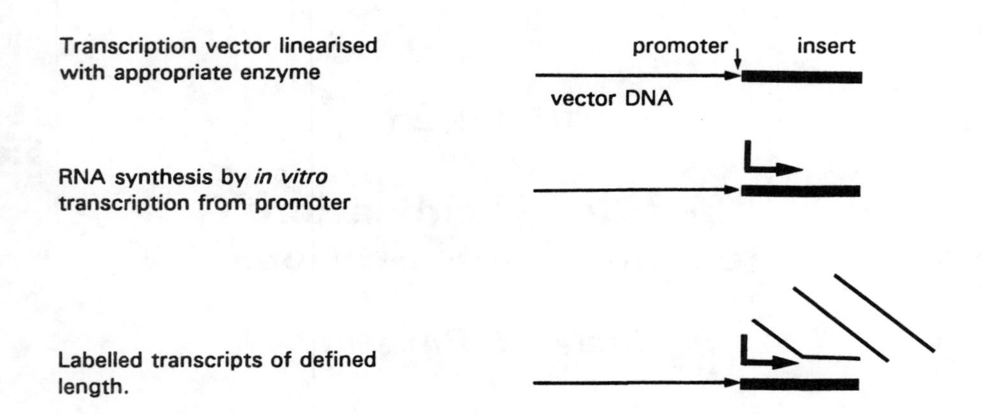

Fig. 1. Production of transcripts of defined length from a transcription vector containing bacteriophage promoter(s) with a cloned insert DNA.

Double-stranded DNA probes were used in earlier studies *(7)*. They do have the advantage of not requiring sub-cloning into transcription vectors, but they tend to reanneal during hybridization *(14)*. Single-stranded DNA probes from bacteriophage vectors have also been used *(15)*.

Oligonucleotides can be designed as very specific hybridization probes and have been successfully used for *in situ* hybridization *(15–17)*, but difficulties in obtaining high-specific activity labeling using end-labeling, and hybrid stability owing to their small size, have so far restricted their use.

Single-stranded RNA probes, synthesized by in vitro transcription are commonly used because of the ease of synthesis using the bacteriophage RNA polymerases and their promoters *(18)*. Vectors, such as pBluescript from Stratagene (La Jolla, CA), have the T3 and T7 polymerase promoters in opposing directions and the gene fragment can be cloned between the promoters using the unique restriction sites. This allows the synthesis of both the sense and the antisense transcripts by linearizing the plasmid with enzymes downstream of each promoter. The transcripts are synthesized from the phage promoter, through to the end of the linearized template, to a defined length depending on the size of the insert DNA (Fig. 1).

Some authors *(7,19,20)* recommend an optimum length of 150 bases to facilitate tissue penetration, but probes of 400 bases have been successfully used (Fig. 2) *(9,18)*. The probes can be shortened by limited hydrolysis and the size of the hydrolyzed transcripts estimated by running samples on agarose gels.

RNA probes are 10–15 times more sensitive than DNA probes. Several factors contribute to their effectiveness.

1. They are free of large stretches of vector sequences thereby reducing nonspecific hybridization. Posthybridization treatment with RNase also reduces background by removing single-stranded RNA.
2. The template can be removed by digestion with RNase-free DNase I.
3. RNA–RNA hybrids are very stable allowing more stringent conditions to be used *(21)*.
4. The use of sense and antisense transcripts on duplicate tissue sections gives a control for hybridization.

Radiolabeling the probe with a [^{35}S]ribonucleotide is the most widely used method and yields high specific activity probes. Labeling with high energy [^{32}P] gives a wider scatter of silver grains that cannot be resolved at the cellular level. This may still be useful for rapid or "whole mount" analyses as used in animal tissues *(22)*. The low energy isotope, ^{3}H, gives the best resolution but will require exposure times of several weeks *(15)*.

Nonradioactively labeled probes are considered to be less sensitive and their use, generally, has been limited to the detection of abundant transcripts. However, they are continually being improved and they are currently being tested in our laboratory. Major advantages of using nonradioactive probes are that they are safer, give results within hours and multiple detection systems can be used for different probes within the same tissue sections *(23–26)*.

A critical stage of the *in situ* hybridization procedure is the tissue preparation. Cryosectioning of tissue and postfixing is used, but the morphological preservation is generally inferior to wax embedding. Sectioning is also more difficult if the tissue does not have a homogeneous density. More often, tissue is fixed and subsequently embedded in a wax medium *(27,28)*. This also allows the handling of larger numbers of samples.

Fixation stabilizes cellular organization and the tissues remain stable for years. The fixative must rapidly penetrate the material to give good tissue preservation and preserve biological activity. The nucleic acids are held in a three-dimensional array with the fixed proteins and other macromolecules, rather than by interaction with the fixative, particularly glutaraldehyde. The retention of nucleic acids should be confirmed by staining sectioned material with a suitable stain, such as Acridine Orange *(29)*.

Good tissue preservation is obtained using the aldehydes, which react rapidly with proteins resulting in crosslinks. Glutaraldehyde is a very effec-

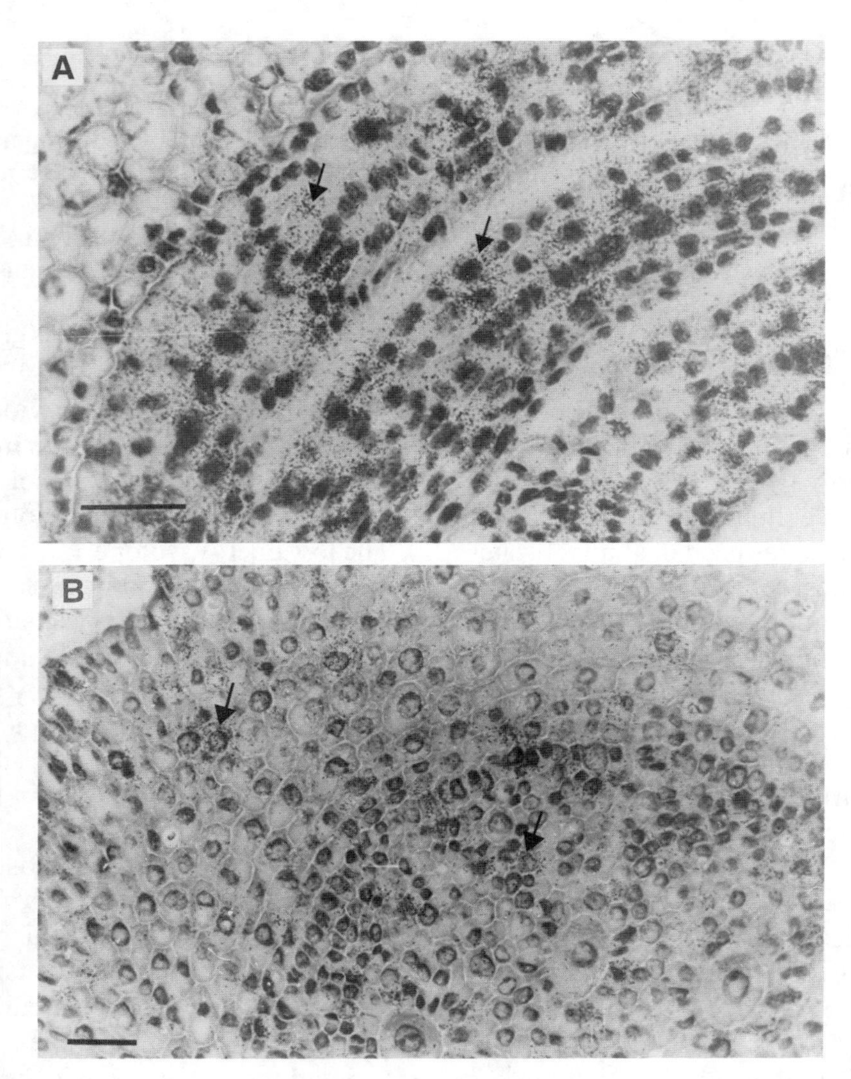

Fig. 2. Transverse sections through **(A)** maize shoot, 3-d after germination, **(B)** maize root, 3-d after germination, and **(C)** coleoptile of maize embryo, 20-d after pollination. The sections were hybridized with the antisense strand of a 400 nt histone H4 riboprobe labeled with [α-^{35}S]CTP, and the slides dipped in (Fig. 2 *opposite page*) emulsion. The black spots (as indicated by arrows) are the hybridization signal obtained after a 5-d exposure at 4°C. An interesting feature of this histone probe is the hybridization to transcripts in small groups of, or individual, cells that appear to specifically accumulate in the S phase of the cell cycle (Bar = 50 μm).

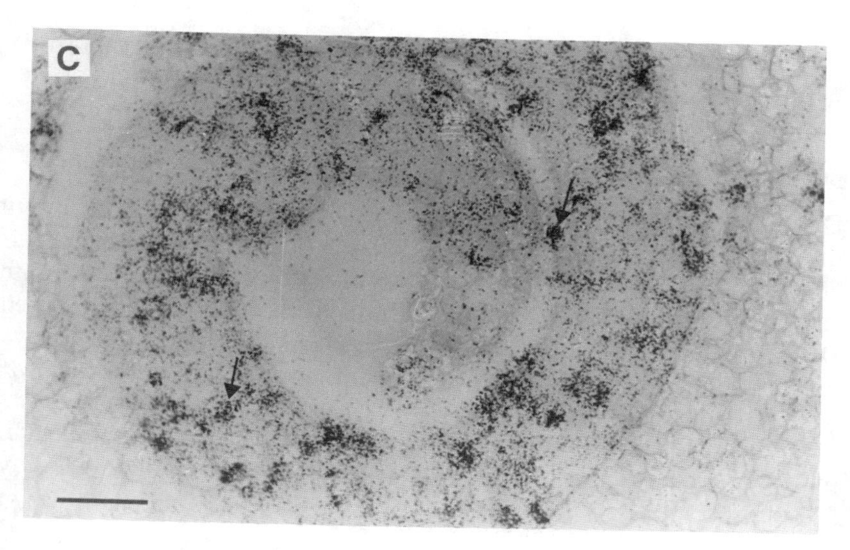

Fig. 2C.

tive crosslinking fixative but, for *in situ* techniques, it can reduce accessibility of the target sequence, and for immunocytochemistry, the proteins may be rendered nonreactive. Formaldehyde has only one aldehyde group, so there is less crosslinking, but this also increases the risk of extraction of cell components. Using low concentrations of glutaraldehyde (0.5–0.1%) with 4% formaldehyde successfully preserves most tissues. Subsequent protease digestion of the tissue sections will improve probe access. Precipitating fixatives, such as ethanol-acetic acid mixtures with or without the addition of formaldehyde, are also used. It is important to optimize the fixation for each specific application. It is not possible to adequately review all of the methodology here, but there are many general texts available that should be consulted *(30,31)*.

After fixation, the tissues are dehydrated and embedded in wax for sectioning. The sections are treated prior to hybridization to remove some of the proteins, allow penetration of the probe, and reduce nonspecific background. Such pretreatment also exposes the RNA to degradation and so should be performed immediately prior to hybridization.

Because of the increasing popularity of *in situ* hybridization, there is a large body of literature available, with many excellent reviews *(32–35)*. Here I have given the details of one particular method that has proved successful.

However, it is worth remembering that there are many steps that must be optimized, for a particular tissue and/or depending on the nature of the probe.

2. Materials

1. Diethylpyrocarbonate (DEPC)-treated water and solutions are required through all stages to remove contaminating RNases. Stir the water or buffer containing DEPC (0.1%) for 1 h at 37°C or overnight at room temperature. If a solution is not DEPC-treated or autoclaved, it should be prepared in DEPC-treated water. DEPC is toxic and combustible and should always be used in a fume cupboard. After treatment autoclave the solution, which will also inactivate the DEPC.
2. All glassware and solutions should be sterilized where possible. Sterile plastic containers are usually RNase free but must be handled with gloves.

2.1. Slide Coating

1. Single-side frosted end glass slides (thoroughly washed).
2. 50 µg/mL poly-D-lysine (Sigma [St. Louis, MO] P1149 mol wt >150,000) in 10 m*M* Tris-HCl, pH 8.0.

2.2. Tissue Preparation

1. Fixatives: The most suitable fixative must be selected for specific tissues. Examples include ethanol:formaldehyde:acetic acid, 80:40:5 v/v (used for cereal embryo tissues), and ethanol:acetic acid, 3:1 (used for leaf tissues).
2. Ethanol solutions for dehydration: 100, 90, 70, 50, and 30%.
3. Tributyl alcohol (2-methyl-propan-2-ol).
4. Embedding wax.

2.3. Probe Labeling

1. Stock 1*M* dithiothreitol (DTT).
2. Ribonuclease inhibitor.
3. RNA polymerases, T3 and T7, 15–20 U/µL (Promega, Madison, WI or Stratagene).
4. Transcription buffer supplied as 5X or 10X stocks (Promega or Stratagene).
5. RNase-free DNase I, 1 U/µL.
6. Carrier RNA: yeast tRNA, 10 mg/mL.
7. Stock 10X DNase buffer: 400 m*M* Tris-HCl, pH 7.5, 600 m*M* $MgCl_2$.
8. 3*M* Sodium acetate, 7.5*M* ammonium acetate, pH 4.8–5.2.
9. Cold 75 and 100% absolute ethanol (–20°C).
10. Digested purified plasmid for transcription, 1 µg/µL in 10 m*M* Tris-HCl, pH 7.4, 1 m*M* EDTA, pH 8.0, or water. Store all solutions at –20°C.

2.3.1. *Radioactive Labeling*

1. Ribonucleotide stocks of 10 m*M* UTP, ATP, GTP, and 100 μ*M* CTP.
2. [^{35}S]α-labeled ribonucleotide (CTP). Activity 370 MBq/mL (Amersham, Arlington Heights, IL).

2.3.2. *Labeling with Digoxigenin-11-UTP*

1. Stock 200 m*M* EDTA, pH 8.0.
2. Stock 10X nucleotide phosphate mix containing 10 m*M* ATP, GTP, CTP, and 6.5 m*M* UTP, 3.5 m*M* Digoxigenin-11-UTP (Boehringer Mannheim, Mannheim, Germany) in 100 m*M* Tris-HCl, pH 7.5.
3. 4*M* LiCl. Store all solutions at –20°C.

2.4. Tissue Section Pretreatment

1. Xylene or nontoxic substitute such as Histoclear.
2. Ethanol series: 100, 90, 70, 50, 30% and DEPC-treated water.
3. 200 m*M* HCl.
4. Stock 20X SSC: 3*M* NaCl, 300 m*M* trisodium citrate, pH 7.0.
5. Proteinase K: 1 μg/mL in 100 m*M* Tris-HCl, pH 8.0, 50 m*M* EDTA.
6 Stock 10X phosphate buffered saline (PBS): 150 m*M* NaCl, 10 m*M* sodium phosphate, pH 7.5.
7. 4% Formaldehyde in 1X PBS. Prepare fresh by heating to 60°C and adjust pH to 9.5 using 1*M* NaOH. Use at room temperature. *Caution:* Toxic.
8. Acetic anhydride: Prepare by mixing 100 m*M* triethanolamine buffer, pH 8.0 (requires vigorous stirring), and glacial acetic acid (500 μL to 500 mL buffer) immediately prior to use. This mixture is very unstable.

2.5. Hybridization

1. Stock 10X salts: 3*M* NaCl, 100 m*M* Tris-HCl, pH 6.8, 100 m*M* sodium phosphate, pH 6.8, 50 m*M* EDTA.
2. Hybridization buffer: Prepare as 1 mL aliquots containing; 1X salts, 50% formamide, 10% dextran sulfate, 1 mg/mL tRNA, 10 m*M* DTT, 0.5 mg/mL poly A. *Caution:* Formamide is toxic.
3. Wash buffer: 50% formamide, 1X salts, 10 m*M* DTT.
4. NTE: 500 m*M* NaCl, 10 m*M* Tris-HCl, pH 8.0, 1 m*M* EDTA, 10 m*M* DTT.
5. RNase A: 10 μg/μL in NTE.
6. Stock 20X SSC (*see* Section 2.4.).
7. Coverslips treated with dimethyldichlorosilane (Sigmacote SL-2 from Sigma). Prepare in a fume cupboard.

2.6. Autoradiography and Development

1. Dipping emulsion: Kodak NTB2 or Ilford K5 emulsions in gel form which melt at 45°C. Avoid creating bubbles in the emulsion as they will increase

the background on the slides. All steps should be performed in complete darkness or with maximum safety light (e.g., using Ilford 904 filter).
2. Kodak D19 developer.
3. Rapid fixer.
4. Suitable counterstains.
5. Ethanol and xylene solutions required for dehydration can be reused from Section 2.4.

2.7. Nonradioactive Localization

1. Antidigoxigenin labeled with alkaline phosphatase (Boehringer Mannheim) diluted in 100 m*M* Tris-HCl, 100 m*M* NaCl, pH 7.5.
2. Stock 5X Tris buffered saline (TBS): 100 m*M* Tris-HCl, pH 7.5, 2.4*M* NaCl.
3. TTBS, Tween–20, 0.5% in TBS.
4. Carbonate buffer: 0.1*M* $NaHCO_3$, 1.0 m*M* $MgCl_2$, pH 9.8.
5. Nitro blue tetrazolium (NBT) (Sigma N-6876) color development solution. Stock 50 mg/mL in dimethylformamide (DMF). Store at 4°C.
6. 5-Bromo-4-chloro-3-indolyl phosphate (BCIP) (Sigma B-8503) color development solution. Stock 25 mg/mL in 70% DMF. Store at –20°C. *Caution:* NBT, BCIP, and DMF are all toxic compounds.

3. Methods

3.1. Slide Preparation

Slides are precoated to improve adhesion of the sections.

1. Glass slides should be clean. Washing them individually with detergent followed by several rinses in distilled water is sufficient.
2. Bake at 180°C for 2 h.
3. Soak in polylysine solution for 30 min at room temperature and then air dry in a dust-free place. Some slide coatings may interfere with the results.

3.2. Fixation and Embedding of Tissue

This must be optimized for different tissues. A sufficient volume of solution should always be used to allow efficient substitution within the specimens.

1. Plant tissue should be selected and cut to 2–3 mm cubes either under, or immediately immersed in, the fixative at room temperature.
2. Fix for 1 h. Overfixation (>4–5 h, or overnight) may cause loss of target signal.
3. Dehydrate the tissues in ethanol solutions (*see* Section 2.2.), and then through tri-butyl alcohol (TBA) in ethanol (50, 70, 85, 100% twice). Leave for at least 1 h in each solution at room temperature.

4. Add an equal volume of molten wax to the final TBA, and infiltrate overnight at the melting temperature of the wax. The wax should not be overheated, as this will reduce its sectioning properties.
5. Replace the wax-TBA with fresh molten wax. Replace after 3 h and repeat. Leave again in fresh wax overnight, then replace the wax with fresh wax for a further 2 h.
6. Pour molten wax into embedding molds. Put specimen into mold gently. Orient the specimen and allow the block to cool. An automatic wax handling system is worth the expense, particularly when handling large numbers of samples.

3.3. Sectioning

1. Put treated slides onto a heating block at 40–45°C and coat with sterile DEPC-treated water.
2. Trim the block to a suitable size and section tissue (5–10 μm).
3. Collect the sections as ribbons and float shiny side down on the slide.
4. Allow the sections to expand and then remove the water using clean paper tissue. Take care not to damage the sections. Leave the sections to dry overnight at 42°C. Protect from dust.
5. Sections may be stored in a desiccator (4°C is best). They can be stored for at least 1 mo, but may tend to fall off the slides during treatment if left for too long.

3.4. Preparation of Riboprobes

The methods for cloning, plasmid isolation, and restriction enzyme digestion are covered in detail elsewhere *(36)* and will not be discussed here.

3.4.1. [^{35}S]Labeled Riboprobes

The DNA fragments should be cloned into an appropriate vector. After linearizing the template DNA it should be resuspended in RNase-free buffer or water. Add 10 m*M* DTT to all solutions containing [^{35}S] and after any step that may inactivate DTT.

1. In two tubes, one for each of the two enzymes, T3 and T7 RNA polymerases, mix together 5.0 μL transcription buffer, 1.0 μL DTT, 1.0 μL RNase inhibitor, 1.0 μL of each 10 m*M* nonlabeled nucleotide (ATP, GTP, UTP), 1.0 μL CTP (100 μ*M*), 1.0 μL linearized plasmid template (equivalent to 1–5 μg), 5.0 μL [^{35}Sa]CTP (50 μCi), 0.5 μL enzyme (T3 or T7), and 9.5 μL water (to a total 20 μL).
2. Incubate at 37°C for 1 h. The yield of transcript can be estimated by comparing the relative staining of DNA and RNA bands on RNase-free agarose gels.

3. To remove the DNA template add the following: 1.0 μL RNase inhibitor, 0.5 μL DTT, 5.0 μL DNase buffer, 1.0 μL RNase-free DNaseI and DEPC water to 50 μL.
4. Incubate at 37°C for 15 min.
5. Use 1 μL of the incubation mixture to quantify the incorporation (*see* Note 1).
6. Phenol/chloroform extract.
7. Precipitate by addition of 25 μL of 7.5*M* ammonium acetate, 1 μL of tRNA and 2–3 vol of ethanol at –70°C for 1 h or at –20°C overnight.
8. Microfuge for 20 min. Wash with 70% ethanol and dry.
9. If the probe needs to be hydrolyzed:
   a. Resuspend the pellet in 0.1*M* sodium carbonate buffer, pH 10.2, 10 m*M* DTT, and hydrolyze at 60°C *(6)*.
   The time required for hydrolysis is calculated using the equation:

$$t = (L_o - L_f)/(kL_oL_f) \quad (1)$$

   Where t = time in minutes, Lo = original length in kb, Lf = final (desired) length in kb, and k = rate constant, approximately 0.11 cuts /kb/min.
   b. Add 1/10 volume of 5% acetic acid, 1/10 volume of 3*M* sodium acetate, pH 4.5 and 2.5 vol of ethanol. Precipitate as in steps 7 and 8.
10. Resuspend pellet in 50% formamide, 10 m*M* DTT at 5X concentration of probe (1X = 0.2–0.3 μg/mL/kb). Estimate activity by counting the sample in a liquid scintillation counter.

3.4.2. Nonradioactive Riboprobes

The procedure outlined here follows the method supplied by Boehringer Mannheim:

1. To a microfuge tube on ice add 1 μL linearized template DNA, 2 μL NTP mix, 2 μL transcription buffer (10X), 40 U enzyme, and DEPC-treated water to 20 μL final volume.
2. Incubate at 37°C for 2 h. Stop the reaction by addition of 2 μL of EDTA and heat to 65°C.
3. Remove the DNA template by DNase treatment as in Section 3.4.1., steps 3–6.
4. Precipitate with 2 μL of 4*M* LiCl and 75 μL of cold ethanol (100%) at –70°C for 30 min or 2 h at –20°C. Microfuge and wash pellet with 70% ethanol.
5. Dry pellet under vacuum and resuspend at 37°C for 30 min in 50 μL sterile DEPC-treated water with 1 μL of RNase inhibitor.

3.5. Section Pretreatment

The sections are treated prior to hybridization in order to increase the accessibility of the nucleic acid to the probe and to reduce nonspecific binding. The pretreament should be carried out at room temperature, except where specified, just prior to hybridization.

1. Deparaffinize sections in xylene or Histoclear (nontoxic) for 2 × 5 min and rehydrate through an ethanol series.
2. Incubate in 0.2*M* HCl for 10 min to denature the proteins and nick the DNA.
3. Rinse in H_2O for 1 min, then incubate in 2X SSC for 5 min. At higher temperatures this is supposed to denature the RNA. Here it is reduced to a rinse to neutralize the acid pH from the previous step.
4. Rinse in PBS for 5 min and treat with proteinase K by incubating at 37°C for 30 min in prewarmed buffer. This improves the probe penetration when crosslinking fixatives, e.g., formaldehyde or glutaraldehyde, are used. It is unnecessary if precipitative fixatives or small oligonucleotide probes are used. Parameters that can be varied are the proteinase concentration or time in order to obtain good hybridization signals without loss of morphology.
5. Stop digestion with 2 mg/mL of glycine in PBS for 2 min, followed by a brief rinse in PBS.
6. Refix tissue for 20 min in freshly prepared 4% formaldehyde. This prevents possible target diffusion.
7. Rinse well in PBS for 2 × 5 min to block the action of the fixative.
8. Place slides in a dish containing triethanolamine buffer (stirring vigorously). Add 500 µL of glacial acetic acid and stir for 10 min. This acetylation of amino groups reduces nonspecific electrostatic binding of the probe.
9. Rinse in H_2O for 1–2min. Dry the slides at 37°C to prevent dilution of the probe.

3.6. Hybridization

3.6.1. Radiolabeled Probe

1. Mix probe with hybridization buffer and add diluted probe to the sections. RNA probes are recommended at about 0.5 ng/µL or 2-5 × 10^5 cpm/µL.
2. Cover slides with silanized cover slips and hybridize at 50–55°C for 12–18 h in a humid container.
3. Place slides in a rack in wash buffer to allow the coverslips to fall off (about 30 min) and then incubate in fresh wash buffer at 50°C for 2–3 h. Do not allow the slides to dry out during the washes.
4. Remove nonhybridized probe by incubating the slides at 37°C for 30 min in RNase A (10 µg/mL) in NTE buffer.

5. Wash in NTE four times for 15 min, then transfer into fresh wash buffer and incubate at 50°C for 4 h. Alternatively, the slides can be left in wash buffer overnight.
6. Dehydrate slides through an ethanol series and leave to air-dry in a dust-free location.

3.6.2. Digoxigenin Labeled Probe

1. Dilute the probe in hybridization buffer to give a concentration of 5–10 ng/μL.
2. Hybridize as described for the radiolabeled probe in Section 3.6.1., step 2, then remove the coverslips by immersing in 2X SSC (about 30 min).
3. Wash the slides for 2 × 15 min in 2X SSC at 42°C, and for 2 × 15 min in 0.2X SSC at 37°C, then leave to dry in a dust-free location.

3.7. Visualization

3.7.1. Autoradiography

The best resolution is obtained by close contact of the specimen and the emulsion, which is achieved by dipping the slides into the emulsion and allowing them to dry to form a thin layer. X-ray film may be used for an assessment of the signal. Work in maximum safelight conditions and make sure that you can locate the back and front faces of the slides using the frosted edge.

1. Melt the emulsion at 45°C and dilute with an equal volume of distilled water. Agitate the liquid as little as possible and stand for at least 15 min to allow bubbles to settle.
2. Dip slides into the emulsion, then drain excess emulsion briefly and wipe the back of the slide. Dip extra slides for control development.
3. Put slide flat onto an ice pack to allow the gel to cool, then transfer to a drying rack and allow to dry in the dark for 2–4 h.
4. Store slides at 4°C in a light-tight box with desiccant to allow formation of the latent image.

3.7.2. Developing

After 3–5 d, depending on the probe and the target concentration, develop 1 or 2 slides. The slides can be left for 1 mo before developing, but longer periods will only increase the background. The image is developed photographically and will produce silver grains. Developing replicate slides allows the assessment of the length of time needed before developing all of the samples.

1. Cool developing solutions to 10–15°C and develop slides for 5–10 min.
2. Rinse in water for 1 min.

3. Fix for 10 min.
4. Wash in water for 10 min (minimum).
5. Allow to air dry.

3.7.3. Development of Digoxigenin Labeling

1. Immerse the slides in TBS for 2 × 5 min.
2. Transfer to alkaline phosphatase labeled anti-digoxigenin (Boehringer Mannhein) diluted 1:300 and incubate for 1–2 h at room temperature then wash the slides for 2 × 5 min in TBS. Longer incubations (12–18 h) can be left at 4°C.
3. Prepare color development solution, immediately before use, using 60 µL of BCIP plus 60 µL of NBT in 10 mL of carbonate buffer.
4. Immerse slides in the color solution and leave for 1–4 h for color development. To stop the reaction, wash with water for 3 × 5 min.

3.7.4. Staining

After development, the sections can be stained to reveal their tissue morphology. There is a wide choice of stains available and it is necessary to refer to specialist texts. Choose one that does not interfere with the interpretation of the signal. Useful stains include hematoxylin, toluidene blue, and fast green.

3.7.5. Microscopy

Under the light microscope the signal from radioactive hybridization will appear as dark grains under bright-field illumination. Viewed under dark-field conditions it will appear as white grains, but the tissue structure is barely visible. Sections hybridized with a nonradioactive probe are viewed under bright-field illumination.

4. Notes

1. To quantify the incorporation of the [^{35}S]ribonucleotide, spot and dry duplicate volumes (1 µL) of probe onto Whatman (Maidstone, UK) DE81 paper. Wash one of the duplicate filters in 0.5*M* phosphate buffer (3 × 5 min) and 90% ethanol (3 × 1 min) before counting. The difference in the dpm will give the percentage incorporated.
2. All of the controls should be used for the first experiment, then they can be reduced as the experimenter becomes more experienced. Controls should include:
   a. A tissue in which the gene of interest should be expressed at high levels (e.g., as measured by a northern assay).
   b. RNase treatment of the sections prior to hybridization to determine binding to nonRNA components.

   c. Using the sense probe to determine background labeling.
   d. Variation of the hybridization times to optimize the signal-to-noise ratio.
   e. Processing at least duplicate slides to determine exposure times for autoradiography.
3. Sample fixation is a crucial step and the fixative should be optimized for the tissue of interest. Over- or under-fixation may lead to poor morphology, artifacts, and possibly loss of signal. Overfixation will also increase background noise.
4. Failure to obtain sections satisfactorily may be owing to poor dehydration or wax impregnation. Sectioning improves with experience.
5. When using riboprobes, all RNase contamination must be eliminated using DEPC where appropriate and always wearing gloves.
6. When applying the dipping emulsion, the slide should not be overloaded. Care must be taken not to expose the emulsion to light. Storing the emulsion near solvents or [^{32}P] sources may increase background. Bubbles give background problems and also any bubbles in the dipping solution will prevent the slides from being smoothly coated with the emulsion.
7. Counterstaining is needed to identify the cellular structure. It is best to check the most appropriate stain and length of time for staining on a few trial sections.

Acknowledgments

The author would like to thank P. Puigdomenech and L. Ruiz-Avila of the Consejo Superior de Investigaciones Cientificas, CID, in Barcelona for introducing me to *in situ* hybridization during my year in Puigdomenech's laboratory.

References

1. Gall, J. G. and Pardue, M. L. (1969) Formation and detection of RNA-DNA hybrid molecules in cytological preparations. *Proc. Natl. Acad. Sci. USA* **63,** 378–383.
2. Pardue, M. L. and Gall, J. G. (1969) Molecular hybridisation of radioactive DNA to the DNA of cytological preparations. *Proc. Natl. Acad. Sci. USA* **64,** 600–604.
3. Brahic, M. and Haase, A. T. (1978) Detection of viral sequences of low reiteration frequency by *in situ* hybridisation. *Proc. Natl. Acad. Sci. USA* **75,** 6125–6129.
4. Wells, M. (1990) The demonstration of viral DNA in human tissues by *in situ* DNA hybridisation, in *In situ* Hybridisation: Application to Developmental Biology and Medicine (Harris, N. and Wilkinson, D. G., eds.), Cambridge University Press, pp. 271–272.
5. Angerer, L. M., Cox, K. H., and Angerer, R. C. (1987) Demonstration of tissue-specific gene expression by *in situ* hybridisation. *Methods Enzymol.* **152,** 649–661.
6. Barker, S. J., Harada, J. J., and Goldberg, R. B. (1988) Cellular localization of soybean storage protein mRNA in transformed tobacco seeds. *Proc. Natl. Acad. Sci. USA* **85,** 458–462.

7. Cox, K. H., DeLeon, D. V., Angerer, L. M., and Angerer, R. C. (1984) Detection of mRNAs in sea urchin embryos, in *in situ* hybridisation using asymetric probes. *Dev. Biol.* **101,** 485–502.
8. Langdale, J. A., Rothermel, B. A., and Nelson, T. (1988) Cellular pattern of photosynthetic gene expression in developing maize leaves. *Genes Dev.* **2,** 106–115.
9. Ruiz-Avila, L., Burgess, S. R., Stiefel, V., Dolors-Ludevid, M., and Puigdomenech, P. (1992) Accumulation of cell wall hydroxyproline-rich glycoprotein mRNA is an early event in maize embryo cell differentiation. *Proc. Natl. Acad. Sci. USA* **89,** 2414–2418.
10. McFadden, G. I. (1989) *In situ* hybridisation in plants: from macroscopic to ultrastructural resolution. *Cell Biol. Int. Repts.* **13,** 3–21.
11. Binder, M., Tourmente, S., Roth, J., Renaud, M., and Gehring,W. J. (1986) *In situ* hybridisation at the electron microscope level: localization of transcripts on ultrathin sections of Lowicryl K4M-embedded tissue using biotinylated probes and protein A-gold complexes. *J. Cell Biol.* **102,** 1646–1653.
12. Harris, N., Grindley, H., Mulchrone, J., and Croy, R. R. D. (1989) Correlated *in situ* hybridisation and immunocytochemical studies of legumin storage protein deposition in pea (*Pisum sativum* L.). *Cell Biol. Int. Repts.* **13,** 23–35.
13. VanDer Loss, C. M., Volkers, H. H., Rook, R., VanDer Berg, F. M., and Houthoff, H. J. (1989) Simultaneous application of *in situ* hybridisation and immunohistochemistry on one tissue section. *Histochem. J.* **21,** 279–284.
14. Wetmur, J. G. and Davidson, N. (1968) Kinetics of renaturation of DNA. *J. Mol. Biol.* **31,** 349–370.
15. Berger, C. N. (1986) *In situ* hybridisation of immunoglobulin-specific RNA in single cells of the B lymphocyte lineage with radiolabeled DNA probes. *EMBO J.* **5,** 85–93.
16. Lewis, M. E., Sherman, T. G., and Watson, S. J. (1985) *In situ* hybridisation histochemistry with synthetic oligonucleotides: Strategies and methods. *Peptides* **6 (Suppl. 2),** 75–87.
17. Taneja, K. and Singer, R. H. (1987) Use of oligodeoxynucleotide probes for quantitative *in situ* hybridisation to actin mRNA. *Anal. Biochem.* **166,** 389–398.
18. Witkiewicz, H, Bolander, M. E., and Edwards, D. R. (1993) Improved design of riboprobes from pBluescript and related vectors for *in situ* hybridisation. *Biotechniques* **14,** 458–463.
19. Moench, T. R., Gendelman, H. F., Clements, J. E., Narayan, O., and Griffin, D. E. (1985) Efficiency of *in situ* hybridisation technique as a function of probe size and fixation technique. *J. Virol. Methods* **11,** 119–130.
20. Smith, L. M., Handley, J., Li, Y., Martin, H., Donovan, L., and Bowles, D. J. (1992) Temporal and spatial regulation of a novel gene in barley embryos. *Plant Mol. Biol.* **20,** 255–266.
21. Casey, J. and Davidson, N. (1977) Rates of formation and thermal stabilities of RNA:DNA and DNA:DNA duplexes at high concentration of formamide. *Nucleic Acids Res.* **4,** 1539–1552.
22. Hemmati-Brivanlou, A., Frank, D., Bolce, M. E., Brown, B. D., Sive, H. L., and Harland, R. M. (1990) Localisation of specific mRNAs in *Xenopus* embryos by whole mount *in situ* hybridisation. *Development* **110,** 325–330.

23. Martin, R, Hoover, C, Grimme, S., Grogan, C., Holtke, J., and Kessler, C. (1990) A highly sensitive, nonradioactive DNA labeling and detection system. *Biotechniques* **9,** 762–768.
24. Leitch, I. J., Leitch, A. R., and Heslop-Harrison, J. S. (1991) Physical mapping of plant DNA sequences by simultaneous *in situ* hybridisation of two differently labeled fluorescent probes. *Genome* **34,** 329–333.
25. Mullink, H., Walboomers, J. M. M., Raap, A. K., and Meyer, C. J. L. M. (1989) Two colour DNA *in situ* hybridisation for the detection of two viral genomes using nonradioactive probes. *Histochemistry* **91,** 195–198.
26. Singer, R. H., Lawrence, J. B., and Villnave, C. (1986) Optimisation of *in situ* hybridisation using isotopic and nonisotopic detection methods. *Biotechniques* **4,** 230–250.
27. Glauert, A. M. (1975) *The Fixation and Embedding of Biological Specimens,* North-Holland Publishers, Amsterdam.
28. McFadden, G. I., Bonig, I., Cornish, E. C., and Clarke, A. E. (1988) A simple fixation and embedding method for use in hybridisation histochemistry of plant tissues. *Histochem. J.* **20,** 575–586.
29. Von Bertalanffy, L. and Bickis, I. (1956) Identification of cytoplasmic basophilia (ribonucleic acid) by fluorescence microscopy. *J. Histochem. Cytochem.* **4,** 481–493.
30. Harris, J. R. (1991) Electron Microscopy in Biology—A Practical Approach. Oxford University Press, Oxford.
31. *Royal Microscopical Society Handbooks Series.* Oxford University Press, Royal Microscopical Society.
32. Coghlan, J. P., Aldred, P., Haralambidis, J., Niall, H. D., Penschow, J. D., and Tregear, G. W. (1985) Hybridisation histochemistry. *Anal. Biochem.* **149,** 1–28.
33. Gee, C. E. and Roberts, J. L. (1983) *In situ* hybridisation histochemistry: a technique for the study of gene expression in single cells. DNA **2,** 155–161.
34. Meyerowitz, E. M. (1987) *In situ* hybridisation to RNA in plant tissue. *Plant Mol. Biol. Reptr.* **5,** 242–250.
35. Tecott, L. H., Eberwine, J. H., Barchas, J. D., and Valentino, K. L. (1987) Methodological considerations in the utilization of *in situ* hybridisation, in *In Situ Hybridisation* (Valentino, K., Eberwine, J., and Barchas, J. D., eds.), Oxford University Press, pp. 3–24.
36. Sambrook, J., Fritsch, E. F., and Maniatis, T. (1989) *Molecular Cloning: A Laboratory Manual.* 2nd ed. Cold Spring Harbor Laboratory Press, Cold Spring Harbor, New York.

CHAPTER 26

Xenopus Oocytes as a Heterologous Expression System

Frederica L. Theodoulou and Anthony J. Miller

1. Introduction

Oocytes of the South African clawed toad, *Xenopus laevis*, have proved a versatile and powerful heterologous expression system for eukaryotic genes *(1)*. *Xenopus* oocytes were first used to express plant genetic material in 1979 *(2)*, and in the past decade it has become increasingly evident that they are an appropriate and useful system to study plant proteins. As several comprehensive reviews of expression in oocytes have been published recently *(1,3–6)*, this chapter will concentrate specifically on those issues which are particularly relevant to the expression of plant genes with particular emphasis on the expression of plant transport proteins.

The ovaries of the adult female *Xenopus* contain thousands of oocytes at all stages of development, which, in the absence of hormonal stimulation, fail to pass into the oviduct and can be readily removed by surgery. Fully mature stage VI oocytes *(7)* are usually used for expression; these are large cells, around 1.2 mm in diameter, with a yellow vegetal pole, and a darker animal pole, under which the nucleus is displaced. The oocyte is surrounded by a glycoprotein matrix, known as the vitelline membrane, and a layer of follicle cells, which are electrically coupled to the oocyte via gap junctions.

During oogenesis, oocytes accumulate large quantities of enzymes and organelles that form a reserve for early embryonic development; these

From: *Methods in Molecular Biology, Vol. 49: Plant Gene Transfer and Expression Protocols*
Edited by: H. Jones Humana Press Inc., Totowa, NJ

include histones, nucleoplasmin, RNA polymerases, tRNAs, and ribosomes, which may be used to transcribe and translate exogenous DNA and RNA, respectively. Foreign genetic material is introduced by microinjection. This may be mRNA extracted from tissue, synthetic RNA prepared by in vitro transcription of a cDNA template, or a cDNA. RNA is microinjected into the vegetal pole of the oocyte, and cDNA may be microinjected into the nucleus. The latter injections are demanding because it is difficult to locate the nucleus but good technical guides have been written (e.g., *3*). Methods that do not require nuclear injection of cDNA have also been described *(8)*.

DNAs injected into the oocyte nucleus are assembled into chromatin and transcribed, and unspliced RNAs are accurately processed to produce mature mRNAs. Injected RNAs are translated by the oocyte, and posttranslational modifications such as precursor processing, phosphorylation, and glycosylation are carried out by endogenous enzymes. Membrane proteins derived from animal RNAs have been shown to be targeted appropriately, and secretory proteins are exported from the cell. If mixtures of RNAs specifying protein subunits are injected, functional multisubunit complexes are assembled by the oocyte. Indeed, oocytes have been used as a model system to study posttranslational modification, sorting, targeting, and assembly of many proteins *(9,10)*. Protocols for the fractionation of oocytes and subsequent detection of protein products have been developed for this purpose. These protocols may also be used to verify expression.

Several plant proteins have been expressed in oocytes (*see* Table 1) and, when appropriate, proteins have been successfully glycosylated and secreted *(12,16)*. A number of plant plasma membrane proteins have also been successfully expressed in the oocyte plasma membrane, and appear to retain their native properties. Boorer et al. (1992) demonstrated that the *Arabidopsis* H^+/hexose transporter, STP1, has similar kinetic properties to those of the same protein expressed in yeast, and similar results were obtained for the *Chlorella* homolog, HUP1 *(23,24)*. However, the question of how plant membrane proteins are targeted in *Xenopus* oocytes is still open: In a recent report, tonoplast intrinsic protein (TIP) was targeted to the oocyte plasma membrane *(25)*, raising the possibility that oocytes may be able to express other tonoplast proteins. Why TIP is targeted to the plasma membrane is not clear; it is possible that TIP lacks the signals for retention on oocyte intracellular membranes and moves to

Table 1
Some Examples of Higher Plant Proteins
That Have Been Successfully Expressed in *Xenopus* Oocytes

Protein	Genetic material	Reference
Soluble proteins		
Maize storage proteins	polyA$^+$ RNA	*2,11,10*
Phaseolus vulgaris phaseolin	mRNA, cRNA,	*12*
	RNA from polysomes	*13*
Soybean glycinins	polyA$^+$ RNA	*14*
Wheat α-amylase	RNA	*15*
Vicia faba storage globulins	polyA$^+$ RNA	*16*
Barley α-amylase	RNA, cDNA, genomic DNA	*17*
Wheat storage proteins	cRNA	*18*
	cRNA	*9*
Tomato ethylene forming enzyme	polyA$^+$ RNA, cRNA	*19*
Membrane proteins		
K^+ channels	cRNA, polyA$^+$ RNA	*20,21*
Nitrate transporter	cRNA	*22*
H^+/hexose cotransporters	cRNA	*23,24*
Tonoplast intrinsic protein (TIP)	cRNA	*25*

the plasma membrane by default. Several sequences responsible for organelle targeting and membrane retention of plant proteins have now been identified *(26)*, whether these are operational in amphibian cells remains to be investigated in detail. Roitsch and Lehle (1991) demonstrated that a yeast vacuolar glycoprotein was delivered to the lysosome of oocytes (the functional equivalent of the vacuole), indicating that a fungal targeting signal may be functional in oocytes *(27)*. However, targeting of plant vacuolar lumen proteins is distinct from that in yeast cells, and little is known about the targeting of vacuolar membrane proteins *(26)*.

These results indicate that the oocyte has great potential as a heterologous expression system, with several main applications:

1. To clone genes. RNA fractions encoding several transporters and receptors have been identified by their functional expression *(1)*. Size-fractionated RNAs that show the desired function are generally used to generate a cDNA library by reverse transcription and packaging in plasmids. RNAs are generated by in vitro transcription, and the cDNA population then subdivided until a single clone is obtained. The advantage of this approach is that a gene is identified by its function, and not by sequence homology or antigenic reactivity, which require further proof of function.

2. To identify the function of genes that have already been cloned. The cDNA is injected into oocytes or, more usually, is inserted in an expression vector containing one or more bacteriophage RNA polymerase promoters, from which RNA may be obtained by in vitro transcription. Activity may then be assayed in oocytes injected with synthetic RNAs, e.g., ethylene forming enzyme *(19)*, TIP *(25)*. The identification can be further refined by using an antisense approach; a gene of known function is expressed in oocytes, and its function is removed by injection of antisense oligonucleotides complementary to a gene of unknown function, e.g., the dihydropyridine-sensitive calcium channel *(28)*.
3. To characterize in detail the protein products of cloned genes, including kinetic analysis of transporters *(29)*, and the effects of specific mutations in structure/function studies. The effects of various second messenger systems can be investigated by coexpression or microinjection of signal transduction pathway components. Also, the oocyte may be used as a common environment to compare properties of homologous proteins from different tissues, or structure/function relationships can be investigated by expression of hybrid proteins, containing subunits from different species.
4. To study posttranslational processing and assembly of proteins.

Other heterologous expression systems are available, and may be used for some of the applications listed here, and each has particular advantages. The baculovirus system will yield large quantities of protein, whereas yeast will provide stable expression which, with the appropriate promoter, can be induced by a particular substrate, but for the study of membrane proteins, particularly by electrophysiology, the oocyte is unsurpassed.

A major advantage of using oocytes over other heterologous expression systems is that it simplifies the handling of single identified cells after transfer of genetic information. For example, a single oocyte can be handled using a Pasteur pipet and is conveniently incubated in the well of a microtiter plate. Another advantage is that a high proportion (>98%) of cells express the genetic information after injection. It is also easy to control the environment of the oocytes by bath perfusion, and in studies of transporters, to control the oocyte membrane potential by voltage clamp *(29)*; the large size of oocytes makes them an easy target for microelectrode impalement. Furthermore, heterologous expression of gradient-coupled transporters in oocytes offers a unique opportunity for kinetic characterization, because parallel electrophysiological and radiometric assays can be performed *(29)*.

The chief disadvantages of oocytes are the seasonal variation in quality and alteration in the ability of individual cells to express a foreign message. For this reason, it is useful to test the protein synthetic ability of each batch of oocytes by injecting a number of the oocytes with a test message, which is known to be translated in oocytes. Further problems are the short transient expression period, which does not last longer than about 2 wk, and the relatively small number of cells that can be handled in a single experiment.

2. Materials

2.1. Isolation and Maintenance of Oocytes

1. Modified barth's saline (MBS): High salt stock: 128 g NaCl, 2 g KCl, 5 g $NaHCO_3$, 89 g HEPES. Dissolve in distilled water, and adjust to pH 7.6 with 1*M* NaOH. Make up to 1 L, and store in 40-mL aliquots at –20°C.

   Divalent cation stock: 1.9 g $Ca(NO_3)_2 \cdot 4H_20$, 2.25 g $CaCl_2 \cdot 6H_2O$, 5 g $MgSO_4 \cdot 7H_20$. Dissolve sequentially in 1 L distilled water. Store in 40-mL aliquots at –20°C. Antibiotic stock: 10 mg/mL sodium penicillin, 10 mg/mL streptomycin sulfate. Store in 1-mL aliquots at –20°C. To prepare MBS, mix 919 mL of distilled water, 40 mL high salt stock, and 40 mL divalent cation stock. Check pH, and adjust to 7.6, if necessary. Sterilize by autoclaving, and add 1 mL antibiotic stock, when cool. Final composition: 88 m*M* NaCl, 1 m*M* KCl, 2.4 m*M* $NaHCO_3$, 15 m*M* HEPES-NaOH, pH 7.6, 0.6 m*M* $Ca(NO_3)_2$, 0.41 m*M* $CaCl_2$, 0.82 m*M* $MgSO_4$, 10 μg/mL sodium penicillin, 10 μg/mL streptomycin sulfate.
2. 0.1% (w/v) ethyl m-aminobenzoate (Tricane, M5222).
3. Collagenase (Sigma [St. Louis, MO] type 1A).
4. Stripping solution: 200 m*M* potassium aspartate, 20 m*M* KCl, 1 m*M* $MgCl_2$, 10 m*M* EGTA, 10 m*M* HEPES, pH 7.4. (Osmolarity: 475 mOsm-hypertonic to MBS).

2.2. RNA Purification Using Magnetic Beads

NB. All reagents should be RNase-free (*see* Note 6).

1. Diethyl pyrocarbonate (DEPC)-treated water.
2. Binding buffer: 20 m*M* Tris-HCl, pH 7.5, 1*M* LiCl, 2 m*M* EDTA.
3. Washing buffer: 10 m*M* Tris-HCl, pH 7.5, 0.15*M* LiCl, 1 m*M* EDTA.
4. Elution buffer: 2 m*M* EDTA.
5. 3*M* sodium acetate.
6. Absolute ethanol. Cooled to –20°C.
7. 70% (v/v) ethanol (made with DEPC-treated water).

2.3. Size-Fractionation of RNA

1. Sucrose, 5 and 25% (w/v). Dissolve in sterile water and treat with 0.1% (v/v) DEPC overnight. Heat at 100°C for 15 min to destroy the DEPC. Cool to room temperature.
2. 1*M* Tris-HCl, pH 7.4. Make up in DEPC-treated water in RNase-free vessels.
3. 0.5*M* EDTA, pH 8. Add 18.61 g EDTA to a small volume of water. Add 10*M* NaOH, with vigorous stirring, until the pH approaches 8 and the salt begins to dissolve. Dilute to about 90 mL, adjust pH to 8, and make up to 100 mL. Sterilize by autoclaving, and treat with DEPC.
4. 20% (w/v) SDS filter sterilized.

2.4. Preparation of Template and In Vitro Transcription

Store reagents at –20°C, unless otherwise stated.

1. Proteinase K buffer: 500 m*M* NaCl, 50 m*M* EDTA, pH 8.0, 100 m*M* Tris-HCl (pH 8.0).
2. Proteinase K, 20 mg/mL.
3. 5% (w/v) SDS filter sterilized. Store at room temperature.
4. 10X transcription buffer: 400 m*M* Tris-HCl (pH 7.5 at 37°C), 60 m*M* $MgCl_2$, 20 m*M* spermidine HCl, 50 m*M* NaCl.
5. RNase-free DNase: Dissolve 10 mg of pancreatic DNase I in 10 mL of 0.1*M* iodoacetic acid, 0.15*M* sodium acetate, pH 5.2. Heat the solution to 55°C for 45 min, cool to 0°C, and add 1*M* $CaCl_2$ to a final concentration of 5 m*M*.
6. Dithiothreitol (DTT).
7. Bovine serum albumin (BSA).
8. 10X rNTPs: 5 m*M* each of rATP, rCTP, and rUTP; 1–5 m*M* rGTP.
9. Cap analog: P^1-5^1-(7-methyl)guanosine-P^3-5^1-guanosine triphosphate. Store at –80°C.
10. Placental RNase inhibitor.
11. Phenol, equilibrated with 0.1*M* Tris-HCl, pH 7.6. Store at 4°C.
12. Chloroform. Store at room temperature.
13. SP6, T7, or T3 RNA polymerase.

2.5. Microinjection

Tributylchlorosilane (TBCS).

2.6. Radiolabeling Oocyte Proteins

1. [^{35}S]methionine, [^{3}H]leucine, or other radiolabeled amino acid.
2. Homogenization buffer: 0.1*M* NaCl, 1% (w/v) Triton X-100, 1 m*M* phenylmethylsulfonyl fluoride (PMSF), 20 m*M* Tris-HCl, pH 7.6. *Caution*: PMSF is highly toxic.

3. 2X SDS-PAGE buffer: 0.125*M* Tris-HCl, pH 6.8, 4% SDS, 20% sucrose, or glycerol, 0.004% bromophenol blue, 10 m*M* DTT.
4. Peroxide solution: 5% (v/v) H_2O_2, 1*M* NaOH, 1 mg/mL unlabeled amino acid (same amino acid as used in Section 1.).
5. 25% (w/v) trichloroacetic acid (TCA).
6. 8% (w/v) TCA.
7. Whatman GF/C glass fiber filters.
8. Suitable scintillant, e.g., BDH Scintran cocktail T.
9. Gentamycin stock: 50 mg/mL. Add to MBS to give 50 μg/L.

2.7. Immunoprecipitation of Oocyte Proteins

1. Immunoprecipitation buffer: 0.1*M* KCl, 5 m*M* $MgCl_2$, 1% Triton X-100, 0.5% SDS, 1% sodium deoxycholate, 1 m*M* PMSF, 0.1*M* Tris-HCl, pH 8.2.
2. Protein A sepharose, or fixed *Staphylococcus aureus.*
3. SDS-PAGE sample buffer.

3. Methods

3.1. Maintenance of Xenopus laevis

Xenopus may be reared in the laboratory, but for most purposes, it is more convenient to obtain groups of animals from a supplier. Blades Biological (Edenbridge, UK) and Nasco Inc. (Fort Atkinson, WI) are suppliers and orders should request large, sexually-mature females. The frogs can be kept in large tanks of tap water, from which the chlorine has been allowed to evaporate. Water should be no deeper than 15 cm, and kept between 15–20°C. A cover, or lid is required to prevent escape (*see* Note 1). Feed the frogs twice weekly with chopped liver, or commercially available Xenopus pellets (same source as frogs), and clean the tank after the frogs have been fed (*see* Note 2).

3.2. Removal of Ovary Tissue

Ovary tissue may be removed in two ways: First, the frog is sacrificed, and the entire ovary removed. However, this is wasteful, as rarely is it possible to use all of the available oocytes. In the second approach, an ovarian lobe is removed under anesthesia, and the frog may be used for further experiments. This has the advantage that a uniform batch of oocytes may be used for several experiments. In both cases, researchers should consult their local government regulations for guidelines on handling animals humanely; a government-issued license may be needed in some countries.

3.2.1. Removal of a Complete Ovary

1. Sterilize surgical instruments by autoclaving, and swab surfaces with ethanol.
2. Kill the frog by a method approved by relevant government regulations, for example, 0.3 mL euthetol (May and Baker) injected under the skin in the lower back. It may be anesthetized prior to handling by immersion in ice for several minutes, until it becomes inactive.
3. Place the frog on a tile or glass plate, ventral side uppermost, and make a small (ca. 1 cm) incision in the loose skin on the lower abdomen above the ovary on one side. Cut through the muscular body wall with a pair of sharp scissors or a scalpel.
4. The ovary should be revealed as a mass of yellow and brown cells that is divided into 16 lobes and fills the abdomen. Carefully pull the ovary through the incision, using a pair of forceps. Check that all the tissue has been removed; place the ovarian lobes in a Petri dish containing MBS.
5. Wash the ovary several times in MBS.

3.2.2. Removal of an Ovarian Lobe

1. Anesthetize the animal by immersion in 0.1% ethyl m-aminobenzoate (also called Tricane or MS222) for 15–30 min. *Caution*: The anesthetic is a potential carcinogen, and latex gloves should be worn throughout the procedure.
2. When the frog is anesthetized, place it, ventral side uppermost, on a damp tissue, and cover it with tissues soaked in anesthetic solution, leaving the lower abdomen exposed. This both prevents drying of the mucous membranes, and maintains anesthesia.
3. Make a 1-cm cut through the skin on one side of the lower abdomen, using a sterile scalpel. Neatly cut through the muscular body wall to reveal the ovary. Avoid the central vein.
4. Gently remove as many ovarian lobes as required, and place immediately in a Petri dish containing MBS.
5. Suture the body muscle using fine thread, and then repeat for the skin.
6. Allow the frog to recover for 24 h in shallow water with its head initially supported to prevent drowning.

3.3. Preparation of Oocytes

If relatively few oocytes are required, they may be manually removed, by careful teasing from the ovarian lobe using Pasteur pipets. It is helpful to "blunt" glass pipet tips in a bunsen flame, to reduce damage to the oocytes. More commonly, oocytes are released by collagenase digestion, which removes the follicle cells, but leaves the vitelline membrane intact. The vitelline membrane may be removed by osmotic shock, but since this renders the oocytes extremely fragile, this is only performed when a clean membrane is required (usually in transport studies using patch-clamp when a very high resistance seal between the membrane and electrode must be formed).

3.3.1. Collagenase Treatment of Oocytes

1. Place the whole ovary or individual ovarian lobes in a dish containing MBS, and tease into smaller clumps using two pairs of watchmaker's forceps. Wash the clumps several times in fresh MBS.
2. Incubate the oocyte clumps in MBS containing 1–2 mg/mL collagenase (Sigma [St. Louis, MO] type 1A), at 18–22°C, with gentle agitation, for 2 h (*see* Notes 4 and 5).
3. Wash the oocytes very thoroughly with several changes of MBS. Store in the dark a temperature-controlled incubator at 18°C.
4. Change the MBS solution daily.

3.3.2. Assessment of Oocyte Quality

1. Inspect the appearance of the oocytes. They should show a clear boundary between the animal and vegetal poles, and have no mottling. They should be firm when injected or impaled.
2. If desired, perform a test translation, using a message that is known to be expressed, and can easily be assayed (e.g., salivary amylase assay *[30]*), or if electrophysiology is to be performed, measure the membrane potential of a few oocytes. Minus 40 mV in pH 7.6 Barths saline is generally considered to be the minimum potential for useful recordings to be made.
3. At this stage it is useful to collect groups of undamaged, healthy-looking oocytes and place them in fresh dishes of MBS. Oocytes keep much better when stored in large volumes of medium (*see* Note 3).

3.3.3. Removal of Vitelline Envelope

1. Remove follicular tissue using watchmaker's forceps or by collagenase digestion (*see earlier*).
2. Place oocytes in "stripping" solution for 5–10 min at 22°C.
3. Peel vitelline envelopes from the oocyte surface with a pair of fine forceps. The oocytes are now extremely fragile, and should be handled with care. This procedure is best performed immediately before the oocyte is required.

3.4. Preparation of Heterologous Genetic Material

3.4.1. Isolation of mRNA

Protocols for the isolation of poly(A)$^+$ mRNA from plants can be found elsewhere in this volume. However, we have found purity of mRNA to be crucial to the survival of the oocyte, and routinely use magnetic separation of poly(A)$^+$ RNA *(31)* (*see* Notes 7 and 9).

3.4.2. Size Fractionation of RNA

1. Prepare two sucrose solutions (e.g., 5 and 25% w/v) using solutions described in Section 2.3.; to give final concentrations of 10 m*M* Tris-HCl, 1 m*M* EDTA, and 0.2% (w/v) SDS.
2. Use a gradient former and peristaltic pump to prepare a sucrose gradient in a sterile 20 mL centrifuge tube.
3. Resuspend poly(A)$^+$ RNA (e.g., 60–120 μg) in 100 μL of 1 m*M* EDTA, 0.5% (w/v) SDS, and heat to 65°C for 10 min. Quickly cool the RNA to room temperature, and apply to the top of the gradient.
4. Centrifuge the gradient at 21,000*g* for 15 h at 4°C.
5. Fractionate the gradient into 0.5-mL aliquots, and measure the absorbance at 260 nm in a spectrophotometer.
6. Add 0.1 vol 3*M* sodium acetate and 2.5 vol absolute ethanol to precipitate the RNA, and incubate at –70°C for 30 min.
7. Centrifuge the fractions at 15,000*g* for 15 min at 4°C, and wash the pellet in 70% (v/v) ethanol.
8. Dry the pellets under vacuum, and recover in sterile, DEPC-treated water.
9. Store at –80°C.
10. Aliquots from each fraction may be tested by translating in a rabbit reticulocyte lysate, if desired.

3.4.3. Preparation of Templates for In Vitro Transcription

1. This requires a cDNA for the gene of interest. This cDNA is subcloned in an expression vector containing one or more bacteriophage RNA polymerase promoters (*see* Note 10).
2. Linearize the template with a restriction enzyme that cuts downstream of the insert (*see* Note 11).
3. If the template has been prepared from minipreps that have been treated with RNase, perform a proteinase K treatment. To the digest, add 0.1 vol proteinase K buffer, 0.1 vol of 5% SDS and proteinase K to a final concentration of 100 μg/mL. Incubate the reaction at 37°C for 1 h.
4. Extract the reaction with phenol/chloroform. Add 0.1 vol 3*M* sodium acetate and 2.5 vol absolute ethanol. Incubate for 15 min on ice, or for 30 min at –20°C.
5. Centrifuge at top speed in a microfuge for 15 min, and wash the pellet in 70% ethanol.
6. Dry the pellet under vacuum, and recover the DNA in DEPC-treated water at a concentration of 0.5–1 μg/mL (*see* Note 12).

3.4.4. Preparation of Capped Synthetic RNA

1. Prepare the transcription mixture by adding the components to a sterile microcentrifuge tube at room temperature, in the following order (*see* Notes 13, 14, and 15):

Template DNA	1 μg
DEPC-treated water	to give final vol of 50 μL
100 m*M* DTT	5 μL
2.5 mg/mL BSA	2.5 μL
5 m*M* rATP, rUTP, rCTP	2.5 μL each
1 m*M* rGTP	2.5 μL
4 m*M* cap analog	2.5 μL
10X transcription buffer	50 μL
Placental RNase inhibitor	25 U
RNA polymerase	30–45 U

2. Centrifuge briefly to collect contents in the bottom of the tube, mix gently, and incubate for 1 h at 37°C.
3. Remove template by digestion with 1 μL RNase-free DNase (1 mg/mL) for 15 min at 37°C (*see* Note 16).
4. Add 1 μL 0.5*M* EDTA (nuclease-free).
5. Phenol/chloroform extract the reaction, and precipitate by adding 0.1 vol 3*M* sodium acetate and 2.5 vol of ethanol for 30 min at –20°C.
6. Centrifuge at 13,000*g* for 15 min and wash the pellet with 70% ethanol.
7. Dry the pellet under vacuum, and resuspend the RNA in DEPC-treated water at a concentration of 1 μg/mL. Store at –80°C.
8. Assess yield by electrophoresis in formaldehyde-containing gels, or by UV spectrophotometry (*see* Note 17).

3.5. Microinjection of Genetic Material into Oocytes

3.5.1. Injection Apparatus

1. Stereomicroscope: Any good dissecting microscope will suffice, but a lens with a good depth of field is desirable, as this permits simultaneous inspection of the oocyte surface and the meniscus within the micropipet with both in reasonable focus.
2. Micromanipulator: A relatively unsophisticated micromanipulator is suitable, but must be readily moved in three dimensions, as this facilitates injection of rows of oocytes. Some workers prefer a high sensitivity manipulator, as the greater precision is less likely to lead to damage of the oocytes. A cold light source is useful as this minimizes drying out of the oocytes during injection.
3. Injection system: This may be a simple manual apparatus, such as a syringe driven by a micrometer screw, or a microprocessor-controlled microinjector. The latter system dispenses predetermined volumes of RNA and is often equipped with a useful "clear" function to eject excess RNA solution when injection is complete.

4. Micropipet-pulling apparatus: A vertical puller equipped with a magnet and heater will give reproducible results. Use of a microforge to produce a sharp tip is unnecessary for the injection of oocytes.

3.5.2. Construction of Micropipets for Injections

1. Prepare micropipets from thin-walled glass capillaries (e.g., Drummond # 3-00–203-G), using a vertical pulling apparatus. A two-stage pull (i.e., under gravity and under a preset magnetic force) produces pipets with sharply-tapering points, which are suitable for injection. Conditions must be determined empirically for each combination of glass and puller.
2. Place the pipets in a drying oven at 180°C. Inject 100 µL of tributylchlorosilane (TBCS) through a port in the top of the oven, and leave to salinize for 30 min. Allow the fumes to disperse, before removing the pipets. *Caution:* TBCS is harmful, and this procedure should be carried out in a fume hood. Take other precautions as local safety rules require.
3. Sterilize the pipets by baking at 180°C overnight, and store in a sterile, dust-free container.
4. A method for calibrating micropipet volume may be found in ref. *4*.

3.5.3. Injection Procedure

1. Leave oocytes in MBS at 18°C for at least 2 h after collagenase treatment, so that damaged oocytes can be identified and discarded. A damaged oocyte can be identified by the leaking of cell contents into the surrounding solution.
2. Wipe down all instruments with ethanol.
3. Attach the micropipet to the injection device and examine the tip under the microscope. Using a pair of sterile forceps, break the tip to a diameter of ca. 10 µm, leaving a bevelled edge.
4. Thaw an aliquot of RNA, and place 1 µL on to a piece of sterile parafilm, placed in a small Petri dish lid.
5. Draw the RNA into the micropipet tip, and place the Petri dish with RNA on ice until needed later.
6. Cover a glass microscope slide with filter paper, and tape a second slide on top with a 2-mm offset from the edge of the other slide and leaving a strip of filter paper exposed. Wet the filter paper with MBS, and transfer 5–10 oocytes onto the filter paper such that they can be supported by the second slide, with their vegetal poles exposed. It is important to keep the oocytes moist at all times; oocyte injection can be done with the oocytes immersed in saline.
7. Adjust the focus of the microscope, such that both the oocyte surface and the RNA meniscus can be observed. Impale the oocyte with the pipet and

inject 10–50 nL RNA (*see* Note 18). As the oocyte is impaled, it will indent, and then "give" as the micropipet tip enters the cell. Slight swelling of the oocyte indicates that the RNA has been injected (*see* Note 19).
8. Gently remove the micropipet, and repeat with the remaining oocytes.
9. Replace the oocytes in MBS without delay (*see* Note 20).
10. Assays for expressed proteins may be performed as soon as 24 h after injection, but the optimal expression period varies between mRNAs and should be determined empirically.

3.6. Transport Assays

Transport activity may be assayed by uptake of radiolabeled solutes, or by electrophysiology. Numerous examples of uptake assays may be found in the oocyte literature (e.g., *23*) as well as accounts of the two electrode voltage clamp and of patch clamp techniques (e.g., *5*) (*see* Notes 21–25).

3.7. Radiolabeling Oocyte Proteins

3.7.1. Oocyte Labeling

1. Inject oocytes with the mRNA of interest as described in Section 3.5.3.
2. Place 10 oocytes in the well of a microtiter plate containing 100 μL MBS. Add 0.1–5 μCi/μL [^{35}S]methionine, [^{3}H]leucine, or other radiolabeled amino acids and incubate overnight at 18°C (*see* Notes 26 and 27).
3. Wash the oocytes several times in ice cold MBS (*see* Note 28).
4. Proceed to a fractionation protocol (*see* Section 3.9.) or homogenize the oocytes in 100 μL of homogenization buffer.
5. Centrifuge at 10,000*g* for 15 min at 4°C to remove yolk and pigment granules.
6. Remove and retain the supernatant, if polyacrylamide gel electrophoresis (PAGE) is to be performed, add an equal volume of 2X SDS PAGE sample buffer. If needed resuspend the pellet in 100 μL homogenization buffer.

3.7.2. Assessment of Incorporation

1. Remove 2–5 μL aliquots and assess for incorporation of radioactivity as follows: Add 0.5 mL distilled water and 0.5 mL of 5% H_2O_2 solution containing 1*M* NaOH and 1 mg/mL unlabeled amino acid. Incubate at 37°C for 10 min.
2. Transfer the solutions to 10 mL tubes and add 2 mL ice cold 25% TCA. Incubate on ice for 30 min.
3. Filter through GF/C Whatman glass filters and wash with 2 × 5 mL 8% TCA.
4. Dry the filters, place in scintillation vial inserts and add 4 mL of a suitable scintillant. If ^{3}H has been used as the label, first wet the filters with a little ethyl acetate.

5. Count the samples to determine suitable loadings for PAGE. Protocols for running and drying gels, together with guidelines for loadings, can be found in ref. *32* (*see* Notes 29 and 30).

3.8. Immunoprecipitation of Expressed Proteins

1. Take 100 μL of protein sample, for example, a crude homogenate, or a membrane-enriched fraction, and add 400 μL of immunoprecipitation buffer. Incubate at 4°C for 30 min.
2. Add the antibody, and incubate at 4°C for a further 30 min. For radiolabeled proteins suitable starting volumes are as follows: 10–100 μL hybridoma supernatant, 0.5–5 μL polyclonal antiserum, 0.1–1 μL ascites fluid (this gives an average of 1 μg heavy chain), but it may be necessary to determine a satisfactory volume empirically. Carry out a parallel reaction with a suitable control antibody to assess nonspecific immunoprecipitation.
3. Recover the immune complex by incubation with either 50 μL of 10% fixed *Staphylococcus aureus*, or 100 μL protein A-sepharose (10% v/v) for 1 h at 4°C, with gentle agitation (*see* Notes 31 and 32). Both of these sources of protein A should be washed three times in immunoprecipitation buffer before use.
4. Collect the immune complexes by centrifugation at 10,000*g* at 4°C for 1 min.
5. Wash the immune complexes three times in immunoprecipitation buffer.
6. Add 50 μL SDS-PAGE sample buffer, and heat to 85°C for 10 min. Centrifuge briefly and load supernatant on to a polyacrylamide gel. *Caution:* Heat treatment will aggregate certain membrane proteins and if this creates problems, incubate at room temperature for 20 min in the presence of protease inhibitors.
7. After electrophoresis, dry the gel and perform autoradiography, if the oocyte proteins have been radioactively labeled.

3.9. Fractionation of Oocytes

Two basic methods are available for fractionation of oocyte components: sucrose density gradient centrifugation *(4)*, or preparation of plasma membrane complexes, followed by differential centrifugation *(33,34)*.

3.10. Antisense Techniques

3.10.1. Protocol 1: General

1. Screen oligonucleotides for their ability to cause RNase H-mediated cleavage of sense RNA *(33)*. Groups of oligonucleotides are usually more effective than single ones and require a lower dose to mediate cleavage (*see* Note 33).

2. Prehybridize the sense RNA with the antisense oligonucleotides. Denature RNA by heating to 65°C for 2 min. Incubate the RNA and oligonucleotides together for 3–4 h at 37°C in 200 m*M* NaCl, 10 m*M* Tris-HCl, pH 7.4. Treat a batch of RNA in the same way, but without oligonucleotides (*see* Notes 34 and 35).
3. Inject oocytes with the hybridized mixture.
4. Assay groups of oocytes, and compare those injected with antisense oligonucleotides to controls injected with only the sense RNA.

3.10.2. Protocol 2: For Transport Proteins

1. Screen oligonucleotides for their ability to cause RNase H-mediated cleavage of sense RNA *(35)*.
2. Inject oocytes with sense RNA(s) of interest.
3. Allow a suitable expression period (e.g., 2–4 d), and observe phenotype, by a nondestructive assay, such as voltage-clamping. Store oocytes in separate Petri dishes or in wells of microtiter plates so that individual oocytes can be identified.
4. Inject a cocktail of antisense oligonucleotides into half the oocytes. Allow 24 h (in the first instance) for ablation of sense RNA.
5. Reassay for phenotype in oligoinjected and control oocytes.
6. A parallel group of oocytes may be prepared for Northern and Western analyses, to demonstrate degradation of message and protein.

3.11. Troubleshooting

The are many reasons for failed expression of genes in *Xenopus*. As there are numerous steps that may be important to the success of expression, and few workers report failed attempts at expression, it is difficult to identify what factors may determine success. One approach is to consider the reasons for failed expression listed below, and systematically work through the most likely possibilities.

1. RNA degraded: Assess quality of RNA on an agarose gel. If necessary, prepare fresh template, and repeat *in vivo* transcription.
2. RNA unstable: This could be owing to the lack of a poly A^+ tail, although the effects of poly A^+ tails are controversial. A cap structure is essential to stability of RNA in oocytes; the proportion of capped RNA may be enhanced by increasing the proportion of cap analog to GTP in the transcription reaction. Northern analysis of injected oocytes may indicate whether degradation is occurring.
3. RNA cannot be translated. This appears to be a common cause of failed expression. Removal of RNA secondary structure in the regions flanking

the coding sequence has resulted in successful expression in at least two cases *(13,36)*. Translation in a cell-free rabbit reticulocyte lysate can give some indication of whether a particular RNA is readily translated. Insertion of coding sequence between the 5' and 3' untranslated regions of a *Xenopus* β-globulin gene may improve expression; the vector pSP64T was constructed for this purpose *(37)* and some authors simultaneously remove flanking sequence and subclone into pSP64T, with some success.

4. More than one protein/RNA species is required for activity. This may be because the functional complex contains more than one protein subunit *(38)*, or because a second protein mediates correct assembly as in the case of the Na^+/K^+ATPase *(39)*.
5. The protein is incorrectly targeted. This may be checked by labeling newly synthesized proteins with radioactive amino acid, and fractionating the oocyte, identifying the location of the protein of interest by immunoprecipitation. Alternatively, immunoblotting of PAGE-separated proteins from membrane fractions may be used.
6. Endogenous activities mask activity of exogenous protein. For example, the solute measured in an uptake study may be simultaneously effluxing from the oocyte.
7. Properties of the protein product may be incompatible with expression. Two *Arabidopsis* K^+ channel clones, AKT1 and KAT1, have been isolated by yeast complementation, but only KAT1 has been subsequently expressed in oocytes (*20*, H. Sentenac, personal communication). Both channels share extensive sequence homology, but only AKT1 contains a hydrophilic carboxyl-terminal domain homologous to the 33-residue ankyrin repeat sequence that is involved in protein to protein interactions. This ankyrin repeat may block functional expression in oocytes.
7. Unexplained reasons: Attempts to express the a subunit of skeletal muscle calcium channels in oocytes both with and without the auxiliary subunits have been unsuccessful *(40)*. In this case, coinjection of heterologous mRNA from heart and *Torpedo* electrical organ led to successful expression. Brain mRNA, however, never produced an L-type Ca^{2+} channel in the oocyte system, suggesting that the oocyte is inadequate for expressing particular Ca^{2+} channels from some tissues.

4. Notes

4.1. Maintenance and Surgery of Xenopus

1. A crystal of potassium permanganate allowed to disperse in the tank water is used in our laboratory to help prevent infection.

2. The frogs feed quickly, but tend to hold food in their mouths without swallowing, which they later spit out. It is therefore advisable to wait several hours after feeding before cleaning the tank. In our experience, Xenopus pellets tend to disintegrate in the water before they can be eaten.
3. Occasionally, a mature female will have very few oocytes or oocytes of poor quality, and another animal will have to be selected. We find that there is little relationship between the appearance of the frog and the quantity or quality of oocytes obtained. Often a rather poor-looking frog will yield large quantities of oocytes with good expression characteristics.
4. Some workers perform collagenase digestion in a calcium-free solution (such as calcium-free Ringer's). We have not found this to be necessary, and use MBS.
5. Collagenase is extremely variable in its activity. It is useful to find a batch that works well, and request this in repeat orders. Some manufacturers, such as Sigma, sell artificial blends with uniform activity.

4.2. Use of mRNA

6. Special precautions must be taken when working with RNA, to reduce the activity of ubiquitous RNases. If possible, a separate working area and set of apparatus should be set aside for RNA work. Water must be of the highest purity available, and should be treated with 0.1% DEPC to inactivate RNase: DEPC is added to a concentration of 0.1%, the solution left overnight in a fume hood, and then autoclaved to inactivate remaining DEPC. Tris and other amine-containing compounds cannot be treated directly with DEPC, but can be prepared with DEPC-treated water. *Caution:* DEPC is a suspected carcinogen, and can also be explosive. Use only in a fume hood, wearing protective clothing, and store in a sealed container at 4°C. All solutions should be prepared and stored in sterile plasticware.
7. Phenotypes are often apparent in oocytes injected with poly A^+ RNA from a specific tissue. However, if the message is of very low abundance, for example in the case of membrane proteins, enrichment for the species of interest may be necessary, and indeed forms part of the cloning strategy. For inducible systems, RNA may be extracted when the species of interest is most abundant. Alternatively size fractionation of mRNA is performed. Sucrose density gradient centrifugation is the method most commonly employed, but separation by nondenaturing agarose gel electrophoresis has also been used *(41)*.
8. Some messenger RNAs are very stable when injected into oocytes, e.g., globulin mRNA was still being efficiently translated 1 wk after the original injection. The role of the poly(A) tail in RNA stability is still controversial *(42)*.
9. Protocols for use of magnetic beads will vary between manufacturers.

4.3. Preparation of Template for In Vitro Transcription

10. Most modern expression vectors contain two promoters in opposing orientation, such that both sense and antisense RNA can be obtained (e.g., pBluescript, Stratagene, La Jolla, CA; pGEM series, Promega, Madison, WI). This is useful for control experiments. To obtain sense template, the plasmid containing the cDNA is then linearized with a restriction enzyme that cleaves on the 3' side of the insert, and to obtain antisense template, the plasmid is linearized with an enzyme that cuts on the 5' side of the insert. It is worth considering this when the initial subcloning is carried out.
11. An enzyme that generates a blunt end or a 5' overhang must be used, as nonspecific transcription can initiate at 3' protruding ends.
12. As an alternative to linearized plasmid containing cDNA, the template may be a PCR-generated DNA fragment, where the phage promoter was incorporated into one of the primers *(43)*.

4.4. Preparation of Capped Synthetic RNA

13. The transcription mix is assembled at room temperature in a specific order to avoid precipitation of DNA by spermidine in the transcription buffer. Commercially available kits supply a buffer that has been optimized for use with the other kit reagents, and often contains RNase inhibitor.
14. Most mRNA molecules have a 5' 7-methyl guanosine residue—the cap structure—that serves to protect the mRNA from degradation by nucleases and also functions in the protein synthesis initiation process. Kreig and Melton (1984) have shown that a 5' cap is essential for stability of synthetic RNAs in *Xenopus* oocytes *(37)*. Capped in vitro transcripts are synthesized by substituting a cap analog for part of the GTP in the transcription reaction. Ratios of 9:1 (cap : GTP) have typically been used; whereas this results in a high proportion of the transcripts being capped, the yield of RNA may be reduced to 20% or lower. Cap analog is also relatively expensive. Therefore, a ratio of 4:1 cap:GTP is often employed as a compromise between capping efficiency and yield of the transcription reaction.
15. In vitro transcription kits are available that employ very high working concentrations of rNTPs, (as high as 7.5 m*M*) that give high yields of RNA (e.g., Ambion [Austin, TX] MEGAscript™, AMS Biotechnology UK Ltd). A correspondingly higher proportion of cap analog is therefore necessary.
16. It is advisable to purchase a preparation of RNase-free DNase, which is recommended for removal of template from in vitro transcription reactions, as the method given in Section 2.4. only removes 98% of contaminating RNase activity.

17. The yield of the transcription reaction varies according to several factors; length of template, age of the enzyme, and so on, but 10 μg or more RNA may be obtained from 1 μg of template DNA. RNA yield can also be assessed by the incorporation of [^{32}P]rNTP in the transcription reaction. Much higher yields may be obtained from commercially available kits, and the use of these is recommended.

4.5. Microinjection

18. A maximum of 100 ng RNA may be injected, without detrimental competition with endogenous "housekeeping" messages *(4)*.
19. A method for checking the volume of solutions injected into oocytes may be found in ref. *44*.
20. Injection of mRNA into the animal pole should be avoided, as this may damage the nucleus, and lead to oocyte death.

4.6. Transport Assays

21. Before attempting to express a transport protein (or other enzyme) in *Xenopus* oocytes, it is important to ascertain whether endogenous activites are present, which may confuse the proposed assays. Oocytes and their follicle cells have several endogenous transporters (reviewed in ref. *6*). Such activites may be subtracted as background, or inhibited (for example by the use of channel blockers). Expression of plant-derived proton cotransporters in oocytes may not be problematic, as cotransport in oocytes is often sodium dependent, and may be distinguished by substituting choline for sodium ions. Usually, however, the activity of a heterologous message expressed in oocytes is manyfold higher than that of an endogenous message and can be easily assayed and distinguished. Interaction between endogenous and heterologous proteins may occur, as in the case of $b^{0,+}$ amino acid transporter *(45)*.
22. As many plant transport proteins are energized by proton gradients, it is important to determine whether the oocyte's internal pH regulating system copes with the acid "load" imposed by a functional proton symport. Changes in the cytosolic pH will also have obvious implications for the functional characterization of H^+ gradient-driven transport. We have investigated this using oocytes expressing the H^+/hexose cotransporter STP1, but find that there is no measurable change in cytosolic pH during H^+/hexose cotransport *(46)*.
23. The lipid composition of membranes is known to modulate the activity of certain transport proteins, and it is possible that this may be important when expressing a plant transporter in *Xenopus* oocytes. However, a hexose transporter from Arabidopsis (STP1) shows similar kinetic properties when expressed both in yeast and oocytes *(23)*.

24. The oocyte is surrounded by several tissue layers, including a layer of follicle cells, a vitelline envelope and an outer theca (layer of connective tissue). The oocyte plasma membrane is highly folded into microvilli that form gap junctions with the macrovilli of the follicle cells. The follicle cells have ion channels and transporter activity and can contribute to the whole oocyte's electrophysiological response, and are therefore removed either manually or by collagenase digestion.
25. When performing electrophysiological assays, it may be important to consider asymmetries in the oocyte, for example the phosphatidylinositol second messenger system is more concentrated in the animal pole, as are certain channels. Also, microvilli are more abundant in the plasma membrane of the animal pole.

4.7. Labeling of Oocyte Proteins

26. Oocytes are freely permeable to amino acids, owing to the presence of several transport proteins in the plasma membrane *(45)*. Thus, labeled amino acids may be added to the medium, rather than injected. Methionine, histidine, proline, and leucine have small pool sizes (e.g., methionine 30 pmol/oocyte, *see* ref. *4*) and are suitable for most purposes.
27. Radiolabeled amino acid may be applied shortly after injection, but some workers prefer to leave a 24-h period before incubation to allow damaged oocytes to be identified.
28. For secreted proteins, add gentamycin and nystatin to MBS. Addition to the medium of 5% fetal calf serum, dialyzed against MBS, prevents reduction of disulfide bonds in the secreted protein. Nonspecific binding sites on the microtiter plate should be blocked by treatment with 0.1–0.5% BSA before use.
29. Abundant polypeptides (e.g., α-amylase—unpublished results) can be visualized with autoradiography of dried SDS-PAGE gels. However, membrane proteins are usually expressed at a much lower level (this is not usually a problem when performing electrophysiological assays) and immunoprepcipitation is routinely used to concentrate labeled protein before running gels. Alternatively, Western blotting is used to identify proteins of interest. If no antibody is available, a protein product of a cloned gene can be "tagged" with an epitope, such as the c-myc oncoprotein, for which commercial antibodies are available.
30. As maternal RNAs are translated by the oocyte in competition with injected RNAs, there will always be a background of endogenous oocyte proteins in radiolabeling experiments.

4.8. Immunoprecipitation

31. Protein A sepharose gives lower backgrounds than fixed *S. aureus*.
32. The affinity of protein A differs for different classes and sources of antibody. If the affinity for the antibody of choice is low, the antibody should be supplemented by anti-immunoglobulin antibodies *(47)*.

4.9. Antisense Techniques

33. Research from several laboratories has shown that expression of RNA may be selectively inhibited by antisense RNA *(48)*, antisense oligonucleotides *(35)*, and partially digested, antisense single-stranded DNA *(35)*. This is thought to be owing to cleavage of the sense RNA by endogenous RNase H.
34. The antisense message may be prehybridized with RNA prior to injection or by injection after the phenotype of the sense message has been demonstrated. The former method has the advantage that rare messages can be selectively suppressed, but it is never clear from these experiments whether the lack of observed phenotype is owing to lack of expression, or to suppression by the antisense message. Noguchi et al. demonstrated function of the Na^+/K^+ATPase and observed removal of both message and protein 24 h after injection of antisense oligonucleotides *(39)*. A cocktail of antisense oligonucleotides is preferable to using a single one because this increases chances of binding to the RNA and overrides problems caused by inaccessible RNA secondary structure *(35)*.
35. Oligonucleotides are convenient to use, as they can readily be obtained in high molar excess of their RNA target. A minimum of 10 residues has been shown to be effective.

References

1. Sigel, E. (1990) Use of *Xenopus* oocytes for the functional expression of plasma membrane proteins. *J. Mem. Biol.* **117,** 201–221.
2. Larkins, B. A., Pedersen, K., Handa, A. K., Hurkman, W. J., and Smith, L. D. (1979) Synthesis and processing of maize storage proteins in *Xenopus laevis* oocytes. *Proc. Natl. Acad. Sci. USA* **76,** 6448–6452.
3. Colman, A. (1984a) Expression of exogenous DNA in *Xenopus* oocytes, in *Transcription and Translation—A Practical Approach* (Hames, B. D. and Higgins, S. J., eds.), IRL, Oxford, pp. 49–69.
4. Colman, A. (1984b) Translation of eukaryotic messenger RNA in *Xenopus* oocytes, in *Transcription and Translation—A Practical Approach* (Hames, B. D. and Higgins, S. J., eds.), IRL, Oxford, pp. 271–302.
5. Rudy, B. and Iverson, L. (eds.) (1992) Ion channels. *Meth. Enzymol.* **207**.
6. Dascal, N. (1987) The use of *Xenopus* oocytes for the study of ion channels. *CRC Crit. Rev. Biochem.* **22,** 317–387.

7. Dumont, J. N. (1972) Oogenesis in *Xenopus laevis* (Daudin). I. Stage of oocyte development in laboratory maintained animals. *J. Morphol.* **136,** 153–180.
8. Yang, X.-C., Karschin, A., Labarca, C., Elroy-Stein, O., Moss, B., Davidson, N., and Lester, H. A. (1991) Expression of ion channels and receptors in *Xenopus* oocytes using vaccinia virus. *FASEB J.* **5,** 2209–2216.
9. Altschuler, Y., Rosenberg, N., Harel, R., and Galolo, G. (1993) The N- and C-terminal regions regulate the transport of wheat gamma-gliadin through the endoplasmic reticulum in *Xenopus* oocytes. *Plant Cell* **5,** 443–450.
10. Wallace, J. C., Galili, G., Kawata, E. E., Cuellar, R. E., Shotwell, M. A., and Larkins, B. A. (1988) Aggregation of lysine-containing zeins into protein bodies in *Xenopus* oocytes. *Science* **240,** 662–664.
11. Hurkman, W. J., Smith, L. D., Richter, J., and Larkins, B. A. (1981) Subcellular compartmentalisation of maize storage proteins in *Xenopus* oocytes injected with zein messenger RNAs. *J. Cell Biol.* **87,** 292–299.
12. Matthews, J. A., Brown, J. W. S., and Hall, T. C. (1981) Phaseolin mRNA is translated to yield glycosylated polypeptides in *Xenopus* oocytes. *Nature* **294,** 175,176.
13. Ceriotti, J. L., Pedrazzini, E., Fabbrini, M. S., Zoppe, M., Bollini, R., and Vitale, A. (1991) Expression of the wild-type and mutated vacuolar storage protein phaseolin in *Xenopus* oocytes reveals relationships between assembly and intracellular transport. *Eur. J. Biochem.* **202,** 959–968.
14. Ereken-Tumer, N., Richter, J. D., and Nielsen, N. C. (1982) Structural characterisation of the glycinin precursors. *J. Biol. Chem.* **257,** 4016–4018.
15. Boston, R. S., Miller, T. J., Mertz, J. E., and Burgess, R. R. (1982) *In vitro* synthesis and processing of wheat α-amylase. Translation of gibberellic acid-induced wheat aleurone layer RNA by wheat germ and *Xenopus laevis* oocyte systems. *Plant Physiol.* **69,** 150–154.
16. Bassüner, R., Huth, A., Manteuffel, R., and Rapoport, T. A. (1983) Secretion of plant storage globulin polypeptides by *Xenopus laevis* oocytes. *Anal. Biochem.* **133,** 321–326.
17. Aoyagi, K., Sticher, L., Wu, M., and Jones, R. L. (1990) The expression of barley α-amylase genes in *Xenopus laevis* oocytes. *Planta* **180,** 333–340.
18. Simon, R., Altschuler, Y., Rubin, R., and Galili, G. (1990) Two closely related wheat storage proteins follow a markedly different subcellular route in *Xenopus laevis* oocytes. *Plant Cell* **2,** 941–950.
19. Spanu, P., Reinhardt, D., and Boller, T. (1991) Analysis and cloning of the ethylene-forming enzyme from tomato by functional expression of its mRNA in *Xenopus laevis* oocytes. *EMBO J.* **10,** 2007–2013.
20. Schachtman, D. P., Schroeder, J. I., Lucas, W. J., Anderson, J. A., and Gaber R. F. (1992) Expression of an inward-rectifying potassium channel by the *Arabidopsis KAT1* cDNA. *Science* **258,** 1654–1657.
21. Cao, Y., Anderova, M., Crawford, N. M., and Schroeder, J. I. (1992) Expression of an outwardly-rectifying potassium channel from maize mRNA and complementary RNA in *Xenopus* oocytes. *Plant Cell* **4,** 961–969.
22. Tsay, Y.-F., Schroeder, J. I., Feldmann, K. A., and Crawford, N. M. (1993) The herbicide sensitivity gene *CHL1* of *Arabidopsis* encodes a nitrate-inducible nitrate transporter. *Cell* **72,** 705–713.

23. Boorer, K. J., Forde, B. G., Leigh, R. A., and Miller, A. J. (1992) Functional expression of a plant plasma membrane transporter in *Xenopus* oocytes. *FEBS Lett.* **302,** 166–168.
24. Aoshima, H., Yamada, M., Sauer, N., Komor, E., and Schobert, C. (1993) Heterologous expression of the H^+/hexose cotransporter from *Chlorella* in *Xenopus* oocytes and its characterization with respect to sugar specificity, pH and membrane potential. *J. Plant Physiol.* **141,** 293–297.
25. Maurel, C., Reizer, J., Schroeder, J. I., and Chrispeels, M. J. (1993) The vacuolar membrane protein γ-TIP creates water specific channels in *Xenopus* oocytes. *EMBO J.* **12,** 2241–2247.
26. Gal, S. and Raikhel, N. V. (1993) Protein sorting in the endomembrane system of plant cells. *Curr. Opinion Cell Biol.* **5,** 636–640.
27. Roitsch, T. and Lehle, L. (1991) The vacuolar protein-targeting signal of yeast carboxypeptidase is functional in oocytes from *Xenopus laevis*. *Eur. J. Biochem.* **195,** 145–150.
28. Lotan, I., Goelet, P., Gigi, A., and Dascal, N. (1989) Specific block of calcium channel expression by a fragment of dihydropyridine receptor cDNA. *Science* **243,** 666–669.
29. Parent, L., Supplisson, S., Loo, D. D. F., and Wright, E. M. (1992) Electrogenic properties of the cloned Na^+/glucose cotransporter: I. Voltage-clamp studies. *J. Mem. Biol.* **125,** 49–62.
30. Urnes, S. E. and Carrol, D. (1990) Amylase synthesis as a simple model system for translation and hybrid arrest in *Xenopus* oocytes. *Gene* **95,** 267–274.
31. Jakobsen, K. S., Brevold, E., and Hornes, E. (1990) Purification of mRNA directly from crude plant tissues in 15 minutes using magnetic oligo dT microspheres. *Nucleic Acids Res.* **18,** 3669.
32. Hames, B. D., and Rickwood, D. (1990) Gel Electrophoresis of Proteins. A Practical Approach. IRL Press, Oxford.
33. Wall, D. A. and Patel, S. (1989) Isolation of plasma membrane complexes from *Xenopus* oocytes. *J. Mem. Biol.* **107,** 189–201.
34. Thomas, H. M., Takeda, J., and Gould, G. W. (1993) Differential targeting of glucose transporter isoforms heterologously expressed in *Xenopus* oocytes. *Biochem. J.* **290,** 707–715.
35. Morgan, R., Edge, M., and Colman, A. (1993) A more efficient and specific strategy in the ablation of mRNA in *Xenopus laevis* using mixtures of antisense oligos. *Nucleic Acids Res.* **21,** 4615–4620.
36. Moorman, J. R., Palmer, C. J., John III, J. A. E., Durieux, M. E., and Jones, L. R. (1992) Phospholemman expression induces a hyperpolarization-activated chloride current in *Xenopus* oocytes. *J. Biol. Chem.* **267,** 14,551–14,554.
37. Kreig, P. A. and Melton, D. A. (1984) Functional messenger RNAs are produced by SP6 *in vitro* transcription of cloned cDNAs. *Nucleic Acids Res.* **12,** 7057–7070.
38. Kroll, B., Emde, M., Jeromin, A., Penner, L., Rechkemmer, G., Kretzschmar, T., Klos, A., Kohl, J., and Bautsch, W. (1991) Functional expression of a human C5a receptor clone in *Xenopus* oocytes requires additional RNA. *FEBS Lett.* **291,** 208–210.

39. Noguchi, S., Higashi, K., and Kawamura, M. (1990) Assembly of the α-subunit of *Torpedo californica* Na^+/K^+-ATPase with its pre-existing β-subunit in *Xenopus* oocytes. *Biochim. Biophys. Acta* **1023,** 247–253.
40. Dascal, N., Lotan, I., Karni, E., and Gigi, A. (1992) Calcium channel currents in *Xenopus* oocytes injected with rat skeletal muscle RNA. *J. Physiol.* **450,** 469–490.
41. Tarnuzzer, R. W., Campa, M. J., Qian, N.-X., Engelsberg, E., and Kilberg, M. S. (1990) Expression of the mammalian system A neutral amino acid transporter in *Xenopus* oocytes. *J. Biol. Chem.* **265,** 13,914–13,917.
42. Sachs, A. and Wahle, E. (1993) Poly(A) tail metabolism and function in eucaryotes. *J. Biol. Chem.* **268,** 22,955–22,958.
43. Milligan, J. F., Groebe, D. R., Witherell, G. W., and Uhlenbeck, O. C. (1987) Oligoribonucleotide synthesis using T7 RNA polymerase and synthetic DNA template. *Nucleic Acids Res.* **15,** 8783–8798.
44. Cupello, A., Baldelli, P., Gheri, A., and Robello, M. (1993) A method for checking the volume of solutions microinjected into *Xenopus* oocytes. *Anal. Biochem.* **213,** 172–174.
45. Van Winkle, L. J. (1993) Endogenous amino acid transport systems and expression of mammalian amino acid transport proteins in *Xenopus* oocytes. *Biochim. Biophys. Acta* **1154,** 157–172.
46. Miller, A. J., Smith, S. J., and Theodoulou, F. L. (1994) The heterologous expression of H^+-coupled transporters in *Xenopus* oocytes, in *Membrane Transport in Plants and Fungi: Molecular Mechanisms and Control* (Blatt, M. R., Leigh, R. A., and Sanders, D., eds.), Company of Biologists, Cambridge.
47. Harlow, E. and Lane, D. (1988) *Antibodies. A Laboratory Manual.* Cold Spring Harbor Laboratory, Cold Spring Harbor, NY.
48. Shuttleworth, J. and Colman, A. (1988) Antisense oligonucleotide-directed cleavage of mRNA in Xenopus oocytes and eggs. *EMBO J.* **7,** 427–434.

CHAPTER 27

Heterologous Expression in Yeast

Laurence J. Trueman

1. Introduction

The complex nature of plant metabolism often can seriously hinder the functional analysis of many proteins; and this has led some researchers to attempt to express such proteins in less sophisticated organisms. Unfortunately, the obvious choice, *Escherichia coli*, is less than ideal with many foreign peptides being misfolded and subsequently sequestered in insoluble protein bodies. The yeast *Saccharomyces cerevisiae*, however, has a combination of properties that make it a particularly attractive system for the functional expression of plant proteins.

Saccharomyces cerevisiae is a single-celled eukaryotic organism able to express functionally many plant genes, correctly folding, modifying, and targeting the protein. Its small genome size (only four times that of *E. coli*) means that, although possessing all the major metabolic pathways of eukaryotes, it lacks many of the minor pathways that give plant metabolism its plastic nature, thus simplifying analysis of the protein. This can be taken a step further by the ability to make absolute mutants of many yeast genes, allowing the analysis of many proteins in a background free of the yeast homolog. This ability to construct absolute mutants is owing to the ability to propagate and manipulate *S. cerevisiae* as a haploid cell. In this state many mutations exhibit a dominant phenotype, especially since the small genome size means that multigene families are rare. Absolute mutants of essential genes can also be constructed. The lethal phenotype is avoided by propagating the mutant gene in the diploid cell. The diploid is then transformed with the plant homolog and

From: *Methods in Molecular Biology, Vol. 49: Plant Gene Transfer and Expression Protocols*
Edited by: H. Jones Humana Press Inc., Totowa, NJ

sporulated to give haploid spores. Absolute haploid yeast mutants can now be isolated by the ability of the plant protein to supply the essential gene function.

A second type of mutant commonly used in yeast genetics is the temperature sensitive mutant. These auxotrophs produce a functional protein at one temperature, but not at another. The yeast is manipulated at the permissive temperature and then switched to the nonpermissive temperature for analysis of the plant protein. Obviously, such mutations occur at a much lower rate than absolute mutants, but the ability to screen billions of yeast cells at a time, means that such mutations occur at an acceptable frequency.

The ability to construct absolute and temperature sensitive mutants in yeast has resulted in the development of one of the most powerful techniques to arise in plant genetics in recent years by allowing the isolation of plant genes by the ability of their product to complement such yeast mutations. This technique has now been used to isolate directly several plant genes that would have taken many years of biochemical analysis to isolate conventionally *(1–4)*.

Finally, the advanced molecular genetics of *S. cerevisiae* developed over the past 20 yr allows it to be manipulated almost as easily as *E. coli*. Coupled with the large amount of knowledge accumulated on this subject (mainly driven by the interest displayed by the pharmaceutical industry in *S. cerevisiae* as a heterologous expression system for proteins intended for human consumption, on the basis that it is a food organism and thus considered safe), *S. cerevisiae* is set to become an invaluable tool to the plant geneticist.

The first consideration for expressing foreign proteins in yeast is the transcriptional unit. Although the genes of yeast and higher eukaryotes are structurally similar, they often differ in detail and this has two implications: First, plant promoters and transcriptional terminators rarely function efficiently in yeast, and second, intron splice sites usually go unrecognized. This means that for the efficient expression of a plant protein, an expression cassette, consisting of a yeast promoter, terminator, and a complementary DNA copy of the gene has to be constructed.

In theory, any yeast promoter can be used for expression; however, the commonly encountered promoters have been selected for their high levels of activity. This ensures that even with the inevitable drop in expression levels compared to the wild-type gene that usually accompanies heterologous expression, reasonable yields of the plant protein can still be achieved. Promoters can be divided into two categories. Constitutive promoters are active under most conditions, and are often taken from glycolytic genes, such as *ADH1*, *PFK*, and *PGK*. Regulated promoters are only active under certain cellular conditions, needing to be activated and/or derepressed. The most commonly used regulated promoters have been taken from the galactose catabolic genes *GAL1* and *GAL10*. These promoters are activated by D-galactose (the activator) in the absence on D-glucose (an absolute repressor), with full induction taking between 4–6 h. Secretion of a protein is achieved by fusing the coding region to a yeast promoter and its associated leader sequence. The most commonly encountered promoter of this type is *MF*α1. This promoter drives the expression of the α mating factor and is thus only active in *MAT*α strains.

Yeast terminators are included to ensure efficient termination and release of the messenger RNA, as well as providing yeast polyadenylation sequences. They are often taken from the same gene as the promoter, although their source is usually unimportant.

The expression cassette is assembled in *E. coli* and carried on a plasmid that also contains sequences that allow it to be selected and maintained in yeast. These *E. coli/S. cerevisiae* shuttle vectors can be divided into two general categories, depending on whether they integrate into the yeast genome or replicate independently as covalently closed circles.

Yeast integrative plasmids (YIps) integrate into the genome by homologous recombination between a plasmid-borne sequence, usually the selective marker, and its chromosomal counterpart. The frequency of transformation is low, but can be significantly increased if the plasmid is first cleaved in the yeast sequence used to target integration. YIps are very stable, being lost at a rate of only 0.1% per generation, but usually only insert in single copy. Multiple insertions can be encouraged by transforming with excessive amounts of DNA, or by targeting the vector to sequences that are iterated in the genome such as the rDNA repeats.

The second type are episomal vectors and these can take several forms. The simplest are termed yeast replicative plasmids or YRps, and contain a selectable marker and a yeast origin of replication *(ARS)*. YRps, however, tend to be mitotically unstable with plasmid-free cells being generated at a rate of 10–20% per generation. This is because of a strong maternal bias in the partitioning of plasmids between the mother and daughter cells during division. Maternal bias can be overcome by the inclusion of other elements in the vector, the simplest way being to add a centromere to give a yeast centromeric plasmid (YCp). YCps are much more stable than YRps being lost at a rate of only 1% per generation, but unfortunately the centromere also reduces the copy number to 1–2 molecules per cell.

A more common solution is to include sequences from the endogenous plasmid of *S. cerevisiae*, the 2 μ circle. The 2 μ plasmid overcomes the problem of maternal bias by a mechanism involving a *cis*-acting locus termed STB. Replicative vectors carrying the STB element (confusingly termed yeast episomal plasmids or YEps) have a copy number of between 10–40 molecules per cell, and are only lost at a rate of 1–3% per generation. To function properly, this type of YEp needs 2 μ to be present in the cell to provide *trans*-acting factors (such strains are called *cir*$^+$). Some vectors, however, also carry the genes for these *trans*-acting factors (*REP1*, *REP2*, and *FLP*) and not only can they be propagated in a *cir*° strain (a strain lacking 2 μ), but also reach higher copy numbers (50–200 molecules per cell) than in cir$^+$ strains owing to the lack of competition with the 2 μ plasmid.

Selection for these vectors in yeast is usually achieved by the ability of a plasmid-borne marker gene to complement an auxotrophic mutation carried by the host strain, allowing it to grow on minimal media. The most commonly encountered marker genes are *URA3*, *TRP1*, *HIS3*, and *LEU2*, allowing mutant strains to grow on media lacking uracil, L-tryptophan, L-histidine, and L-leucine, respectively.

Expression vectors can be obtained from several sources. A wide variety can be acquired through the literature offering almost every combination of promoter, selectable marker and vector type. There are also several commercial companies offering a limited range of general purpose expression vectors, but often with useful features such as multiple cloning sites. A review of commonly used vectors has been compiled by Parent et al. *(5)*.

Next, a suitable strain is required. General purpose strains can be obtained through the literature, commercial outlets, or through culture collections such as American Type Culture Collection (ATCC) or the yeast genetic stock center (*see* Note 1). Once obtained, yeast cells should be revived on complete media (YPD) at 30°C, and then their phenotype checked by plating on minimal medium (SD) with and without the addition of any required supplements (*see* Section 2.1.). Most strains are cir^+ and are thus suitable for use with vectors carrying the STB locus. Strains that are termed petite (*pet*, ρ^- or ρ°) should be avoided as they lack functional mitochondria. If a mutant is required then the first place to look is the literature. The long history of *S. cerevisiae* as a model eukaryotic organism has resulted in the isolation of a wide range of mutants, many of which should be available. Occasionally a useful mutation may reside in an unsuitable genetic background; for example, many older yeast strains possess mutations in the galactose permease gene *(gal2)* making the use of galactose inducible promoters impossible. Section 3.1. details how to change the genetic background of a mutation. This is achieved by crossing the mutant with a strain having the desired genotype, sporulating the diploids generated, and then selecting for recombinant cells that possess the mutation in a more useful genetic background.

Often, however, a convenient mutant is not available and one has to be constructed. If a clone of the yeast gene is available then mutants can be generated by one-step gene replacement as follows. A mutant copy of the gene is constructed, usually by introducing a deletion and/or by the insertion of a selectable marker into the coding sequence. A fragment containing the disruption flanked by gene sequences is then isolated and used to transform an appropriate yeast strain. These flanking sequences direct the integration of the fragment into the genomic locus by homologous recombination, replacing the wild-type gene with the mutant. Mutants are then selected using either loss of gene function or the fragment-borne marker.

If only sequence data is available, then Baudin et al. *(6)* have published a simple protocol that allows the generation of a disrupted gene fragment using the polymerase chain reaction (PCR). This method relies on the design of bipartite primers, the 3' half of which are complementary to flanking sequences of a yeast selectable marker, whereas the 5' ends are complementary to regions of the gene to be mutated. These primers are used to amplify, by PCR, a copy of the selectable marker flanked by target gene sequences. This PCR fragment is then used to disrupt the wild-type gene as described earlier. With the *S. cerevisiae* genome sequencing project estimated to be completed by the year 1997 (personal communication from S. G. Oliver), this method should allow the disruption of all yeast genes.

If none of the described approaches are possible, then the only solution is to mutagenize a strain using DNA modifying agents and then screen for the desired mutant. Since the frequency of some mutations is often dependent on the method used, and many of the mutagens used are hazardous and require more respect than I can give them here, I refer the reader to the complete set of protocols described by Lawrence *(7)* for further information. Once obtained, yeast strains can be stored for 6 mo or more at 4°C on plates or for several years in YPD stabs. For long-term storage, strains can be frozen in glycerol, as described in Section 3.2.

When the expression cassette has been constructed and a suitable strain obtained, transformation of the yeast can be achieved using the protocol of Dohmen et al. *(8)* described in Sections 3.3. and 3.4. Although this protocol does not offer the highest transformation frequencies published for *S. cerevisiae* (*see* refs. *9* and *10*), transformation efficiencies of up to 1×10^5 transformants/μg DNA can be achieved but with the advantage that the competent cells can be stored for several months at –70°C. Transformant colonies should appear after several days, and the presence of the foreign gene in the cells can be verified using the colony hybridization protocol described in Section 3.5. (adapted from ref. *11*). Analysis of the transformants can now begin, and it is recommended that several cell lines are investigated since stable variations in the copy number of both integrative and episomal vectors can occur. If required, cellular fractionation of *S. cerevisiae* into its component organelles for protein localization or biochemical analysis can be achieved using the protocols described by Franzusoff et al. *(12)*. Plasmids identified by their ability to complement a yeast mutation can be isolated using the protocol of Lorincz *(13)* described in Section 3.6. This latter protocol will only work with *E. coli/S. cerevisiae* shuttle vectors, since it involves the rescue of the plasmid into *E. coli*, from which it is then amplified and isolated using standard techniques *(14)*. This is necessary since the direct isolation of plasmid DNA from yeast yields insufficient quantities to be of use. This protocol can also be used to isolate small amounts of total genomic DNA; however, for large scale isolation the reader is referred to the protocol of Cryer et al. *(15)*.

Sadly, the insertion of a plant gene into an expression vector and its subsequent transformation into yeast does not always guarantee success. Expression is a complex multistep process and many problems can occur. Sometimes yeast is unable to process the protein correctly. In the case of the plasma membrane H^+-ATPase from *Arabidopsis thaliana*, the protein is fully functional, but appears to become trapped in the endoplasmic reticulum *(16)*. Occasionally, a cofactor or activator may be absent. For instance, many plant enzymes derive their reducing power from ferridoxin; *S. cerevisiae*, however, cannot synthesize ferridoxin, deriving most of its reducing power from NADPH.

Often, however, the problem is one of low yield, but this can frequently be improved by changing the culture conditions. Temperature in particular can have a significant effect, since the optimal temperature to culture yeast in the laboratory (30°C) is often not the ideal temperature for maximal expression of a foreign protein. In many cases a culture temperature between 20–24°C gives better results. Strain dependent variations in expression levels can also occur, but often the problem is owing to the expression cassette itself being a poor imitation of a yeast gene. Fortunately, many things can be altered to improve transcription from the cassette and the reader is directed to the review by Romanos et al. *(17)* for further information. The isolation of high molecular weight RNA from yeast for determining the level of transcription by Northern analysis can be achieved using the method of Köhrer and Domdey *(18)*. Doubts about the vector itself can be tested by expressing the reporter gene β-glucoronidase *(gus)*. This also has the added advantage of allowing the induction kinetics of regulated promoters to be determined under the conditions used.

2. Materials

2.1. Media

1. Complete medium (YPD): 1%(w/v) Bacto yeast extract, 2% (w/v) Bacto peptone, 2% (w/v) D-glucose (*see* Notes 2–4).
2. Minimal medium (SD): 0.17% (w/v) yeast nitrogen base without ammonium sulfate and amino acids (Difco, Detroit, MI), 0.5% (w/v) ammonium sulfate, 2% (w/v) D-glucose (*see* Notes 2–6). Sterilize by filtration.
3. SD supplements: The following supplements should be added to minimal medium as appropriate. To a final concentration of: 20 µg/mL of adenine sulfate, L-arginine-HCl, L-histidine-HCl, L-methionine, L-tryptophan, and uracil; 30 µg/mL of L-isoleucine, L-lysine-HCl, and L-tyrosine; 50 µg/mL of L-phenylalanine; 60µg/mL of L-leucine; 150 µg/mL of L-valine.

2.2. Mating and Sporulation

1. Presporulation medium: 0.8% (w/v) Bacto yeast extract, 0.3% (w/v) Bacto peptone, and 10% D-glucose (*see* Note 2).

2. Sporulation medium: 1% potassium acetate, 0.1% Bacto yeast extract, and 0.05% D-glucose (*see* Note 2).
3. 100 mg/mL Novozyme 234 (*see* Note 7).

2.3. Long-Term Storage of Yeast Cells

15% (v/v) glycerol. Autoclave to sterilize.

2.4. Preparation of Competent Cells

Solution A: 1*M* sorbitol, 10 m*M* bicine-NaOH pH 8.35, 3% ethylene glycol, and 5% DMSO (*see* Note 8).

2.5. Transformation of Competent Yeast Cells

1. Carrier DNA: 10 mg/mL salmon sperm DNA, sonicated, and heat denatured (*see* Note 9).
2. Solution B: 40% (w/v) polyethylene glycol (PEG) 1000 dissolved in 0.2*M* bicine-NaOH pH 8.35 (*see* Note 10).
3. Solution C: 150 m*M* NaCl, 10m*M* bicine-NaOH pH 8.35.

2.6. Colony Hybridization

1. Solution I: 1*M* sorbitol, 0.1*M* sodium citrate, 50 m*M* EDTA pH 8.0, 50 m*M* dithiothreitol, 2 mg/mL Novozyme 234 (*see* Note 7).
2. 10% SDS.
3. 0.5*M* NaOH.
4. Solution II: 0.2*M* Tris-HCl, pH7.5, 2X SSC (0.3*M* NaCl, 0.03*M* sodium citrate).

2.7. Isolation of Plasmids from Yeast

1. STE: 100 m*M* NaCl, 10 m*M* Tris-HCl, pH 8.0, 1 m*M* EDTA, 0.1% sodium dodecylsulfate.
2. Tris-buffered phenol, pH 8.0.
3. Chloroform/isoamyl alcohol (24:1 v/v).
4. 3*M* NaOAc pH 5.0.
5. Absolute ethanol.
6. 70% Ethanol.
7. TE: 10 m*M* Tris-HCl, pH 8.0, 1 m*M* EDTA.
8. Glass beads 0.45–0.5 mm in diameter.

3. Methods

3.1. Mating and Sporulation

1. Select two suitable strains (*see* Note 11) and streak to single colony on a YPD plate and incubate at 30°C for 2 d.

2. Using a sterile loop or toothpick, streak six parallel lines of strain one onto a YPD plate. Next streak six parallel lines of strain two onto the plate at right angles to strain one. Incubate overnight at 30°C.
3. Isolate diploids by replica plating the cells onto selective medium and incubate at 30°C. Diploid colonies should appear within 2–3 d.
4. Streak the diploid to single colony on a plate of presporulation medium and incubate for 2 d at 30°C (*see* Note 12).
5. Using a sterile loop, streak a single diploid colony from the presporulation medium on a plate of sporulation medium. Incubate at 30°C for 3–7 d on sporulation medium (*see* Note 13). Check for asci formation using a light microscope (*see* Note 14).
6. Scrape a loopful of sporulation culture off the sporulation plate and suspend in 180 µL of sterile distilled water. Add 20 µL of 100 mg/mL Novozyme 234 (*see* Note 7) and 100 µL of glass beads. Incubate for 1 h at 30°C with shaking (*see* Note 15).
7. Add 800 µL of sterile distilled water and agitate vigorously for 1 min on a vortex mixer. Check for the disruption of the asci using the light microscope and revortex if necessary.
8. Spread a range of dilutions (typically 10^{-2} to 10^{-4} in sterile distilled water) on selective plates containing 80 µg/mL canavanine sulfate and incubate at 24°C for 2–3 d.

3.2. Long-Term Storage of Yeast Cells

1. Streak the yeast strain onto a YPD plate or onto a plate of SD medium containing all supplements not required for selection. Incubate at 30°C until the colonies are 2–3 mm in diameter.
2. Scrape several loopfulls of cells off the plate and suspend in 15% glycerol.
3. Store the cells at –70°C or below (*see* Note 16).
4. Revive the cells by spreading an aliquot onto a YPD plate and incubate at 30°C for 2–3 d.

3.3. Preparation of Competent Cells

1. Inoculate 5 mL of YPD with an appropriate yeast strain (*see* Note 17) and grow overnight at 30°C with shaking.
2. Inoculate 100 mL of YPD with 1 mL of the overnight culture and incubate at 30°C with shaking, until the cells reach an OD of 0.6–1.0.
3. Harvest the cells at 3000*g* for 5 min at room temperature and resuspend the cells in 50 mL (0.5 vol) of solution A.
4. Repellet the cells as described in step 3 and resuspend in 2 mL (0.02 vol) of solution A.

5. Transfer 200-µL aliquots into 1.5-mL microcentrifuge tubes and place directly into a –70°C freezer (*see* Note 18). The cells can be stored at –70°C for several months.

3.4. Transformation of Competent Yeast Cells

1. To a frozen aliquot of yeast cells, add 50 µg of single-stranded carrier DNA (*see* Note 9) and 0.1–5 µg of a plasmid or DNA fragment in a total volume of up to 20 µL.
2. Thaw cells at 37°C for 5 min with rapid agitation in an Eppendorf shaker (*see* Notes 19 and 20).
3. Add 1.4 mL of solution B, gently mix by inversion, and incubate at 30°C for 60 min.
4. Harvest the cells at 3000*g* for 5 min at room temperature and resuspend in 500 µL (2.5 vol) of solution C.
5. Repellet the cells as described in step 4 and resuspend in 100 µL of solution C.
6. Spread an appropriate volume onto selective plates and incubate at 24–30°C (*see* Note 21). Transformant colonies usually appear after 2–5 d and should be purified from the background of untransformed cells by streaking to single colonies on fresh selective medium.

3.5. Colony Hybridization

1. Carefully place a circular piece of nitrocellulose or nylon membrane onto a selective plate. Ensure that the membrane is in complete contact with the medium.
2. Using a sterile loop or toothpick streak the yeast cells onto the filter. Alternatively, replica plate cells from a master plate directly onto the filter using a piece of velvet. Incubate the filter on the plate overnight at 30°C.
3. The filter should be removed from the plate and placed on 3MM paper saturated with the solutions listed in the following. Ensure that the membrane is in complete contact with each solution.
   a. Solution I overnight at 30°C (*see* Note 22).
   b. 10% SDS for 5 min at room temperature.
   c. 0.5*M* NaOH for 10 min at room temperature.
   d. Solution II for 5 min at room temperature. Repeat twice.
4. The DNA is now fixed to the membrane according to the manufacturer's recommendations and probed using standard techniques *(14)* (*see* Note 23).

3.6. Isolation of Plasmids from Yeast

1. Pick a single, medium-sized, plasmid containing yeast colony (2–3 mm diameter) from a selective plate and suspend in 200 µL of STE in a microcentrifuge tube.

2. Add 0.9 vol of glass beads (0.45–0.5 mm diameter) and mix at full speed on a vortex mixer for 1 min to lyse the cells.
3. Add 200 µL of Tris-saturated phenol (pH 8.0) and mix at full speed for 30 s on a vortex mixer.
4. Centrifuge for 1 min at 12,000*g* in a microfuge to separate phases and remove the upper aqueous phase to a fresh microcentrifuge tube.
5. Repeat the phenol extraction (steps 3 and 4).
6. Add 400 µL of chloroform/isoamyl alcohol (24:1) and mix thoroughly for 30 s using a vortex mixer and separate phases as described in step 4. Again transfer the upper aqueous phase to a fresh microcentrifuge tube.
7. Add 40 µL of 3*M* NaOAc (pH 5.0) and 1 mL of absolute ethanol. Leave at –20°C for 20 min.
8. Pellet the nucleic acid by centrifugation at 12,000*g* at 4°C for 15 min.
9. Wash the nucleic acid pellet by adding 1 mL of cold 70% alcohol and mix briefly using a vortex mixer. Repellet the nucleic acid by centrifugation as described in step 8.
10. Discard the supernatant and dry the nucleic acid pellet under vacuum. Resuspend the pellet in 100 µL of TE buffer.
11. Transform *E. coli* using varying amounts (1–20 µL) of the DNA solution (*see* Note 24) and spread onto selective plates.

4. Notes

1. The American type culture collection (Rockville, MD). The yeast genetic stock center, Department of Biophysics and Medical Physics, University of California, Berkeley, CA.
2. All media can be made solid by the addition of 2% (w/v) Bacto agar.
3. D-glucose is the preferred carbon source and should be used whenever possible since other substrates often result in a noticeable decrease in growth rate. *GAL* promoters can be induced by the addition of 2% (w/v) D-galactose.
4. Media (except SD) should be sterilized, in the absence of the carbon source, using an autoclave at 121°C and 1.05 kg/cm^2 for 15 min. The carbon source, which is sterilized by filtration and is usually stored as a 10X stock solution, is added prior to use.
5. Synthetic medium can be stored as a 10X concentrate at 4°C. In practice, however, it is better to store the carbon source separately since complete concentrates easily become contaminated.
6. Enhanced growth may be achieved by growing the cells on medium containing all supplements not required for selection. Growth of *TRP1* and *URA3* plasmid-containing strains can be aided by the addition of casamino acids or similar acid protein hydrolyzates.

7. Several other enzyme preparations can be substituted for the Novozyme 234 (Novo Biolabs) such as Lyticase (Zymolase 60,000) or Glusulase (β-glucuronidase and sulfatase).
8. The DMSO is optional.
9. Schiestl and Gietz *(10)* have shown the addition of single-stranded carrier DNA to increase transformation efficiency significantly. The source and size of the DNA, however, is important. For optimum transformation frequencies they recommend the addition of Salmon sperm DNA that has been sonicated to give a fragment size range of between 2–15 kb (as assessed on an agarose gel) and then denatured by boiling.
10. The source and purity of the PEG 1000 can have a significant effect on transformation efficiency. When high transformation frequencies are required, it is recommended that PEG 1000 from several sources are tested.
11. The strains must be of opposite mating type, which in *S. cerevisiae* is determined by the two alleles *MAT*α and *MAT***a**, of the mating type locus. It is also desirable to select the phenotypes of the strains carefully since this can simplify the protocol considerably. First, choose two stains that have different auxotrophic markers as this will allow positive selection for the diploid on selective medium. For example, take the following cross: strain 1 *(MAT***a***, ura3, leu2)* X strain 2 *(MAT*α*, ura3, trp1)*. The diploid *(MAT***a***/ MATa, ura3/ura3, leu2/LEU2, TRP1/trp1)* is phenotypically only *ura*3 and thus can be selected from a background of parental haploids by its ability to grow on medium lacking leucine and tryptophan.

    Second, one of the strains should carry the recessive marker *can*^{R}, which confers resistance to the toxic arginine analog canavanine. This will allow selection against nonsporulated dipolids (since no diploid strain sporulates completely) when screening the recombinant haploids for the desired phenotype. It works because the diploid (*can*^{R}/*can*^{S}) carries the dominant *can*^{S} marker and is thus sensitive to canavanine. Only 50% of the haploids, however, will be *can*^{S}, the rest being *can*^{R} and thus canavanine resistant. It is from this latter population that the mutant is isolated. Spontaneous *can*^{R} mutants of most strains can be isolated by growing the strain on YPD plates containing 80 μg/mL canavanine sulfate for several days at 30°C.
12. This step can be omitted and the diploid taken directly from a fresh YPD or SD plate, but with an accompanying decrease in sporulation efficiency.
13. Diploids vary in their ability and the time taken to sporulate. If a strain proves difficult to sporulate, try incubating the cells at 24°C instead of 30°C or by performing the sporulation in liquid medium with shaking. Sporulation of auxotrophic diploids can be aided by the addition of supplements to 25% of the concentration quoted in materials Section 2.3. Note, however, that higher supplement concentrations can inhibit sporulation.

14. An ascus consists of four spherical ascospores arranged as a tetrahedron. Under the light microscope, however, the asci often appear triangular owing to the seclusion of one spore behind the other three.
15. Alternatively, the glass beads can be omitted and step 7 substituted as follows: Add 4.8 mL of sterile-distilled water and sonicate at full power for 15 s, while cooling the spores with a jacket of ice. Check for the disruption of the asci using the light microscope and repeat the sonication if necessary.
16. The viability of yeast cells tend to decrease significantly if stored above –55°C.
17. Strains often differ in their ability to be transformed. It is therefore suggested that several strains are tested.
18. Alternatively, the cells can be frozen in a dry ice/ethanol bath and then stored at –70°C. Freezing the cells in liquid nitrogen should be avoided as this reduces the efficiency of transformation.
19. A 5-μL aliquot of 1*M* histamine may be added at this point.
20. Thawing and mixing the cells at room temperature gives slightly lower transformation frequencies.
21. The carbon source can often have an effect on the transformation frequency. Maximal efficiencies are usually obtained by selecting on plates containing D-glucose.
22. It is important not to let the membrane dry out by placing the membrane and filter paper into a covered container (a Petri dish is ideal). Incubation of the filter in solution I for a shorter period of time (i.e., 2–3 h) at 37°C may also give satisfactory results.
23. Some researchers like to try and remove some of the cell debris from the surface of the filter before fixing the DNA. This can be achieved by either lightly scratching the surface of the filter with a gloved hand, or by placing the membrane between two pieces of 3MM paper (Whatman, Maidstone, UK) and then applying pressure by rolling a glass rod over the surface.
24. The number of colonies obtained per plate can depend on a number of factors including: the volume of nucleic acid suspension used, the competence of the *E. coli* cells, and the copy number of the shuttle vector.

References

1. Frommer, W., Hummel, S., and Riesmeier, J. (1993) Expression cloning in yeast of a cDNA encoding a broad specificity amino acid permease from *Arabidopsis thaliana. Proc. Natl. Acad. Sci. USA* **490,** 5944-5959.
2. Leustek, T., Murillo, M., and Cervantes, M. (1994) Cloning of a cDNA encoding ATP sulfurylase from *Arabidopsis thaliana* by functional expression in *Saccharomyces cerevisiae. Plant Physiol.* **105,** 897–902.

3. Riesmeier, J., Willmitzer, L., and Frommer, W. (1992) Isolation and characterisation of a sucrose carrier cDNA from spinach by functional expression in yeast. *EMBO* **11,** 4705–4713.
4. Sentenac, H., Bonneaud, N., Minet, M., Lacroute, F., Salmon, J., Gaymard, F., and Grignon, C. (1992) Cloning and expression in yeast of a plant potassium ion transport system. *Science* **256,** 663–665.
5. Parent, S., Fenimore, C., and Bostain, K. (1985) Vector systems for the expression, analysis and cloning of DNA sequences in *S.cerevisiae. Yeast* **1,** 83–138.
6. Baudin, A., Ozier-Kalogeropoulos, O., Denouel, A., Lacroute, F., and Cullin, C. (1993) A simple and efficient method for direct gene deletion in *Saccharomyces cerevisiae. Nucleic Acids Res.* **421,** 3329,3330.
7. Lawrence, C. (1991) Classical mutagenesis techniques. *Methods Enzymol.* **194,** 273–281.
8. Dohmen, R., Strasser, A., Höner, C., and Hollenberg, C. (1991) An efficient transformation procedure enabling longterm storage of competent cells of various yeast genera. *Yeast* **7,** 691,692.
9. Manivasakam, P. and Schiestl, R. (1993) High efficiency transformation of *Saccharomyces cerevisiae* by electroporation. *Nucleic Acids Res.* **421,** 4414,4415.
10. Schiestl, R. and Gietz, R. (1989) High efficiency transformation of intact yeast cells using single-stranded nucleic acids as a carrier. *Curr. Genet.* **16,** 339–346.
11. Brownstein, B., Silverman, G., Little, R., Burke, D., Korsmeyer, S., Schlessinger, D., and Olson, M. (1989) Isolation of single-copy human genes from a library of yeast artificial chromosome clones. *Science* **244,** 1348–1351.
12. Franzusoff, A., Rothblatt, J., and Schekman, R. (1991) Analysis of polypeptide transit through yeast secretory pathway. *Methods Enzymol.* **194,** 662–674.
13. Lorincz, A. (1984) Quick preparation of plasmid DNA from yeast. *Focus* **6,** 11.
14. Sambrook, J., Fritsch, E. F., and Maniatis, T. (1989) *Molecular Cloning: A Laboratory Manual.* Cold Spring Harbor Laboratory, Cold Spring Harbor, New York.
15. Cryer, D. R., Eccleshall, R., and Marmur, J. (1975) Isolation of Yeast DNA. *Methods Cell Biol.* **412,** 39–44.
16. Villalba, J., Palmgren, M., Berberian, G., Ferguson, C., and Serrano, R. (1992) Functional expression of plant plasma membrane H^+-ATPase in yeast endoplasmic reticulum. *J. Biol. Chem.* **267,** 12,341–12,349.
17. Romanos, M., Scorer, C., and Clare, J. (1992) Foreign gene expression in yeast: a review. *Yeast* **8,** 423–488.
18. Köhrer, K. and Domdey, H. (1991) Preparation of high molecular weight RNA. *Methods Enzymol.* **194,** 398–405.

PART VI

Techniques for Studying Chloroplast Gene Expression

CHAPTER 28

The Isolation of Intact Chloroplasts

Johnathan A. Napier
and Simon A. Barnes

1. Introduction

The isolation of intact chloroplasts from plant tissue allows the study of processes carried out in this subcellular compartment. Such processes may be enzymatic, e.g., photosynthetic oxygen-evolution, or mechanistic, like the import and targeting of precursor proteins synthesized in the cytosol. The latter is important in the case of the plant molecular biologist wishing to study the subcellular localization of a given protein.

The isolation of intact chloroplasts was first carried out by workers interested in the study of photosynthetic processes *(1)*. Although the methods used gave reasonable yields it became clear that the isolated plastids were not totally intact and that different isolation procedures yielded different classes of "intact" chloroplasts *(2)*. Subsequent advances in techniques such as the use of modified silica gels such as Percoll to provide gradients have greatly aided the isolation of intact chloroplasts *(3)*. However, the isolation appears to be somewhat species-dependent; high yields can be obtained from young leaves of pea and spinach, whereas other species have proved to be more difficult to work with. An excellent review of the various approaches to chloroplast isolation can be found in ref. *4*. Recently, tobacco, a species widely used in transgenic plant studies, has been shown to be capable of yielding intact chloroplasts suitable for study of protein import and targeting *(5)*.

The isolation of good chloroplast preparations is dependent on a number of factors to be outlined later, but the principal procedure is very

From: *Methods in Molecular Biology, Vol. 49: Plant Gene Transfer and Expression Protocols*
Edited by: H. Jones Humana Press Inc., Totowa, NJ

straightforward. Plant tissue (usually young green leaves) is homogenized in a buffered osmoticum. A low speed spin is employed to pellet the chloroplasts along with the cell wall debris and other contaminants (such as nuclei). The intact chloroplasts are separated from broken organelles and debris using a Percoll step gradient; intact chloroplasts penetrate the gradient more readily than broken ones. The intact organelles can then simply be recovered from the gradient, washed to remove the Percoll, and resuspended in the appropriate osmotically active buffer.

Intactness can be measured in a number of different ways including in vivo labeling, microscopic examination, and oxygen evolution. The simplest method is to view the chloroplasts under an oil-immersion lens at 1000X magnification. Intact chloroplasts are characterized by a "halo" surrounding them.

A more precise method (and the most widely used) is the measurement of ferricyanide-dependent oxygen evolution by intact and broken chloroplasts *(6)*. Ferricyanide is an electron acceptor, which may be reduced by water via the chloroplast thylakoid electron transport chain. However, ferricyanide cannot cross the intact chloroplast envelope and comparison of the rates of oxygen evolution on addition of ferricyanide to intact and broken chloroplast preparations is therefore a measure of the intactness of the chloroplast envelope. For the intactness assay, NH_4Cl is added as an uncoupler of electron transport from ATP synthesis allowing higher, and therefore more measurable, rates of oxygen evolution. In order to decrease the rate of oxygen evolution caused by the fixation of CO_2 by intact chloroplast preparations, D,L-glyceraldehyde is added to the assay as an inhibitor of the Calvin cycle. It is this method that will be described in detail.

2. Materials

1. 1X GRM: 330 m*M* sorbitol, 50 m*M* HEPES-KOH, pH 8.0, 2 m*M* EDTA, 1 m*M* $MgCl_2$, 1 m*M* $MnCl_2$. The 1X GRM is for the isolation of intact chloroplasts and should be frozen to a slurry prior to use.
2. 2X GRM: 660 m*M* sorbitol, 100 m*M* HEPES-KOH, pH 8.0, 4 m*M* EDTA, 2 m*M* $MgCl_2$, 2 m*M* $MnCl_2$. The 2X GRM is for the intactness assay and should be prewarmed to 20°C. 1X GRM and 2X GRM can be prepared fresh from the following stock solutions. These are also used in the preparation of the Percoll gradient solutions (PGS, *see* Section 3.1.5.).
3. 1*M* HEPES-KOH pH 8.0.
4. 0.5*M* EDTA pH 8.0.

5. 1*M* $MgCl_2$.
6. 1*M* $MnCl_2$.
7. PBF-Percoll: Percoll (Sigma, St. Louis, MO) containing 3% PEG 6000, 1% BSA, and 1% Ficoll.
8. Resuspension buffer: 330 m*M* sorbitol, 50 m*M* HEPES-KOH, pH 8.0.
9. 0.5*M* Potassium ferricyanide.
10. 2*M* D,L-glyceraldehyde.
11. 0.5*M* NH_4Cl. Solutions 9–11 are prepared weekly, store at 4°C.
12. Liquid phase oxygen electrode apparatus with control box (Hanstatech Instruments, Kings Lynn, Norfolk, UK), water bath with circulating pump, chart recorder, tungsten halogen lamp capable of 1500 μmol m^2/s PAR.
13. Polytron tissue homogenizer, muslin cloth, plastic beaker, funnels, centrifuge tubes, all chilled on ice.
14. 7–10-d-old pea seedlings grown in compost under greenhouse conditions.

3. Methods

3.1. Chloroplast Isolation

1. Harvest about 50 g of young nonwilted pea seedlings (*see* Note 1). Wash briefly in cold water to remove soil, and so on.
2. Working in the cold room from this point on, add 300 mL of 1X GRM slurry to the seedlings in a suitable plastic container and homogenize with a Polytron PT20 tissue homogenizer (Kinematica Ag, Lucerne, Switzerland) set at speed 7 for 7 s (*see* Note 2).
3. Pour the homogenate through eight layers of muslin cloth into chilled centrifuge bottles and spin at 4°C in a prechilled rotor at 2000*g* for 2 min.
4. Remove the supernatant gently (but quickly) and resuspend the large green pellet in a small volume (10 mL) of ice cold 1X GRM. Keeping the tubes on ice at all times, use an orbital shaker to resuspend the pellet gently (*see* Note 3). While this is happening, make up the Percoll step gradients.
5. Make the two solutions required for the Percoll gradient (40 and 85% Percoll gradient solution, PGS) as detailed later.
   a. 40% PGS: 10 mL PBF-Percoll in a 25 mL solution containing 330 m*M* sorbitol, 5 m*M* EDTA, 1 m*M* $MgCl_2$, 1 m*M* $MnCl_2$, 50 m*M* HEPES-KOH, pH 8.0 (made up using stock solutions).
   b. 85% PGS: 21.25 mL PBF-Percoll in a 25 mL solution with the same ingredients and concentrations as for 40% PGS.
6. Prepare Percoll steps as follows. To a Corex 30 mL centrifuge tube add 10 mL of 40% PGS. Using a syringe and a large bore needle, "inject" 10 mL of 85% PGS underneath the 40% PGS. Start this by placing the needle on the bottom of the tube. Add the 85% PGS slowly and carefully, checking that a clear interface can be seen.

7. Load the resuspended chloroplasts onto the top of the gradient using a cutoff 5 mL Gilson (Paris) tip; allow approx 5 mL per gradient made as above (*see* Note 4). Centrifuge for 20 min in a swing-out rotor at 5000*g*, 4°C. After centrifugation distinct bands should be present at the interfaces between the 40/85% PGS and the GRM/40% PGS.
8. Aspirate off the broken chloroplasts present in the GRM/40% PGS. Remove the intact chloroplasts, found at the interface between the 40/85% PGS, using a Pasteur pipet with the tip snapped off and place in a Corex tube on ice. Add approx 25 mL of GRM (i.e., fill the tube) and mix gently. Spin at 2000*g* for 5 min, 4°C. Repeat this process several times, each time with fresh GRM and gentle resuspension of the chloroplast pellet using the Pasteur pipet. This step is essential for the removal of the Percoll (*see* Note 5).
9. After the final wash, resuspend the pellet in a small volume of resuspension buffer. Again, this may be aided by orbital shaking.
10. Measure the chlorophyll concentration by adding 5 µL of the chloroplast solution to 995 µL of 80% acetone solution and vortexing. Spin in a microfuge to remove debris and read the absorbance at 652 nm. Calculate chlorophyll concentration by multiplying the absorbance by 2.9 to give the concentration in mg/mL. The chloroplasts are now ready to be assayed for intactness.

3.2. Measurement of Intactness

1. Assemble and calibrate the oxygen electrode apparatus as directed by the manufacturer's instructions. The water bath should be set to a temperature of 20°C.
2. The lamp for the illumination of the chloroplast suspension should be placed 20–30 cm from the O_2 electrode chamber to yield an incident light intensity of 1500 µmol m^2/s PAR. The response of the O_2 electrode to the heating effect of the lamp may be assessed by adding 2 mL water to the electrode chamber and measuring the drift in the signal from the electrode caused when the light is switched on. If the drift is significant, then the lamp may be moved further from the chamber or the drift may be subtracted from the rates of ferricyanide-dependent O_2 evolution in the calculation of intactness.
3. Rinse the O_2 electrode chamber carefully with distilled water prior to each intactness assay, and in between the measurements of the rates for the intact and broken preparations.
4. To measure intactness, add 1 mL 2X GRM and 0.9 mL distilled water to the O_2 electrode chamber.
5. Add 100 µL of chloroplast preparation (equivalent to 50 µg chlorophyll).
6. Add 10 µL of 0.5*M* potassium ferricyanide, 10 µL D,L-glutaraldehyde and 10 µL 0.5*M* NH_4Cl.

7. Switch on the light and record the rate of O_2 evolution by the intact chloroplasts until a steady state is reached *(a)*.
8. Remove the chloroplast suspension and reagents from the O_2 electrode chamber and rinse thoroughly with distilled water.
9. Add 0.9 mL distilled water to the O_2 electrode chamber, and add 100 μL chloroplast preparation. Incubate for 1 min to allow the chloroplasts to rupture fully.
10. Add 1 mL 2X GRM, and reagents as in step 6.
11. Switch on the light and record the rate of O_2 evolution by the broken chloroplasts until a steady state is reached *(b)*.
12. Calculate the intactness of the chloroplast preparation:

$$\text{Intactness (\%)} = 100\ (b - a)/b.$$

Intactness values of greater than 95% should be expected routinely with the preparation procedure described for pea chloroplasts (*see* Note 6).

4. Notes

1. The condition of the plant material will determine the yield of intact chloroplasts. Old leaves or water-stressed tissue will invariably give much lower yields. Plants exposed to high temperatures such as greenhouse conditions in summer are also prone to decreased yields.
2. All manipulations should be carried out in a cold room. The need to be gentle with the isolated plastids cannot be overemphasized. Homogenization by the Polytron should be brief, otherwise the initial pellet will be mainly thylakoid membranes, not intact chloroplasts.
3. The resuspension of the chloroplasts after the initial spin will take about 30 min. Do not attempt to rush the resuspension of the organelles by vortexing; the process may be aided by very gentle use of a Pasteur pipet.
4. Take care not to overload the Percoll gradients, otherwise incomplete separation will occur. Usually a pellet containing starch and debris is seen at the bottom of the tube.
5. The complete removal of the Percoll solution by washing in GRM is essential for reproducible results.
6. Isolated chloroplasts have a relatively short shelf life; they should be used within a few hours of isolation.

Acknowledgments

Simon Barnes was the recipient of a Gatsby studentship in J. C. Gray's lab in the Department of Plant Sciences, Cambridge, UK.

References

1. Kalberer, P. P., Buchanan, B. B., and Arnon D. I. (1967) Rates of photosynthesis by isolated chloroplasts. *Proc. Natl. Acad. Sci.* **57,** 1542–1549.
2. Hall, D. O. (1972) Nomenclature for isolated chloroplasts. *Nature New Biol.* **235,** 125,126.
3. Mills, W. R. and Joy, K. W. (1980) A rapid method for isolation of purified, physiologically active chloroplasts, used to study the intracellular distribution of amino acids in pea leaves. *Planta* **148,** 75–83.
4. Walker, D. A. (1980) Preparation of higher plant chloroplasts. *Methods Enzymol.* **69,** 94–104.
5. Madueno, F., Napier, J. A., Cejudo, F. J., and Gray, J. C. (1992) Import and processing of the precursor of the Rieske FeS protein of tobacco chloroplasts. *Plant. Mol. Biol.* **20,** 289–299.
6. Walker, D. A. (1987) *The Use of the Oxygen Electrode and Flourescence Probes in Simple Measurements of Photosynthesis,* Oxygraphics Limited, Sheffield, UK, pp. 121–126.

CHAPTER 29

In Vitro Protein Import by Isolated Chloroplasts

Johnathan A. Napier

1. Introduction

In Chapter 28, a method for the isolation of intact chloroplasts was described. The availability of this technique has allowed researchers the opportunity to investigate the mechanisms of chloroplast biogenesis. The chloroplast is a complex organelle, being made up of six subcompartments. The thylakoid membrane, which makes up about half the chloroplast, is the site of photosynthetic electron transport, a process involving four multimeric protein complexes *(1)*. The majority of chloroplast proteins are nuclear-encoded and synthesized in the cytosol; therefore, there must be some mechanism that allows proteins to cross the chloroplast envelope (the membrane that delineates the organelle) yet also discriminates between the plastid and other organelles, like mitochondria *(2)*. Early work in the then embryonic field of chloroplast targeting demonstrated that nuclear-encoded chloroplast proteins were synthesized as precursor proteins. This precursor was processed to the mature form when the precursor was taken up by isolated chloroplasts. This lead to the proposal that the precursor protein interacted with a carrier protein (i.e., a receptor) on the chloroplast envelope that allowed import into the organelle *(3)*. Subsequent research demonstrated that an N-terminal extension, termed the transit sequence, present in the precursor, was absolutely required for chloroplast import (reviewed in *2*; *see also* Chapter 30, this volume).

From: *Methods in Molecular Biology, Vol. 49: Plant Gene Transfer and Expression Protocols*
Edited by: H. Jones Humana Press Inc., Totowa, NJ

However, the initial experiments relied on (and were hindered by) the generation of [^{35}S]methionine-labeled precursor proteins by in vitro translation of total mRNA populations in a cell-free system (such as wheat germ or rabbit reticulocyte lysates) with the gene product of interest identified by immunoprecipitation *(4)*. The obvious drawback of this approach was the need for a monospecific antibody; also, the amount of target protein was only ever likely to be a small percentage of the total mRNA pool.

With the development of in vitro transcription vectors *(5)*, the researcher was afforded the opportunity to make RNA to a specific cDNA or gene simply by cloning the DNA behind a bacteriophage RNA polymerase promoter. This in vitro synthesized RNA could then be translated in the cell-free system to yield unique [^{35}S]methionine-labeled proteins. Such transcription vectors are now widely available from a number of commercial sources and have made it much easier to set up an assay to investigate the targeting of a specific protein. In the case of targeting to the chloroplast, the assay takes the form of posttranslational import into isolated pea chloroplasts. The chloroplasts can be added to the in vitro synthesized protein in the presence of ATP, and the chloroplasts then protease-treated and refractionated to determine whether or not the protein of interest is protease-protected by the chloroplasts as a result of import into the organelle. For recent experimental examples of this technique, *see* refs. *6* and *7*.

This chapter will describe the in vitro synthesis of RNA from a cloned cDNA template and the production of [^{35}S]labeled protein in a cell-free translation system. The use of this protein for in vitro import into isolated chloroplasts will also be detailed. The procedure can be split into three sections: in vitro synthesis of the RNA, in vitro translation of the RNA in the presence of [^{35}S]methionine, and import into isolated chloroplasts followed by analysis on SDS-PAGE and autoradiography (the latter will not be described in this chapter since they are covered in detail elsewhere in this volume; *see* Chapters 34 and 35). It is usually best to check the efficiency of the in vitro transcription/translation prior to attempting chloroplast imports; efficient in vitro synthesis and labeling of the protein is a prerequisite for successful chloroplast import assays.

2. Materials

2.1. In Vitro Transcription

1. 5X Transcription buffer: 200 m*M* Tris-HCl, pH 7.5, 30 m*M* $MgCl_2$, 10 m*M* spermidine, 50 m*M* NaCl.

2. 100 m*M* Dithiothreitol (DTT).
3. 1 mg/mL Acetylated BSA.
4. RNase inhibitor; RNasin @ 40 U/μL.
5. rNTP mix: 5 m*M* ATP, CTP, UTP, 0.5 m*M* GTP.
6. rGTP chase: 5 m*M* GTP.
7. Cap: 1 m*M* $m^7G(5')ppp(5')G$ (from Ambion, Austin, TX). All solutions should be sterilized by filtration and stored at –20°C.
8. RNA polymerase: SP6/T7 depending on vector.
9. Sterile distilled water.
10. DNA template: cDNA cloned into a vector such as pGEM11Zf(-) (Promega, Madison, WI), at ~1 mg/mL and ultrapure.
11. Water bath set at 40°C. Sterile Eppendorf tubes, pipet tips.

2.2. *In Vitro* Translation

1. Wheat germ cell-free extract (available from Promega/Amersham, Arlington Heights, IL).
2. 1*M* potassium acetate.
3. 1 m*M* amino acid mix minus methionine (this and step 2 are usually supplied with the wheat germ extract).
4. $[^{35}S]$methionine (15μCi/μL).
5. Water bath set at 25°C. Sterile Eppendorf tubes, pipet tips.

2.3. Chloroplast Import

1. Isolated chloroplasts, 1X GRM, Percoll gradient solutions, described in Chapter 28.
2. Mg^{2+} ATP: Stock solution 100 m*M* in Tris-HCl, pH 7.0.
3. Thermolysin (Protease X, from Sigma, St. Louis, MO) 10 mg/mL in 1X GRM.
4. Standard SDS-PAGE reagents and equipment.
5. 5% Trichloroacetic acid (TCA) solution to fix gel, 2*M* Tris (unbuffered) to neutralize TCA-treated gel.
6. Autoradiograph cassette, X-ray film.

3. Methods

3.1. *In Vitro* Transcription

1. Make up the following master mix in a sterile tube on ice (*see* Note 1): 10 μL 5X reaction buffer, 5 μL 0.1*M* DTT, 5 μL 1 mg/ml BSA, 1 μL RNasin, 5 μL rNTP mix, 1 μL 1 m*M* Cap, 13 μL sterile distilled water. This is enough for five 10-μL reactions. The reaction can be scaled up if required. It is important to use sterile and clean components for the synthesis of the RNA.

2. For each reaction, aliquot out 8 μL of the master mix into sterile Eppendorf tubes on ice. Add 1 μL of plasmid DNA (transcription vector containing cDNA of interest) and 1 μL of the appropriate RNA polymerase. Spin briefly, mix gently, and incubate at 40°C for 20 min (*see* Note 2).
3. Add 1 μL of rGTP chase, mix, and incubate for a further 40 min at 40°C.
4. The RNA is now ready for testing by in vitro translation. No further purification is needed. Store the RNA at –80°C if not being used immediately.

3.2. In Vitro Translation

1. Make up the following master mix in a sterile tube on ice: 25 μL wheat germ extract, 4 μL 1 m*M* amino acid mix minus methionine, 0–3 μL 1*M* potassium acetate, 2 μL [^{35}S]methionine (15 μCi/μL), 1 μL RNasin, sterile distilled water to 45 μL. The amount of potassium acetate can be varied to enhance the translation of a given RNA. It may be worthwhile determining optimal concentrations of potassium acetate for individual RNAs. Obviously, care should be taken when using the radiolabel. [^{35}S]methionine is labile and should not be allowed to stand at room temperature for any length of time.
2. For each reaction, aliquot 9 μL of the master mix into a sterile tube on ice. Add 1 μL of RNA and mix. Incubate at 25°C for 1–2 h.
3. Analyze the translation products by SDS-PAGE. To 1 μL of translation product add 7 μL of SDS-PAGE sample loading buffer, mix, and incubate at 95°C for 2 min. Analyze on an appropriate percentage SDS-PAGE gel, with either radioactive or prestained molecular weight markers.
4. When the gel is run to completion, dismantle apparatus and fix the gel by pouring boiling 5% TCA (~200 mL) over it (*see* Note 3). Allow to cool for 10 min, carefully remove the TCA, and wash with water. Remove the water and add ~ 200 mL of 2*M* unbuffered Tris. Allow to stand for 10 min. Remove the Tris, wash briefly in water, and then dry the gel using a gel drier. Expose the gel to preflashed X-ray film at room temperature. The labeled protein band should be clearly seen after an overnight exposure. Assuming this is the case, import of the in vitro translation product into intact chloroplasts may now be attempted.

3.3. Import into Isolated Chloroplasts

1. Set up the wheat germ in vitro translation as described earlier. You may wish to increase the total reaction volume/precursor protein to 25 μL final volume.
2. While the labeled precursor is being synthesized, isolate intact chloroplasts as described in the previous chapter (*see* Note 4). Remember to keep the 40% PGS solution, you will need it later. Determine the chlorophyll con-

centration. Import is carried out in a reaction volume of ~300 μL with chloroplasts equivalent to 25 μg chlorophyll. Therefore, add the appropriate volume of chloroplasts (depending on your chlorophyll concentration) to a tube on ice. Make up to volume with 1X GRM containing 1 m*M* methionine and 5 m*M* Mg^{2+} ATP (added from stock solutions; *see* Note 5). To start the import reaction add ~25 μL of freshly synthesized in vitro translation product. Incubate at 25°C for 20 min on the bench; there is no need for supplementary lighting of the samples.

3. Split the import into two equal volumes and place on ice. Treat one sample with thermolysin to final concentration of 100 μg/mL and allow both to stand on ice for 20 min. While this is incubating, aliquot 1 mL of 40% PGS (from chloroplast isolation) into two Eppendorfs.
4. Gently layer the + and - thermolysin treated reactions on top of the two Percoll-containing tubes, and spin in a microfuge at low speed setting (~6000*g*) for 5 min. Intact chloroplast will pellet to the bottom of the tube. Discard supernatant (which may be radioactive—*caution!*) and resuspend the pellet in 1 mL of 1X GRM.
5. Pellet the washed chloroplasts by microfuging at high speed (~13,000*g*) for 2 min. Resuspend in a small volume (~20 μL) of SDS-PAGE sample buffer. Heat to 95°C for 2 min and run ~5 μL on an SDS-PAGE gel. Also run a small amount (0.5 μL) of the in vitro translation product alongside the two (-/+ thermolysin) import reactions. Autoradiograph the dried gel and compare the three tracks. Usually, if import has occurred a protease-protected band will be seen in the + thermolysin treatment. This band would also be expected to be of a smaller size than the in vitro translation product of the precursor owing to proteolytic processing by the chloroplast stromal processing peptidase. An example of a typical experiment is shown in Fig. 1. It is clear that the precursor form of the protein has been processed to a smaller size and that this smaller form is protected from thermolysin digestion by being inside the chloroplast envelope. It can, therefore, be considered to have been imported by the isolated chloroplasts (*see* Notes 6 and 7).

4. Notes

1. All the outlined procedures can be carried out over 2 d. On the first day, it is best to synthesize in vitro the RNA and check it by translation. The gel can be exposed overnight and if the protein of interest is clearly visible on the autoradiograph the next morning, then the isolation of intact chloroplasts and import can be carried out. It is important to ensure that the translation product is suitably "hot"; if methionine residues are infrequent in the protein, then either [^{35}S]cysteine or ^{14}C amino acid mixtures can be used.

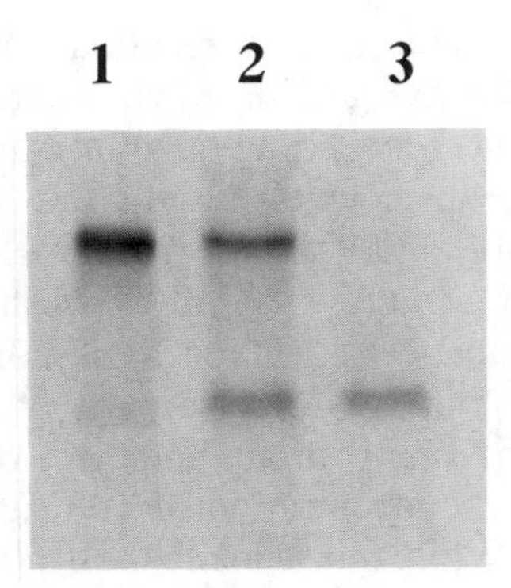

Fig. 1. The import and processing of the precursor for the light-harvesting complex II (LHCII) into isolated pea chloroplasts. The precursor is synthesized as a polypeptide of ~31 kDa. On import (as defined by protease-protection) the precursor of LHCII is processed to the mature form of ~27 kDa. The in vitro synthesis of the precursor and its import into chloroplasts was carried out as described in the text. The tracks are: 1. precursor; 2. import, - thermolysin treatment; 3. import, + thermolysin treatment. Napier, J. A. and Gray, J. C. (University of Cambridge), unpublished results.

2. The point where most problems occur is at the transcription stage; the wheat germ extract is usually very efficient at translating any RNA template. Transcription can be limited by a number of factors, although amount and quality of DNA template seem very important, as are sterile reagents. Using an uncut plasmid template (as described here) appears to improve transcription. The capping of the RNA molecule is thought to increase its translatability in vitro. If problems persist with the transcription there are a number of technical manuals available; Promega's "Protocols and Applications" has a large section on transcription.
3. Boiling 5% TCA is used to fix the gels since this will remove any *t*-RNA molecules that have picked up the [^{35}S]methionine. If these radioactive *t*-RNAs are not removed they tend to run in the middle of the gel as a large fuzzy band.
4. It is important to use "fresh" in vitro synthesized protein for the import to maintain a high level of translocation. The use of wheat germ extracts rather than rabbit reticulocyte is owing to the observation of increased chloroplast lysis in the presence of rabbit reticulocyte lysates. It also might be expected that wheat germ extracts would be more likely to contain any cytosolic factors required for chloroplast import.

5. ATP, which is required to drive import should be made up in small aliquots and frozen. Each aliquot should be used only once, i.e., not subject to freeze–thaw cycles.
6. An additional control to verify import is to have a third treatment, thermolysin in the presence of 1% Triton X100. This will solubilize the chloroplasts and allow the protease access to the translocated and processed protein; the result of this treatment should be the absence of any band on the autoradiograph. The treatment will demonstrate the protease-sensitive nature of the protein under study, i.e., determine if the resistance to thermolysin is owing to chloroplast protection or inherent for the protein itself.
7. Once import into isolated chloroplasts has been clearly demonstrated then it is possible to subfractionate the organelle into stroma, thylakoids, and envelope fractions to gain further information about targeting. The next chapter (Chapter 30) deals with targeting to the different compartments present in the chloroplast.

References

1. Knight, J. S., Madueno, F., and Gray, J. C. (1993) Import and sorting of proteins by chloroplasts. *Biochem. Soc. Trans.* **21,** 31–35.
2. Keegstra, K., Olsen, L. J., and Theg, S. M. (1989) Chloroplast precursors and their transport across the envelope membrane. *Ann. Rev. Plant Physiol. Plant Mol. Biol.* **40,** 471–501.
3. Highfield, P. E. and Ellis, R. J. (1978) Synthesis and transport of the small subunit of chloroplast ribulose bisphosphate carboxylase. *Nature* **271,** 420–424.
4. Dobberstein, B., Blobel, G., and Chua, N.-H. (1978) *In vitro* synthesis and processing of a putative precursor for the small subunit of ribulase-1,5-bisphosphate carboxylase of *Chlamydomonas reinhardtii. Proc. Natl. Acad. Sci. USA* **74,** 1082–1085.
5. Melton, D. A., Krieg, P. A., Rebagliati, M. R., Maniatis, T., Zinu, K., and Green M. R. (1984) Efficient *in vitro* synthesis of biologically active RNA and RNA hybridisation probes from plasmids containing a bacteriophage SP6 promoter. *Nucleic Acids Res.* **12,** 7035–7056.
6. Napier, J. A., Hoglund, A.-S., Plant, A. L., and Gray J. C. (1992) Chloroplast import of the precursor of the gamma subunit of pea chloroplast ATP synthase. *Plant Mol. Biol.* **20,** 737–741.
7. Moore, T. and Keegstra, K. (1993) Characterisation of a cDNA clone encoding a chloroplast-targeted Clp homologue. *Plant Mol. Biol.* **21,** 525–537.

CHAPTER 30

Targeting of Foreign Proteins to the Chloroplast

Johnathan A. Napier

1. Introduction

The chloroplast is the product of two genomes with the majority of its polypeptide complement encoded by nuclear DNA and synthesized on cytosolic ribosomes. Proteins destined for the chloroplast must therefore contain targeting information to allow their correct localization. It has been clear now for a number of years that nuclear-encoded chloroplast proteins are synthesized as precursors with a cleavable N-terminal signal sequence *(1)*. This N-terminal extension, termed the transit sequence, is removed during or just after translocation into the organelle by a soluble peptidase resident in the chloroplast stroma *(2)*. The transit sequence has been shown to be necessary and sufficient to direct proteins across the chloroplast envelope and into the stroma; the transit sequence of the small subunit of Rubisco (abbreviated SSU) has been shown to direct import into chloroplasts of cytosolic, mitochondrial and bacterial proteins when placed at their N-terminus *(3)*. Conversely, SSU devoid of its transit sequence cannot be imported into isolated chloroplasts *(4)*.

A large number of transit sequences of precursor proteins destined for the chloroplast have now been deduced by molecular cloning of cDNAs. The amino acid sequences of these domains show little homology at the level of primary or secondary structure *(5)*, although the cleavage site for the removal of the transit peptide by the stromal processing peptidase (SPP) does appear to be a conserved motif V/I-X-A/C-A *(6)*.

From: *Methods in Molecular Biology, Vol. 49: Plant Gene Transfer and Expression Protocols*
Edited by: H. Jones Humana Press Inc., Totowa, NJ

It is also clear that although the chloroplast transit sequence is sufficient and necessary for chloroplast import, it does not direct the insertion of proteins into the thylakoid membrane or translocation across this membrane into the thylakoid lumen. This requires additional targeting information that may take the form of either a second cleavable N-terminal domain or a noncleavable signal present in the mature protein. For example, the 33 kDa protein of the photosynthetic oxygen-evolving complex has an N-terminal chloroplast import domain that is removed by the SPP to expose (at the N-terminus) a thylakoid transfer domain that serves to target the intermediate form to the thylakoid. The thylakoid transfer domain is then cleaved (to produce the mature form of the protein) by a thylakoid processing peptidase that shows target sequence specificity similar to bacterial leader peptidase *(7)*. However, some integral thylakoid membrane proteins such as the light harvesting complex II (LHCII) or the Rieske iron-sulfur protein of the *b/f* complex are targeted by noncleavable signals. These domains take the form of hydrophobic membrane spans, but there appears to be no homology between different proteins *(5)*.

Targeting to the chloroplast envelope was until recently poorly understood; however, ongoing studies have started to indicate the route of targeting and the type of signal sequences involved *(8)*. The chloroplast envelope, although only comprising ~2% of the chloroplast, is in fact made up of three distinct compartments: the outer envelope, the inner envelope, and the intermembrane space. The outer envelope acts as the interface between the cytosol and the chloroplast and is the presumed site of any receptor molecules for the transit sequence. Targeting of proteins to the outer envelope is different from all other organellar compartments in that it does not require a cleavable signal sequence. Moreover, the process itself appears not to be receptor-mediated or ATP-requiring *(9)*, unlike targeting to all other chloroplast compartments *(3,8)*. No analysis has been carried out on outer envelope proteins to determine domains involved in targeting. A cDNA encoding an intermembrane space protein has yet to be isolated so targeting to this compartment remains undefined.

Targeting to the inner envelope is better understood. Initially it was thought that targeting occurred via import into the stroma followed by "export" into the inner envelope. This idea was given weight by the fact that the inner envelope protein, the phosphate translocator (PiT), had a cleavable transit sequence that was shown to be processed by the SPP *(10)*. However, recent work has shown that the mature portion of the PiT contains targeting information that may serve as a stop-transfer domain; a chimera of the PiT transit sequence and the mature portion of SSU was targeted to the stroma, indicating that the envelope targeting information for the PiT did not reside in the transit sequence. If, however, a chimera of the transit sequence and the first putative membrane span of the mature PiT was fused to the mature form of SSU, then the SSU was targeted to the envelope *(3)*. Similar results have been obtained for the targeting of the plastid ATP/ADP carrier encoded by the *Bt1* locus, i.e., envelope targeting information resides in the mature portion of the protein *(11)*. Thus, the transit sequence serves perhaps to engage the precursor with the chloroplast import machinery and the protein's translocation across the envelope is halted by stop-transfer signals. However, it is likely that the chloroplast envelope stop-transfer signals are significantly different from those of other systems since the addition of ER stop-transfer signals to SSU failed to prevent its accumulation in the stroma *(12)*. The removal of an inner envelope protein's transit sequence by the SPP may be owing to the entry of only this portion of the protein into the stroma. This might imply that the SPP would be expected to be in close contact with the import machinery and the chloroplast envelope.

Thus it can be seen that there is a considerable body of information regarding the targeting domains of various chloroplast proteins. The advent of modern molecular biology techniques such as PCR means that creating fusions or chimeras is a relatively simple process, whereas previously it would have been a daunting prospect. It is therefore possible to manipulate the targeting information of a given protein and so alter its subcellular location. Such a proposition may be attractive in terms of genetic engineering for crop improvement.

2. Materials and Methods

There is no single "method" that can be described for targeting foreign proteins to chloroplast compartments since each example will be different, both in terms of the targeting requirements and the passenger protein. A chimeric fusion between the targeting domain of a chloroplast-localized protein and the protein that you wish target to the chloroplast (i.e., the passenger) has to be generated. It is not the aim of this chapter to describe PCR methods, since they are covered in detail in other chapters and in a

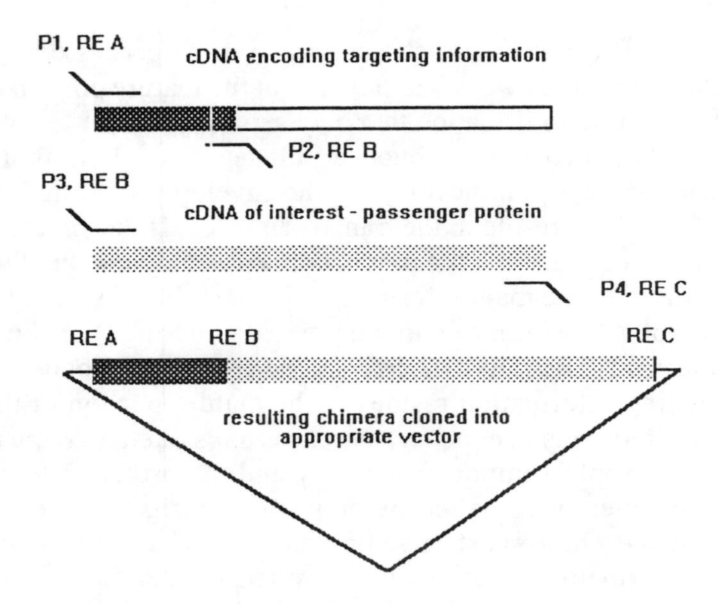

Fig. 1. Schematic representation of the formation of a chloroplast-targeted chimera using PCR. P = oligonucleotide primer specific for cDNA under amplification. RE = restriction site built into 5' end of primer. Two separate PCR products are generated from the two templates indicated; the two PCR products can be subcloned or directly ligated together using the restriction site B introduced into both fragments. The chimera can then be directionally cloned into the vector of choice using restriction sites A and C.

number of technical volumes *(13–15)*. It is the intention of this chapter to indicate the possible fusions that can be created to potentially alter the targeting of a given protein. However, a basic outline of the principles of using PCR to create a chimera will be given later. The precise domains to be used to target the passenger protein also are indicated later.

Two PCR products are generated, one encoding the targeting information and the second the passenger protein. The generation of the first PCR product can either be from available cDNA clones or from ds-cDNA synthesized from the appropriate mRNA template. A chimera is most simply created by the ligation of the two PCR products together. This is done by the incorporation of restriction sites into the 5' end of the oligonucleotide primers used for the PCR reaction (a schematic representation is shown in Fig. 1). Thus, by the resulting addition of compatible sites to the 3' end of the first PCR product and the 5' end of the second it is relatively easy to create a fusion. Obviously, the choice of restriction sites will be dictated by any internal sites present in either of the PCR products; therefore, it is important to have determined any restriction sites present in either cDNA before embarking on the PCR-mediated formation of chimeras. It is also equally vital to ensure that the chimera formed by the ligation of the two PCR products together produces an uninterrupted reading frame (*see* Note 1).

2.1. Targeting to the Stroma

The fusion of the transit sequence of any chloroplast targeted protein to the N-terminus of the passenger protein should result in localization in the stroma. A particularly well-characterized transit sequence is that of SSU and it is this that is recommended for stromal targeting. A PCR product should be generated that contains the initiating methionine of the SSU transit sequence, up to just beyond the SSP processing site in the mature portion of the protein. This should yield a PCR product of about 200 bp including the additional restriction sites at the 5' end of the primers. A suitable template for this reaction is the cDNA encoding the precursor of SSU from pea described in ref. *4*. This can then be subcloned into an appropriate vector, and then ligated to the similarly cloned PCR product encoding the passenger protein. The ligated chimera can then be cloned into, e.g., a transcription vector. The resulting chimera should be sequenced to check for any errors incorporated by the Taq polymerase misreading (this is true for all fusions; *see* Notes 2 and 3).

2.2. Targeting to the Thylakoid Membrane

The information for thylakoid targeting is contained as a noncleavable signal in the mature portion of proteins such as LHCII or the Rieske iron-sulfur protein. In the case of the Rieske protein this information is conveniently positioned at the N-terminus of the mature protein *(16)*; the precursor protein contains an N-terminal transit sequence that directs import into the chloroplast stroma, and once the transit is removed, the noncleaved membrane insertion sequence directs thylakoid integration. For the purpose of targeting a passenger protein across the chloroplast envelope and stroma and into the thylakoid membrane, then a PCR product from the first methionine of the transit sequence up to the end of the hydrophobic portion of the mature sequence (about 55 residues of the mature Rieske protein) should be synthesized. The template for this reaction should be the cDNA described in *(17)* encoding the precursor of the tobacco Rieske iron-sulfur protein. Since the transit sequence of Rieske is about 50 amino acids long, the PCR product should be about 330 bp, including restriction sites. This can then be ligated to the N-terminus of the passenger protein to make the appropriate chimera.

2.3. Targeting to the Thylakoid Lumen

Proteins targeted to the thylakoid lumen (thylakoid-translocating proteins) are synthesized as precursor proteins with a bipartite signal sequence; the N-terminal region allows chloroplast envelope transfer and is cleaved by the SPP to reveal a second signal sequence; this directs integration and translocation across the thylakoid and is in turn removed by the thylakoid processing peptidase. A well defined bipartite transit peptide is that of the 33-kDa protein of the oxygen-evolving complex of photosystem II (abbreviated to OEC33). The generation of a fusion between the OEC33 transit (81 amino acids long) and the passenger should result in lumenal targeting. Therefore a PCR product spanning the first methionine of the precursors up to just beyond the start of the mature protein should be synthesized, giving a PCR product of about 270 bp. The template suitable for the generation of the targeting information is the precursor of the pea OEC33 protein described in ref. *18*. This should then be fused to the passenger protein. *See also* ref. *19* for further details about sorting pathways for OEC33.

2.4. Targeting to the Chloroplast Inner Envelope

Proteins targeted to the inner envelope require an N-terminal chloroplast import signal such as the transit sequence of SSU and also an inner envelope target sequence. In the case of PiT the inner envelope target information is at the N-terminal of the mature protein, with the protein being synthesized as a precursor with a cleaved transit sequence. Therefore to target a protein to the chloroplast inner envelope, a PCR product spanning the first methionine of the PiT precursor up to the second putative membrane span of the mature protein should be generated. Since the transit sequence is 72 residues and up to the second membrane span is a further 75 residues, a PCR product of about 450 bp should be generated. The template for this reaction is the pea PiT cDNA described in ref. *10*. This can be joined to the passenger protein to give an envelope targeted chimera (*see* Note 4).

2.5. Targeting to the Other Chloroplast Compartments

At present no information is available regarding targeting to the two other plastidic locations (the outer envelope and the intermembrane space). However, it might be possible to target a passenger protein to the outer envelope by fusing it to the small outer envelope protein OE7 described in ref. *9*; a PCR product encoding the entire protein (62 residues ~190 bp) could be fused to a passenger protein. This, however, is speculation.

3. Notes

1. It is vital to sequence all chimeras synthesized by PCR.
2. The simplest way of testing the efficiency of targeting of the chimera is by in vitro import of the radiolabeled precursor into intact chloroplasts as described in Chapter 29. The intact chloroplasts can be subfractionated and the imported protein analyzed as described in ref. *10*.
3. *Caution:* The effect of the passenger protein on the targeting information of the chimera is an unknown variable. In some cases it may be that inefficient targeting or even mistargeting may occur simply as a result of the choice of passenger protein. This may be owing to cryptic targeting information present in the passenger protein. The results of various different passenger proteins on the targeting efficiencies of thylakoid-translocating bipartite signal sequences are detailed in ref. *19* and *20*.
4. All the cDNAs described for the PCR-mediated generation of targeting chimeras should be available on request from the research groups that isolated and characterized them.

References

1. Keegstra, K., Olsen, L. J, and Theg, S. M. (1989) Chloroplastic precursors and their transport across the envelope membranes. *Annu. Rev. Plant Physiol. Plant Mol. Biol.* **40,** 471–501.
2. Robinson, C. and Ellis, R. J. (1984) Transport of proteins into chloroplasts. Partial purification of chloroplast protease involved in the processing of imported precursor polypeptides. *Eur. J. Biochem.* **142,** 337–342.
3. Knight, J. S., Madueno, F., and Gray, J. C. (1993) Import and sorting of proteins by chloroplasts. *Biochem. Soc. Trans.* **21,** 31–35.
4. Anderson, S. and Smith, S. M. (1986) Synthesis of the small subunit of ribulose-bisphosphate carboxylase from genes cloned into plasmids containing the SP6 promoter. *Biochem. J.* **240,** 709–715.
5. Theg, S. M. and Scott, S. V. (1993) Protein import into chloroplasts. *Trends Cell Biol.* **3,** 186–190.
6. Von Heijne, G., Steppuhn, J., and Hermann, R. G. (1989) Domain structure of mitochondrial and chloroplast targeting peptides. *Eur. J. Biochem.* **180,** 535–545.

7. Gray, J. C. (1989) Targeting and assembly of chloroplast proteins. *Curr. Opin. Cell Biol.* **1,** 706–711.
8. Soll, J. and Alefsen, H. (1993) The protein import apparatus of chloroplasts. *Physiol. Plant* **87,** 433–440.
9. Saloman, M., Fischer, K., Flugge, U.-I., and Soll, J. (1990) Sequence analysis and protein import studies of an outer chloroplast envelope polypeptide. *Proc. Natl. Acad. Sci. USA* **87,** 5778–5782.
10. Willey, D. L., Fischer, K., Wachter, E., Link, T. A., and Flugge, U.-I. (1991) Molecular cloning and structural analysis of the phosphate translocator from pea chloroplasts and its comparison to the spinach phosphate translocator. *Planta* **183,** 451–461.
11. Li, H.-M., Sullivan, T. D., and Keegstra, K. (1992) Information for the targeting to the chloroplast inner envelope membrane is contained in the mature region of the maize *Bt-1*-encoded protein. *J. Biol. Chem.* **267,** 18,999–19,004.
12. Lubben, T. H., Bansberg, J., and Keegstra, K. (1987) Stop-transfer regions do not halt translocation of proteins into chloroplasts. *Science* **238,** 1112,1113.
13. Horton, R. M. and Pease, L. R. (1991) Recombination and mutagenesis of DNA sequences using PCR, in *Directed Mutagenesis* (McPherson, M. J., ed.), Oxford University Press, Oxford, UK, pp. 217–247.
14. McPherson, M. J., Quirke, P., and Taylor, G. R., eds. (1991) *PCR—A Practical Approach.* Oxford University Press, Oxford, UK.
15. White, B. A., ed. (1993) *Methods in Molecular Biology,* vol. 15: *PCR Protocols.* Human, Totowa, NJ.
16. Madueno, F., Bradshaw, S. A., and Gray, J. C. (1994) The thylakoid-targeting domain of the chloroplast Rieske iron-sulphur protein is located in the N-terminal hydrophobic region of the mature protein. *J. Biol. Chem.* **269,** 17,458–17,463.
17. Madueno, F., Napier, J. A., Cejudo, F. J., and Gray, J. C. (1992) Import and processing of the precursor of the Rieske FeS protein of tobacco chloroplasts. *Plant Mol. Biol.* **20,** 289–299.
18. Wales, R., Newman, B. J., Pappin, D., and Gray, J. C. (1990) The extrinsic 33 kDa polypeptide of the oxygen-evolving complex of photosystem II is a putative calcium-binding protein and is encoded by a multi-gene family in pea. *Plant Mol. Biol.* **12,** 439–451.
19. Henry, R., Kapazoglou, A., McCaffery, M., and Cline, K. (1994) Differences between lumen targeting domains of chloroplast transit peptides determine pathway specificity for thylakoid transport. *J. Biol. Chem.* **269,** 10,189–10,192.
20. Clausmeyer, S., Klosgen, R. B., and Herrmann, R. G. (1993) Protein import into chloroplasts; the hydrophilic lumenal proteins exhibit unexpected import and sorting specificities in spite of structurally conserved transit peptides. *J. Biol. Chem.* **268,** 13,869–13,876.

PART VII

Techniques for Studying Mitochondrial Gene Expression

CHAPTER 31

Isolation of Mitochondria

Stefan Binder and Lutz Grohmann

1. Introduction

Plant cells contain different subcellular compartments that serve distinct physiological functions. One of these organelles, the mitochondrion, provides the nonphotosynthetic energy required in the cell. Mitochondria are of endosymbiotic origin and contain their own genome with essential genes *(1,2)*. However, during evolution most of the genetic information has been transferred and was integrated into the nuclear genome *(3)*. Only a limited number of polypeptides are still expressed within the mitochondria, whereas most of the mitochondrial proteins are synthesized in the cytoplasm and have to be imported *(2)*.

A prerequisite for the detailed molecular analysis of the plant mitochondrial genome and its expression (*see* Chapters 32 and 33) is an isolation and purification procedure that yields an organelle preparation almost free from contamination by other subcellular structures. The efficient isolation of large quantities of mitochondria is hindered by the rigid cell wall of plant cells. It is therefore necessary to use relatively strong mechanical forces to disrupt the plant cells. The required rough handling of the plant tissue might, however, lead to the collapse of subcellular compartments, resulting in a decreased yield of intact mitochondria. It has to be kept in mind that disruption of other different types of organelles, i.e., plastids and nuclei, might cause contamination of the mitochondrial nucleic acid preparation.

The high content of secondary metabolites in some plant tissues is another factor that influences the yield of mitochondria obtained in a

From: *Methods in Molecular Biology, Vol. 49: Plant Gene Transfer and Expression Protocols*
Edited by: H. Jones Humana Press Inc., Totowa, NJ

Table 1
Preparation of Solutions for Percoll Step Gradients[a]

% Medium per gradient	Percoll	2X res.	Double distilled water	Total volume	Volume per gradient
14%	10.5 mL	37.50 mL	27.0 mL	75.0 mL	12 mL
28%	21.0 mL	37.50 mL	16.5 mL	75.0 mL	12 mL
45%	16.9 mL	18.75 mL	1.9 mL	37.5 mL	6 mL

[a]The volumes given are sufficient for the preparation of six Percoll step gradients in 36-mL Beckman [Fullerton, CA] ultraclear centrifuge tubes.

routine isolation procedure. In some cases, e.g., for mtRNA extraction *(4)* (Chapter 32) or *in organello* protein synthesis *(5)* (Chapter 33) it is necessary to purify mitochondria by Percoll step gradient centrifugation. This additional purification step is, however, not essential for the isolation of mtDNA. We describe here a Percoll step gradient based protocol *(6)* for the purification of mitochondria.

2. Materials

A wide range of plant tissues has been successfully used as a source for the isolation of plant mitochondria. However, since preparations from green tissues are generally contaminated with chloroplasts it is recommended to use tissues like dark grown callus cultures, etiolated seedlings, or storage organs like potato tubers or beets (*see* Note 1).

2.1. Stock Solutions

1. Stock 1: 0.1*M* EGTA, pH 7.2.
2. Stock 2: 0.2*M* Tricine, pH 7.2.
3. Stock 3: Percoll (Pharmacia, Uppsala, Sweden).
4. Grinding buffer: 0.4*M* mannitol, 1 m*M* EGTA, 25 m*M* MOPS, pH 7.8. Add 40 m*M* β-mercaptoethanol and 0.1% BSA (final concentrations) immediately before use.
5. Wash medium: 0.4*M* mannitol, 1 m*M* EGTA, 5 m*M* MOPS, pH 7.5. Add 0.1% BSA (final concentration) immediately before use.
6. Resuspending medium: 0.4*M* mannitol, 1 m*M* EGTA, 20 m*M* Tricine, pH 7.2.
7. 2X Resuspending medium: Ingredients are added in doubled concentrations of those given in step 6.
8. The above media *(4–7)* are autoclaved and stored at 4°C.
9. Prepare solutions for Percoll step gradients freshly, as outlined in Table 1.

10. Equipment: Waring blender, Miracloth, muslin cloth, 400 mL centrifuge bottles, Dounce-homogenizer (calibrated borosilicate glass, tight fitting Teflon™ pestle; Braun-Melsungen, Melsungen, Germany).

3. Methods

All following steps are done either at 4°C or in the cold room (*see* Note 2).

3.1. Cell Disruption

1. Harvested callus cultures, etiolated shoots, or peeled potatoes and cut into appropriate pieces. Wash once with double-distilled water.
2. Fill the Waring blender with the washed plant material and use 1–2 vol of grinding buffer per vol of tissue. Disrupt cells by 3 × 5 s strokes at high speed interrupted by at least 30-s breaks (*see* Note 3).
3. Remove the cell debris by filtering the cell homogenate through four layers of muslin cloth and one layer of miracloth.

3.2. Differential Centrifugation

1. Spin at 3500*g* for 5 min at 4°C to pellet cell debris and other organelles.
2. Transfer the supernatant containing the mitochondria to a fresh centrifuge bottle and spin at 18,000*g* for 30 min at 4°C.
3. Discard supernatant and carefully resuspend mitochondria with a smooth brush in a small volume of resuspending medium (*see* Note 4).
4. For further purification by Percoll step gradient centrifugation, resuspend the mitochondria in 3–4 mL (or less) of wash medium per gradient.
5. Transfer the suspension to a Dounce homogenizer and carefully disperse the mitochondria by applying 3–5 strokes (*see* Note 5).

3.3. Percoll Density Gradient Centrifugation

1. Prepare Percoll step gradients according to the scheme outlined in Table 1. We recommend that you prepare the gradients during the differential centrifugation step. For pouring the gradient, take a 10-mL glass pipet and carefully layer the solutions under each other (Table 1).
2. Carefully layer the suspension of mitochondria (3–4 mL) onto the freshly prepared Percoll step gradients, and spin in a swing-out rotor (Beckman SW28 rotor) at 70,000*g* for 45 min at 4°C.
3. Use a Pasteur pipet to remove mitochondria from the 28%/45% interface (*see* Fig. 1), and transfer the organelles to a 250-mL centrifuge bottle and dilute the suspension with at least 5 vol of wash medium.
4. Pellet the mitochondria at 18,000*g* for 30 min at 4°C. Carefully remove the supernatant (the first pellet is very fluffy), dilute again with 5 vol of wash medium and centrifuge as above. This step is repeated until the Percoll is completely removed and a tight mitochondrial pellet is obtained.

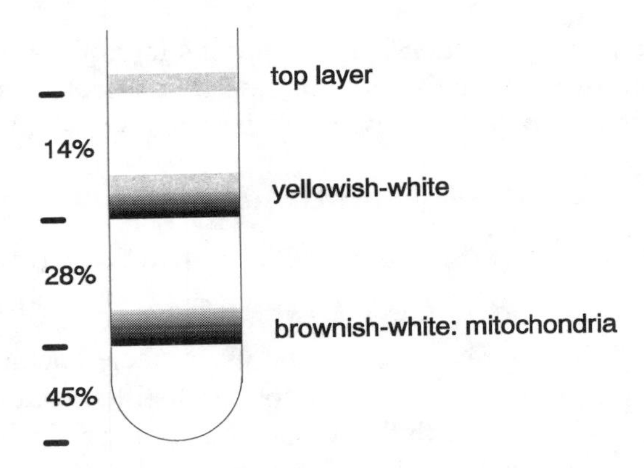

Fig. 1. Purification of plant mitochondria by Percoll density gradient centrifugation. The schematic drawing shows the distribution of different fractions and the relative position of plant mitochondria.

5. Resuspend the final mitochondrial pellet in a maximum total volume of 1.0 mL wash medium and transfer to a 1.5-mL Eppendorf tube. The Eppendorf tube is weighed before filling with suspension in order to estimate the yield of wet weight mitochondria.
6. Pellet the mitochondria in an Eppendorf centrifuge at 4°C, freeze in liquid nitrogen, and store at –80°C. As a rule of thumb, 1 g of wet mitochondria corresponds to approx 100 mg mitochondrial protein.

4. Notes

1. For the isolation of sufficient quantities of mitochondria we recommend the following amounts of starting material: etiolated seedlings, at least 100 g; tissue cultures, at least 1000 g; storage organs, at least 500 g.
2. The isolation procedure is carried out at 4°C in order to minimize the degradation of mitochondrial nucleic acids by nucleases.
3. Avoid excessive cell disruption, since this might lead to a disruption of organelles and therefore decrease the yield of intact mitochondria during the isolation procedure. On the other hand, incomplete disruption of the plant tissue could have the same effect.
4. The actual volume depends on the chosen procedure for further purification. For mtDNA preparation, resuspend in a total volume of 20–25 mL of DNase I buffer (*see* Chapter 32).
5. Careful dispersion of mitochondria is essential, since otherwise aggregated mitochondria would pellet through the Percoll gradient and decrease the yield.

References

1. Gray, M. W. (1989) The evolutionary origins of organelles. *Trends Genet.* **5,** 294–299.
2. Gray, M. W., Hanic-Joyce, P. J., and Covello, P. S. (1992) Transcription, processing and editing in plant mitochondria. *Annu. Rev. Plant Physiol. Plant Mol. Biol.* **43,** 145–175.
3. Brennicke, A., Grohmann, L., Hiesel, R., Knoop, V., and Schuster, W. (1993) The mitochondrial genome on its way to nucleus: different stages of gene transfer in higher plants. *FEBS Lett.* **325,** 140–145.
4. Binder, S. and Brennicke, A. (1993) Transcription initiation sites in mitochondria of *Oenothera berteriana*. *J. Biol. Chem.* **268,** 7849–7855.
5. Leaver, C. J., Hach, E., and Forde, B. G. (1983) Protein synthesis by isolated plant mitochondria. *Methods Enzymol.* **97,** 476–484.
6. Boutry, M., Faber, A. M., Charbonnier, M., and Briquet, M. (1984) Micro-analysis of plant mitochondrial protein synthesis products. *Plant Mol. Biol.* **3,** 445–452.

Chapter 32

Mitochondrial Nucleic Acid Purification and Analysis

Stefan Binder

1. Introduction

During the last years the structure and gene content of the mitochondrial genome of land plants have been subject of intensive molecular investigations. Because of the large size (200–2500 kb), and the complex, multipartite organization, the actual configuration of the mitochondrial genome and its coding capacity is still unclear *(1)*. A currently accepted model suggests that the mitochondrial genome of higher plants is organized as multiple circular molecules. A master chromosome is assumed to represent the entire mitochondrial genome, although its existence in vivo has not been shown. However, subgenomic circles, which are formed by intra- and/or intergenomic homologous recombination via direct and/or inverted repeats, have been found in almost all land plants *(2,3)*. In addition, circular or linear DNA plasmids of low molecular weight have been reported in plant mitochondria *(4,5)*.

Sequence analysis of the complete *Marchantia* mitochondrial genome showed about 94 genes to be encoded by the organellar DNA. These include three genes for rRNAs, 29 genes for tRNAs, 30 protein coding genes, and 28–42 unidentified open reading frames *(6)*.

Since plant cells contain three organelles with independent genetic systems, the extraction of total cellular nucleic acids generally results in a mixture of nucleic acids from these different compartments. Although for some applications total nucleic acids might be satisfactory, detailed analysis of DNA and/or RNA of mitochondria requires the extraction of nucleic acids from enriched or highly purified mitochondrial fractions.

From: *Methods in Molecular Biology, Vol. 49: Plant Gene Transfer and Expression Protocols*
Edited by: H. Jones Humana Press Inc., Totowa, NJ

Whereas Southern or Northern blot analysis usually can be done with relatively crude fractions of mitochondrial nucleic acids, the construction of organelle specific genomic or cDNA libraries needs highly purified mitochondrial nucleic acids, since this is the only possibility to avoid or reduce contamination of the libraries by nonmitochondrial clones. Therefore, mitochondrial nucleic acids have to be extracted from highly enriched mitochondrial fractions.

For the isolation of mitochondrial DNA (mtDNA), it is sufficient to enrich mitochondria by differential centrifugation (*see* Chapter 31). An important step during the extraction procedure is the digestion of the isolated organelles with DNase I prior to the extraction of the DNA. This treatment removes nuclear and chloroplast-derived DNA sticking to the outer membrane of the mitochondria. However, before lysis of the organelles, the DNase activity has to be completely inactivated to avoid degradation of the mtDNA. Final purification of the mtDNA is performed by isopycnic centrifugation on CsCl density gradients, which results in an almost complete separation of mitochondrial DNA and RNA.

Mitochondrial RNA (mtRNA) is usually extracted from Percoll density gradient purified mitochondria. This isolation procedure guarantees high purity of the mitochondrial fraction and diminishes the possibility of degradation of the RNA by contaminating agents. To further suppress residual degradation of the mtRNA, mitochondria are lysed in the presence of guanidinium thiocyanate, a strong RNase inhibitor *(7)*. The author describes here protocols for the efficient isolation of nondegraded mtDNA and mtRNA from plant tissue.

2. Materials

2.1. Plant Mitochondria

For the isolation of mtDNA use mitochondria enriched by differential centrifugation. MtRNA is extracted from organelles that have been purified on Percoll gradients (*see* Chapter 31).

2.2. Buffers and Solutions for mtDNA Isolation

All buffers should be autoclaved and stored at 4°C; β-mercaptoethanol should be added immediately before use.

1. DNase I buffer: 300 m*M* mannitol, 5 m*M* $MgCl_2$, 10 m*M* KH_2PO_4, 50 m*M* Tris-HCl, pH 7.5, 4 m*M* β-mercaptoethanol.
2. 0.5*M* EDTA, pH 8.0.

3. STE: 10 m*M* Tris-HCl, pH 8.0, 100 m*M* NaCl, 1 m*M* EDTA.
4. 10% SDS.
5. Phenol (water saturated).
6. Chloroform/isoamyl alcohol (24:1).
7. Ethidium bromide (10 mg/mL in double-distilled water).
8. Paraffin.
9. Isopropanol saturated over CsCl-saturated TE (10 m*M* Tris-HCl, pH 8.0, 1 m*M* EDTA).
10. 100% Ethanol.
11. 70% Ethanol.
12. Solid Cesium chloride.
13. DNase I.

2.3. Buffers and Solutions for the Extraction of mtRNA

For precautions for working with RNA please see notes.

1. Lysis buffer: 4*M* guanidinium thiocyanate, 25 m*M* sodium citrate pH 7.0, 0.5% (w/v) sarcosyl, 0.1*M* β-mercaptoethanol (add immediately before use).
2. 2*M* Sodium acetate (NaOAc), pH 4.0.
3. Phenol (water saturated).
4. Chloroform/isoamyl alcohol (24:1).
5. Isopropanol.
6. 70% Ethanol.

3. Methods

3.1. Extraction of Mitochondrial DNA

1. After differential centrifugation, use a fine brush to resuspend mitochondria in 3–5 mL of DNase I buffer in a 250-mL centrifugation flask. It is generally recommended to keep the volume as small as possible to ensure a complete resuspendation of the mitochondria. Adjust the volume to 25 mL and add 3–4 mg of DNase I. Digestion is performed for 1 h at 0°C.
2. DNase I is then inactivated by the addition of 5 mL of 0.5*M* EDTA. Add 230 mL of DNase I buffer and centrifuge at 18,000*g* for 30 min at 4°C.
3. Discard supernatant and resuspend mitochondria again in 250 mL of DNase I buffer containing 5 m*M* EDTA. Centrifuge as indicated earlier and repeat this wash step again to ensure that DNase I is washed out completely.
4. Disperse mitochondria in 7 mL STE and transfer the suspension to a 13 mL tube. Add 350 µL of 10% SDS to a final concentration of 0.5%.

5. Remaining protein is removed by extraction with phenol/chloroform. Add 1/2 vol of phenol (3.675 mL) and the same amount of chloroform/ isoamyl alcohol (24:1), mix gently, and centrifuge at 12,000*g* for 10 min at 4°C. Carefully remove upper phase, transfer to a new 13-mL centrifuge tube, add 1 vol of chloroform/isoamyl alcohol (24:1), mix gently, and centrifuge again.
6. Transfer supernatant to a fresh 13-mL ultracentrifuge tube. Add 200 µL EtBr (10 mg/mL) and solid CsCl in a ratio of 0.95 g/mL of final suspension. Carefully dissolve the CsCl and avoid vigorous shaking and excess exposure to light since this might damage the mtDNA. Fill up with paraffin and close the tube. Centrifuge CsCl gradients at 100,000*g* for 48 h at 18°C.
7. A thin band containing the mtDNA should be visible in the gradient under UV illumination. An additional second broad band (above the mtDNA) containing nuclear DNA might be visible in the case of an incomplete digestion of mitochondria with DNase I. Recover mtDNA from the CsCl gradient by piercing the tube from the side and transfer the mtDNA (500–1000 µL solution) to a 13-mL tube.
8. Add an equal amount of isopropanol saturated with CsCl-saturated TE buffer to remove EtBr. Mix gently, and after phase separation, recover upper phase into a new tube. Repeat this extraction twice. Complete removal of EtBr can be monitored by a brief inspection under UV illumination.
9. Add 2–3 vol of double-distilled water and precipitate DNA with at least 2 vol of absolute ethanol. After incubation for at least 1 h at –20°C, pellet the mtDNA by centrifugation at 12,000*g* for 20 min at 4°. Remove supernatant and add 1 mL 70% ethanol. Mobilize the pellet and transfer to a 1.5 mL Eppendorf tube. Spin at 12,000*g* for 5 min in an Eppendorf centrifuge, remove supernatant, and air dry the mtDNA pellet.
10. Dissolve the mtDNA in 20–100 µL of TE and photometrically determine the concentration of the mtDNA by absorption measurements at 260 and 280 nm. The ratio 260:280 should be 2:1 for good quality mtDNA. The quality of the mtDNA preparation might be examined by restriction analysis (*see* Fig. 1).

3.2. Isolation of Mitochondrial RNA

Because of the ubiquitous presence of RNase, appropriate precautions must be taken for preparation of the solutions as well as during the extraction procedure (*see* Note 1).

1. Dissolve Percoll gradient purified mitochondria in 2–3 vol (maximum 5 mL) of lysis buffer in a 13-mL centrifuge tube.

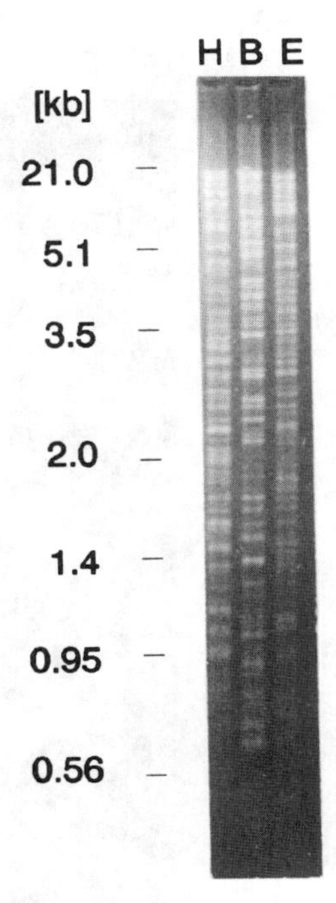

Fig. 1. Restriction digests of about 2 μg of *Oenothera berteriana* mtDNA isolated from dark grown callus cultures. Restriction enzymes used are *Hin*dIII (H), *Bam*HI (B), and *Eco*RI (E).

2. Incubate for 5 min at room temperature and add 0.1 vol of 2*M* NaOAc, pH 4.0 (*see* Note 2). Mix gently and extract the suspension by addition of 1 vol of water saturated phenol and 0.2 vol of chloroform/isoamyl alcohol (24:1). Mix gently and keep on ice for 15 min.
3. Separate phases by centrifugation at 12,000*g* for 15 min at 4°C (*see* Note 3). Transfer the aqueous phase to a fresh tube and extract two times with 1 vol of chloroform/isoamyl alcohol (24:1).
4. Precipitate mtRNA by addition of 1 vol of isopropanol and overnight incubation at –20°C. Pellet mtRNA by centrifugation at 12,000*g* for 15 min at 4°C. Wash the mtRNA two times with 70% ethanol, air dry, and resuspend the RNA in 50–100 μL double-distilled water.
5. In some cases it might be necessary to repeat the phenol/chloroform extraction steps (steps 2 and 3) to completely remove protein contamination.
6. The quality of the mtRNA can be evaluated by size fractionation on a 1.3% (w/v) denaturing agarose gel (*see* Fig. 2).

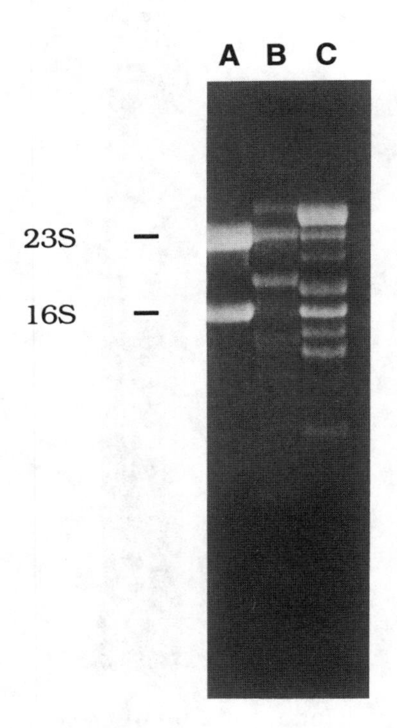

Fig. 2. Formaldehyde agarose gel (1.3% w/v) of *E. coli* 16S and 23S rRNA as size markers **(A)**, mtRNA **(B)**, and total cellular RNA **(C)** from *Solanum tuberosum*. RNAs were digested with DNase I in order to remove residual DNA.

4. Notes

1. Precautions for working with RNA. Because of the overall presence of RNase, it is recommended to follow these precautions:
   a. All water used should be of highest purity, i.e., double distilled and autoclaved. However, treatment of the water with DEPC is not necessary.
   b. Tubes and all other plastic and glassware used during the isolation procedure should be sterile.
   c. Wear gloves while working with RNA.
2. Precipitation of nucleic acids in the presence of 200 m*M* NaOAc, pH 4.0 leads to a preferred precipitation of the mtRNA; however, at least trace amounts of mtDNA are still present in this mtRNA preparation. For some applications it is recommended to eliminate DNA contamination by digestion with RNase-free DNase I.
3. If phases are not separated after the centrifugation, add another 0.2 vol of chloroform/isoamyl alcohol (24:1), mix briefly, and centrifuge again.

References

1. Newton, K. J. (1988) Plant mitochondrial genomes: Organization, expression and variation. *Ann. Rev. Plant Physiol. Plant Mol. Biol.* **39,** 503–532.
2. Palmer, J. D. and Shields, C. R. (1984) Tripartite structure of the *Brassica campestris* mitochondrial genome. *Nature* **307,** 437–440.
3. Lonsdale, D. M., Hodge, T. P., and Fauron, C. M.-R. (1984) The physical map and organization of the mitochondrial genome from fertile cytoplasm of maize. *Nucleic Acids Res.* **12,** 9249–9261.
4. Pring, D. R., Levings, C. S., III, Hu, W. W. L., and Timothy, D. H. (1977) Unique DNA associated with mitochondria in the "S"-type cytoplasm of male-sterile maize. *Proc. Natl. Acad. Sci. USA* **74,** 2904–2908.
5. Schardl, C. L., Lonsdale, D. M., Pring, D. R., and Rose K. R. (1984) Linearization of maize mitochondrial chromosomes by recombination with linear episomes. *Nature* **310,** 292–296.
6. Oda, K., Katsuyuki, Y., Ohta, E., Nakamura, Y., Takemura, M., Nozato, N., Akashi, K., Kanegae, T., Ogura, Y., Kohchi, T., and Ohyama, K. (1992) Gene organization deduced from the complete sequence of liverwort *Marchantia polymorpha* mitochondrial DNA. *J. Mol. Biol.* **223,** 1–7.
7. Chomczynski, P. and Sacchi, N. (1987) Single step method of RNA isolation by acid guanidinium thiocyanate-phenol-chloroform extraction. *Anal. Biochem.* **162,** 156–159.

CHAPTER 33

In organello Protein Synthesis

Lutz Grohmann

1. Introduction

Protein synthesis by isolated plant mitochondria under in vitro conditions is a direct method to study the translational products of this organellar genetic system. Compared to fungal and animal mitochondrial systems, a considerably higher number of proteins is expected to be encoded by the plant mitochondrial genome (for details, *see* the introduction to Chapter 32). Since no in vitro translation system using isolated mitochondrial ribosomes, e.g., in an S100 extract, has yet been established, protein synthesis with intact organelles is the best way to analyze the products of the mitochondrial translation apparatus. The *in organello* synthesized proteins should in fact reflect the number of polypeptides expressed by plant mitochondria. The analysis relies on incubation of isolated mitochondria with radiolabeled amino acids such as [^{35}S]methionine. After a period of continuing the *in organello* synthesis the radiolabeled translation products are subsequently separated by SDS polyacrylamide gel electrophoresis and analyzed by autoradiography (Fig. 1). Such an experimental procedure was developed and has been described in detail by Leaver et al. *(1)*. The system established by these authors has reproducibly worked well with mitochondria from different tissues of various plants and here we describe this method with minor changes.

An application of this method is the identification of products whose expression has been analyzed only at the nucleic acid level (i.e., Northern hybridization). The immunological analysis of *in organello* synthe-

From: *Methods in Molecular Biology, Vol. 49: Plant Gene Transfer and Expression Protocols*
Edited by: H. Jones Humana Press Inc., Totowa, NJ

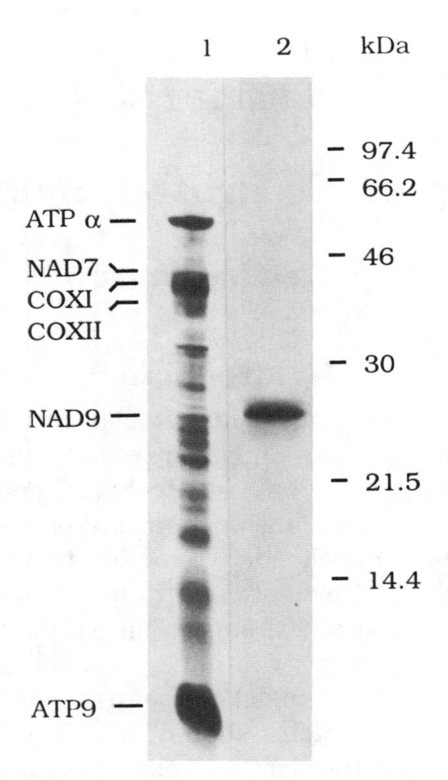

Fig. 1. Autoradiograph of *in organello* [^{35}S]labeled translation products from plant mitochondria separated on a 15% SDS-polyacrylamide gel. Lane (1), total translation products of potato mitochondria were directly loaded on the gel. Lane (2), after incubation with an antibody directed against the overexpressed wheat mitochondrial NAD9 protein and subsequent immunoprecipitation with Protein-A Sepharose. Assignments of individual labeled bands are given on the left side and the molecular mass standards are on the right hand margin.

sized products will give information about the expression of a known or unidentified open reading frame at the protein level. A prerequisite for this analysis is an antiserum specifically reacting with the polypeptide encoded by the open reading frame of interest. This can be a heterologous antibody *(2)*, an antibody directed against the gene product that has been overexpressed in *E. coli (3)* or raised against synthetic oligopeptides made from the predicted protein sequence of the ORF *(4)*. The immunoprecipitation of an *in organello* radiolabeled mitochondrial protein will indicate whether the protein is actually expressed by the respective mitochondrial open reading frame (Fig. 1).

Particularly, the maternally inherited trait of cytoplasmatic male sterility (CMS) has been shown to be associated with rearrangements in the mitochondrial genomes and the creation of novel open reading frames by a new arrangement of different DNA segments. Identification and assignment of these CMS related proteins expressed by the mitochondrial genome of such sterile plant lines can be performed by comparing the SDS-PAGE pattern of the *in organello* labeled protein products synthesized by isolated mitochondria from sterile and fertile lines. The unique presence of a labeled polypeptide in the pattern of the respective CMS plant lines and its absence in the labeled translation products of the fertile (wild-type) and, if available, of a restored plant line is a way to illustrate the involvement of these mitochondrially expressed proteins in this specific phenotype *(4)*.

2. Materials

All buffers and stock solutions are prepared with double-distilled water.

2.1. Chemicals

Protein A-Sepharose (Pharmacia, Uppsala, Sweden); L-[^{35}S]methionine (in vivo cell labeling grade, specific activity 37 TBq/m*M*, Amersham, Arlington Heights, IL); amino acid mixture without L-methionine (Amersham).

2.2. Plant Material

Use freshly isolated plant mitochondria purified on Percoll density step gradients (*see* Chapter 31).

2.3. Buffers and Solutions for In organello Protein Synthesis

1. Resuspension-buffer: 0.4*M* mannitol, 10 m*M* Tricine, 1 m*M* EGTA, adjust pH to 7.2 with KOH, autoclave, and store at 4°C.
2. Stock solutions: 54 m*M* ADP; 50 m*M* GTP; adjust pH of both to 7–8 with NaOH, filter sterilize, and store small aliquots at –20°C (do not refreeze GTP); 1.0*M* DTT, filter sterilize, and store aliquots at –20°C; 1.0*M* sodium acetate, autoclave, and store at 4°C.
3. Stock solutions for salt mix: 0.8*M* mannitol; 1.0*M* KCl; 0.2*M* Tricine, pH 7.2; 0.2*M* potassium phosphate, pH 7.2; 0.5*M* $MgCl_2$; 0.1*M* EGTA, pH 7.2. Autoclave each stock solution and store at 4°C.

4. Salt mix: Mix 4.7 mL of 0.8*M* mannitol, 1.875 mL of 1.0*M* KCl, 1.05 mL of 0.2*M* Tricine, pH 7.2, 0.5 mL of 0.2*M* phosphate buffer, pH 7.2, 0.41 mL of 0.5*M* $MgCl_2$, 0.165 mL of 0.1*M* EGTA, pH 7.2, and add 1.3 mL of double-distilled H_2O to a final volume of 10 mL. Adjust pH to 7.2 with KOH, filter sterilize, and store aliquots at –20°C.

2.4. Buffers and Solutions for Immunoprecipitation

1. Stock solution: 20% (w/v) SDS.
2. Tris-buffer: 0.1*M* Tris-HCl, pH 8.4.
3. TNET: 50 m*M* Tris-HCl, pH 7.4, 5 m*M* EDTA, 150 m*M* NaCl, 1% (v/v) Triton X-100.

3. Methods

3.1. *In organello* Protein Synthesis

1. Before setting up the labeling reaction, freshly prepare the following amounts of working solutions:
   a. Incubation medium. For each labeling reaction, mix 120 µL of salt solution with 2.5 µL of 0.05*M* GTP, 0.5 µL of 1.0*M* DTT, 12.5 µL of amino acid mix, 2.0 µL of L-[^{35}S]methionine (10 µCi/µL), and 12.5 µL H_2O. Prepare an excess of incubation mix (75 µL should be enough) to provide samples for background and total cpm counting.
   b. Energy mix. Prepare fresh 1.0*M* sodium succinate solution by dissolving 0.675 g of sodium succinate and 125 µL of 0.2*M* phosphate buffer pH 7.2 in a final volume of 2.5 mL H_2O. Take 2.5 µL of the 1.0*M* sodium succinate solution and mix with 5 µL of 54 m*M* ADP and 42.5 µL of H_2O.
   c. Sodium acetate. Mix 5 µL of 1.0*M* sodium acetate and 45 µL H_2O. This substrate is used in a control reaction. Sodium acetate is a nonoxidizable substrate that can be utilized by bacteria but not by mitochondria, hence amino acid incorporation in its presence gives an indication of bacterial contamination.
2. For each mitochondrial preparation, set up at least two labeling reactions in 1.5-mL microcentrifuge tubes, one for the acetate control, to estimate the bacterial contamination, and one for the recommended energy mix (for an alternative energy mix, *see* Note 1). The final labeling reactions are performed in volumes of 250 µL. Mix 50 µL of energy mix, or 50 µL of the acetate solution for the control reaction, with 150 µL of the incubation mix and then add 50 µL of mitochondrial suspension (2 mg/mL mitochondrial protein in resuspension-buffer).
3. Incubate at 25°C for 60–90 min.

4. After incubation remove 2 × 5 μL aliquots from each reaction and 4 × 5 μL from the left over incubation mix, spot separately onto 1.5-cm Whatman (Maidstone, UK) 3MM filter paper discs and estimate the trichloroacetic acid (TCA) precipitable counts (*see* Notes 2 and 3).
5. Stop the reaction by adding 1 mL of ice cold resuspension-buffer containing 10 m*M* unlabeled L-methionine and pellet the mitochondria at 12,000*g* for 5 min in a microcentrifuge at 4°C. Remove the supernatant (containing the free [^{35}S]methionine) and wash the mitochondrial pellet for two more times in 250 μL resuspension-buffer. The pellet may be frozen on dry ice and stored at –80°C or suspended in 20–30 μL of electrophoresis sample buffer for immediate SDS-PAGE analysis.

3.2. SDS-Polyacrylamide Gel Electrophoresis

1. Solubilize the samples in a buffer containing 2% (v/v) sodium dodecyl sulfate (SDS), 10% (w/v) sucrose, and 60 m*M* Tris-HCl, pH 6.8 by heating for 2 min at 90°C. Cool to room temperature and add DTT to 20 m*M* or β-mercaptoethanol to 1% (v/v).
2. Analyze the samples on a 15%T/2%C or 12–20%T/2%C linear gradient SDS-polyacrylamide gel *(5)* and stain the gel with Coomassie blue. If using colored or radiolabeled molecular weight markers staining is not necessary.
3. Destain the gel, dry it onto a Whatman 3MM paper, and expose for up to 2 wk to a Kodak X-OMAT AR X-ray film (*see* Note 4).

3.3. Immunoprecipitation

1. Combine 5–8 separately performed *in organello* labeling-reactions, pellet the mitochondria, and resuspend in a final volume of 0.5–1.0 mL of cold 0.1*M* Tris-HCl, pH 8.4.
2. Freeze (in liquid nitrogen) and thaw the mitochondrial suspension, at least three times, in order to break the mitochondria. Pellet the proteins in an ultracentrifuge at 430,000*g* for 10 min at 4°C, or at 140,000*g* for 30 min, and resuspend in 125 μL of 0.1*M* Tris-HCl, pH 8.4 using a small Teflon™ pestle in a glass homogenizer (or equivalent) for efficient suspension.
3. Add 22 μL of 20% (w/v) SDS, vortex, and boil the mixture for 5 min at 100°C, then add 2.7 mL TNET buffer and centrifuge for 10 min at 100,000*g* to pellet nonsolubilized proteins.
4. Take the supernatant, add 10–50 μL of the antiserum, and incubate overnight at 4°C. The optimum antiserum concentration has to be experimentally determined.
5. For preswelling of Protein A-Sepharose, mix 5 mg in 50–100 μL TNET buffer, and place at 4°C for 1–2 h with gentle agitation.

6. Shake the Protein A-Sepharose into suspension, add 50 µL with a pipet tip (chopped off at the end) to the mitochondrial protein-antibody suspension, and incubate with continuous mixing for 1 h at 4°C.
7. Wash the antibody-protein A-Sepharose complexes four times with 1.0 mL TNET buffer and centrifuge each time for 10 s.
8. After the last wash, resuspend the beads in 50 µL SDS-PAGE sample buffer (without DTT or β-mercaptoethanol), boil for 5 min, and cool to room temperature, add DTT to 20 m*M* or β-mercaptoethanol to 1% (v/v), and analyze on an SDS-polyacrylamide gel (*see* Section 3.2.).

4. Notes

1. The succinate/ADP energy mix used in this protocol provides ATP in a self-generating system of respiratory chain-linked phosphorylation of ADP. An alternative to the succinate/ADP energy mix is to supply ATP externally. Mix 12.5 µL of 0.24*M* ATP (pH adjusted to 7–8, filter sterilized, and stored in small aliquots at –20°C) with 37.5 µL H_2O. This often gives a higher rate of incorporation than when using succinate/ADP (C. J. Leaver and A. Liddell, personal communication).
2. The time course of the reaction can be followed by removing 5-µL aliquots every 5–10 min. Aliquots are spotted onto Whatman 3MM paper disks (1.5 cm), air-dried, and immediately immersed in ice cold 10% TCA for at least 15 min. Unincorporated radioactivity is removed by washing in 5% TCA at 90°C for 10 min, followed by four washes in 5% TCA at room temparature for 5 min and one wash in 95% EtOH or acetone. The filter discs are finally washed in ether at 37°C for 15 min and thoroughly air-dried. The radioactivity incorporated into protein is estimated by scintillation counting in the presence of 4–5 mL of scintillation fluid (e.g., Pharmacia Optiphase HiSafe 3).
3. Optimum rates of incorporation depend on the isolation of intact mitochondria free of bacterial contamination. In the presence of a nonoxidizable substrate, such as acetate, the incorporation should be less than 5% of the rates obtained with the oxidizable substrate. Problems owing to bacterial contamination can be circumvented by reinforced sterile working conditions. Inhibitors of bacterial protein synthesis cannot, however, be used since plant mitochondria contain ribosomes similar to those of eubacteria, and are also sensitive to chloramphenicol, tetracyclin, or puromycin.
4. The relatively low translational activity of isolated mitochondria, and as a consequence low rates of incorporation of the radioactive amino acid, can result in relatively long exposure times necessary for autoradiography of the labeled protein products. Exposure times can be shortened by increasing the $[^{35}S]$methionine concentration to 60–80 µCi per labeling reaction

(C. J. Leaver and A. Liddell, personal communication). However, this can be very costly for routine work.

Acknowledgments

The author is grateful to C. J. Leaver and A. Liddell (Department of Plant Sciences, University of Oxford, UK) for communication of unpublished results and comments on the procedure, and to Axel Brennicke for his continuous support.

References

1. Leaver, C. J., Hack, E., and Forde, B. G. (1983) Protein synthesis by isolated plant mitochondria. *Methods Enzymol.* **97,** 476–484.
2. Gäbler, L., Herz, U., Liddell, A., Leaver, C. J., Schröder, W., Brennicke, A., and Grohmann, L. (1994) The 42.5 kDa subunit of the NADH:ubiquinone oxidoreductase (complex I) in higher plants is encoded by the mitochondrial *nad7* gene. *Mol. Gen. Genet.* **244,** 33–40.
3. Prioli, L., Huang, J., and Levings III, C. S. (1993) The plant mitochondrial open reading frame *orf221* encodes a membrane-bound protein. *Plant Mol. Biol.* **23,** 287–295.
4. Monéger, F., Smart, C. J., and Leaver, C. J. (1994) Nuclear restoration of cytoplasmic male sterility in sunflower is associated with the tissue-specific regulation of a novel mitochondrial gene. *EMBO J.* **13,** 8–17.
5. Laemmli, U. K. (1970) Cleavage of structural proteins during assembly of the head of bacteriophage T4. *Nature* **227,** 680–685.

PART VIII

Immunological Detection of Proteins

CHAPTER 34

Separation of Plant Proteins by Electrophoresis

Peter R. Shewry, Arthur S. Tatham, and Roger J. Fido

1. Introduction

Direct identification of the product of a transferred gene is clearly an important part of the characterization of a transgenic plant. It is possible to detect proteins using highly sensitive ELISA techniques (Chapter 36, this volume), and to determine their precise histological or subcellular locations using immunohistochemistry (Chapter 37, this volume). However, it is essential that the expressed protein should also be separated by electrophoresis, and detected either by staining or Western blot analysis. Selection of the most suitable electrophoresis system will allow the researcher to determine the M_r of the protein in the native and denatured states, its isoelectric point and charge at various pH values, its interactions (via disulfide bonds) with other subunits, and the presence or absence of glycosylation. This will, in fact, confirm whether the protein has been correctly processed, or whether any proteolysis or incorrect post-translational processing has occurred. Furthermore, it is possible to confirm that the protein has the correct N-terminal amino acid sequence by transfer to a PVDF based membrane for microsequencing. The present chapter describes methods for the extraction, electrophoresis, and detection of proteins, and for their transfer to membranes for microsequencing. These can be used for the direct analysis of expressed proteins, or combined with Western blotting as described in the following chapter (Chapter 35).

From: *Methods in Molecular Biology, Vol. 49: Plant Gene Transfer and Expression Protocols*
Edited by: H. Jones Humana Press Inc., Totowa, NJ

Detailed practical instructions are not included as these vary with the type of apparatus that is available.

2. Materials

2.1. Protein Extraction and Sample Preparation

2.1.1. Extraction of Seed Tissues

Total protein extraction buffer: 0.0625*M* Tris-HCl, pH 6.8, 2% (w/v) SDS, 5% (v/v) 2-mercaptoethanol or 1.5 % (w/v) dithiothreitol, 10% (v/v) glycerol, 0.002% (w/v) Bromophenol blue.

2.1.2. Extraction of Leaf Tissues

Homogenization buffer: 0.1*M* Tris-HCl, pH 8.0, 0.01*M* $MgCl_2$, 18% (w/v) sucrose, 40 m*M* 2-mercaptoethanol (+2% (w/v) SDS to extract membrane proteins and other insoluble proteins).

2.2. SDS-PAGE Systems

2.2.1. The Laemmli (1970) System (1)

1. Solution A: 30% (w/v) acrylamide, 0.8% (w/v) bisacrylamide.
2. Solution B: 1.5*M* Tris-HCl, pH 8.8.
3. Solution C: 0.5*M* Tris-HCl, pH 6.8.
4. Solution D: 10% (w/v) SDS.
5. Solution E: 10% (w/v) ammonium persulfate (freshly prepared).
6. Solution F: 0.004% (w/v) riboflavin.
7. Running buffer: 0.025*M* Tris, 0.192*M* glycine, pH 8.3, 0.1% (w/v) SDS.
8. Sample buffer: Same as total protein extraction buffer (*see* Section 2.1.1.).

2.2.2. The Tricine Gel System

1. Anode buffer: 0.2*M* Tris-HCl, pH 8.9.
2. Cathode buffer: 0.1*M* Tris, 0.1*M* Tricine, 0.1% (w/v) SDS. No adjustment of the pH, which is about pH 8.25, is required.
3. Gel buffer: 3.0*M* Tris-HCl, pH 8.45, 0.3% (w/v) SDS.
4. Acrylamide stock: 48% (w/v) acrylamide, 1.5% (w/v) bisacrylamide.
5. Ammonium persulfate: 10% (v/v).

2.2.3. The Tris-Borate System

1. Separating gel buffer: 1.25*M* Tris-borate, with 1% (w/v) SDS prepared by dissolving 378 g Tris, 15 g SDS, and 95 g boric acid to a volume of 2.5 L. The pH should be 8.9 with no adjustment required.
2. Stacking gel buffer: 1.0*M* Tris-HCl, pH 6.8, 10% (w/v) SDS.
3. Ammonium persulfate (prepared fresh), 10% (w/v).

4. Sample buffer: Mix 6.55 mL stacking gel buffer (pH 6.8), 3.3% (w/v) SDS, 10 mL (v/v) glycerol, and 1.54% (w/v) DTT (100 m*M* final concentration). Make up to 100 mL with water.
5. Running buffer: Use separating gel buffer diluted 10X. This may be reused by mixing the buffer from the upper and lower chambers.
6. Acrylamide (40% w/v) and *NN'*-methylenebisacrylamide (2% w/v) solutions are purchased from BDH.

2.3. Nondenaturing Gel Systems

2.3.1. The Davis (pH 8.9) System (2)

1. Solution A: 48 mL 1*M* HCl, 36.6 g Tris, 230 µL TEMED. Water to 100 mL. pH should be 8.9.
2. Solution B: ≃48 mL 1*M* HCl, 5.98 g Tris, 460 µL TEMED. Water to 100 mL. Adjust pH to 6.7 with 1*M* HCl.
3. Solution C: 28.0% (w/v) acrylamide, 0.735% (w/v) bis.
4. Solution D: 10.0% (w/v) acrylamide, 2.5% (w/v) bis.
5. Solution E: 0.004% (w/v) riboflavin.
6. Solution F: 40% (w/v) sucrose.
7. All solutions can be stored in the dark at 4°C.

2.3.2. The pH 3.2 Lactate System for Cereal Seed Storage Proteins

1. Solution A: 32% (w/v) acrylamide, 1% (w/v) bis.
2. Solution B: 0.032% (w/v) ferrous sulfate.
3. Upper reservoir buffer (1 L): 0.1% (w/v) aluminum lactate, 0.16% (v/v) lactic acid (85%), pH should be 3.1.
4. Lower reservoir buffer (4 L): 0.4% (v/v) lactic acid (85%), adjust pH to 2.6 with further lactic acid.
5. Sample buffer: 35% (v/v) ethanol, 10% (v/v) glycerol, 3*M* urea, with one crystal of basic fuchsin.

2.4. Isoelectric Focusing (IEF)

1. Stock solution A: 10% (w/v) acrylamide, 0.5% (w/v) bisacrylamide. Store in the dark at 4°C and mix well before use.
2. Catalyst: 10% (w/v) ammonium persulfate, make up fresh before use.
3. Anode buffer: 1*M* orthophosphoric acid. Store at 4°C.
4. Cathode buffer: 1*M* sodium hydroxide. Make up fresh before use.
5. Sample buffer (10X stock): 100 m*M* glycine, adjusted to pH 8 by the addition of solid Tris base. Store at 4°C.
6. Sample buffer: Dilute 10X stock with 6*M* urea. Add 1% (v/v) 2-mercaptoethanol. Make up fresh before use.

2.5. Two-Dimensional Systems

2.5.1. The O'Farrell (1975) System (3)

1. Lysis buffer to dissolve proteins: 9.5*M* urea, 2% (w/v) Nonidet NP-40, 2% (w/v) ampholytes, 5% (v/v) 2-mercaptoethanol.
2. Solution A: 28.38% (w/v) acrylamide, 1.62% (w/v) bis.
3. Solution B: 10% (w/v) Nonidet NP40.
4. Solution C: 10% (w/v) ammonium persulfate (fresh prepared).
5. Anode solution: 0.01*M* H_3PO4.
6. Cathode solution: 0.02*M* NaOH (degassed and stored under vacuum).
7. Sample overlay solution: 9*M* urea, 1% (w/v) ampholytes.
8. Solution D: (Separating gel buffer): 1.5*M* Tris-HCl, pH 8.8, 0.4% (w/v) SDS.
9. Solution E: (Stacking gel buffer): 0.5*M* Tris-HCl, pH 6.8, 0.4% (w/v) SDS.
10. Solution F: 29.2% (w/v) acrylamide, 0.8% (w/v) bis.
11. SDS sample buffer: 0.0625*M* Tris-HCl, pH 8.8, 10% (w/v) glycerol, 5% (v/v) 2-mercaptoethanol, 2.3% (w/v) SDS.
12. Running buffer: 0.025*M* Tris, 0.192*M* glycine, 0.1% (w/v) SDS, or 2.0% (w/v) SDS.

2.5.2. Alternative Two-Dimensional System

Equilibration buffer: 0.12*M* Tris-HCl, pH 6.8, 2.5% (w/v) SDS, 5% (v/v) 2-mercaptoethanol + 10% (v/v) glycerol for storage.

2.6. Gel Staining Systems

2.6.1. Coomassie Blue-Based Stains for Total Proteins

2.6.1.1. Standard Stain for Total Proteins

1. Stain solution: Dissolve 1 g of Coomassie brilliant blue R250 in 400 mL of methanol. Carefully mix with 100 g of TCA dissolved in water and make up to 1 L. Filter only if necessary.
2. Destain solution: 10% (w/v) TCA.

2.6.1.2. Blakesley Stain

This method is based on Blakesley and Boezi *(4)*.

Stain solution: Prepare a 2% (w/v) aqueous solution of Coomassie brilliant blue G. Add an equal volume of 1*M* sulfuric acid. Mix and stir for 3 h. Filter through Whatman (Maidstone, UK) No.1 paper. Measure the volume and slowly add with stirring 1/9 vol (111 mL/L) of 10*M* KOH, leading to the development of a purple color. Add 100% (w/v) TCA to a final concentration of 12% (w/v).

2.6.2. Silver Stain

The method below is based on Morrissey *(5)*.

1. Prefixative: 50% (v/v) methanol, 10% (v/v) acetic acid for 30 min, followed by 5% (v/v) methanol, 7% (v/v) acetic acid for 30 min.
2. Fixative: 30 min in 10% (v/v) glutaraldehyde.
3. Wash in distilled water, preferably overnight.
4. Soak: 5 µg/mL dithiothreitol for 30 min. Pour off solution and without rinsing add 0.1% (w/v) silver nitrate. Leave for 30 min.
5. Rinse: Once rapidly in a small amount of distilled water and then twice rapidly in a small amount of developer (50 µL of 37% [v/v] formaldehyde in 100 mL 3% [w/v] sodium carbonate). Soak in developer until the desired level of staining is obtained.
6. Stop solution: Add 5 mL of 2.3*M* citric acid directly to the developer and agitate for 10 min. Discard this solution and wash with distilled water.

2.6.3. Glycoprotein Stains

Fuchsin-sulfite stain: Dissolve 2 g of basic fuchsin in 400 mL distilled water. Cool and filter the solution and add 10 mL of 2*M* HCl and 4 g of potassium metabisulfite. Leave overnight in a stoppered bottle in a cool dark place and then add 1 g activated charcoal. Stir, filter, and add 2*M* HCl (10 mL +) until a small aliquot does not turn red when dried on a glass slide. Store in a stoppered bottle in a cool dark place and discard when the solution turns pink.

2.7. Microsequencing from Gels

1. 10X stock CAPS (100 m*M*, pH11): Dissolve 22.13 g of CAPS (3-[cyclohexylamino]-1-propanesulfonic acid) in 900 mL of deionized water. Add 2*M* NaOH to pH11 (approx 15 mL) and make up to 1L with deionized water. Store at room temperature.
2. Electroblotting buffer: Mix 200 mL of 10X stock CAPS with 200 mL of methanol and 1600 mL of deionized water.

3. Methods

3.1. Protein Extraction and Sample Preparation

The initial step in analyzing a transgenic plant for expression of a foreign protein is to extract total proteins and separate them by SDS-PAGE. The simplest and most effective way to do this is to extract with a buffer that will allow direct application of the sample to an SDS-PAGE gel.

3.1.1. Extraction of Seed Tissues

The following method is based on direct extraction with the Laemmli *(1)* SDS-PAGE sample buffer.

1. Grind in a mortar with 25 μL of total protein extraction buffer per mg of material (*see* Section 2.1.1.). Transfer to an Eppendorf tube and allow to stand for 2 h.
2. Suspend in a boiling water bath (in a fume hood if 2-mercaptoethanol is used) for 2 min, allow to cool, and spin in a microfuge.
3. Apply appropriate aliquots of the supernatant (usually 10–20 μL) to the SDS-PAGE gel.

3.1.2. Extraction of Leaf Tissues

The method discussed in Section 3.1.1. gives poor resolution of proteins from leaves owing to the high proportion of chloroplast proteins. The method that follows, based on Nelson et al. *(6)*, is recommended for such tissues.

1. Freeze tissue in liquid N_2.
2. Grind in mortar with 3 mL/g of homogenization buffer (*see* Section 2.1.2.).
3. Filter through cheesecloth (if necessary) and centrifuge (15 min at 10,000*g* in microfuge) to remove debris.
4. If SDS is omitted from the homogenization buffer a second fraction containing membrane proteins can be obtained by re-extracting the pellet with about 0.05 vol (relative to original homogenate) of 2% (w/v) SDS, 6% (w/v) sucrose, and 40 m*M* 2-mercaptoethanol.
5. Dilute samples to about 2 mg/mL, ensuring that the final solution contains 2% (w/v) SDS, 0.002% (w/v) Bromophenol blue, and not less than 6% (w/v) sucrose.

3.1.3. Specific Extraction Methods

Analysis of the total extractable proteins is often sufficient to demonstrate protein expression, either by gel staining or Western blot analysis (*see* Chapter 35). If not it may be necessary to use specific extraction methods, in order to produce enriched fractions. In establishing such methods it is important to consider the solubility properties expected of the expressed protein, but to be aware that the actual properties may differ from these (for example, owing to incorrect folding or assembly). Similarly, the solubility properties of the proteins present in the host tissue should also be considered. Detailed discussions of extraction methods for individual proteins are outside the scope of the present chapter.

3.1.4. Sample Preparation by Microdialysis

Although it is often possible to apply crude extracts to SDS-PAGE gels, all the other gel systems discussed here are more sensitive to the pH and salt content of the samples. When working with such samples it is usually necessary to remove contaminating salts, and so on, and adjust the buffer concentration and pH by dialysis prior to separation. A simple and convenient system for multiple samples is to dialyze in 1.5-mL Eppendorf tubes, which are covered with a piece of dialysis membrane secured with an elastic band and suspended upside down (in a perspex or polystyrene rack) in a stirred buffer solution (Fig. 1A). It is, of course, necessary to avoid the use of SDS when the samples are to be separated on systems other than one-dimensional SDS-PAGE or the O'Farrell two-dimensional IEF/SDS-PAGE.

3.2. SDS-PAGE Systems

SDS-PAGE is the most widely used system for protein electrophoresis and is particularly suitable for the analysis of crude extracts because of wide tolerance of variation in the salt content and pH of the samples applied. The proteins are initially complexed with SDS (an anionic detergent) that binds at a ratio of about 1.4 g/g protein. This gives an extended conformation and a high net negative charge at pH 8.9. Because of this high charge (which masks the charge of the protein) all proteins migrate rapidly toward the anode, and separation is achieved by molecular sieving through the polyacrylamide matrix. Although mobility is in theory independent of the protein sequence, single amino acid substitutions have been shown to affect the mobility of some proteins *(7)*. Mobility is also affected by the presence of residual elements of secondary structure, which may be eliminated by the inclusion of urea at concentrations of up to 6*M*. Similarly the presence of intrachain disulfide bonds will result in a more compact conformation and faster mobility. Although it is usual to carry out separations under reducing conditions, comparisons of reduced/native separations can provide information on the presence of intrachain disulfide bonds and thus help to confirm the identity and correct folding of the expressed protein.

Intrachain disulfide bonds can reform during electrophoresis of some proteins (for example, cereal prolamins), and this can be prevented by inclusion of 2 m*M* DTT in the gels. Comparisons of reduced and nonreduced protein samples should be made on separate gels or leave at

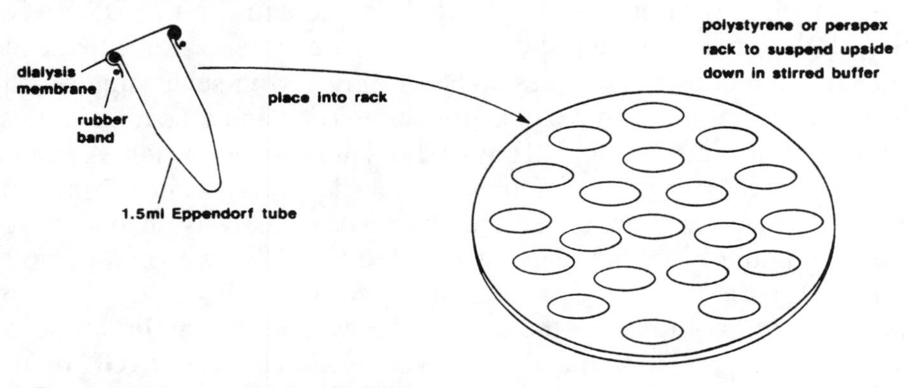

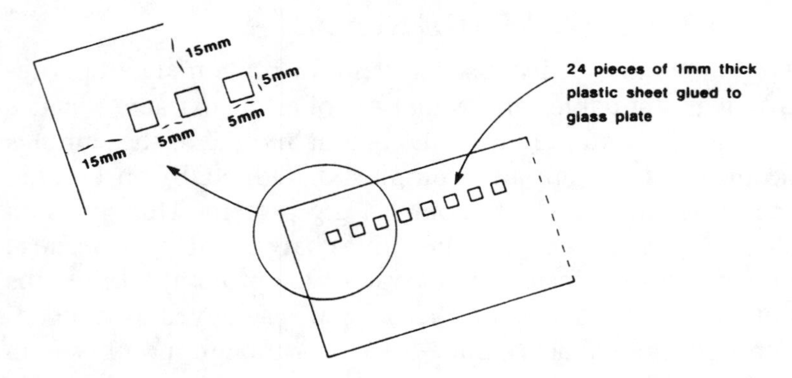

Fig. 1. **A.** Simple apparatus for microdialysis of protein samples. **B.** Template for casting wells in IEF gel.

least one or two tracks empty between samples as 2-mercaptoethanol present in sample buffer can rapidly diffuse between tracks.

3.2.1. The Laemmli (1970) System (1)

Since its introduction the Laemmli system has been widely used for a range of samples from animal, plant, and microbial origin. We have provided recipes for a range of acrylamide concentrations with volumes calculated for two widely used types of apparatus: the Hoefer SE600 (which is identical to the BioRad [Richmond, CA] Protean I) with 14 cm × 16 cm × 1.5 mm gels, and the BioRad Mini Protean II with 8 cm ×

Table 1
Volumes Required for Casting Two Laemmli (1970) Separating Gels[a]

	Percent acrylamide				
	5	7.5	10	12.5	15
A, mL	1.66	2.5	3.32	4.16	5.0
B, mL	2.5	2.5	2.5	2.5	2.5
D, µL	100	100	100	100	100
H_2O, mL	5.7	4.8	4.0	3.2	2.3
E, µL	50	50	50	50	50
TEMED, µL	5	5	5	5	5

[a]Using the BioRad Mini Protein II gel apparatus.

Table 2
Volumes Required for Casting Two Laemmli (1970) Separating Gels[a]

	Percent acrylamide				
	5	7.5	10	12.5	15
A, mL	10	15	20	25	30
B, mL	15	15	15	15	15
D, µL	600	600	600	600	600
H_2O, mL	34	29	24	19	14
E, µL	300	300	300	300	300
TEMED, µL	30	30	30	30	30

[a]Using the Hoefer SE 600 gel apparatus.

7.3 cm × 0.75 mm gels (*see* Tables 1 and 2). Whereas the Mini Protean II apparatus is invaluable for rapid routine analyses, the larger gel size of the Hoefer SE600 can give better resolution of complex mixtures (*see* Fig. 2).

1. Cast separating gels according to the manufacturer's instructions, using the volumes given in Tables 1 and 2. The solutions are mixed before the addition of E and TEMED but not degassed. The gel is carefully overlaid with water or water-saturated butan-1-ol and allowed to set for about 1 h.
2. Pour off water-saturated butan-1-ol and gently wash gel surface with distilled water. Dry gel surface and inside of cassette by inserting a folded No.1 filter paper to just above surface.
3. Place combs into gel cassettes and pipet stacking gel to form wells. Stacking gel is prepared as follows: 1.3 mL of sol A, 2.5 mL sol C, 100 µL of sol D, 50 µL of sol E, and 4.7 mL of water. The solution is well mixed before the addition of 1.3 mL sol F and 10 µL of TEMED.

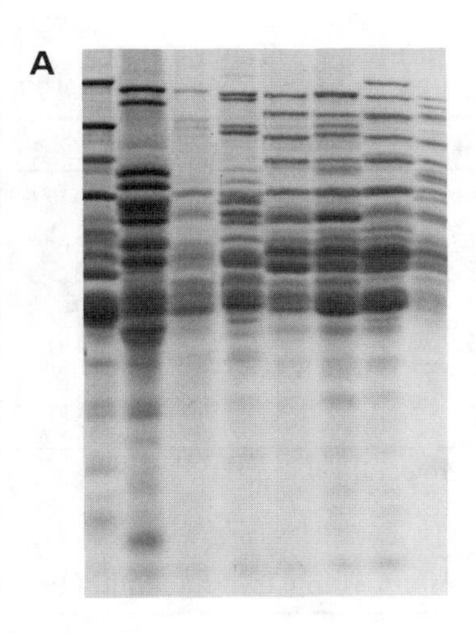

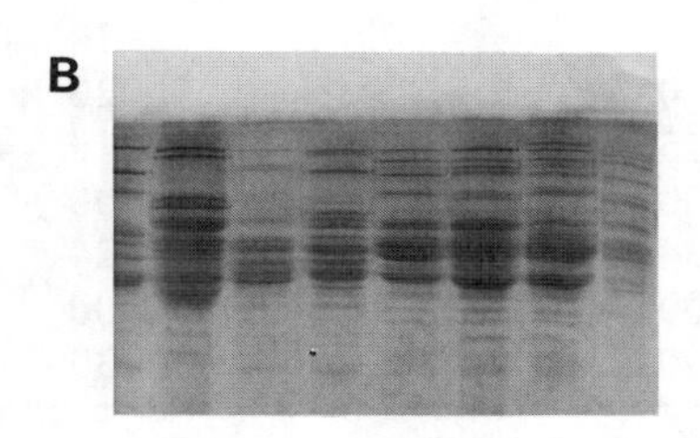

Fig. 2. Comparison of separations of total prolamin extracts of different wheat varieties on large **(A)** and small **(B)** Laemmli gels.

4. After about 30 min remove combs and wash sample wells with running buffer to remove nonpolymerized acrylamide (*see* Note 1).
5. Load about 10–20 μL of sample per well for the Hoefer SE600 (20 wells per gel) or 10 μL per well for the Bio-Rad Mini Protean II (10 wells per gel).
6. Run gels at 20°C with water cooling for Hoefer (San Francisco, CA) SE600 (*see* Notes 2–4). The Hoefer SE600 has an initial power setting of 20 mA/gel (for 30 min) followed by 30 mA/gel for about 3 h. The BioRad Mini Protean II is run at constant 200V (approx 70 mA at start and 35 mA at finish) for 50 min.

3.2.2. The Tricine Gel System

This is recommended for the analysis of low M_r proteins (between 5–20,000) and peptides. It is based on Schägger and von Jagow *(8)*.

Table 3
Volumes Required for Casting Two Tricine SDS-PAGE Gels[a]

	Stacking gel	Separating gel, 10%	Separating gel, 16%
Acrylamide, mL	0.80	2.03	3.30
Gel buffer, mL	2.48	3.30	3.30
Glycerol, g	—	1.3	1.3
ammonium persulfate, µL	80	50	50
TEMED, µL	8	5	5
water, mL	6.7	3.62	2.35

[a]Using the Bio-Rad Mini Protean II gel apparatus.

Table 4
Volumes Required for Casting Two Tris-Borate SDS-PAGE Gels[a]

	Stacking gel, 3%	Separating gel, 10%
Acrylamide, mL	1.5	15
Bisacrylamide, mL	0.8	3.12
Resolving gel buffer, mL	—	6
Stacking gel buffer, mL	2.48	—
10% SDS, w/v	0.2	—
Water, mL	14.8	35.3
Ammonium persulfate, 10% w/v, mL	0.2	0.56
TEMED, µL	20	25

[a]Using the Hoefer SE600 gel system.

1. Volumes for 2 Bio-Rad Mini Protean II gels are given in Table 3 (*see* Note 5).
2. Gels are run initially at 30 V for about 60 min, then increased to 120 V for a further 90 min, all at 20°C.

3.2.3. The Tris-Borate System

This system, based on Koenig et al. *(9)*, is a useful alternative to the Laemmli *(1)* system, providing better separation of some protein mixtures (for example, the high molecular weight subunits of wheat glutenin).

1. Volumes for two Hoefer SE600 Gels (60 mL total volume) are given in Table 4.
2. Follow the method given for the Laemmli *(1)* system (*see* Section 3.2.1.).
3. 1.5-mm thick gels are run at 30 mA/gel for about 3 h (*see* Note 6).

3.3. Nondenaturing Gel Systems

A number of nondenaturing gel systems have been developed, although these are less widely used than SDS-PAGE. The earliest was the pH 8.9 system of Ornstein *(10)* and Davis *(2),* which is described in detail by Hames *(11)*. This system has proved to be the most durable, and is still used with various modifications. The basic system is described in Section 3.3.1. Other systems have been developed for specific uses and one example of these is given (*see* Section 3.3.2.). This is a low pH system, based on aluminum lactate buffers, which was developed for analysis of the storage proteins of wheat and other cereals. With all nondenaturing gel systems it is essential that the samples should be dialyzed to remove excess salts, and so on. In addition SDS must be avoided during extraction and purification. Traces of SDS on the apparatus can also interfere with separation, and it is advisable to use a separate apparatus to that used for SDS-PAGE.

3.3.1. The Davis (pH 8.9) System

This was the original gel electrophoresis system *(2)*, and has formed the basis for many modified systems. The details that follow are taken from the original paper.

1. Prepare the separating gel by mixing 1 part A, 2 parts C, and 1 part distilled water. Add an equal volume of freshly prepared 0.14% (w/v) ammonium persulfate.
2. Prepare the stacking gel by mixing 1 part B, 2 parts D, 1 part E, and 4 parts F. Polymerization is initiated by exposure to fluorescent light.
3. The running buffer is repared by mixing 28.8% (w/v) glycine, 6% (w/v) Tris. pH should be 8.3.

3.3.2. The pH 3.2 Lactate System for Cereal Storage Proteins

The gliadin storage proteins of wheat are classified into groups on the basis of their mobility at low pH. A number of low pH electrophoretic systems have been used, initially with starch gels but more recently using polyacrylamide gels. Most are based on lactate/lactic acid buffer systems, with special catalysts to allow polymerization at low pH (3.0–3.2). The system described here is based on Clements *(12)* and is used in our laboratory with a dedicated BioRad Protean II gel apparatus.

1. Buffers should be precooled to 4°C.
2. Volumes are given for the BioRad Protean II System. Mix the following (sufficient for 2 gels), 25 mL solution A, 10 mL solution B, 45 mL water, 0.4 mL lactic acid (85%), and 80 mg ascorbic acid.

3. Precool at –20°C until ice crystals start to appear. Add 0.2 mL of 0.6% (v/v) hydrogen peroxide and pour rapidly at room temperature. The mixture will polymerize in 2–3 min (*see* Note 7).
4. Dissolve protein samples in sample buffer at about 1 mg/mL, centrifuge, and load up to 100 μL.
5. Run with reverse polarity (compared to SDS-PAGE) at 40–45 mA/gel, either at 4°C in a cold room or using a cooling system (*see* Note 8). Run for 30 min longer than the time taken for the tracking dye to reach the front (a total of 2–2$^{1}/_{2}$ h).
6. Carefully remove gels and place into 350 mL of 10% (w/v) TCA for 30 min to fix the proteins. 5 mL of 0.5% Coomassie brilliant blue R250 is then added and left overnight. No destaining is required.

3.4. Isoelectric Focusing (IEF)

Although more widely used as the first dimension of two-dimensional analyses (*see* Section 3.5.), the high resolving power of IEF means that one-dimensional systems can also be used to separate highly complex mixtures. It is, of course, of particular value if the expressed protein has an unusually high or low pI. The system described below (based on Righetti et al. *[13]*), uses an LKB Multiphor or similar flat bed apparatus, and can be used to separate up to 24 denatured protein samples in a single run.

In order to facilitate sample application the gel is cast with 24 sample wells (each 5 mm^2 × 1 mm deep and 5 mm apart) positioned about 10 mm from one edge. These are made by cementing 5mm^2 × 1 mm thick squares of plastic to the upper (thick) plate of the gel sandwich (Fig. 1B).

1. Dissolve samples at a concentration of about 10 mg/mL (*see* Note 9). If necessary dialyze overnight (*see* Section 3.1.4.).
2. Dissolve 21.6 g urea in 30 mL of Stock A and 3 mL of LKB 40% pH 3.5–10 carrier ampholyte and make up to 60 mL with distilled deionized water. Degas under vacuum, add 300 μL of catalyst and 20 μL of TEMED, and mix.
3. Pour into assembled gel sandwich with the wells uppermost to avoid air bubbles, and allow to polymerize. Remove top plate from gel.
4. Add 25 μL of 1% pH 3.5–10 ampholyte to each sample well and prerun for 1 h at 4°C with 10 W power (*see* Note 10).
5. The wells are then blotted dry and 15–25 μL of sample added to each (*see* Note 11). The gel is then run for 3 h at 4°C with 13 W power.
6. The glass plate bearing the gel is then placed for 2 h in 10% (w/v) TCA. The gel is then washed three times for 2 h each with distilled water. This is essential to remove ampholyte prior to staining (*see* Notes 12–14).

7. Stain for 2 h in 0.05% (w/v) Coomassie blue G dissolved in 7% (v/v) acetic acid/25% (v/v) ethanol, and destain in 7% (v/v) acetic acid/25% (v/v) ethanol. Alternatively stain with Blakesley stain (*see* Section 3.6.1.2.).
8. In order to measure the pH gradient a gel track containing separated proteins can be cut into 0.5-mm slices and mascerated with deionized water. The pH of the slurry is then measured with a micro pH electrode.

3.5. Two-Dimensional Systems

Only one type of two-dimensional system has been widely adopted—the combination of IEF with SDS-PAGE. This has the advantages of wide application, adaptability (pH range, gel concentration, and so on), and high resolution. We will describe two alternative systems later: the widely used O'Farrell *(3)* system and a simpler system based on combining the one-dimensional IEF systems described in Section 3.4. with the Laemmli SDS-PAGE system (*see* Section 3.2.1.).

A major advantage of the O'Farrell system is that samples can be applied to the first dimension IEF gel in the presence of SDS, which comes off the protein and forms anionic mixed micelles with the detergent Nonidet NP40. Because of this it is suitable for the analysis of proteins that are not readily solubilized.

O'Farrell et al. *(14)* subsequently reported a novel system in which nonequilibrium pH gradient electrophoresis (NEPHGE) is combined with SDS-PAGE to give better resolution of highly basic proteins. This is not described in detail here.

3.5.1. The O'Farrell (1975) System

1. Prepare IEF gel recipe (10 mL) by mixing 5.5 g urea, 1.33 mL solution A, 2.0 mL solution B, 1.97 mL H_2O, and 0.4 mL ampholytes (to 2% w/v final concentration). Dissolve urea then add 10 µL solution C (degassed) and 7 µL TEMED (*see* Note 15).
2. Prepare SDS separating gel by mixing 0.25 vol solution D, 0.75 vol solution F, water (varied to give required gel concentration), and 0.0033 vol solution C (degassed). Degas, add 0.0005 vol TEMED and pour.
3. Prepare SDS stacking gel by mixing 0.25 vol solution E, 0.15 vol solution F, 0.6 vol H_2O, and 0.003 vol solution C (degassed). Degas, add 0.001 vol TEMED.
4. Gels are cast in 130 × 2.5 mm (id) glass tubes sealed at the bottom with parafilm, and overlaid with 8*M* urea (*see* Note 16).
5. After 1–2 h the overlay is removed and replaced with 20 µL lysis buffer overlaid with water.

6. After a further 1–2 h the tubes are loaded into the apparatus and the gel overlaid with 20 µL fresh lysis buffer followed by cathode solution. The upper reservoir is also filled with cathode solution (extensively degassed) and the lower reservoir with anode solution. The tubes are prerun at 200 V for 15 min, 300 v for 30 min, and 400 v for 30 min.
7. The upper reservoir is then emptied and the gels loaded with samples overlaid first with 9*M* urea/1% ampholyte and then with cathode solution. The upper reservoir is refilled and the gels run at 400 v for 12 h followed by 800 v for 1 h (*see* Note 17).
8. The gels are removed from the tubes into 5 mL of SDS sample buffer and shaken at room temperature for 2 h (*see* Note 19).
9. The SDS-PAGE slab gel is poured in an apparatus with notched plates to hold the IEF gel. The separating gel is poured to 25 mm below the notch and overlaid with water. After polymerization (≃1 h) the water and unpolymerized gel are moved from the surface, replaced with fourfold diluted solution D, and the gel allowed to stand overnight (*see* Note 18).
10. Remove the solution from the gel surface and pour stacking gel to the base of the notch. Insert a Teflon™ strip 1.2 mm below the base of the notch to form a flat surface. Allow to polymerize 30–60 min.
11. Remove the Teflon™ strip. Pour on 1 mL of melted agarose solution (1% in SDS sample buffer) and transfer IEF gel into this. Allow to set for 5 min.
12. Load gel into apparatus. Add 40 µL of 1% (w/v) Bromophenol blue to upper reservoir. Run at 20 mA till dye has reached front.
13. Fix and stain as appropriate (*see* Section 3.6.; *see* Note 20).

3.5.2. Alternative Two-Dimensional System

1. Run IEF gel as in Section 3.4.
2. Place the glass plate supporting the gel on to graph paper and cut between the sample wells into 10-mm wide strips.
3. Either place strips in 10 mL of equilibration buffer for about 30 min, or place strips in plastic vials containing 10 mL of equilibration buffer with 10% (v/v) glycerol. Place vials at –20°C after initially frosting by suspending in liquid nitrogen. Thaw before use by placing the vials in warm water for 10 min.
4. The gel strip is placed 1 cm above a Laemmli separating gel in a Hoefer SE600 slab unit and surrounded with stacking gel.
5. Electrophoresis is carried out at 30 mA for about 6 h, using Bromophenol blue as a tracking dye.
6. Gels are washed for 3 h in 5% (v/v) acetic acid containing 30% (v/v) ethanol to remove carrier ampholytes (*see* Note 21). They may then be stained with acetic acid/ethanol or TCA/methanol based systems.

3.6. Gel Staining Systems

3.6.1. Coomassie Blue-Based Stains for Total Proteins

The most commonly used total protein stain is Coomassie brilliant blue. This has essentially replaced most other dyes used for protein staining, although copper staining is an alternative method (Lee et al. *[15]*). Coomassie blue is available as R-250 (R for reddish hue) or G-250 (G for greenish hue). Numerous protocols have been reported for the use of these two dyes, some of which are said to have similar sensitivities to silver staining (Neuhoff et al. *[16]*). In some protocols the Coomassie blue is dissolved in solutions including TCA. Because it is relatively insoluble under these conditions it forms dye–protein complexes very rapidly. Alternatively, the dye can be prepared in the leuco form that is nearly colorless in solution, but binds to protein to give blue color against an almost colorless background. An advantage of this method is that the gel can be left in solution for as long as is necessary to develop the required intensity and no destaining is required.

A typical stain solution includes an alcohol as a solvent (either methanol or ethanol) and an acid as a fixative (either acetic acid or TCA). TCA is often used rather than acetic acid because it is a much stronger fixative and is less likely to dissolve the proteins. The time required for staining depends on the gel thickness and concentration as well as temperature.

Gels can be destained by simple diffusion using the solvent used for staining or with 10% (w/v) TCA. A small piece of foam rubber can also be added to the destain solution to absorb the stain lost from the gel. Care must be taken with destaining as the stain can be completely removed from the proteins. The destaining solution can be recycled by pouring through a charcoal filter to absorb any remaining dye.

3.6.1.1. Standard Stain for Total Proteins

1. Place the gel in a shallow tray and immerse in stain. Cover the tray and shake gently for several hours or more conveniently leave overnight at ambient temperature (*see* Note 22).
2. Recover the stain solution and rinse the gel in water before adding destain solution. Add foam rubber and shake for several hours until background clears. Recover destain solution.

3.6.1.2. Blakesley Stain

A stain based on that of Blakesley and Boezi *(4)* is particularly suited for IEF and two-dimensional IEF/SDS-PAGE gels, as it gives minimum background without destaining. It can also be used after staining with Coomassie blue R to intensify protein staining and to reduce the background (for one-dimensional or two-dimensional gels).

1. Immerse the gel in the stain until desired intensity is reached.
2. Wash gel in water. No destaining is required.

3.6.2. Silver Stain

The original silver staining method for the detection of proteins after gel electrophoresis (Switzer et al. *[17]*) has subsequently been developed and modified by a large number of workers including Oakley et al. *(18)* and Morrissey *(5)*. In addition, some methods combine Coomassie blue staining with silver staining to give enhanced sensitivity (Moreno et al. *[19]*). Silver staining alone is reported to be between 10 and 100 times more sensitive than Coomassie blue, and because of the increased sensitivity great care must be taken to ensure all glassware is thoroughly cleaned and gloves worn at all times. It is advisable not to touch the gels at all, even with gloved hands.

It is common to prepare reagents for silver staining following published methods. However, if only a limited amount of silver staining is done it is often more convenient to purchase kits such as the BioRad "Silver Stain Plus" Kit (based on Gottleib and Chavko, *20*) which is reported to detect 2 ng protein with low background staining.

Although silver staining is certainly the most sensitive chromogenic detection method there are a number of problems associated with its use. Not all proteins will stain, some only very weakly and others not at all. Some proteins that do not contain cysteine residues stain negatively, whereas others that do contain cysteine residues stain orange, red, brown, or green depending on cysteine content and on whether they contain covalently attached lipids (Chuba and Palchaudhuri *[21]*). Also, not only proteins are detected using silver stain, but also DNA and the lipid portions of lipopolysaccharides, thus confusing protein identification.

Silver staining is reported to be very rapid, but may in fact be time consuming as numerous steps are required that need careful timing and handling. We therefore recommend the use of Coomassie blue staining for most applications, including protein blotting for sequencing (*see* Section 3.8.).

The method below is based on Morrissey *(5)*. All steps should be carried out with agitation.

1. Prefix in 50% (v/v) methanol, 10% (v/v) acetic acid for 30 min, followed by 5% (v/v) methanol, 7% (v/v) acetic acid for 30 min.
2. Fix for 30 min in 10% (v/v) glutaraldehyde.
3. Rinse in distilled water, preferably overnight.
4. Soak in 5 μg/mL dithiothreitol for 30 min.
5. Pour off solution and without rinsing add 0.1% (w/v) silver nitrate. Leave for 30 min.
6. Rinse once rapidly in a small amount of distilled water and then twice rapidly in a small amount of developer (50 μL of 37% [v/v] formaldehyde in 100 mL 3% [w/v] sodium carbonate). Soak in developer until the desired level of staining is obtained.
7. Stop the staining by adding 5 mL of 2.3*M* citric acid directly to the developer and agitate for 10 min. Discard this solution and wash with distilled water.

3.6.3. Glycoprotein Stains

Many plant proteins are glycosylated and it is routine to test for this using gel staining. Most glycoprotein staining systems are based on the periodic acid–Schiff method, including the system described below which is taken from Sargent and George *(22)* based on Zaccharius et al. *[23]*).

1. Fix in 12.5% (w/v) TCA (*see* Note 23).
2. Rinse with distilled water and soak in 1% (w/v) periodic acid, 3% (v/v) acetic acid for 50 min.
3. Wash in several changes of distilled water for 12–18 h.
4. Immerse in fuchsin-sulfite stain in the dark for 50 min.
5. Wash three times for 10 min each with freshly prepared 0.5% (w/v) metabisulfite.
6. Wash in distilled water to remove excess stain.

3.7. Gel Preservation

Gels are most readily preserved by drying. A quick, economical, and very simple method is to use a commercially available gel drying frame such as that supplied by Promega. No pretreatment is required except for soaking the gel in water. The gel is then placed between wetted sheets of gel drying film and clamped between the frames. Four minigels or one large gel can be dried at one time, the minigels drying overnight at ambient temperature.

Once dry the gels are readily handled and can be stored simply by punching holes and filing. They can also be scanned or displayed using an overhead projector. Both acrylamide and agarose gels can be dried in this way.

For SDS-polyacrylamide gels:

1. The stained gel is soaked overnight in water.
2. Two cellulose sheets are wetted in an open tray.
3. Place one drying frame on a clean bench and carefully lay one cellulose sheet uniformly over the frame so as to remove any wrinkles or folds.
4. Holding the gel from one edge carefully position on to the wetted sheet.
5. Holding the second sheet by one edge carefully and evenly position it over the gel, avoiding the creation of air pockets. If air bubbles form lift the sheet and reposition.
6. Place the second frame over the first and slide the assembly to the edge of the bench. Fold the cellulose sheets over the frame and clamp the entire frame.
7. Place the frame over an open tray in a horizontal position and allow to dry overnight. Large gels often require 1 or 2 d to dry completely.

3.8. Microsequencing from Gels

The ultimate confirmation of the identity and integrity of an expressed protein is to determine its amino acid sequence. It is, of course, unrealistic to determine the full sequence, but the development of highly sensitive gas and pulsed liquid phase sequencers coupled with high efficiency membranes means that it is often possible to determine N-terminal sequences of proteins transferred from one- or two-dimensional gels. Using this approach, it is possible to obtain sequence information from less than 20 pg of protein, although the procedure is more successful with proteins of low M_r. Whereas early studies used chemically activated filters, most workers now prefer PVDF-based membranes such as Immobilon™ (Millipore, Bedford, MA) or Pro-Blot™ (Applied Biosystems Inc., Foster City, CA). The latter has been developed specifically for use with the Applied Biosystems Pulsed Liquid Sequencer, with a specific cycle (BLOTT 1) to give improved efficiency in the absence of polybrene.

The following protocol has given excellent results in our laboratory, and is taken from the Applied Biosystems Protein Sequencer Users Bulletin (Yuen et al. *[24]*).

1. Wet two sheets of PVDF membrane with methanol for a few seconds, then place in a dish containing blotting buffer (*see* Note 24).
2. Remove the gel from the electrophoresis cell and soak in blotting buffer for 5 min.
3. Assemble the transblotting sandwich and electroblot at 90 V (300 m*M*) at room temperature for 10–30 min.

4. Remove the membranes from the transblotting sandwich and rinse with deionized water.
5. Visualize by staining with Coomassie blue R250.
   a. Rinse in distilled deionized water.
   b. Saturate with 100% methanol for a few seconds.
   c. Stain briefly with 0.1% (w/v) Coomassie blue R250 in 1% (v/v) acetic acid/40% (v/v) methanol. Spots should appear within 1 min.
   d. Destain with 50% (v/v) aq. methanol.
6. Rinse thoroughly with distilled deionized water.
7. Excise bands or spots for microsequencing (*see* Note 25).

4. Notes

4.1. The Laemmli (1970) System

1. The inclusion of up to 4*M* urea in the separating and stacking gels can result in improved resolution, but also leads to decreased mobility and, in some cases, changes in the relative mobilities of the different components of a mixture (*see* Bunce et al., *25*).
2. It is often convenient to run large gels overnight with a constant setting of 50 V. This is adjusted to 30 mA/gel for the final hour.
3. The choice of gel size can have a great effect on the separation, as illustrated by the Hoefer SE600 and BioRad Mini Protean II gels shown in Fig. 2.
4. The Laemmli system gives good resolution of proteins over a wide range of M_r s, but the Tricine system (later) may be preferred for proteins of M_r below about 20,000. The difference in resolution given by two systems is illustrated in Fig. 3.

4.2. The Tricine System

5. Urea can be added to 6*M* by dissolving 3.6 g solid urea in the separating gel mixture before adding catalyst.

4.3. The Tris-Borate System

6. A Tris-borate transfer buffer can be used to transfer proteins from Tris-borate gels to membranes: 50 m*M* boric acid, 50 m*M* Tris, pH 8.5 (*see* Baker et al. *[26]*). The transfer buffer can be reused several times without loss of efficiency.

4.4. The pH 3.2 Lactate System for Cereal Seed Storage Proteins

7. The apparatus and especially the plates must be scrupulously clean and free of SDS.
8. Efficient cooling is essential as the gel is run at a high voltage.

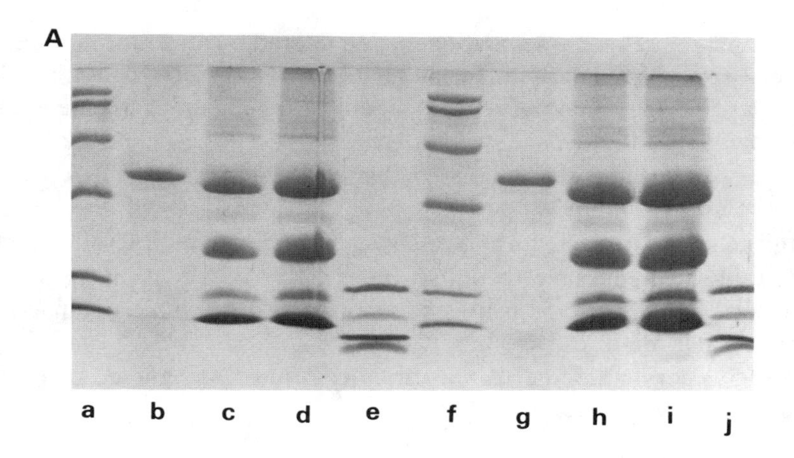

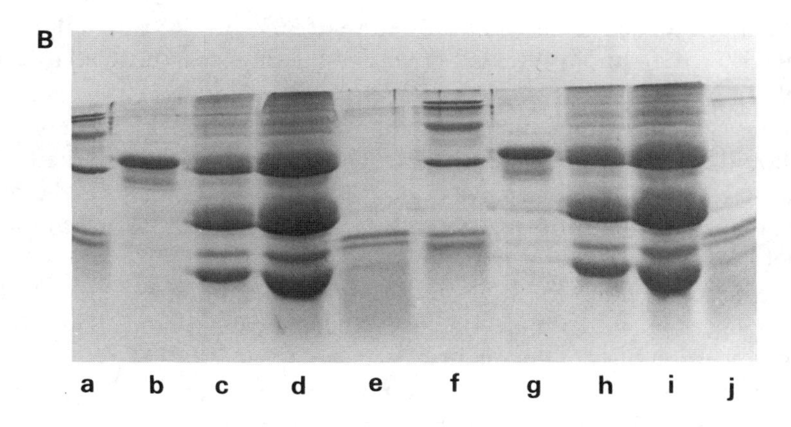

Fig. 3. Comparison of Tricine **(A)** and Laemmli **(B)** gel systems for separation of small proteins. Molecular weight marker proteins in tracks a and f are, from the top of gel: ovotransferrin (76–78,000); bovine serum albumin (66,250); ovalbumin (42,700); carbonic anhydrase (30,000); myoglobin (17,200); cytochrome c (12,300). In tracks e and j are cyanogen bromide cleaved products of myoglobin in the molecular weight range of 16,949–2512 (BDH Ltd.). Tracks c, d, h, and i show cyanogen bromide cleaved products of purified B1 Hordein (tracks b and g).

4.5. Isoelectric Focusing

9. It is essential to avoid high concentrations of salts in the samples as they affect the pH gradient.

10. The pH range of the separation can be altered by substituting mixtures of narrow range ampholytes for the pH 3.5–10 ampholyte. Similarly low or high pH range ampholyte can be mixed with wide range to extend the ends of the gradient.
11. The protein sample may be loaded at either side of the gel, depending on its solubility and pI.
12. It is essential to wash the gel well before staining to remove all ampholytes.
13. It may be necessary to substitute a TCA-based staining system as described in Section 3.6.1. as some proteins are soluble in ethanol/acetic acid/water mixtures.
14. The 5% acrylamide gel is very fragile. It must be handled carefully and shaken only gently.

4.6. Two-Dimensional Systems

15. O'Farrell recommends using a mixture of 80% pH 5–7 ampholyte to 20% pH 3.5 to 10 ampholyte, but other combinations can be used to vary the pH range.
16. It is essential that the plates and tubes are scrupulously clean. The tubes should be cleaned in chromic acid, rinsed in water, and soaked in alcoholic KOH prior to extensive rinsing and air drying. Tubes and plates can be rinsed with a proprietary wetting agent (e.g., photoflo) to facilitate gel removal.
17. The time schedule for IEF can be varied, maintaining 5,000–10,000 V hours per gel, with a final voltage of 400 at the maximum. This will affect spot position but not gel quality.
18. The acrylamide concentration in the separating gel can be varied from 5–22.5% by varying the proportions of solution F and water. Gradient gels can be run using 75% (v/v) glycerol instead of water to prepare the high concentration acrylamide solution, and reducing the amounts of solution C (ammonium persulfate) in both mixtures to 0.00145 volumes.
19. Equilibration time can be reduced in duration or omitted to prevent protein losses. In the latter case specific modifications to the procedure are required depending on the solubility of the protein (*see* O'Farrell *[13]*).
20. IEF gels can be frozen in 5 mL SDS sample buffer using an ethanol/dry ice bath, and stored at –70°C.
21. As with one-dimensional IEF, care must be taken to remove ampholytes before staining. However, TCA must also be avoided in the initial washing as an insoluble precipitate is formed when SDS is also present. The SDS and the bulk of the ampholytes can be removed by washing with acetic acid/ethanol before staining.

4.7. Gel Staining Systems

22. The methanol is lost from the stain solution during use and fresh stain should be prepared regularly.
23. It is essential to run control proteins (glycosylated and nonglycosylated) on all gels in order to reduce the risk of false positives or negatives.

4.8. Microsequencing from Gels

24. It is essential to wear gloves at all stages to avoid contaminating the membrane.
25. It is usually only worth attempting sequencing if the spots are clearly visible on staining with Coomassie blue and often necessary to bulk several spots from two-dimensional gels. One-dimensional gels can be streaked or bands from several tracks bulked.

References

1. Laemmli, U. K. (1970) Cleavage of structural proteins during the assembly of the head of bacteriophage T4. *Nature* **227,** 680–685.
2. Davis, B. J. (1964) Disc Electrophoresis—II. Method and application to human serum proteins. Ann. NY *Acad. Sci.* **121**, 404–427.
3. O'Farrell, P. H. (1975) High resolution two-dimensional electrophoresis of proteins. *J. Biol. Chem.* **250**, 4007–4021.
4. Blakesley, R. W. and Boezi, J. A. (1977) A new staining technique for proteins in polyacrylamide gels using Coomassie brilliant blue G250. *Anal. Biochem.* **82**, 580–582.
5. Morrissey, J. H. (1981) Silver stain for proteins in polyacrylamide gels: A modified procedure with enhanced uniform sensitivity. *Anal. Biochem.* **117**, 307–310.
6. Nelson, T., Harpster, M. H., Mayfield, S. P., and Taylor, W. C. (1984) Light regulated gene-expression during maize leaf development. *J. Cell Biol.* **98,** 558–564.
7. de Jong, W. W., Zweers, A., and Conen, L. H. (1978) Influence of single amino acid substitutions on electrophoretic mobility of sodium dodecyl sulphate-protein complexes. *Biochem. Biophys. Res. Commun.* **82**, 532–539.
8. Schägger, H. and von Jagow, G. (1987) Tricine-sodium dodecyl sulphate-polyacrylamide gel electrophoresis for the separation of proteins in the range from 1 to 100 KDa. *Anal. Biochem.* **166**, 368–379.
9. Koenig, R., Stegemann, H., Francksen, H., and Paul, H. L. (1970) Protein subunits in the potato virus N group. Determination of the molecular weights by polyacrylamide electrophoresis. *Biochem. Biophys. Acta.* **207**, 184–189.
10. Ornstein, L. (1964) Disc Electrophoresis—I. Background and theory. *Ann. NY Acad. Sci.* **121**, 321–346.
11. Hames, B. D. (1990) In Gel electrophoresis of Proteins: A practical approach, 2nd ed. (Hames, B. D., and Rickwood, D., ed.), IRL Press, Oxford, pp. 1–147.
12. Clements, R. L. (1987) A study of gliadins of soft wheats from the Eastern United States using a modified polyacrylamide gel electrophoresis procedure. *Cereal Chem.* **64**, 442–448.

13. Righetti, P. G., Gianazza, E., Viotti, A., and Soave, C. (1977) Heterogeneity of storage proteins of maize. *Planta* **136**, 115–123.
14. O'Farrell, P. Z., Goodman, H. M., and O'Farrell, P. H. (1977) High resolution two dimensional electrophoresis of basic as well as acidic proteins. *Cell* **12**, 1133–1142.
15. Lee, C., Levin, A., and Branton, D. (1987) Copper staining: A five-minute protein stain for sodium dodecyl sulphate-polyacrylamide gels. *Anal. Biochem.* **166**, 308–312.
16. Neuhoff, V., Arold, N., Taube, D., and Ehrhardt, W. (1988) Improved staining of proteins in polyacrylamide gels including isoelectric focusing gels with clear background at nanogram sensitivity using Coomassie brilliant blue G-250 and R-250. *Electrophoresis* **9**, 255–262.
17. Switzer, R. C., Merril, C. R., and Shifrin, S. (1979) A highly sensitive silver stain for detecting proteins and peptides in polyacrylamide gels. *Anal. Biochem.* **98**, 231–237.
18. Oakley, B. R., Kirsch, D. R., and Morris, N. R. (1980) A simplified ultrasensitive silver stain for detecting proteins in polyacrylamide gels. *Anal. Biochem.* **105**, 361–363.
19. Moreno, De M. R., Smith, J. F., and Smith, R. V. (1985) Silver staining of proteins in polyacrylamide gels: Increased sensitivity through a combined Coomassie blue-silver stain procedure. *Anal. Biochem.* **151**, 466–470.
20. Gottlieb, M. and Chavko, M. (1987) Silver staining of native and denatured eucaryotic DNA in agarose gels. *Anal. Biochem.* **165**, 33–37.
21. Chuba, P. J. and Palchaudhuri, S. (1986) Requirement for cysteine in the color silver staining of proteins in polyacrylamide gels. *Anal. Biochem.* **156**, 136–139.
22. Sargent, J. R. and George, S. G. (1975) in *Methods in Zone Electrophoresis,* BDH Chemicals Ltd., Poole, UK.
23. Zaccharius, R. J., Zell, T. E., Morrison, J. H., and Woodlock, J. J. (1969) Glycoprotein staining following electrophoresis on acrylamide gels. *Anal. Biochem.* **31**, 148–152.
24. Yuen, S. W., Chui, A. H., Wilson, K. J., and Yuan, P. M. (1988) Microanalysis of SDS-PAGE electroblotted proteins, in *Applied Biosystems Protein Sequences User Bulletin No. 36.*
25. Bunce, N. A. C., White, R. P., and Shewry, P. R. (1985) Variation in estimation of molecular weights of cereal prolamins by SDS-PAGE. *J. Cereal Sci.* **3**, 131–142.
26. Baker, C. S., Dunn, M. J., and Yacoub, M. H. (1991) Evaluation of membranes used for electroblotting of proteins for direct automated microsequencing. *Electrophoresis* **12**, 342–348.

CHAPTER 35

Western Blotting Analysis

Roger J. Fido, Arthur S. Tatham, and Peter R. Shewry

1. Introduction

The immunodetection of proteins bound to a membrane has widespread applications in plant biochemistry and molecular biology, including the identification and semiquantitative determination of foreign proteins expressed in transgenic plants. The approach is usually applied to protein transferred from electrophoretic separations (*see* Chapter 34), which allows positive identification to be combined with the provision of information about the protein M_r, charge, pI, and so on. This is commonly called Western blotting, and provides the operator with a wide range of options for choice of membrane type, transfer system, and detection system. We will initially discuss these options, and then provide detailed step-by-step instructions for a well-established method of protein transfer and identification using an enzyme-labeled second antibody. In addition, we will discuss two other applications of protein blotting that provide rapid but less precise results. These are analysis of tissue extracts using dot blotting, and "squash" blots of whole plants or plant organs.

These different applications of protein blotting and immunochemical detection facilitate the rapid screening of multiple samples for protein expression, and detailed analysis of the amount and properties of proteins expressed in individual plants or plant organs. An example of the use of Western blotting to study the expression and processing of a protein encoded by a transgene is shown by Higgins and Spencer *(1)*, (*see also* Fig. 1).

From: *Methods in Molecular Biology, Vol. 49: Plant Gene Transfer and Expression Protocols*
Edited by: H. Jones Humana Press Inc., Totowa, NJ

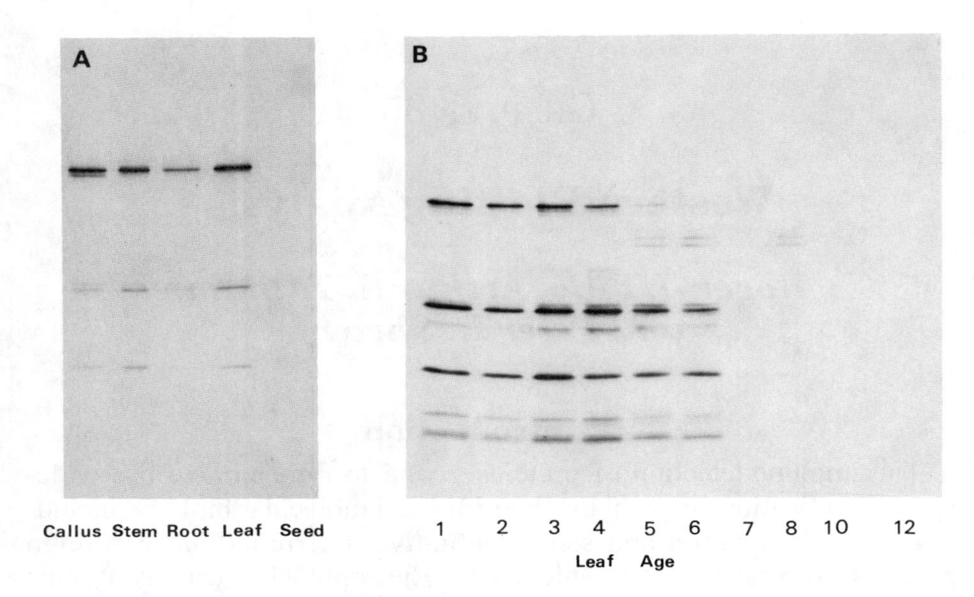

Fig. 1. The distribution of polypeptides related to pea vicilin in transgenic tobacco plants (14/2) transformed with a 35S-vicilin chimeric gene. **(A)** In whole tissue extracts of callus, stem, root, leaf, and seed; and **(B)** In immunopurified fractions from leaves of various ages. The vicilin related polypeptides were detected after fractionation by SDS-PAGE, blotted onto nitrocellulose and detected by alkaline phosphatase-coupled antibodies (from Higgins and Spencer, 1991 *[1]*).

1.1. General Principles of Western Blotting

Protein blotting developed from the original work of DNA blotting as described by Southern *(2)*. However, instead of using capillary action as the driving force to transfer protein from gel to membrane, Towbin et al. *(3)* used electroelution in a system that has developed into the standard method of protein transfer. Burnett *(4)* slightly modified the system and adopted the term "Western blotting" in accordance with the previously used nomenclature for nucleic acid blotting procedures *(2)*.

Western blotting enables proteins to be immobilized and made more accessible for detection and identification by probing after various types of electrophoretic separation using a wide variety of methods. The procedure is highly sensitive in that discrete components of a complex mixture can be identified and convenient in that the filters can be processed immediately after transfer or stored.

1.2. Transfer Membranes

Several types of membrane are available for blotting, the choice depending on the protein under study. The most commonly used is nitrocellulose (NC), which is available either pure or supported, and in a range of formats and pore sizes. Although a 0.45-μm pore size is widely used, low molecular weight proteins (i.e., below 20 kDa) require a smaller pore size, with 0.2 μm being recommended. Suppliers such as Schleicher and Schuell (Keene, NH) provide pure NC membrane in six pore sizes. Because pure nitrocellulose membrane is fragile and will tear and crack easily, it must be handled carefully, and always with gloved hands.

Although more suitable for nucleic acid techniques, nylon membranes are an alternative to NC and are available in charged (e.g., Zeta-Probe from Bio-Rad, Richmond, CA) or uncharged (e.g., Hybond-N from Amersham International, Arlington Heights, IL) forms. They are resistant to tearing, very pliable, and have advantages over NC in certain circumstances, including a higher protein-binding capacity. However, there are also some limitations to their use. Because of their increased binding capacity they can give high nonspecific binding and consequently require increased concentrations of blocking agent. In addition, nylon membranes are incompatible with the most commonly used anionic protein stains, such as Amido Black and Coomassie blue. The polyvinyldifluoride (PVDF) membranes, such as Immobilon (Millipore Corp., Bedford, MA) and ProBlott (Applied Biosystems Inc., Foster City, CA) can be used for automated microsequencing of proteins directly after transfer. They can be used in the same way as NC after initially prewetting in 100% methanol for a short time (*see* Immobiline Technical Bulletin, Millipore). In a recent comparison of different membrane types it was shown that the PVDF-based membranes were the most efficient for microsequencing analysis *(5)*.

Many companies, including Bio-Rad (Poole, UK) and BDH, provide free guides to blotting membranes.

2. Materials

2.1. Electroblotting of Gels

1. Transfer buffer: 25 m*M* Tris, 192 m*M* glycine, and 20% methanol, pH 8.3, with or without 0.02% (w/v) SDS.
2. Nitrocellulose transfer membrane cut to size of gel.
3. Filter papers (Whatman [Maidstone, UK] 3MM) cut to the size of the gel.

2.2. Membrane Blocking

1. Tris-buffered saline (TBS): 20 m*M* Tris-HCl, 500 m*M* NaCl, pH 7.5.
2. Blocking agent: 20 m*M* Tris-HCl, 500 m*M* NaCl, pH 7.5, 5% (w/v) nonfat dried milk powder (Marvel type).

2.3. Periodate Treatment of Western Blots

1. 50 m*M* Sodium acetate, pH 4.5.
2. 10 m*M* Periodic acid in sodium acetate (prepare fresh before use).
3. TBS containing 1% (w/v) glycine (GTBS).

2.4. Total Protein Staining

The stain is available in powder form from a number of suppliers and as a 2% (w/v) solution in 30% (w/v) TCA/30% sulfosalicylic acid from Sigma (St. Louis, MO). Dilute 10X for use.

2.5. Alkaline Phosphatase Detection System

1. Tris-buffered saline (TBS): 20 m*M* Tris-HCl, 500 m*M* NaCl, pH 7.5.
2. Tween-Tris buffered saline (TTBS): TBS containing 0.05% v/v Tween 20.
3. Blocking agent: 5% (w/v) solution of Marvel (nonfat powdered milk) in TBS.
4. Antibody buffer: TTBS containing 1% BSA (w/v).
5. Solution A: 50 mg/mL solution of nitro blue tetrazolium prepared in 70% (v/v) of N,N-dimethylformamide (DMF).
6. Solution B: 25 mg/mL solution of 5-bromo-4-chloro-3-indolyl phosphate in 100% DMF.
7. Solution C: Carbonate buffer: 0.1*M* $NaHCO_3$, 1.0 m*M* $MgCl_2$ adjusted to pH 9.8 with NaOH.

3. Methods

3.1. Methods of Protein Transfer

There are essentially four methods of protein transfer from polyacrylamide gels to membranes, although in practice the method most commonly used is electroblotting. These methods, in terms of increasing efficiency, are simple diffusion, capillary action, vacuum blotting, and electroblotting.

An extremely simple system for transferring proteins is diffusion, in which two membranes (one on each side of the gel) are sandwiched between filter papers and foam pads. The entire assembly is clamped between an open frame (to allow free movement of buffer), and submerged under transfer buffer *(6,7)* for 24 h or longer if necessary.

An assembly for capillary blotting was described by Southern *(2)* and used extensively for DNA and RNA blotting. Mass flow was achieved by placing the gel on paper wicks soaked in transfer buffer, with the movement of fluid acting as the driving force to transfer the proteins from the gel into the membrane. McDonald *(8)* also reported the use of a simple transfer method for nondenatured proteins from isoelectric-focusing (IEF) gels to a PVDF-type membrane. The membrane was placed directly on the gel, followed by three layers of blotting paper and a 5 kg weight to assure intimate contact. Proteins were transferred in 3 min, after which a second membrane was blotted for 7 min. However, this system proved to be slow and inefficient for transferring proteins from other types of gel, and more efficient transfer was achieved by using negative pressure (i.e., low vacuum) to increase the mass flow of buffer. The early development of vacuum blotting was made by Peferoen et al. *(9)* who adapted a slab gel drier.

In 1979 Towbin et al. *(3)* described the use of electroelution to transfer proteins from polyacrylamide gels to nitrocellulose membranes. The method gave much improved transfer and has subsequently developed into the standard method for protein transfer. The term "Western blotting" *(4)* now covers essentially all forms of protein blotting.

Many blotting systems are available commercially from suppliers such as Bio-Rad, Hoefer (San Francisco, CA), and Pharmacia-LKB (Uppsala, Sweden), who produce both wet-tank and semidry blotters. The choice of system will depend on numerous factors. Although the semidry method does have some advantages including lower buffer volumes and shorter blotting times, the wet-tank method can give better transfer, especially of high molecular weight proteins.

3.2. Electroblotting of Gels

Proteins can be blotted from any type of gel (*see* Chapter 34), although the conditions of transfer must be adjusted. IEF and low pH (non-SDS) gels can both be transferred to NC using 0.7% (v/v) acetic acid as transfer buffer, with the proteins migrating toward the cathode. Alternatively these gels can be equilibrated in SDS-PAGE blotting buffer containing high concentrations (2% w/v) of SDS before transferring as standard SDS gels (the transfer buffer containing 0.01–0.02% w/v SDS *(see later)*. Matthaei et al. *(10)* also blotted hydrophobic integral membrane proteins from IEF gels to NC, using octyl glucoside detergent. This detergent did not interfere with transfer, whereas CHAPS and Triton X-100 did.

Two-dimensional IEF/SDS-PAGE gels are frequently used because of their high resolution (*see* Chapter 34). These can be treated as standard one-dimensional SDS-PAGE gels, and equilibrated in transfer buffer for between 30 and 40 min prior to electroblotting.

The protocol described is for SDS-PAGE using the Bio-Rad Trans-Blot cell.

1. Prior to transfer, the membrane must be fully wetted to remove all air pockets. This is best achieved by carefully wetting one edge and allowing capillary action to drive air out while slowly lowering into buffer. Leave under buffer until required. Disposable gloves must be worn at all times when handling the membrane.
2. When polyacrylamide-gel electrophoresis is complete the gel is marked by removing one corner and equilibrated in transfer buffer (up to 30 min for a 1.5-mm-thick gel).
3. The gel holder of the Trans-Blot cell is placed cathode (grey) side down into a shallow tray containing sufficient buffer to maintain all further steps just under liquid.
4. A wetted Scotch-Brite pad is placed onto the gel holder and a wetted sheet of Whatman 3MM paper onto the pad.
5. The equilibrated gel is carefully placed onto the filter paper so as to avoid trapping air bubbles (*see* Note 1).
6. The transfer membrane is held by two sides and carefully laid onto the gel from the centre outward. It is then gently rolled with a sterile 10 mL graduated pipet to remove any air pockets, ensuring good contact (*see* Note 2).
7. The sandwich is completed by placing further filter paper(s) over the membrane and a second pad over the filter paper. The gel holder is held firmly and closed before being placed into the transfer tank with sufficient buffer to cover the blot (*see* Notes 3 and 4).

3.3. Membrane Blocking

In order to minimize nonspecific binding during the immunoreactive detection of transferred protein, all reactive sites on the filter must be blocked. This is commonly done with protein, often bovine serum albumin (BSA) (up to 5% w/v), fetal calf serum (10% w/v), or gelatin (3% w/v). For work with NC we have found that the cheapest and most efficient blocking agent is a 5% (w/v) suspension of nonfat dried milk powder (Marvel type) in Tris-buffered saline (TBS). Nonionic detergents such as Tween-20 can also be used, but it has been reported that Tween-20 can give false positive reactions when used alone *(11)*. A comparison of four commonly used blocking agents showed that defatted milk powder was the most powerful, but also demonstrated the importance of using the correct blocker for any given monoclonal antibody in order to give maximum immunoreactivity *(12)*.

Following transfer, the membrane is washed briefly in TBS before carefully immersing in sufficient blocking solution to cover the membrane. Shake gently for 1 h at room temperature (*see* Note 5).

3.4. Periodate Treatment of Western Blots

A further problem of high background owing to nonspecific binding of antibodies to glycoproteins can be overcome by treatment with periodate *(13)*.

1. Wet blot in TTBS before adding blocking agent (*see* Section 2.2.).
2. Wash the membrane in sodium acetate for 3 × 15 min before transferring to periodic acid solution for 1 h.
3. Wash the membrane twice with sodium acetate briefly and then twice with GTBS (for 10 and 30 min).
4. Wash twice with TBS briefly and leave for 15 min in blocking agent before continuing as usual (*see* Note 6).

3.5. Detection Methods

A wide range of methods can be used, including general staining methods for total proteins or highly specific detection systems for individual proteins within a complex mixture.

3.5.1. Total Protein Stains

Total proteins in NC and PVDF membranes can be detected using a variety of anionic dyes which are also used to stain proteins in polyacrylamide gels. However, these dyes are not suitable for use with nylon membranes as they bind irreversibly.

Staining is useful to determine the efficiency of transfer and can be done in combination with immunodetection (*see* Section 3.5.3.). The most commonly used stains are Coomassie blue R-250 *(4)*, which is very suitable for PVDF membrane but gives high background with NC, Amido black *(3)*, Fastgreen *(7)*, India ink *(14)*, and Aurodye *(15)*, the latter being sensitive in the low nanogram range.

It is also possible to identify the position of molecular weight marker proteins using prestained markers such as those supplied by Amersham or Bio-Rad. These are clearly visible on the gel while running and also on the membrane following transfer. However, the mobilities of the prestained proteins may not correspond precisely to those of unstained markers. An alternative method is to use biotinylated marker proteins that are available from BioRad and have the added advantage of allowing processing as for a normal immunoblot. Avidin conjugated with horseradish peroxidase (avidin-HRP) or alkaline phosphatase (avidin-AP) can be added to the conjugated second antibody prior to incubation. The second antibody will detect the immobilized protein (antigen), and the labeled avidin will detect the biotinylated standards.

When staining for total protein it is often convenient to use a reversible staining dye such as Amido black *(16)* or the red dye Ponceau S *(17)*. Although not as sensitive as Amido black, Ponceau S is very rapid and simple to use. However, before using a reversible stain prior to immunodection, any possible interference with the immunoreactivity of proteins should be carefully checked. The following is a protocol for total protein staining with Ponceau S.

1. Wash the membrane briefly with water before adding the dilute stain. Leave in contact for several minutes, with shaking, before pouring off. Retain for reuse.
2. Remove background stain by washing the membrane in distilled water (*see* Note 7). The stain can then be removed by washing twice with TTBS before blocking the membrane.

3.5.2. Labeling Systems

Detection methods are usually indirect, requiring a specific ligand or antibody to bind to the immobilized protein, followed by a labeled second antibody or ligand.

Probably the most commonly used method is indirect immunological detection using a specific labeled second antibody. These second antibodies are available commercially with a wide choice of labels. They are species-specific, and will recognize whole primary antibodies of different classes, antibody subtypes, or immunoglobulin fragments. The labels may be radioactive, fluorescent, or enzymic, the choice depending on convenience and the degree of sensitivity required. Much of the early work used radiolabels, especially [^{125}I], with detection by autoradiography. However, enzymic labels are much easier to use in terms of handling and safety, producing insoluble reaction products at the binding site. Horseradish peroxidase (HRP) was the first enzyme used for detection, but is not as sensitive as other systems and the insoluble chromogen fades on exposure to light. A more sensitive system was alkaline phosphatase (AP), which produces a stable end color when used with one of several substrates, the most sensitive being 5-bromo-4-chloro-3-indolyl phosphate (BCIP) and nitro blue tetrazolium (NBT) *(18)*.

Alternative methods are available, including enhanced chemiluminescence (ECL) (developed by Amersham) in which the primary antibody is detected by HRP-labeled second antibodies. A Luminol substrate is oxidized by HRP to emit light that is enhanced 1000-fold. Gold-labeled second antibodies, Protein A or Protein G can also be used, with high sensitivity that can be increased by silver enhancement (BioRad). Both protein A and Protein G bind specifically to the Fc region of the antibody, but do not bind equally well to immunoglobulins of different subclasses or from different species. They are also not as sensitive as species-specific antibodies and are less sensitive because only one ligand molecule binds to each antibody. The following is a protocol based on alkaline phosphatase detection.

1. The blotted membrane is washed briefly in TBS.
2. The membrane is blocked (*see* Section 3.3.) by immersion in blocking solution with gentle shaking for 1 h at room temperature.
3. Discard the blocking solution and wash the membrane for 2 × 5 min with TTBS.
4. Transfer the membrane to the antibody diluted in antibody buffer. Shake for 1–2 h or overnight if more convenient. The dilution factor for the antibody must be worked out empirically, but in our experience can vary from only tenfold to several thousandfold.
5. The primary antibody is removed and the membrane washed for 2 × 5 min with TTBS.
6. Add the alkaline phosphatase-labeled second antibody (anti-species), diluted in 1% (w/v) BSA in TTBS for 1–2 h with shaking. The dilution factor is given by the supplier, but may require adjustment for optimum results.
7. The second antibody is discarded and the membrane washed for 2 × 5 min with TTBS followed by 1 × 5 min with TBS to remove residual Tween 20.
8. The color reagents are prepared just prior to use by diluting 0.6 mL of each of solutions A and B into 100 mL of solution C.
9. Incubate at 37°C until full color development occurs with minimal background color.
10. Stop the reaction by washing the membrane in distilled water followed by air drying.

3.5.3. Double Staining Method

The stains used to detect total proteins on membranes are often prepared in methanolic solution that will shrink NC membrane. This and other factors makes direct comparisons of the positions of proteins immunologically detected to those on total protein-stained membranes difficult, especially when comparing two-dimensional gels that may contain hundreds of individual spots. A simple way to overcome this problem is to use a double staining method *(19)* (*see also* Fig. 2).

1. The transferred membrane is immunoreacted as described (*see* Section 3.5.2.), using AP-labeled second antibody and NBT/BCIP color reagents to produce an insoluble colored end product (*see* Note 8).
2. While the membrane is wet, or following rewetting, add sufficient volume of 0.1% (v/v) India ink (or equivalent) in TTBS to cover the membrane.
3. Rinse the membrane in water to remove background stain and air-dry.

3.6. Dot and Tissue Blotting

3.6.1. Dot Immunoblotting

Dot blotting is a very rapid and simple method to detect antibody binding to a protein (antigen) of interest, and is 10- to 1000-fold more sensitive than electroblotting *(20)*. The protein can be applied directly to dry NC by adding small volume (1–5 μL) dots directly to the membrane. If the antigen is too dilute, multiple loadings can be made allowing the membrane to dry in between. The positions of applied samples can be marked with a soft pencil, or NC printed with a grid can be used. Alternatively, it is possible to use a commercially available apparatus such as those produced by Bio-Rad and Schleicher and Schuell. The Bio-Dot (from Bio-Rad) is available with 96-well dots (microtiter plate type) or with 48-well slots, which are attached to a common manifold base for connection to a vacuum system for rapid application and washing. Following sample application the membrane is blocked and the proteins detected as described for immunoblotting membranes (*see* Section 3.5.2.).

Using radiolabeled Protein A for detection, Jahn et al. *(21)* developed a quantitative dot-immunobinding assay that was able to process a large number of samples and was also very sensitive (being able to measure 10 pmol of synapsin I protein and 50 ng of total vesicle membrane protein).

3.6.2. Squash Blots

Specific proteins can be detected in whole plant tissues by using so-called squash blots. In this, whole tissue Sections are pressed onto NC and the

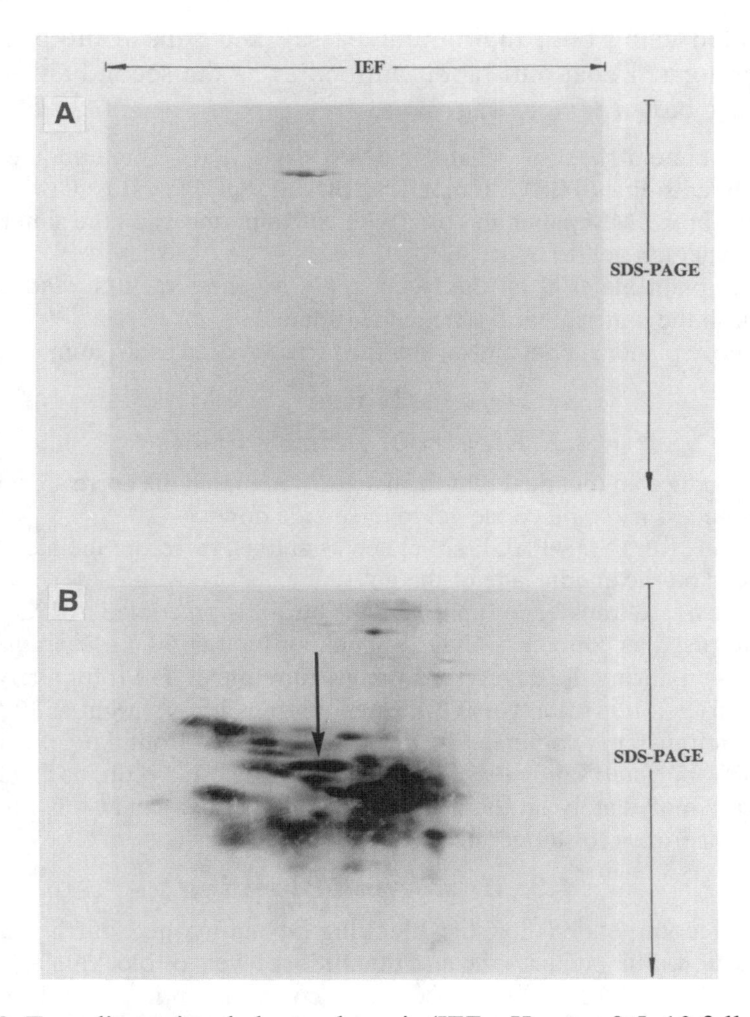

Fig. 2. Two-dimensional electrophoresis (IEF, pH range 3.5–10 followed by SDS-PAGE) of total gliadin extract from wheat cv. Chinese Spring and immunoblotted onto nitrocellulose. **(A)** Immunoblot developed with MAb IFRN 0067. **(B)** Immunoblot stained for total protein using India ink, followed by immunodetection with IFRN 0067. Alkaline phosphatase-labeled second antibody used for detection. The two immunoreactive proteins are arrowed.

bound proteins detected using labeled antibody *(22)*. The activities of numerous enzymes can also be assayed *(23)*. For example, the presence and pattern of expression of pea lectin in transgenic potato plants were determined using both radioimmunoassay and squash blotting, using [^{125}I]conjugated goat anti-rabbit antibodies as the second antibody for the squash blots *(24)* (*see* Fig. 3).

1. Equilibrate a sheet of Whatman 3MM paper and NC membrane in Tris-buffered saline (TBS) (20 m*M* Tris, pH 7.2, 0.9% (w/v) NaCl).
2. Place both 3MM paper and NC (with NC uppermost) on the sintered plate of a vacuum gel drier.
3. Place plant material on the filter, apply a gentle vacuum, and carefully squash the material with a flat ended spatula.
4. Remove plant material and air-dry the filter. Proceed as for immunoblotting.

4. Notes

4.1. Electroblotting of Gels

1. To produce an identical blot from a gel, i.e., one in the correct orientation, it is necessary to place the gel reverse-side down.
2. Proteins from SDS-PAGE are eluted as anions, therefore the membrane is placed on the anodic side of the gel.
3. For standard transfers of up to 5 h, the buffer is precooled to 4°C and cold water used as coolant. Vôltage is held constant at 60 V. When gels have been run during the day, with a typical run time of 3–4 h for a large (16 × 12 cm) gel, it is often very convenient to transfer overnight at 30 V.
4. When transfer is complete the membrane is lifted from the gel and rinsed in Tris-buffered saline (*see* Section 3.3.). The bound proteins can be visualized immediately or, the membrane stored, either wet at 4°C for several days or frozen for longer periods.

4.2. Membrane Blocking

5. It is recommended that the blocking be maintained during immunodetection. This requires the addition of low levels of blocking agent (routinely 1% (w/v) BSA) to the solutions containing primary and secondary antibodies. Over 60% of blocking protein is reported to be lost from the membrane during washing *(19)*.
6. This method works well with NC but gives a grainy appearance with ProBlot when dry.

4.3. Detection Methods

7. The stained membrane can be photographed using a green filter. The proteins of interest and standards as well as individual tracks can be marked using a soft grade pencil.
8. It is possible to stain with India ink first, followed by immunostaining. However, it is preferable to immunostain first in order to eliminate the

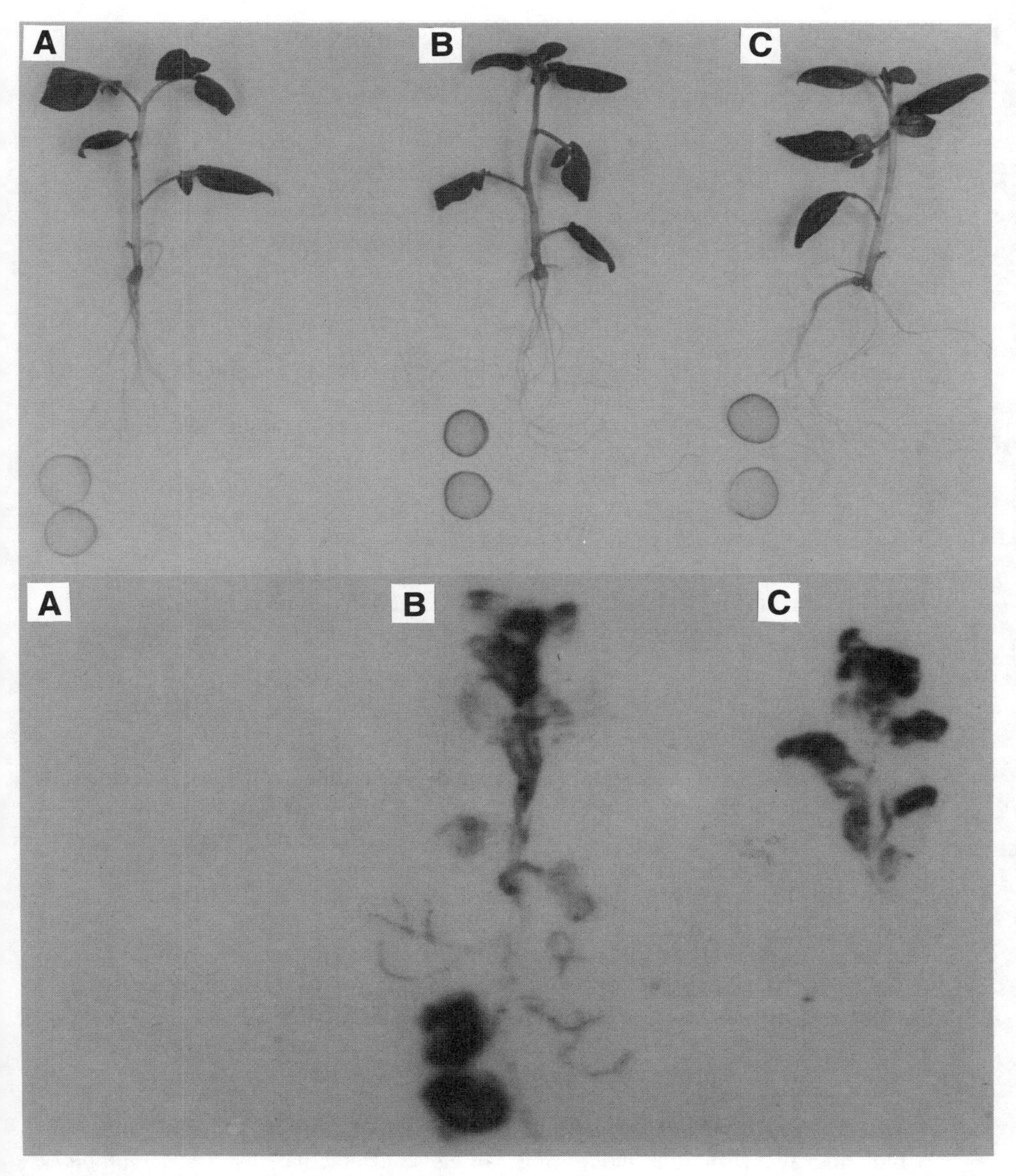

Fig. 3. The localization of pea lectin in various tissues of transformed plants using a squash-blot immunolocalization technique. **(A)** Control plant and tuber Section transformed using pGV3850::pDUB126a; **(B)** CaMV-Lec transformant; **(C)** ssRubisco-Lec transformant. Upper panel, plants and tuber sections before squash; lower panel, autoradiograph demonstrating localization of pea lectin (from Edwards et al., 1991 *[24]*).

possibility that the dye will affect the immunoreactivity of the proteins. An example of the use of double staining to detect two immunoreactive proteins in a complex mixture (total wheat seed storage proteins) is shown in Fig. 2.

Acknowledgments

The authors are grateful to Glyn Edwards (Shell, Sittingbourne) for providing Fig. 3, and to T. J. Higgins (CSIRO, Canberra) for providing Fig. 1.

References

1. Higgins, T. J. V. and Spencer, D. (1991) The expression of a chimeric cauliflower mosaic virus (CaMV-35S)-pea vicilin gene in tobacco. *Plant Sci.* **74,** 89–98.
2. Southern, E. M. (1975) Detection of specific sequences among DNA fragments separated by gel electrophoresis. *J. Mol. Biol.* **98,** 503–517.
3. Towbin, H., Staehelin, T., and Gordon, J. (1979) Electrophoretic transfer of proteins from polyacrylamide gels to nitrocellulose: Procedure and some applications. *Proc. Natl. Acad. Sci. USA* **76,** 4350–4354.
4. Burnette, W. N. (1981) "Western Blotting"; Electrophoretic transfer of proteins from sodium dodecylsulphate-polyacrylamide gels to unmodified nitrocellolose and radiographic detection with antibody and radioiodinated Protein A. *Anal. Biochem.* **112,** 195–203.
5. Baker, C. S., Dunn, M. J., and Yacoub, M. H. (1991) Evaluation of membranes used for electroblotting of proteins for direct automated microsequencing. *Electrophoresis* **12,** 342–348.
6. Bowen, B., Steinberg, J., Laemmli, U. K., and Weintraub, H. (1980) The detection of DNA-binding proteins by protein blotting. *Nucleic Acids Res.* **8,** 1–20.
7. Reinhart, M. P. and Malamud, D. (1982) Protein transfer from isoelectric focusing gels: The native blot. *Anal. Biochem.* **123,** 229–235.
8. McDonald, M. B. (1991) Blotting of seed proteins from isoelectrically focused gels for cultivar identification. *Seed Sci. Technol.* **19,** 33–40.
9. Peferoen, M., Huybrechts, R., and De Loof, A. (1982) Vacuum-blotting: a new simple and efficient transfer of proteins from sodium dodecylsulphate-polyacrylamide gels to nitrocellulose. *FEBS Letts.* **145,** 369–372.
10. Matthaei, S., Baly, D. L., and Horuk, R. (1986) Rapid and effective transfer of integral membrane proteins from isoelectric focusing gels to nitrocellulose membranes. *Anal. Biochem.* **157,** 123–128.
11. Bird, C. R., Gearing, A. J. H., and Thorpe, R. (1988) The use of Tween 20 alone as a blocking agent for immunoblotting can cause artefactual results. *J. Immunol. Methods* **106,** 175–179.
12. Hauri, H.-P. and Bucher, K. (1986) Immunoblotting with monoclonal antibodies: Importance of the blocking solution. *Anal. Biochem.* **159,** 386–389.
13. Hahn, M. G., Lerner, D. R., Fitter, M. S., Norman, P. M., and Lamb, C. J. (1987) Characterisation of monoclonal antibodies to protoplast membranes of *Nicotiana tabacum* identified by an enzyme-linked immunosorbent assay. *Planta* **171,** 453–465.

14. Hancock, K. and Tsang, V. C. W. (1983) India ink staining of proteins on nitrocellulose paper. *Anal. Biochem.* **133,** 157–162.
15. Moeremans, M., Daneels, G., and De Mey, J. (1985) Sensitive colloidal metal (gold and silver) staining of protein blots on nitrocellulose membranes. *Anal. Biochem.* **145,** 315–321.
16. Harper, D. R., Liu, K.-M., and Kangro, H. O. (1986) The effect of staining on the immunoreactivity of nitrocellulose-bound proteins. *Anal. Biochem.* **157,** 270–274.
17. Salinovich, O. and Montelaro, R. C. (1986) Reversible staining and peptide mapping of proteins transferred to nitrocellulose after separation by sodium dodecylsulphate-polyacrylamide gel electrophoresis. *Anal. Biochem.* **156,** 341–347.
18. Blake, M. S., Johnston, K. H., Russel-Jones, G. J., and Gotschlich, E. C. (1984) A rapid, sensitive method for detection of alkaline phosphatase-conjugated anti-antibody on Western blots. *Anal. Biochem.* **136,** 175–179.
19. Ono, T. and Tuan, R. S. (1990) Double staining of immunoblot using enzyme histochemistry and india ink. *Anal. Biochem.* **187,** 324–327.
20. Bernstein, J. M., Stokes, C. E., and Fernie, B. (1987) Comparative sensitivity of ^{125}I-Protein A and enzyme-conjugated antibodies for detection of immunoblotted proteins. *J. Clin. Microbiol.* **25,** 72–75.
21. Jahn, R., Schiebler, W., and Greengard, P. (1984) A quantitative dot-immunobinding assay for proteins using nitrocellulose membrane filters. *Proc. Natl. Acad. Sci. USA* **81,** 1684–1687.
22. Jung, J.-L. and Hahne, G. (1992) A simple method to increase resolution in whole leaf blotting. *Plant Sci.* **82,** 125–132.
23. Spruce, J., Mayer, A. M., and Osborne, D. J. (1987) A simple histochemical method for locating enzymes in plant tissue using nitrocellulose blotting. *Phytochemistry* **26,** 2901–2903.
24. Edwards, G. A. Hepher, A., Clerk, S. P., and Boulter, D. (1991) Pea lectin is correctly processed, stable and active in leaves of transgenic potato plants. *Plant Mol. Biol.* **17,** 89–100.

CHAPTER 36

ELISA Detection of Foreign Proteins

E. N. Clare Mills, Geoffrey W. Plumb, and Michael R. A. Morgan

1. Introduction

Immunoassays were first developed over 30 yr ago. The radioimmunoassay for insulin described by Yalow and Berson *(1)* heralded a new era in the use of antibody reagents for the quantification of proteins and peptides. The Nobel Prize-winning research revolutionized analysis by virtue of much improved specificity and sensitivity coupled with ease of application to large numbers of samples. Subsequently, a number of further innovations have increased the power and utility of immunoassay. Of particular importance was the methodology for generating monoclonal antibodies described by Kohler and Milstein *(2)*, theoretically allowing the production of unlimited amounts of identical antibodies (as opposed to the mixed populations making up polyclonal preparations). Second, the use of enzyme-linked immunosorbent assays (ELISAs) has obviated the need for handling radioisotopes *(3)* and consequently widened the utilization of immunoassays from clinical research to virtually all areas of biological analysis *(4)*. More recently, the production of recombinant antibody fragments *(5)*, using a variety of expression systems, offers the next leap forward. Such technical innovations will offer a faster and more reliable means to synthesize reagents of the desired affinity and specificity, taking immunotechnology on into the next century *(6)*.

As immunoassay offers one of the few means of quantifying a particular protein in the presence of a mixture of other proteins, it is not surpris-

From: *Methods in Molecular Biology, Vol. 49: Plant Gene Transfer and Expression Protocols*
Edited by: H. Jones Humana Press Inc., Totowa, NJ

ing that it has been used extensively to analyze crude recombinant protein preparations. In addition, antibodies can give information on both the folded and glycosylated state of a protein *(7)*. The region of the protein sequence to which an antibody binds is known as an epitope, and its availability for antibody binding can be affected by the conformational state of the protein, particularly so for an epitope comprised of disparate sections of sequence brought together by the way in which the polypeptide chain is folded. Similarly, if the epitope spans a glycosylation site, the presence or absence of carbohydrate will also be indicated by antibody recognition.

This chapter will focus on the use of microtitration plate-based ELISAs for the analysis of expressed proteins, as this format is rapid, easy to use for large sample numbers, and most laboratories possess the basic equipment required for processing the microtitration plates. Those readers requiring information on methods for generating antibodies, their purification and conjugation to enzymes should consult texts such as Godding *(8)*. When using ELISAs, some consideration must be given to assay design. Figures 1 and 2 show the principles behind the two most common types of assay used, which are often referred to by a baffling array of acronyms. Both types may be suitable for determining the level of an expressed protein or probing its folded state.

The first is known as an inhibition ELISA (Fig. 1), where a limited number of antibody binding sites are available to bind either to protein immobilized to the microtitration plate surface, or to protein in solution. Thus standards or sample proteins are added to microtitration plate wells coated with native protein, and mixed with a suitable dilution of primary antibody. After an incubation step the plate is washed to remove all unbound material before adding a second antibody. This is specific for the IgG of the species in which the primary antibody was raised (usually rabbit or mouse), and has been labeled with an enzyme such as alkaline phosphatase or horseradish peroxidase. It is possible to omit this step if enzyme-labeled primary antibody is available, although this means exposing the sensitive enzyme-label directly to components present in the sample, which may modify its activity. Thus when analyzing crude samples it is generally preferable to use a second enzyme-labeled antibody. Following another incubation and washing step, substrate is added, the color allowed to develop, and the optical density of each well determined. A typical standard curve for this type of format obtained for an

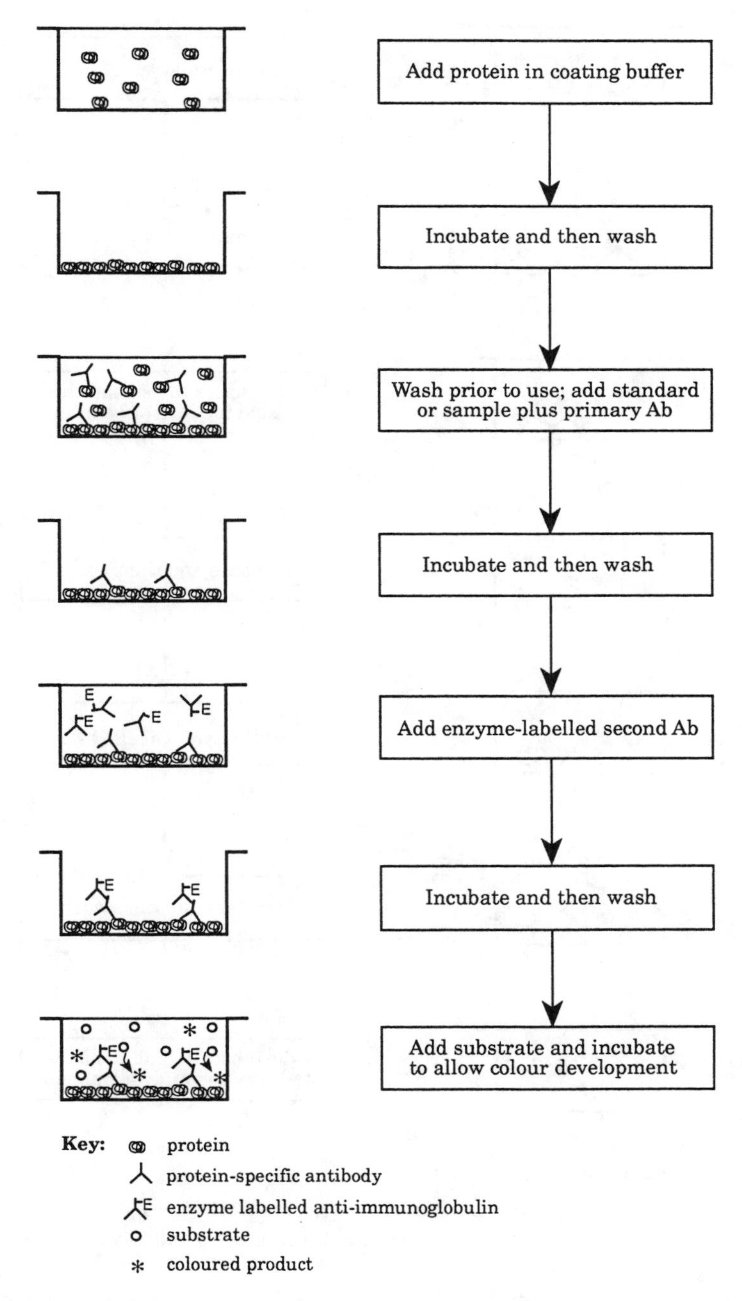

Fig. 1. Flowchart describing the steps involved in performing an inhibition ELISA.

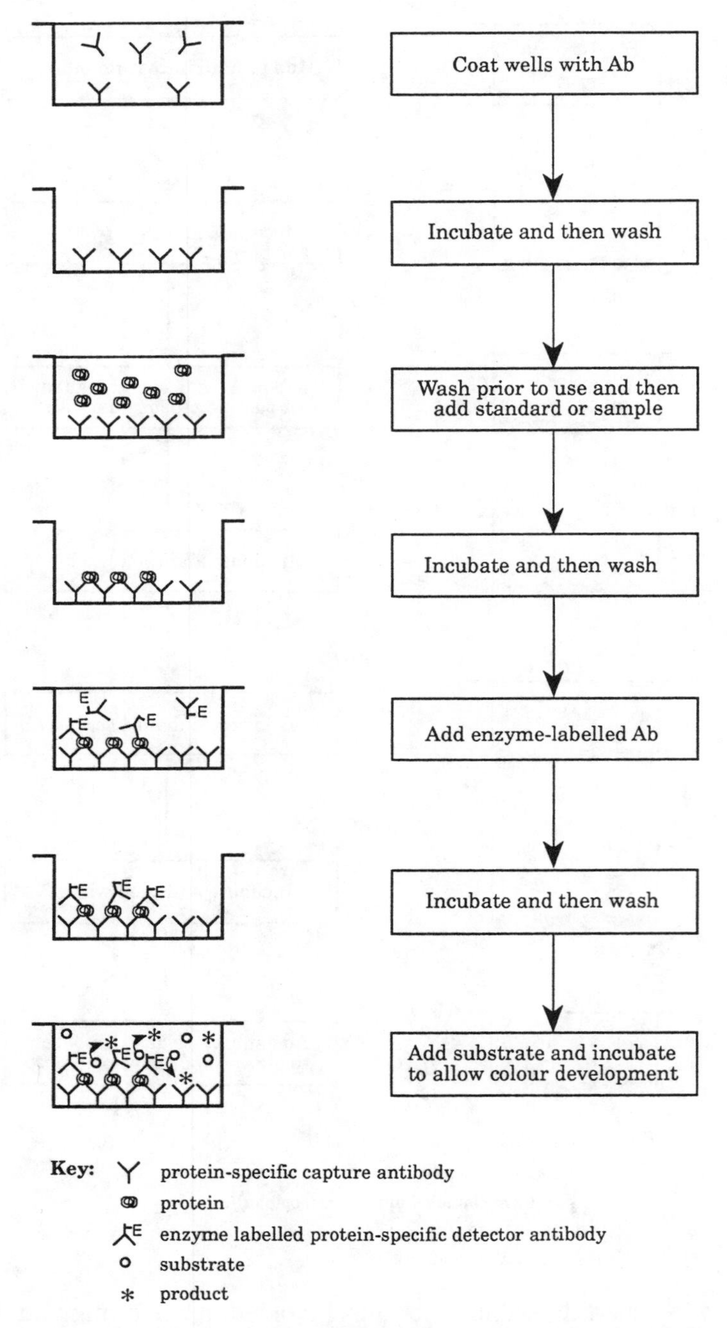

Fig. 2. Flowchart describing the steps involved in performing a two-site ELISA.

ELISA for legumin, the pea 11S globulin, is shown in Fig. 3A. The absorbance values obtained with an inhibition ELISA such as this are inversely proportional to the concentration of protein in solution.

A second format (Fig. 2) employs an excess of one antibody immobilized to the plate which binds, or "captures," all the target protein present in a sample. After a wash step the adsorbed protein is labeled by incubating with a second antibody (known as a "detector") conjugated to an enzyme. If the capture and detector antibodies have been developed in different species of animals, it may be possible to use enzyme-conjugated anti-IgG to label the detector, thereby introducing another incubation step. Further washing and substrate incubation steps are as described for the inhibition ELISA. From the example standard curve, also obtained for pea legumin (Fig. 3B), it can be seen that with this format the color development is directly proportional to the concentration of protein. This type of ELISA has been termed a two-site or sandwich ELISA as a consequence of employing two specific antibodies to capture and to detect the sample protein.

Following are given the protocols used to extract and analyze for vicilin, the 7S globulin from pea expressed in yeast, using an inhibition ELISA. The Notes Section contains the additional information required to set up a two-site ELISA.

2. Materials

Store all buffers at room temperature unless otherwise indicated.

2.1. Production of a Yeast Soluble Extract

1. TMS buffer: 50 m*M* Tris-HCl, pH 8.2, containing 10 m*M* $MgCl_2$ and 1.2*M* sorbitol. Make up fresh.
2. TPL buffer: 50 m*M* Tricine, pH 8.2, containing 1 m*M* PMSF and 0.1 mg/mL leupeptin. Make up fresh and store at 1°C.
3. 10% (w/v) SDS solution.

2.2. ELISAs

1. PBS: 0.0015*M* KH_2PO_4, 0.14*M* NaCl, 0.008*M* Na_2HPO_4, pH 7.4, containing 0.001% (v/v) Kathon (Rhom and Haas, Ltd., Philadelphia, PA) as preservative.
2. PBST: PBS containing 0.05% (v/v) Tween-20.
3. Coating solution: 1 µg/mL pea vicilin in 0.05*M* sodium carbonate/bicarbonate buffer pH 9.6. Make up fresh.

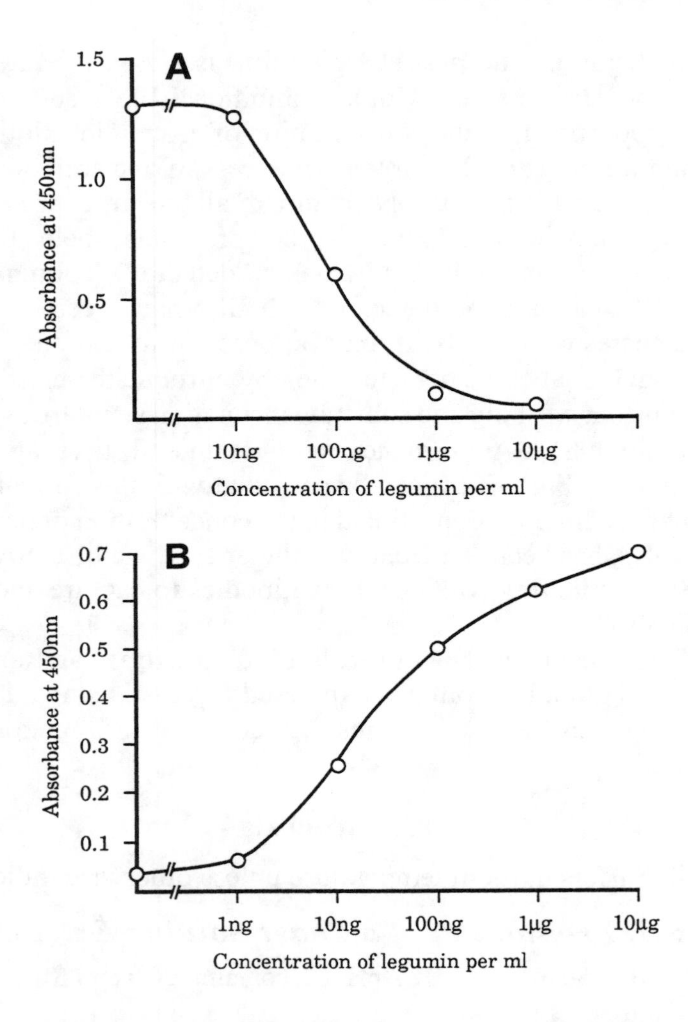

Fig. 3. ELISA calibration curves for pea legumin: **(A)** inhibition ELISA; **(B)** two site ELISA. ELISAs were as described in the text but using horseradish peroxidase as the enzyme label. A (inhibition ELISA): This employed a legumin specific Mab, IFRN 0039, and gave a sensitivity of 12 ng/mL and an ED_{50} = 100 ng/mL. B (two-site ELISA): this employed a rabbit anti-legumin antiserum, $R131b_4$ as the capture antibody, and IFRN 0039 as the detector and gave a calibration curve with a limit of detection of 1 ng/mL and an ED_{50} = 23 ng/mL.

4. Stock vicilin solution (2 mg/mL) for standards. Make up fresh. Substrate solution: *p*-nitrophenol phosphate (1 mg/mL) added as tablets (Sigma, Poole, UK) 0.05*M* sodium carbonate/bicarbonate buffer containing 5 m*M* $MgCl_2$.

5. Primary and alkaline phosphatase labeled secondary antibodies (affinity purified) should be stored aliquoted in PBS at –20 and 4°C, respectively. Make up fresh.
6. 2*M* H_2SO_4.

3. Methods

3.1. Production of a Soluble Extract from Yeast

1. Harvest 1 L of yeast culture by centrifugation.
2. Resuspend the pellet in 25 mL of TMS buffer, add 1 mg/mL zymolyase-20T, then incubate at 25°C for 1 h.
3. Check the amount of cell breakage before the addition of zymolyase-20T and after the 1 h incubation period (*see* Section 3.2.; *see* Note 1).
4. Centrifuge gently at 2000*g* for 5 min, then resuspend the pellet in 40 mL TPL buffer. All subsequent steps to be performed at 1°C.
5. Homogenize the pellet in a 50 mL glass/Teflon™ homogenizer, then centrifuge for 30 min at 12,000*g*. The supernatant comprises the soluble extract.

3.2. Measurement of Cell Breakage

1. Remove 50 µL of extract before the addition of the enzyme zymolyase-20T and dilute 30:1 with TMS buffer.
2. Remove another 50 µL of extract and dilute 30:1 with 10% (w/v) SDS.
3. Measure the absorbance of each fraction from steps 1 and 2 above at 600 nm.
4. Repeat steps 1–3 after the 1 h incubation with the enzyme. Cell breakage is said to be complete when the absorbance in SDS is approximately 20% of the absorbance in TMS buffer.

3.3. ELISA: Preparation of Coated Plates

1. Add 0.3 mL of a 1 µg/mL solution of the relevant protein to the inner 60 wells of 96-well polystyrene microtiter plates (Nunc Immunoplate I, Gibco, Paisley, UK).
2. Incubate overnight at 4°C.
3. Wash three times with PBST using a plate washer (Titertek Microplate Washer 120 supplied by ICN-Flow, Thame, UK) and blot on tissue and dry in air.
4. Store dry at –20°C until required for use.

3.4. ELISA: Inhibition ELISA

1. Wash a coated ELISA plate three times with PBST using a plate washer.
2. Prepare a range of vicilin standards for the assay. Take 50 µL of the stock vicilin solution and add to 950 µL of PBS and mix well. This gives the top standard of 100 µg/mL. Take 100 µL of the 100 µg/mL standard and add to 900 µL of PBS and mix well to give the 10 µg/mL standard. Repeat this

dilution with the 10 μg/mL standard to give the 1 μg/mL standard and so on to give a set of standards ranging from 100 μg–1 ng/mL.

3. Add 0.1 mL of both the protein fractions under test and the standard solutions in triplicate (or as required) to designated wells on the assay plate.
4. Add 0.1 mL of rabbit anti-vicilin diluted 1:10,000 (v:v) in PBST to these wells and incubate at 37°C for 2 h (*see* Notes 2 and 3).
5. Wash five times with PBST using a plate washer, then add 0.2 mL of second antibody (anti-rabbit IgG labeled with alkaline phosphatase) diluted 1:1000 (v:v) in PBST to the relevant microtiter wells (*see* Note 4). Incubate at 37°C for 2 h.
6. Wash five times as described in step 5, then add 0.2 mL of substrate solution to the relevant wells. Incubate at 37°C for 15 min.
7. Stop the reaction by the addition of 50 μL of 2*M* H_2SO_4 to each well and measure the absorbance of each well at 405 nm using a plate reader (Titertek Multiscan MCC supplied by ICN-Flow) (*see* Note 5).

4. Notes

1. Yeast cell breakage is required to release a protein expressed in the cytosol thus allowing further purification and analysis. This is usually achieved by sonication or enzymic hydrolysis of the cell wall. However, yeast cells harvested after the late log phase of growth (i.e., in stationary phase) tend to possess thicker cell walls, and hence a longer incubation with zymolyase is required to achieve adequate levels of cell-lysis. Consequently, the risk of proteolysis to the expressed protein by yeast proteases is increased. Care must therefore be taken in harvesting yeast cells at the correct time and, using effective cell breakage techniques, to maximize the amount of recombinant protein released for analysis.
2. Antibody preparations used in the analysis of recombinant proteins must be chosen with care. For quantification it is preferable to use an antibody preparation that recognizes the native and expressed protein equally well, or to use purified expressed protein as a standard. Figure 4 illustrates this problem for the analysis of recombinant vicilin, using the ELISA described earlier. It can be seen clearly that the recombinant protein was recognized very poorly compared with the native protein to which the polyclonal antiserum had been raised. Extensive characterization of the expressed protein showed that it had not been correctly folded *(9)*, a fact clearly detected by the ELISA. Both polyclonal and monoclonal antibody preparations can be used to develop immunoassays, the former often giving more sensitive assays, a characteristic that can be useful when trying to overcome sample interference (*see* Note 5). When using antibodies to probe the conforma-

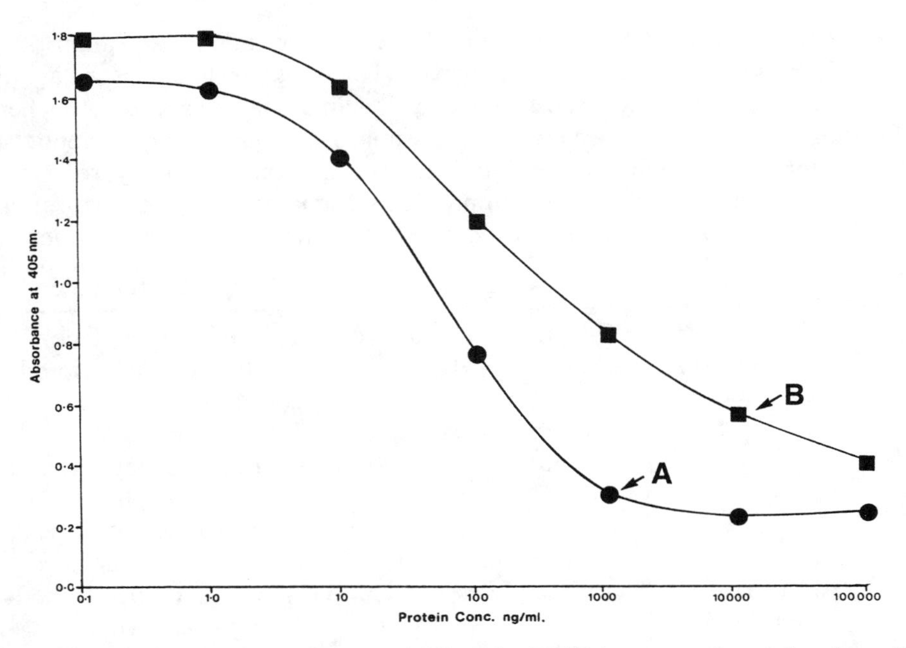

Fig. 4. Inhibition ELISA for pea vicilin. The ELISA was performed as described in the Methods Section; **A:** calibration curve for native vicilin; **B:** calibration curve for purified expressed vicilin.

tion or glycosylation state of the expressed protein, well characterized monoclonal antibodies known to recognize either conformationally sensitive epitopes or glycosylation sites are an advantage, although polyclonal antisera such as the one used in the vicilin study can yield some limited information.

3. The assay protocol described above may need refining for a given protein and antiserum. Thus the optimal dilution of the primary antibody must be determined that will give a maximal absorbance reading of around 2.0 after a substrate incubation time of 10–20 min. It is also possible to employ overnight incubations with specific antibody at 4°C, instead of 2–3 h at 37°C.
4. There are a number of enzyme-labels and substrates available that are suitable as ELISA endpoint detection methods. Horseradish peroxidase (HRP) or alkaline phosphatase are the most common labels employed, for which substrate buffer tablets can be purchased from commercial sources, such as Sigma. Many of the substrates used for HRP are mutagenic and in our experience one of the safest, which also performs very well, is 3,3',5,5'-tetramethylbenzidine (TMB). This can be purchased in, for example, a liq-

uid, ready-to-use form from Cambridge Veterinary Sciences (Cambridge, UK) or as substrate tablets from Sigma. When using alkaline phosphatase as the enzyme-label longer substrate incubation times may be needed and when using *p*-nitrophenyl phosphate as the substrate background color development is often higher, owing to the significant rates of nonenzymic hydrolysis.

5. There are a number of common problems encountered when performing ELISAs, a few of which are listed below, together with some remedies.

Problem	Possible cause	Suggested remedies
Lack of color development	Enzyme label inactive	Replace lot of enzyme label; ensure it is stored correctly
	Incorrect substrate or buffer	Check substrate; check pH of buffers and, if alkaline phosphatase is used, that mg^{2+} is present
Background adsorbance is high (≥0.3)	Insufficient washing of the plates	Make sure wells are emptied and filled properly during wash cycles
	Problems with incomplete coating of plates	Try blocking with BSA either by recoating at 1 µg/mL or adding 0.1% (w/v) BSA to buffers used for diluting samples and standards
	Color development too fast (less than 5 min)	Reoptimize assay using specific antibody at a higher dilution
Poor replication	Inconsistent coating of plates	Ensure that the protein used for coating is not aggregated or insoluble
	Bad pipeting	Check the accuracy and reproducibility of pipets; follow guidelines for good pipeting technique

6. One difficulty encountered particularly when analyzing crude extracts is assay interference. This arises from components present in the sample affecting the antibody recognition reaction. It can be readily detected by preparing standard in an extract that does not contain expressed protein. The absence of interference is demonstrated by parallelism between standards prepared in buffer and those prepared in blank sample. The relation-

Table 1
Matrix Interference in the Analysis of Two Preparations of Recombinant Vicilin[a]

Dilution of yeast extract assayed	μg Vicilin/mL, ELISA determination		Percentage of purified recombinant vicilin detected	
	Sample, 1	Sample, 2	Sample, 1	Sample, 2
Neat	0.35	0.10	0.01	0.004
1:10 (v:v)	3.50	1.20	1.30	0.50
1:100 (v:v)	15.00	14.00	53	62

[a]The levels of recombinant vicilin, calculated from purification data, were 2.8 (sample 1) and 2.3 (sample 2) mg/mL of yeast extract.

ship between the volume or dilution of sample analyzed and the assay response should also be linear. Table 1 shows the lack of linear response obtained for expressed vicilin, showing considerable matrix interference by the yeast extracts. Such deviation may be owing to a number of causes such as:

a. Spurious cross-reactions with proteins present in extracts which share homology with the recombinant protein;
b. High concentrations of protein may interfere as a result of nonspecific protein–protein interactions with the microtitration plate surface or the antibody preparation;
c. The presence of high concentration of salt and/or low pH (less than 4) will dissociate antibody–protein complexes; and
d. Chaotropic agents, detergents, and high levels of reductants or proteases that will denature or digest the antibody preparations.

Many of these factors may be removed by dialysis, gel filtration (e.g., using Pharmacia PD-10 columns), or if the ELISA is sufficiently sensitive, they may simply be diluted away. Should the problem of interference persist it can be avoided by analyzing partially purified material.

7. With a few modifications the assay protocol described earlier can be used to set up a two-site ELISA. Instead of coating with the protein of interest, microtitration plates are coated with protein-specific antibody. When using a polyclonal antiserum a dilution of 1:1000 is often appropriate but this should be optimized for each assay and antibody preparation. It may also improve the assay to dilute the antibody in PBS in case the alkaline pH of the coating buffer has a detrimental effect on binding capacity. A greater volume (0.2 mL) of sample or standard can be added to each well, thus maximizing assay sensitivity. Otherwise incubation and washing steps are

exactly as described for the inhibition ELISA (Section 3.4.), the dilution of enzyme-labeled protein specific antibody used also requiring optimization to give a maximal absorbance of around 2.0.

8. As with all analytical methodology, precision and accuracy with continued performance must be ensured. The shape of a calibration curve influences precision, the steepness of the curves generally giving better precision. Parameters that should be monitored include assay sensitivity, frequently defined as the absorbance reading obtained at zero analyte plus (for a sandwich ELISA) or minus (an inhibition ELISA) two standard deviations. Another measure is the ED_{50}, the dose of analyte that will give an absorbance reading 50% of the maximum, and is affected by changes in curve shape. For those researchers wishing to use more rigorous quality control measures, precision profile analysis undoubtedly offers a very powerful technique for assessing standard curve performance. Detailed information on this methodology can be found in Ekins and Edwards *(10)* and Price and Newman *(4)*. PC-based software is available from Beckman (Fullerton, CA) and Pharmacia that is capable of performing detailed analysis of ELISA results, including precision profiles.
9. Another ELISA format, in addition to those mentioned earlier, can be used to detect expressed proteins. Essentially all that is involved is immobilization of sample or standard protein to a microtitration plate surface (*see* Section 3.3.), followed by blocking (either with skim milk powder or BSA at 0.5% [w/v] in PBST), the subsequent washing and incubation steps with antibody reagents and substrate being as described in Section 3.4. This type of ELISA can also be performed using a membrane, such as nitrocellulose, as the solid-phase, in what is frequently termed as a "dot-blot." Samples are applied in 2–20 μL of buffer (without detergents present) and allowed to dry in air before being developed as described in Chapter 35 of this book. Although easy to perform, such blots can only be analyzed qualitatively in terms of color development, and depending on the substrates used, be at least tenfold less-sensitive than a microtitration-plate based ELISA. Whatever the surface used to immobilize the protein, the noncovalent forces involved in adsorption to the solid-phase can cause significant conformational changes in a protein *(11)*. It may be that in using this type of format expressed proteins incorrectly folded may not be recognized to the same degree as the native protein, although as mentioned earlier, this is also dependent on the characteristics of the protein-specific antibody preparation.

References

1. Yalow and Berson, S. (1959) Assay of plasma insulin in human subjects by immunological methods. *Nature* **184,** 1648,1649.

2. Kohler, G. and Milstein, C. (1975) Continuous cultures of fused cells secreting antibody of predefined specificity. *Nature* **256,** 495–497.
3. Engvall, E. and Perlman, P. (1971) Enzyme-linked immunosorbent assay (ELISA). Quantitative assay of immunoglobulin G. *Immunochemistry* **8,** 871–874.
4. Price, C. P. and Newman, D. J. (1991) *Principles and Practice of Immunoassay*, Macmillan, London.
5. Winter, G. and Milstein, C. (1991) Man-made antibodies. *Nature* **349,** 293–299.
6. Lee, H. A. and Morgan, M. R. A. (1993) Food Immunoassay: applications of polyclonal, monoclonal and recombinant antibodies. *Trends Food Sci. Technol.* **4,** 129–134.
7. Katchalski-Katzir, E. and Kenett, D. (1988) Use of monoclonal antibodies in the study of the conformation and conformational alterations in proteins. *Bull. Chem. Soc. Jpn.* **61,** 133–139.
8. Godding, J. W. (1989) *Monoclonal Antibodies: Principles and Practice. Production and Application of Monoclonal Antibodies in Cell Biology, Biochemistry and Immunology.* Academic, London.
9. Watson, M. D., Lambert, N., Delainey, A., Yarwood, J. N., Croy, R. R. D., Gatehouse, J. A., Wright, D. J., and Boulter, D. (1988) Isolation and expression of a pea vicilin cDNA in the yeast *Saccharomyces cerevisiae. Biochem. J.* **251,** 857–864.
10. Ekins, R. P. and Edwards, P. R. (1983) The precision profile: its use in assay design, assessment and quality control, in *Immunoassays for Clinical Chemistry* (Hunter, W. M. and Corrie, J. E. T., eds.), Churchill Livingstone, Edinburgh, UK, pp. 76–105.
11. Friguet, B., Djavadi-Ohaniance, L., and Goldberg, M. (1984) Some monoclonal antibodies raised with a native protein bind preferentially to the denatured antigen. *Mol. Immunol.* **21,** 673–677.

CHAPTER 37

Immunocytochemical Localization of Proteins

John Davies

1. Introduction

The localization of proteins within plant tissues is readily accomplished using the techniques of immunocytochemistry: the identification of a cell-bound antigen *in situ* by means of a specific antigen–antibody reaction, tagged microscopically by a visible label *(1–3)*.

This chapter assumes that the investigator has an antibody available that is specific to the antigen under study. Such an antibody should preferably have been characterized using Western blot analysis (detailed in Chapter 35)—to enable specificity and optimal antibody dilution to be assessed.

In essence, the technique is as follows: The tissue is first fixed to prevent target diffusion, embedded in a solid matrix to allow thin sectioning, then incubated with the antibody against the target antigen. After washing, the specifically bound antibody is localized using an antibody, raised against the species in which the primary antibody was raised, conjugated to a marker. Such markers include fluorochromes *(4)*, colloidal gold *(5)*, and enzymes whose catalytic reaction can precipitate an insoluble colored residue in the presence of appropriate substrates *(6)*.

Each detection method has its advantages—the most important of which are summarized in Table 1. For the purposes of this chapter we will assume that a light microscopical investigation is appropriate and that colloidal gold will be employed as the marker. Although colloidal gold is not the most sensitive marker it has the advantages of relative sim-

From: *Methods in Molecular Biology, Vol. 49: Plant Gene Transfer and Expression Protocols*
Edited by: H. Jones Humana Press Inc., Totowa, NJ

Table 1
A Comparison of Three Common Antibody Labeling/Detection Techniques

Features	Secondary (2°) antibody marker		
	Fluorescent	Enzymatic	Colloidal gold
Resolution	+++	+	+++
Sensitivity	+	+++	++
Signal/noise ratio	++	+	+++
Safety	+++	++	+++
Permanence	+	++	+++
Potential for dual-labeling	+++	++	++
Requirement for special equipment	Yes	No	No

plicity, permanence, and high resolution. The sensitivity of immunogold staining is greatly increased by silver enhancement—the immunogold-silver, or IGSS, technique *(7–9)*—in which the gold particles act as nucleation sites for the precipitation of silver. The strong signal obtained using this technique may be visualized using conventional bright-field microscopy, though sensitivity is significantly enhanced when epipolarized illumination is utilized *(10)*.

Although it is possible to label the antigen-specific antibody directly; the more common, more sensitive, and flexible approach, is to detect the bound primary antibody using labeled antibodies directed against the species used to derive the primary. This "indirect labeling" approach is the one employed in the following protocol and illustrated schematically in Fig. 1.

2. Materials

2.1. Tissue Fixation and Embedding

The fixatives preferably should be made fresh just prior to use. Fixative stored at 4°C will, however, remain usable for several weeks.

1. Phosphate-buffered saline (PBS): 10 m*M* Na_2HPO_4/NaH_2PO_4, 150 m*M* NaCl, pH 7.4.
2. Paraformaldehyde.
3. 10, 25, 50, 75, 90% and absolute ethanol.
4. Dewaxing solvent (xylene, Inhibisol, or Histoclear).
5. Paraffin wax (Paraplast Plus or equivalent).
6. Clean glass microscope slides.
7. Acetone.
8. 3-aminopropyltriethoxysilane (APES).

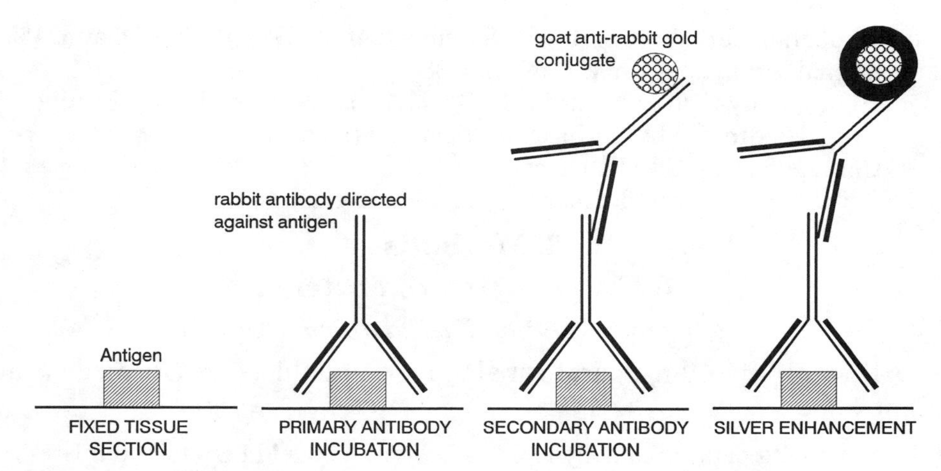

Fig. 1. Schematic diagram illustrating the stages of a typical indirect immunogold-silver staining (IGSS) technique.

2.2. Tissue Pretreatments

1. Trypsin.
2. Trypsin dilution buffer: 0.1% $CaCl_2$.

2.3. Immunolocalization

Antibodies should not be subjected to repeated freeze–thaw cycles and are best stored at 2–8°C (sodium azide may be added to 20 m*M* to prevent microbial growth) or frozen in small aliquots. The protein-containing buffers (blocking, incubation, and wash) should be stored at 2–8°C. Care should be taken when disposing of azide-containing solutions since, in addition to its considerable toxicity, sodium azide may form explosive metallic complexes in copper and lead plumbing.

1. Antibody against the antigen of interest (primary, 1°, antibody) either specific antiserum, ascites fluid, or purified antibody.
2. Antibody gold (5 nm) conjugate against species used to derive primary antibody (secondary, 2°, antibody).
3. Phosphate-buffered saline (PBS): 10 m*M* Na_2HPO_4/NaH_2PO_4, 150 m*M* NaCl, pH 7.4.
4. Blocking buffer: 0.8% Bovine Serum Albumin (BSA), 0.1% gelatin, 5% normal serum, and 2 m*M* NaN_3 in PBS.
5. Washing buffer: 0.8% Bovine Serum Albumin (BSA), 0.1% gelatin, 2 m*M* NaN_3 in PBS.

6. Incubation buffer: 0.8% Bovine Serum Albumin (BSA), 0.1% gelatin, 1% normal serum, and 2 m*M* NaN_3 in PBS.
7. High quality distilled water (ideally such as that provided by Millipore Corp. [Bedford, MA] Milli-Q purification system).
8. Commercial silver enhancement kit.
9. Suitable permanent mountant (DPX, Merckoglas).

3. Methods

3.1. Preparation of Material

3.1.1. Preparation of 4% Paraformaldehyde

All steps involving paraformaldehyde should be performed in a fume hood.

1. Weigh 4 g paraformaldehyde into a beaker, then add PBS to 100 mL.
2. Heat with constant stirring to 60°C.
3. Add a dilute sodium hydroxide solution dropwise until the paraformaldehyde dissolves.
4. Allow solution to cool to room temperature.
5. Check pH—if necessary, adjust to pH 7.4 using H_2SO_4 (not HCl).
6. Store at 4°C until required (*see* Notes 1 and 2).

3.1.2. Fixation and Embedding

1. Cut slices of the plant tissue of interest such that the pieces are no larger than 5–10 mm in any dimension (*see* Notes 3 and 4).
2. Incubate in an excess (>10-fold) 4% paraformaldehyde overnight at 4°C.
3. Dehydrate sections through an ethanol series to 100% ethanol over a period of 1–2 d depending on the size of the tissue pieces.
4. Incubate sections in a 1:1 (v/v) mixture of ethanol:Histoclear at room temperature for 2–4 h, changing the solution twice, then incubate the tissue pieces in 100% Histoclear for 2–4 h.
5. Incubate the tissue pieces in 1:1 mixture Histoclear:molten paraffin wax at 56–60°C overnight, followed by incubation in molten paraffin wax at 56–60°C until the sections are fully infiltrated with the molten wax (*see* Notes 6–8).
6. The infiltrated pieces of tissue are allowed to solidify in suitable molds. Blocks solidified at temperatures below ambient (0–4°C) may exhibit finer grain structure and, therefore, better cellular preservation (*see* Note 5).

3.1.3. Preparation of APES-Coated Slides

All the procedures involving APES should be performed in a fume hood. The APES should be allowed to warm to room temperature before opening to avoid reaction with condensed water (*see* Notes 9–11).

1. Dip slides for 30 s in a freshly prepared 2% solution of APES in dry acetone.
2. Dip slides for 30 s in acetone, and repeat acetone wash.
3. Rinse with distilled water and allow slides to air dry (*see* Note 12).

3.1.4. Sectioning

1. Trim wax block square such that 2–3 mm of wax surround the tissue (*see* Note 13).
2. Place in microtome and section at 7–10 µm according to the manufacturer's instructions (*see* Notes 14 and 15).
3. Float sections onto distilled water at 45–50°C and pick up the sections onto APES-coated slides.
4. Allow slides to dry down on a heated plate or in an oven (~45°C) for a minimum of 4 h, preferably overnight.

3.1.5. Pretreatments

1. Incubate slides (+ adhered sections) in Histoclear for 5 min. Repeat this step (*see* Note 16).
2. Rehydrate sections in an ethanol series, and rinse in distilled water.
3. Incubate sections with 0.1% trypsin in trypsin dilution buffer for 10 min at room temperature. This protease treatment is optional—serving to reveal antigenic sites potentially masked by the fixation process (*see* Note 17).

3.2. Immunolocalization

The washing and incubation times specified in the following are cautious optima. Shorter washes and incubation times may have no adverse affect on the results.

3.2.1. Primary Antibody Incubation

1. Dilute primary antibody to the required concentration in incubation buffer. The optimal concentration must be determined empirically, but 1:50–1:250 dilutions are appropriate for initial experiments (*see* Note 18).
2. Pipet 50–100 µL of the diluted antibody solution onto each slide and cover with a piece of Parafilm so that the sections are covered. Great care must be taken to ensure that no air bubbles are trapped.
3. Incubate the slides at room temperature for 2 h in a humid chamber (*see* Notes 19 and 20).

3.2.2. Secondary Antibody Incubation

1. Wash the sections with washing buffer for 3 × 10 min.
2. Incubate sections with the gold-conjugated secondary antibody at the manufacturer's recommended dilution (normally 1/50–1/100) in incubation buffer for 2 h. Shorter incubation times at a higher antibody concen-

tration may give equivalent results and a prolonged incubation at the recommended dilution can give a stronger signal (*see* Note 21).

3.3. Detection of Bound Antibody

1. Wash the slides with washing buffer for 3 × 10 min.
2. Wash the slides with PBS for 3 × 5 min.
3. If desired, bound antibody may be fixed to the sections by incubation with 2% glutaraldehyde in PBS for 10 min (*see* Note 22).
4. Rinse twice with an excess pure water. Milli-Q is ideal, but high-quality distilled water will suffice (*see* Note 23).
5. Perform silver enhancement of the gold signal according to the manufacturer's recommendations. Monitor the progress of the reaction carefully (*see* Notes 24 and 25).
6. Rinse sections in excess distilled water for 3X 5 min.
7. Counterstain and mount. Bound, silver-enhanced, gold is seen as a dark brown/black precipitate under bright field illumination and as bright particles under dark-field or epipolarized illumination (*see* Notes 26 and 27).

4. Notes

4.1. Preparation of 4% Paraformaldehyde

1. If a faint cloudy precipitate remains in suspension once the solution has cooled to room temperature, the paraformaldehyde may be filtered through Whatman (Maidstone, UK) No. 1 filter paper.
2. If desired, a 16% stock solution of paraformaldehyde may be made—such a stock solution exhibits reasonable stability if stored at 4°C.

4.2. Fixation

3. The fixation process is crucial to the success of the procedure as a whole and must be optimized alongside other stages. Fixatives fall into two broad classes: predominantly precipitative fixatives such as FAA (formalin/acetic acid/ethanol, 1:1:3) and crosslinking fixatives containing aldehydes. Both types are suitable for immunocytochemistry, but a useful starting point is 4% paraformaldehyde, as described here. Microwave irradiation greatly increases the rate of fixation *(10)*, although its utility in immunocytochemical studies is not yet clear.
4. It should be noted that even mild fixation will affect protein structure and may result in total or partial loss of antigenicity. In such cases, limited proteolysis can be used to unmask antigenic sites. Alternatively, flash freezing of the tissue followed by cryosectioning may be appropriate if the facilities are available.

4.3. Embedding

5. In order to section the plant material it must be supported during the sectioning procedure. The tissue is embedded in an inert solidified matrix. Paraffin is the most common such matrix, although water soluble alternatives such as polyethylene glycol (PEG 1000) also have great utility. Molten paraffin wax is allowed to infiltrate the tissue fully over a period of days or hours, depending on the size of the tissue fragments being processed.
6. Modern paraffin preparations contain plastic polymers that aid sectioning but are destroyed by heating above 63°C. The temperature must, therefore, be carefully controlled throughout infiltration. The infiltration stage is commonly performed in a laboratory oven in which the temperature can vary significantly.
7. The time required for complete infiltration of the tissue varies with the tissue size—small pieces may be completely infiltrated in a matter of hours, whereas larger pieces can require 4–5 d.
8. The elevated temperatures and presence of DMSO in the paraffin preparations can affect polypropylene containers and it is recommended that all preparation stages be performed in glass apparatus. Glass scintillation vials are cheap and about the right size.

4.4. Preparation of APES-Coated Slides

9. Prior to the immunocytochemical localization, the cut sections must be adhered to glass microscope slides. Slides may be treated with a number of compounds to ensure that the sections remain attached throughout the lengthy procedures that follow—these include coating with poly L-lysine, gelatin/chrome alum, or a variety of silane derivatives *(11)*. In our hands the use of APES (3-aminopropyltriethoxysilane) has proved both convenient and effective.
10. We have not found it necessary to clean high-quality microscope slides prior to coating. Indeed, cleaning can result in uneven coating.
11. A variety of commercial preparations are available that perform well.
12. The APES-coated slides are stable for several weeks if stored dry.

4.5. Sectioning

13. Thin sections of the plant material, embedded within the paraffin matrix, are cut using a bench microtome according to the manufacturer's instructions.
14. It may be found to be easier to cut a complete ribbon of sections if the wax block is first cooled, using a freezing spray (freon).
15. It is essential that the knife used to cut the sections be clean and sharp. Microtome stages that are capable of using disposable razor blades may be found preferable to those that utilize traditional "cut-throat" blades.

4.6. Pretreatments

16. To allow penetration of the antibodies, the paraffin matrix surrounding the tissue must first be dissolved away and, if necessary, antigenic sites within the tissue exposed by protease treatment. The paraffin may be removed by washes in a number of organic solvents (e.g., xylene) or less toxic commercial preparations such as Histoclear.
17. The requirements will depend on the conditions determined in previous experiments to be optimal for the particular tissue under investigation. In many, if not most, cases, a signal will be visible even if no treatments were undertaken to expose or renature the antigen. However, such pretreatments can greatly increase the signal achieved.

4.7. Primary Antibody Incubation

18. The sections are incubated with the primary antibody in a solution containing blocking agents to minimize nonspecific interactions. These blocking agents may be relatively pure proteins, such as BSA, or complex proteinaceous mixtures, such as reconstituted dried milk powder. Inadequate blocking results in high nonspecific binding of the antibodies with concomitant high background and low signal:noise ratio (giving rise to a signal in the negative, preimmune, control). If unsatisfactory results cannot be improved by changing the primary or secondary antibody concentration, changing the blocking concentration or blocking agent may be considered.
19. Antibody binding is affected by antibody concentration, incubation time, and incubation temperature. Short incubation times at higher temperatures (~37°C) giving results equivalent to prolonged (overnight) incubation at 4°C. Once again the choice is up to the investigator. Often a 2-h incubation at room temperature will give good results. Techniques have evolved to decrease incubation and wash times by the application of microwaves.
20. Negative controls are essential to confirm the validity of immunocytochemical localizations. These usually consist of a preimmune control—where the primary antibody is replaced by normal (preimmune) serum from the same species used to derive the primary antibody and controls in which the secondary antibody conjugate is omitted. A positive control, using a tissue known to express high levels of the antigen, is also useful; particularly while optimizing the conditions for the technique.

4.8. Secondary Antibody Incubation

21. The secondary antibody conjugate will come with a suggested dilution for its use in immunocytochemical localization. This dilution provides a good basis for subsequent optimization. Gold conjugates are available with the

antibody coupled to gold particles of a variety of sizes—typically 10, 5, and, more recently, 1 nm. For general light microscope localizations, where the bound signal is silver enhanced, the 5-nm conjugates are a good starting point.

4.9. Detection of Bound Antibody

22. It is reported that the antibody:antigen complex is unstable in low ionic strength solutions and that the water wash prior to silver enhancement may result in a significant loss of the antibody. Brief crosslinking of the antibodies bound to the tissue sections can alleviate such problems and is readily accomplished by incubating the sections in a 2% solution of glutaraldehyde in PBS just prior to the water rinse.
23. Impurities in the water, heavy metals, and halide salts in particular, can adversely affect the silver enhancement process and give rise to high background signals. The purity of most laboratory distilled water is quite adequate for most purposes and should not give rise to any problems. If difficulties are encountered, Milli-Q water (Millipore Corp.) may be used confidently. Contact of any of the solutions, or slides themselves, with metallic objects should be avoided.
24. A wide range of silver enhancement kits are available commercially (e.g., Amersham, Arlington Heights, IL; BioCell, Cardiff, UK; and Boehringer, Mannheim, Germany), all of which should give satisfactory results. These kits generally consist of two solutions, stable for about 6 mo at 4°C, which are mixed in equal quantities just prior to use.
25. The silver enhancement process may be monitored microscopically—check at 30-s intervals after 2 min. Optimal enhancement generally occurs from 2–15 min, although this is strongly temperature dependent. After about 20 min at room temperature self nucleation of the reagents begins, leading to a high background signal.
26. Once the sections have been silver enhanced, they may be permanently mounted using any of the wide range of commercial mounting agents. Certain mountants are acidic, however, and this can result in an appreciable loss of signal with time.
27. A strong silver enhanced gold signal may be visible to the naked eye if it covers sufficient tissue area. Microscopic examination of the tissue sections under conventional bright-field illumination will show bound antibody:gold complexes as a black precipitate. A significant increase in sensitivity (5- to 10-fold) is achieved if the sections are examined under epipolarizing illumination and the use of oil-immersion objectives provides a similar improvement in signal visibility.

References

1. Harris, N. (1994) Immunocytochemistry for light and electron microscopy, in *Plant Cell Biology: A Practical Approach* (Harris, N. and Oparka, K., eds.), Oxford University Press, Oxford, pp. 157–176.
2. Wang, T. (1986) *Immunology in Plant Science.* Cambridge University Press, Cambridge, UK.
3. Dewey, M., Evans, D., Coleman, J., Priestly, R., Hull, R., Horsley, D., and Hawes, C. (1991) Antibodies in plant science. *Acta Bot. Neerland.* **40,** 1–27.
4. Lloyd, C. W., Slabas, A. R., Powell, A. J., and Lowe, S. B. (1980) Microtubules, protoplasts and plant cell shape: an immunofluorescent study. *Planta* **147,** 500–506.
5. Shaw, P. J. and Henwood, J. A. (1985) Immuno-gold localization of cytochrome f, light harvesting complex, ATP synthase and ribulose 1,5-bisphosphate carboxylase/oxygenase. *Planta* **165,** 333–339.
6. Davies, J. T., Shewry, P. R., and Harris, N. (1993) Spatial and temporal patterns of B hordein synthesis in developing barley (*Hordeum vulgare* L.) caryopses. *Cell Biol. Int.* **17,** 195–203.
7. Holgate, C., Jackson, P., Cowen, P., and Bird, C. (1983) Immunogold silver staining: a new method of immunostaining with enhanced sensitivity. *J. Histochem. Cytochem.* **31,** 938–944.
8. Danscher, G. (1981) Localization of gold in biological tissue. A photochemical method for light and electron microscopy. *Histochemistry* **8,** 1081–1083.
9. Beesley, J. E. (1989) *Colloidal Gold: A New Perspective for Cytochemical Marking.* Royal Microscopical Society Handbook 17. Oxford Science Publications, Oxford University Press, Oxford, UK.
10. Walsh, G. E., Bohannon, P. M., and Wessinger-Duvall, P. B. (1989) Microwave irradiation for rapid killing and fixing of plant tissue. *Can. J. Bot.* **67,** 1272–1274.
11. Maddox, P. H. and Jenkins, D. (1987) 3-aminopropyltriethoxysilane (APES): a new advance in section adhesion. *J. Clin. Pathol.* **40,** 1256,1257.

Index